Computational Topology for Data Analysis

Topological data analysis (TDA) has emerged recently as a viable tool for analyzing complex data, and the area has grown substantially in both its methodologies and applicability. Providing a computational and algorithmic foundation for techniques in TDA, this comprehensive, self-contained text introduces students and researchers in mathematics and computer science to the current state of the field. The book features a description of mathematical objects and constructs behind recent advances, the algorithms involved, computational considerations, as well as examples of topological structures or ideas that can be used in applications. It provides a thorough treatment of persistent homology together with various extensions – like zigzag persistence and multiparameter persistence – and their applications to different types of data, like point clouds, triangulations, or graph data. Other important topics covered include discrete Morse theory, the mapper structure, optimal generating cycles, as well as recent advances in embedding TDA within machine learning frameworks.

TAMAL KRISHNA DEY is Professor of Computer Science at Purdue University. Before joining Purdue, he was a faculty member in the CSE Department of the Ohio State University. He has held academic positions at Indiana University–Purdue University at Indianapolis, Indian Institute of Technology Kharagpur, and Max Planck Institute. His research interests include computational geometry, computational topology, and their applications to geometric modeling and data analysis. He has (co)authored two books, *Curve and Surface Reconstruction: Algorithms with Mathematical Analysis* (Cambridge University Press) and *Delaunay Mesh Generation* (CRC Press), and (co)authored more than 200 scientific articles. Tamal is a fellow of the IEEE, ACM, and Solid Modeling Association.

YUSU WANG is Professor in the Halıcıoğlu Data Science Institute at the University of California, San Diego. Prior to joining UCSD, she was Professor of Computer Science and Engineering at the Ohio State University and postdoctoral fellow at Stanford University. Yusu primarily works in topological and geometric data analysis, developing effective and theoretically justified algorithms for data analysis using geometric and topological ideas, as well as in applying them to practical domains. She received the DOE Early Career Principal Investigator Award in 2006 and NSF Career Award in 2008.

Computational Topology for Data Analysis

TAMAL KRISHNA DEY
Purdue University

YUSU WANG
University of California, San Diego

CAMBRIDGE
UNIVERSITY PRESS

University Printing House, Cambridge CB2 8BS, United Kingdom

One Liberty Plaza, 20th Floor, New York, NY 10006, USA

477 Williamstown Road, Port Melbourne, VIC 3207, Australia

314–321, 3rd Floor, Plot 3, Splendor Forum, Jasola District Centre,
New Delhi – 110025, India

103 Penang Road, #05–06/07, Visioncrest Commercial, Singapore 238467

Cambridge University Press is part of the University of Cambridge.

It furthers the University's mission by disseminating knowledge in the pursuit of
education, learning, and research at the highest international levels of excellence.

www.cambridge.org
Information on this title: www.cambridge.org/9781009098168
DOI: 10.1017/9781009099950

© Tamal Krishna Dey and Yusu Wang 2022

First published 2022

A catalogue record for this publication is available from the British Library

Library of Congress Cataloging-in-Publication data
Names: Dey, Tamal K. (Tamal Krishna), 1964- author. | Wang, Yusu, 1976- author.
Title: Computational topology for data analysis / Tamal Krishna Dey,
Department of Computer Science, Purdue University, Yusu Wang, Halıcıoğlu
glu Data Science Institute University of California, San Diego.
Description: First edition. | New York : Cambridge University Press, 2022.
| Includes bibliographical references and index.
Identifiers: LCCN 2021041859 (print) | LCCN 2021041860 (ebook) | ISBN
9781009098168 (hardback) | ISBN 9781009099950 (ebook)
Subjects: LCSH: Topology.
Classification: LCC QA611 .D476 2022 (print) | LCC QA611 (ebook) | DDC
514/.7–dc23/eng/20211029
LC record available at https://lccn.loc.gov/2021041859
LC ebook record available at https://lccn.loc.gov/2021041860

ISBN 978-1-009-09816-8 Hardback

Contents

Preface

In recent years, the area of topological data analysis (TDA) has emerged as a viable tool for analyzing data in applied areas of science and engineering. The area started in the 1990s with the computational geometers finding an interest in studying the algorithmic aspect of the classical subject of algebraic topology in mathematics. The area of computational geometry flourished in the 1980s and 1990s by addressing various practical problems and enriching the area of discrete geometry in the course of doing so. A handful of computational geometers felt that, analogous to this development, computational topology has the potential to address the area of shape and data analysis while drawing upon and perhaps developing further the area of topology in the discrete context; see, for example, [26, 116, 119, 188, 292]. The area gained momentum with the introduction of persistent homology in early 2000 followed by a series of mathematical and algorithmic developments on the topic. The book by Edelsbrunner and Harer [149] presents these fundamental developments quite nicely. Since then, the area has grown in both its methodology and applicability. One consequence of this growth has been the development of various algorithms which intertwine with the discoveries of various mathematical structures in the context of processing data. The purpose of this book is to capture these algorithmic developments with the associated mathematical guarantees. It is appropriate to mention that there is an emerging sub-area of TDA which centers more around statistical aspects. This book does not deal with these developments, though we mention some of them in the last chapter where we describe the recent results connecting TDA and machine learning.

We have 13 chapters in the book listed in the table of contents. After developing the basics of topological spaces, simplicial complexes, homology groups, and persistent homology in the first three chapters, the book is then devoted to presenting algorithms and associated mathematical structures in various contexts of topological data analysis. These chapters present materials mostly not covered in any book on the market. To elaborate on this claim, we briefly give an overview of the topics covered by the present book. Chapter 4 presents a generalization of the persistence algorithm to extended settings such as to simplicial maps (instead of inclusions), and zigzag sequences with both inclusions and simplicial maps. Chapter 5 covers algorithms on computing optimal generators for both persistent and nonpersistent homology. Chapter 6 focuses on algorithms that infer homological information from point cloud data. Chapter 7 presents algorithms and structural results for Reeb graphs. Chapter 8 considers general graphs, including directed ones. Chapter 9 focuses on various recent results on characterizing nerves of covers, including the well-known mapper and its multiscale version. Chapter 10 is devoted to the important concept of discrete Morse theory, its connection to persistent homology, and its applications to graph reconstruction. Chapters 11 and 12 introduce multiparameter persistence. The standard persistence is defined over a one-parameter index set such as $\mathbb{Z}$ or $\mathbb{R}$. Extending this index set to a poset such as $\mathbb{Z}^d$ or $\mathbb{R}^d$, we get d-parameter or multiparameter persistence. Chapter 11 focuses on computing indecomposables for multiparameter persistence that are generalizations of bars in the one-parameter case. Chapter 12 focuses on various definitions of distances among multiparameter persistence modules and their computations. Finally, we conclude with Chapter 13, which presents some recent developments of incorporating persistence into the machine learning (ML) framework.

This book is intended for an audience comprising researchers and teachers in computer science and mathematics. Graduate students in both fields will benefit from learning the new materials in topological data analysis. Because of the topics, the book plays the role of a bridge between mathematics and computer science. Students in computer science will learn the mathematics in topology that they are usually not familiar with. Similarly, students in mathematics will learn about designing algorithms based on mathematical structures. The book can be used for a graduate course in topological data analysis. In particular, it can be part of a curriculum in data science which has been/is being adopted in universities. We are including exercises for each chapter to facilitate teaching and learning.

There are currently a few books on computational topology/topological data analysis on the market to which our book will be complementary. The materials covered in this book predominantly are new and have not been covered in any of the previous books. The book by Edelsbrunner and Harer [149] mainly focuses on early developments in persistent homology and do not cover the materials in Chapters 4–13 in this book. The recent book of Boissonnat et al. [39] focuses mainly on reconstruction, inference, and Delaunay meshes. Other than Chapter 6, which focuses on point cloud data and inference of topological properties, and Chapters 1–3, which focus on preliminaries about topological persistence, there is hardly any overlap. The book by Oudot [250] mainly focuses on algebraic structures of persistence modules and inference results. Again, other than the preliminary Chapters 1–3 and Chapter 6, there is hardly any overlap. Finally, unlike ours, the books by Tierny [286] and by Rabadán and Blumberg [260] mainly focus on applying TDA to specific domains of scientific visualizations and genomics, respectively.

This book, as any other, is not created in isolation. Help coming from various corners contributed to its creation. It was seeded by the class notes that we developed for our introductory course on Computational Topology and Data Analysis which we taught at the Ohio State University. During this teaching, the class feedback from students gave us the hint that a book covering the increasingly diversified repertoire of topological data analysis was necessary at this point. We thank all those students who had to bear with the initial disarray that was part of freshly gathering coherent material on a new subject. This book would not have been possible without our own involvement with TDA, which was mostly supported by grants from the National Science Foundation (NSF). Many of our PhD students worked through these projects, which helped us consolidate our focus on TDA. In particular, Tao Hou, Ryan Slechta, Cheng Xin, and Soham Mukherjee gave their comments on drafts of some of the chapters. We thank all of them. We thank everyone from the TGDA@OSU group for creating one of the best environments for carrying out research in applied and computational topology. Our special thanks go to Facundo Mémoli, who has been a great colleague (who has collaborated with us on several topics) as well as a wonderful friend at OSU. We also acknowledge the support of the Department of CSE at the Ohio State University where a large amount of the contents of this book were planned and written. The finishing came to fruition after we moved to our current institutions.

Finally, it is our pleasure to acknowledge the support of our families who kept us motivated and engaged throughout the marathon of writing this book, especially during the last stretch overlapping the 2020–2021 Coronavirus pandemic. Tamal recalls his daughter Soumi and son Sounak asking him continually about the progress of the book. His wife Kajari extended all the help necessary to make space for extra time needed for the book. Despite suffering from the reduced attention to family matters, all of them offered their unwavering support and understanding graciously. Tamal dedicates this book to his family and his late parents Gopal Dey and Hasi Dey without whose encouragement and love he would not have been in a position to take up this project. Yusu thanks her husband Mikhail Belkin for his never-ending support and encouragement throughout writing this book and beyond. Their two children Alexander and Julia contributed in their typical ways by making every day delightful and unpredictable for her. Without their support and love, she would not have been able to finish this book. Finally, Yusu dedicates this book to her parents Qingfen Wang and Jinlong Huang, who always gave her space to grow and encouraged her to do her best in life, as well as to her great aunt Zhige Zhao and great uncle Humin Wang, who kindly took her into their care when she was 13. She can never repay their kindness.

Prelude

We make sense of the world around us primarily by understanding and studying the "shape" of the objects that we encounter in real life or in a digital environment. Geometry offers a common language that we usually use to model and describe shapes. For example, the familiar descriptors such as distances, coordinates, angles, and so on from this language assist us to provide detailed information about a shape of interest. Not surprisingly, people have used geometry for thousands of years to describe objects in their surroundings.

However, there are many situations where detailed geometric information is not needed and may even obscure the really useful structure that is not so explicit. A notable example is the *Seven Bridges of Königsberg* problem, where, in the city of Königsberg, the Pregel river separates the city into four regions, connected by seven bridges, as shown in Figure 1 (map and description taken from the Wikipedia page for "Seven bridges of Königsberg"). The question is to find a walk through the city that crosses each bridge exactly once. The story goes that the mathematician Leonhard Euler observed that factors such as the precise shape of these regions and the exact path taken are not important. What is important is the *connectivity* among the different regions of the city as connected by the bridges. In particular, the problem can be modeled abstractly using a graph with four nodes, representing the four regions in the city of Königsberg, and seven edges representing the bridges connecting them. The problem then reduces to what's later become known as finding the Euler tour (or Eulerian cycle) in this graph, which can be easily solved.

For another example, consider animation in computer graphics, where one wants to develop software that can continuously deform one object into another (in the sense that one can stretch and change the shape, but cannot break and add to the shape). Can we continuously deform a frog into a prince this way?[1]

[1] Yes, according to Disney movies.

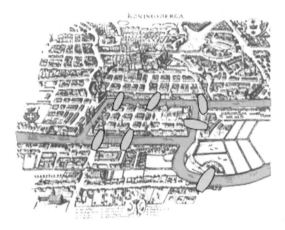

Figure 1 "Map of Königsberg in Euler's time showing the actual layout of the seven bridges, highlighting the river Pregel and the bridges" (the drawing by Bogdan Giuşcă is licensed under CC BY-SA 3.0).

Is it possible to continuously deform a tea cup into a bunny? It turns out the latter is not possible.

In these examples, the core structure of interest behind the input object or space is characterized by the way the space is connected, and the detailed geometric information may not matter. In general, topology intuitively models and studies properties that are invariant as long as the connectivity of space does not change. As a result, topological language and concepts can provide powerful tools to characterize, identify, and process essential features of both spaces and functions defined on them. However, to bring topological methods to the realm of practical applications, we need not only new ideas to make topological concepts and the resulting structures more suitable for modern data analysis tasks, but also algorithms to compute these structures efficiently. In the past two decades, the field of applied and computational topology has developed rapidly, producing many fundamental results and algorithms that have advanced both fronts. This progress has further fueled the significant growth of *topological data analysis* (TDA), which has already found applications in various domains such as computer graphics, visualization, materials science, computational biology, neuroscience, and so on.

In Figure 2, we present some examples of the use of topological methodologies in applications. The topological structures involved will be described later in the book.

An important development in applied and computational topology in the past two decades centers around the concept of *persistent homology*, which

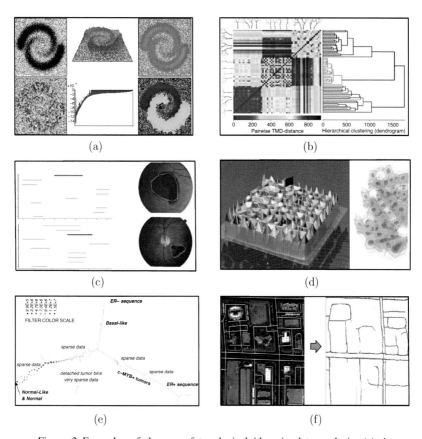

(a)   (b)

(c)   (d)

(e)   (f)

Figure 2 Examples of the use of topological ideas in data analysis. (a) A persistence-based clustering strategy. The persistence diagram of a density field estimated from an input noisy point cloud (shown in the top row) is used to help group points into clusters (bottom row). Reprinted by permission from Springer Nature: Springer Nature, Discrete and Computational Geometry, "Analysis of scalar fields over point cloud data," Chazal et al. [86], © 2011. (b) Using persistence diagram summaries to represent and cluster neuron cells based on their tree morphology. Image taken from [206] licensed by Kanari et al. (2018) under CC BY 4.0 (https://creativecommons.org/licenses/by/4.0/). (c) Using the optimal persistent 1-cycle corresponding to a bar (red) in the persistence barcode, defects in diseased eyes are localized. Image taken from [127]. (d) Topological landscape (left) of the 3D volumetric silicium dataset from [299]. A volume rendering of the silicium dataset is on the right. However, note that it is hard to see all the structures forming the lattice of the crystal, while the topological landscape view shows clearly that most of them have high function values and are of similar sizes. Image taken from [299], reprinted by permission from IEEE: Weber et al. (2007). (e) Mapper structure behind the high-dimensional cell gene expression dataset can show not only the cluster of different tumor or normal cells, but also their connections. Image taken from [245], reprinted by permission from Nicolau et al. (2011, figure 3). (f) Using a discrete Morse-based graph skeleton reconstruction algorithm to help reconstruct road networks from satellite images even with few labeled training data. Image taken from [138].

generalizes the classic algebraic structure of homology groups to the multiscale setting aided by the concept of so-called *filtration* and *persistence modules* (discussed in Chapters 2 and 3). This helps significantly to broaden the applications of homological features to characterizing shapes/spaces of interest.

Figure 2(a) gives an example where persistent homology of a density field is used to develop a clustering strategy for the points [86]. In particular, at the beginning, each point is in its own cluster. Then, these clusters are grown using persistent homology, which identifies their importance and merges them according to this importance. The final output captures key clusters which may look like "blobs" or "curvy strips" – intuitively, they comprise dense regions separated by sparse regions.

Figure 2(b) gives an example where the resulting topological summaries from persistent homology have been used for clustering a collection of neurons, each of which is represented by a rooted tree (as neuron cells have tree morphology). We will see in Chapter 13 that persistent homology can serve as a general way to vectorize the features of such complex input objects.

In Figure 2(c), diseased parts of retinal degeneracy in two eyes are localized from image data. Algorithms for computing optimal cycles for bars in the persistent barcode as described in Chapter 5 are used for this purpose.

In Figure 2(d), we present an example where the topological object of a contour tree (the special loop-free case of the so-called Reeb graph as discussed in Chapter 7) has been used to give a low-dimensional terrain metaphor of a potentially high-dimensional scalar field. To illustrate further, suppose that we are given a scalar field $f : X \rightarrow \mathbb{R}$ where X is a space of potentially high dimension. To visualize and explore X and f in $\mathbb{R}^2$ and $\mathbb{R}^3$, just mapping X to $\mathbb{R}^2$ can cause significant geometric distortion, which in turn leads to artifacts in the visualization of f over the projection. Instead, we can create a 2D terrain metaphor $f' : \mathbb{R}^2 \rightarrow \mathbb{R}$ for f which preserves the contour tree information as proposed in [299]; intuitively, this preserves the valleys/mountain peaks and how they merge and split. In this example, the original scalar field is in $\mathbb{R}^3$. However, in general, the idea is applicable to higher-dimensional scalar fields (e.g., the protein energy landscape considered in [184]).

In Figure 2(e), we give an example of an alternative approach of exploring a high-dimensional space X or functions defined on it via the mapper methodology (introduced in Chapter 9). In particular, the mapper methodology constructs a representation of the essential structure behind X via a pullback of a covering of Z through a map $f : X \rightarrow Z$. This intuitively captures the continuous structure of X at coarser level via the discretization of Z. See Figure 2(e), where the one-dimensional skeleton of the mapper structure behind a breast cancer microarray gene expression dataset is shown [245]. This continuous space representation not only shows "clusters" of different groups of tumors and of normal cells, but also how they connect in the space of cells, which are typically missing in standard cluster analysis.

Finally, Figure 2(f) shows an example of combining topological structures from the discrete Morse theory (Chapter 10) with convolutional neural networks to infer road networks from satellite images [138]. In particular, the so-called 1-unstable manifolds from discrete Morse theory can be used to extract hidden graph skeletons from noisy data.

We conclude this prelude by summarizing the aim of this book: introduce recent progress in applied and computational topology for data analysis with an emphasis on the algorithmic aspect.

1
Basics

Topology, mainly algebraic topology, is the fundamental mathematical subject on which topological data analysis is based. In this chapter, we introduce some of the very basics of this subject that are used in this book. First, in Section 1.1, we give the definition of a topological space and other notions such as open and closed sets, covers, and subspace topology that are derived from it. These notions are quite abstract in the sense that they do not require any geometry. However, the intuition of topology becomes more concrete to nonmathematicians when we bring geometry into the mix. Section 1.2 is devoted to make the connection between topology and geometry through what is called metric spaces.

Maps such as homeomorphism and homotopy equivalence play a significant role to relate topological spaces. They are introduced in Section 1.3. At the heart of these definitions sits the important notion of continuous functions which generalizes the concept mainly known for Euclidean domains to topological spaces. Certain categories of topological spaces become important for their wide presence in applications. Manifolds are one such category which we introduce in Section 1.4. Functions on them satisfying certain conditions are presented in Section 1.5. They are well known as Morse functions. The critical points of such functions relate to the topology of the manifold they are defined on. We introduce these concepts in the smooth setting in this chapter, and later adapt them for the piecewise-linear domains that are amenable to finite computations.

1.1 Topological Space

The basic object in a topological space is a ground set whose elements are called points. A topology on these points specifies how they are *connected* by listing what points constitute a neighborhood – the so-called *open set*.

The expression "rubber-sheet topology" commonly associated with the term "topology" exemplifies this idea of connectivity of neighborhoods. If we bend and stretch a sheet of rubber, it changes shape but always preserves the neighborhoods in terms of the points and how they are connected.

We first introduce basic notions from point set topology. These notions are prerequisites for more sophisticated topological ideas – manifolds, homeomorphism, isotopy, and other maps – used later to study algorithms for topological data analysis. Homeomorphisms, for example, offer a rigorous way to state that an operation preserves the topology of a domain, and isotopy offers a rigorous way to state that the domain can be deformed into a shape without ever colliding with itself.

Perhaps it is more intuitive to understand the concept of topology in the presence of a metric because then we can use the metric balls such as Euclidean balls in a Euclidean space to define neighborhoods – the open sets. Topological spaces provide a way to abstract out this idea without a metric or point coordinates, so they are more general than metric spaces. In place of a metric, we encode the connectivity of a point set by supplying a list of all of the open sets. This list is called a *system* of subsets of the point set. The point set and its system together describe a topological space.

Definition 1.1. (Topological space) A *topological space* is a point set $\mathbb{T}$ endowed with a *system of subsets* T, which is a set of subsets of $\mathbb{T}$ that satisfies the following conditions:

- $\varnothing, \mathbb{T} \in T$.
- For every $U \subseteq T$, the union of the subsets in U is in T.
- For every finite $U \subseteq T$, the common intersection of the subsets in U is in T.

The system T is called a *topology* on $\mathbb{T}$. The sets in T are called the *open sets* in $\mathbb{T}$. A *neighborhood* of a point $p \in \mathbb{T}$ is an open set containing p.

First, we give examples of topological spaces to illustrate the definition above. These examples have the set $\mathbb{T}$ finite.

Example 1.1. *Let $\mathbb{T} = \{0, 1, 3, 5, 7\}$. Then, $T = \{\varnothing, \{0\}, \{1\}, \{5\}, \{1, 5\}, \{0, 1\}, \{0, 1, 5\}, \{0, 1, 3, 5, 7\}\}$ is a topology because $\varnothing$ and $\mathbb{T}$ are in T as required by the first axiom, the union of any sets in T is in T as required by the second axiom, and the intersection of any two sets is also in T as required by the third axiom. However, $T = \{\varnothing, \{0\}, \{1\}, \{1, 5\}, \{0, 1, 5\}, \{0, 1, 3, 5, 7\}\}$ is not a topology because the set $\{0, 1\} = \{0\} \cup \{1\}$ is missing.*

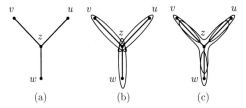

Figure 1.1 Example 1.3: (a) a graph as a topological space, stars of the vertices and edges as open sets; (b) a closed cover with three elements; and (c) an open cover with four elements.

Example 1.2. *Let* $\mathbb{T} = \{u, v, w\}$. *The power set* $2^{\mathbb{T}} = \{\varnothing, \{u\}, \{v\}, \{w\}, \{u, v\},$ $\{u, w\}, \{v, w\}, \{u, v, w\}\}$ *is a topology. For any ground set* $\mathbb{T}$, *the power set is always a topology on it which is called the* discrete topology.

One may take a subset of the power set as a ground set and define a topology, as the next example shows. We will recognize later that the ground set here corresponds to simplices in a simplicial complex and the "stars" of simplices generate all open sets of a topology.

Example 1.3. *Let* $\mathbb{T} = \{u, v, w, z, (u, z), (v, z), (w, z)\}$; *this can be viewed as a graph with four vertices and three edges as shown in Figure 1.1. Let*

- $T_1 = \{\{(u, z)\}, \{(v, z)\}, \{(w, z)\}\}$ *and*
- $T_2 = \{\{(u, z), u\}, \{(v, z), v\}, \{(w, z), w\}, \{(u, z), (v, z), (w, z), z\}\}$.

Then, $T = 2^{T_1 \cup T_2}$ *is a topology because it satisfies all three axioms. All open sets of T are generated by the union of elements in* $B = T_1 \cup T_2$ *and there is no smaller set with this property. Such a set B is called a basis of T. We will see later in the next chapter (Section 2.1) that these are open stars of all vertices and edges.*

We now present some more definitions that will be useful later.

Definition 1.2. (Closure; Closed sets) A set Q is *closed* if its complement $\mathbb{T} \setminus Q$ is open. The *closure* Cl Q of a set $Q \subseteq T$ is the smallest closed set containing Q.

In Example 1.1, the set $\{3, 5, 7\}$ is closed because its complement $\{0, 1\}$ in $\mathbb{T}$ is open. The closure of the open set $\{0\}$ is $\{0, 3, 7\}$ because it is the smallest closed set (complement of open set $\{1, 5\}$) containing 0. In Example 1.2, all sets are both open and closed. In Example 1.3, the set $\{u, z, (u, z)\}$ is closed, but the set $\{z, (u, z)\}$ is neither open nor closed. Interestingly, observe that

$\{z\}$ is closed. The closure of the open set $\{u, (u, z)\}$ is $\{u, z, (u, z)\}$. In all examples, the sets $\varnothing$ and $\mathbb{T}$ are both open and closed.

Definition 1.3. Given a topological space $(\mathbb{T}, T)$, the *interior* Int A of a subset $A \subseteq \mathbb{T}$ is the union of all open subsets of A. The *boundary* of A is Bd $A = \text{Cl } A \setminus \text{Int } A$.

The interior of the set $\{3, 5, 7\}$ in Example 1.1 is $\{5\}$ and its boundary is $\{3, 7\}$.

Definition 1.4. (Subspace topology) For every point set $\mathbb{U} \subseteq \mathbb{T}$, the topology T induces a *subspace topology* on $\mathbb{U}$, namely the system of open subsets $U = \{P \cap \mathbb{U} : P \in T\}$. The point set $\mathbb{U}$ endowed with the system U is said to be a *topological subspace* of $\mathbb{T}$.

In Example 1.1, consider the subset $\mathbb{U} = \{1, 5, 7\}$. It has the subspace topology

$$U = \{\varnothing, \{1\}, \{5\}, \{1, 5\}, \{1, 5, 7\}\}.$$

In Example 1.3, the subset $\mathbb{U} = \{u, (u, z), (v, z)\}$ has the subspace topology

$$\{\varnothing, \{u, (u, z)\}, \{(u, z)\}, \{(v, z)\}, \{(u, z), (v, z)\}, \{u, (u, z), (v, z)\}\}.$$

Definition 1.5. (Connected) A topological space $(\mathbb{T}, T)$ is *disconnected* if there are two disjoint non-empty open sets $U, V \in T$ so that $\mathbb{T} = U \cup V$. A topological space is *connected* if it is not disconnected.

The topological space in Example 1.1 is connected. However, the topological subspace (Definition 1.4) induced by the subset $\{0, 1, 5\}$ is disconnected because it can be obtained as the union of two disjoint open sets $\{0, 1\}$ and $\{5\}$. The topological space in Example 1.3 is also connected, but the subspace induced by the subset $\{(u, z), (v, z), (w, z)\}$ is disconnected.

Definition 1.6. (Cover; Compact) An *open (closed) cover* of a topological space $(\mathbb{T}, T)$ is a collection C of open (closed) sets so that $\mathbb{T} = \bigcup_{c \in C} c$. The topological space $(\mathbb{T}, T)$ is called *compact* if every open cover C of it has a finite *subcover*, that is, there exists $C' \subseteq C$ such that $\mathbb{T} = \bigcup_{c \in C'} c$ and C' is finite.

In Figure 1.1(b), the cover consisting of $\{\{u, z, (u, z)\}, \{v, z, (v, z)\}, \{w, z, (w, z)\}\}$ is a closed cover whereas the cover consisting of $\{\{u, (u, z)\}, \{v, (v, z)\},$

$\{w, (w, z)\}, \{z, (u, z), (v, z), (w.z)\}\}$ in Figure 1.1(c) is an open cover. Any topological space with finite point set $\mathbb{T}$ is compact because all of its covers are finite. Thus, all topological spaces in the discussed examples are compact. We will see examples of noncompact topological spaces where the ground set is infinite.

In the above examples, the ground set $\mathbb{T}$ is finite. It can be infinite in general and a topology may have uncountably infinitely many open sets containing uncountably infinitely many points.

Next, we introduce the concept of *quotient topology*. Given a space $(\mathbb{T}, T)$ and an equivalence relation $\sim$ on elements in $\mathbb{T}$, one can define a topology induced by the original topology T on the quotient set $\mathbb{T}/\sim$ whose elements are equivalence classes $[x]$ for every point $x \in \mathbb{T}$.

Definition 1.7. (Quotient topology) Given a topological space $(\mathbb{T}, T)$ and an equivalence relation $\sim$ defined on the set $\mathbb{T}$, a quotient space $(\mathbb{S}, S)$ induced by $\sim$ is defined by the set $\mathbb{S} = \mathbb{T}/\sim$ and the *quotient topology S* where

$$S := \big\{ U \subseteq \mathbb{S} \mid \{x : [x] \in U\} \in T \big\}.$$

We will see the use of quotient topology in Chapter 7 when we study Reeb graphs.

Infinite topological spaces may seem baffling from a computational point of view, because they may have uncountably infinitely many open sets containing uncountably infinitely many points. The easiest way to define such a topological space is to inherit the open sets from a metric space. A topology on a metric space excludes information that is not topologically essential. For instance, the act of stretching a rubber sheet changes the distances between points and thereby changes the metric, but it does not change the open sets or the topology of the rubber sheet. In the next section, we construct such a topology on a metric space and examine it from the concept of limit points.

1.2 Metric Space Topology

Metric spaces are a special type of topological space commonly encountered in practice. Such a space admits a *metric* that specifies the scalar *distance* between every pair of points satisfying certain axioms.

Definition 1.8. (Metric space) A *metric space* is a pair $(\mathbb{T}, \mathsf{d})$ where $\mathbb{T}$ is a set and d is a distance function $\mathsf{d} \colon \mathbb{T} \times \mathbb{T} \to \mathbb{R}$ satisfying the following properties:

- $d(p, q) = 0$ if and only if $p = q$ for all $p \in \mathbb{T}$;
- $d(p, q) = d(q, p)$ for all $p, q \in \mathbb{T}$;
- $d(p, q) \leq d(p, r) + d(r, q)$ for all $p, q, r \in \mathbb{T}$.

It can be shown that the three axioms above imply that $d(p, q) \geq 0$ for every pair $p, q \in \mathbb{T}$. In a metric space $\mathbb{T}$, an open *metric ball* with center c and radius r is defined to be the point set $B_o(c, r) = \{p \in \mathbb{T} : d(p, c) < r\}$. Metric balls define a topology on a metric space.

Definition 1.9. (Metric space topology) Given a metric space $\mathbb{T}$, all metric balls $\{B_o(c, r) \mid c \in \mathbb{T}$ and $0 < r \leq \infty\}$ and their union constituting the open sets define a topology on $\mathbb{T}$.

All definitions for general topological spaces apply to metric spaces with the above defined topology. However, we give alternative definitions using the concept of limit points which may be more intuitive.

As we have mentioned already, the heart of topology is the question of what it means for a set of points to be *connected*. After all, two distinct points cannot be adjacent to each other; they can only be connected to one another by passing through uncountably many intermediate points. The idea of *limit points* helps express this concept more concretely, specifically in the case of metric spaces.

We use the notation $d(\cdot, \cdot)$ to express minimum distances between point sets $P, Q \subseteq \mathbb{T}$:

$$d(p, Q) = \inf\{d(p, q) : q \in Q\},$$
$$d(P, Q) = \inf\{d(p, q) : p \in P, q \in Q\}.$$

Definition 1.10. (Limit point) Let $Q \subseteq \mathbb{T}$ be a point set. A point $p \in \mathbb{T}$ is a *limit point* of Q, also known as an *accumulation point* of Q, if for every real number $\epsilon > 0$, however tiny, Q contains a point $q \neq p$ such that $d(p, q) < \epsilon$.

In other words, there is an infinite sequence of points in Q that gets successively closer and closer to p – without actually being p – and gets arbitrarily close. Stated succinctly, $d(p, Q \setminus \{p\}) = 0$. Observe that it does not matter whether $p \in Q$ or not.

To see the parallel between the definitions given in this subsection and the definitions given before, it is instructive to define limit points also for general topological spaces. In particular, a point $p \in \mathbb{T}$ is a limit point of a set $Q \subseteq \mathbb{T}$ if every open set containing p intersects Q.

Definition 1.11. (Connected) A point set $Q \subseteq \mathbb{T}$ is called *disconnected* if Q can be partitioned into two disjoint non-empty sets U and V so that there is no

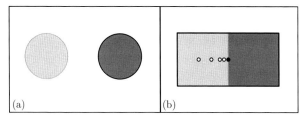

Figure 1.2 (a) The point set is disconnected; it can be partitioned into two connected subsets shaded differently. (b) The point set is connected; the black point at the center is a limit point of the points shaded lightly.

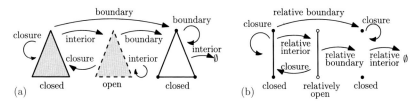

Figure 1.3 Closed, open, and relatively open point sets in the plane. Dashed edges and open circles indicate points missing from the point set.

point in U that is a limit point of V, and no point in V that is a limit point of U. (See Figure 1.2[a] for an example.) If no such partition exists, Q is *connected*, like the point set in Figure 1.2(b).

We can also distinguish between closed and open point sets using the concept of limit points. Informally, a triangle in the plane is *closed* if it contains all the points on its edges, and *open* if it excludes all the points on its edges, as illustrated in Figure 1.3. The idea can be formally extended to any point set.

Definition 1.12. (Closure; Closed; Open) The *closure* of a point set $Q \subseteq \mathbb{T}$, denoted Cl Q, is the set containing every point in Q and every limit point of Q. A point set Q is *closed* if $Q = $ Cl Q, that is, Q contains all its limit points. The *complement* of a point set Q is $\mathbb{T} \setminus Q$. A point set Q is *open* if its complement is closed, that is, $\mathbb{T} \setminus Q = $ Cl $(\mathbb{T} \setminus Q)$.

For example, consider the open interval $(0, 1) \subset \mathbb{R}$, which contains every $r \in \mathbb{R}$ so that $0 < r < 1$. Let $[0, 1]$ denote a *closed interval* $(0, 1) \cup \{0\} \cup \{1\}$. The numbers 0 and 1 are both limit points of the open interval, so Cl $(0, 1) = [0, 1] = $ Cl $[0, 1]$. Therefore, $[0, 1]$ is closed and $(0, 1)$ is not. The numbers 0 and 1 are also limit points of the complement of the closed interval, $\mathbb{R} \setminus [0, 1]$, so $(0, 1)$ is open, but $[0, 1]$ is not.

The definition of *open set* of course depends on the space being considered. A triangle τ that is missing the points on its edges is open in the two-dimensional affine Euclidean space supporting τ. However, it is not open in the Euclidean space $\mathbb{R}^3$. Indeed, every point in τ is a limit point of $\mathbb{R}^3 \setminus \tau$, because we can find sequences of points that approach τ from the side. In recognition of this caveat, a simplex $\sigma \subset \mathbb{R}^d$ is said to be *relatively open* if it is open relative to its affine hull. Figure 1.3 illustrates this fact where, in this example, the metric space is $\mathbb{R}^2$.

We can define the interior and boundary of a set using the notion of limit points also. Informally, the boundary of a point set Q is the set of points where Q meets its complement $\mathbb{T} \setminus Q$. The interior of Q contains all the other points of Q.

Definition 1.13. (Boundary; Interior) The *boundary* of a point set Q in a metric space $\mathbb{T}$, denoted $\mathrm{Bd}\, Q$, is the intersection of the closures of Q and its complement; that is, $\mathrm{Bd}\, Q = \mathrm{Cl}\, Q \cap \mathrm{Cl}\, (\mathbb{T} \setminus Q)$. The *interior* of Q, denoted $\mathrm{Int}\, Q$, is $Q \setminus \mathrm{Bd}\, Q = Q \setminus \mathrm{Cl}\, (\mathbb{T} \setminus Q)$.

For example, $\mathrm{Bd}\,[0, 1] = \{0, 1\} = \mathrm{Bd}\,(0, 1)$ and $\mathrm{Int}\,[0, 1] = (0, 1) = \mathrm{Int}\,(0, 1)$. The boundary of a triangle (closed or open) in the Euclidean plane is the union of the triangle's three edges, and its interior is an open triangle, illustrated in Figure 1.3. The terms *boundary* and *interior* have similar subtlety as open sets: the boundary of a triangle embedded in $\mathbb{R}^3$ is the whole triangle, and its interior is the empty set. However, relative to its affine hull, its interior and boundary are defined exactly as in the case of triangles embedded in the Euclidean plane. Interested readers can draw the analogy between this observation and the definition of interior and boundary of a manifold that appear later in Definition 1.23.

We have seen a definition of the compactness of a point set in a topological space (Definition 1.6). We define it differently here for a metric space. It can be shown that the two definitions are equivalent.

Definition 1.14. (Bounded; Compact) The *diameter* of a point set Q is $\sup_{p,q \in Q} \mathsf{d}(p, q)$. The set Q is *bounded* if its diameter is finite, and is *unbounded* otherwise. A point set Q in a metric space is *compact* if it is closed and bounded.

In the Euclidean space $\mathbb{R}^d$ we can use the standard Euclidean distance as the choice of metric. On the surface of a coffee mug, we could choose the Euclidean distance too; alternatively, we could choose the *geodesic distance*, namely the length of the shortest path from p to q on the mug's surface.

Example 1.4. (Euclidean ball) *In $\mathbb{R}^d$, the Euclidean d-ball with center c and radius r, denoted $B(c, r)$, is the point set $B(c, r) = \{p \in \mathbb{R}^d : \mathsf{d}(p, c) \leq r\}$. A 1-ball is an edge, and a 2-ball is called a disk. A unit ball is a ball with radius 1. The boundary of the d-ball is called the Euclidean $(d − 1)$-sphere and denoted $S(c, r) = \{p \in \mathbb{R}^d : \mathsf{d}(p, c) = r\}$. The name expresses the fact that we consider it a $(d − 1)$-dimensional point set – to be precise, a $(d − 1)$-dimensional manifold – even though it is embedded in d-dimensional space. For example, a circle is a 1-sphere, and a layman's "sphere" in $\mathbb{R}^3$ is a 2-sphere. If we remove the boundary from a ball, we have the open Euclidean d-ball $B_o(c, r) = \{p \in \mathbb{R}^d : \mathsf{d}(p, c) < r\}$.*

The topological spaces that are subspaces of a metric space such as $\mathbb{R}^d$ inherit their topology as a subspace topology. Examples of topological subspaces are the Euclidean d-ball $\mathbb{B}^d$, Euclidean d-sphere $\mathbb{S}^d$, open Euclidean d-ball $\mathbb{B}^d_o$, and Euclidean half-ball $\mathbb{H}^d$, where

$$\mathbb{B}^d = \{x \in \mathbb{R}^d : \|x\| \leq 1\},$$

$$\mathbb{S}^d = \{x \in \mathbb{R}^{d+1} : \|x\| = 1\},$$

$$\mathbb{B}^d_o = \{x \in \mathbb{R}^d : \|x\| < 1\},$$

$$\mathbb{H}^d = \{x \in \mathbb{R}^d : \|x\| < 1 \text{ and } x_d \geq 0\}.$$

1.3 Maps, Homeomorphisms, and Homotopies

The equivalence of two topological spaces is determined by how the points that comprise them are connected. For example, the surface of a cube can be deformed into a sphere without cutting or gluing it because they are connected the same way. They have the same topology. This notion of topological equivalence can be formalized via functions that send the points of one space to points of the other while preserving the connectivity.

This preservation of connectivity is achieved by preserving the open sets. A function from one space to another that preserves the open sets is called a *continuous function* or a *map*. Continuity is a vehicle to define topological equivalence, because a continuous function can send many points to a single point in the target space, or send no points to a given point in the target space. If the former does not happen, that is, when the function is injective, we call it an *embedding* of the domain into the target space. True equivalence is given by a *homeomorphism*, a bijective function from one space to another which has continuity as well as a continuous inverse. This ensures that open sets are preserved in both directions.

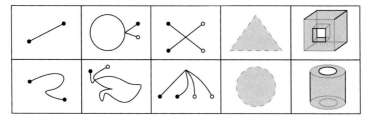

Figure 1.4 Each point set in this figure is homeomorphic to the point set above or below it, but not to any of the others. Open circles indicate points missing from the point set, as do the dashed edges in the point sets second from the right.

Definition 1.15. (Continuous function; Map) A function $f : \mathbb{T} \to \mathbb{U}$ from the topological space $\mathbb{T}$ to another topological space $\mathbb{U}$ is *continuous* if for every open set $Q \subseteq \mathbb{U}$, $f^{-1}(Q)$ is open. Continuous functions are also called *maps*.

Definition 1.16. (Embedding) A map $g : \mathbb{T} \to \mathbb{U}$ is an *embedding* of $\mathbb{T}$ into $\mathbb{U}$ if g is injective.

A topological space can be *embedded* into a Euclidean space by assigning coordinates to its points so that the assignment is continuous and injective. For example, drawing a triangle on paper is an embedding of $\mathbb{S}^1$ into $\mathbb{R}^2$. There are topological spaces that cannot be embedded into a Euclidean space, or even into a metric space – these spaces cannot be represented by any metric.

Next we define a homeomorphism that connects two spaces that have essentially the same topology.

Definition 1.17. (Homeomorphism) Let $\mathbb{T}$ and $\mathbb{U}$ be topological spaces. A *homeomorphism* is a bijective map $h : \mathbb{T} \to \mathbb{U}$ whose inverse is continuous too.

Two topological spaces are *homeomorphic* if there exists a homeomorphism between them.

Homeomorphism induces an equivalence relation among topological spaces, which is why two homeomorphic topological spaces are called *topologically equivalent*. Figure 1.4 shows pairs of homeomorphic topological spaces. A less obvious example is that the open d-ball $\mathbb{B}_o^d$ is homeomorphic to the Euclidean space $\mathbb{R}^d$, given by the homeomorphism $h(x) = x/(1 - ||x||)$. The same map also exhibits that the half-ball $\mathbb{H}^d$ is homeomorphic to the Euclidean half-space $\{x \in \mathbb{R}^d : x_d \geq 0\}$.

For maps between compact spaces, there is a weaker condition to be verified for homeomorphisms because of the following property.

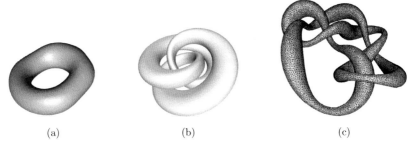

(a) (b) (c)

Figure 1.5 Two tori knotted differently, one triangulated (c) and the other not (b). Both are homeomorphic to the standard unknotted torus in (a), but not isotopic to it.

Proposition 1.1. *If $\mathbb{T}$ and $\mathbb{U}$ are compact metric spaces, every bijective map from $\mathbb{T}$ to $\mathbb{U}$ has a continuous inverse.*

One can take advantage of this fact to prove that certain functions are homeomorphisms by showing continuity only in the forward direction. When two topological spaces are subspaces of the same larger space, a notion of similarity called *isotopy* exists which is stronger than homeomorphism. If two subspaces are isotopic, one can be continuously deformed to the other while keeping the deforming subspace homeomorphic to its original form all the time. For example, a solid cube can be continuously deformed into a ball in this manner.

Homeomorphic subspaces are not necessarily isotopic. Consider a torus embedded in $\mathbb{R}^3$, illustrated in Figure 1.5(a). One can embed the torus in $\mathbb{R}^3$ so that it is knotted, as shown in Figure 1.5(b) and (c). The knotted torus is homeomorphic to the standard, unknotted one. However, it is not possible to continuously deform one to the other while keeping it embedded in $\mathbb{R}^3$ and homeomorphic to the original. Any attempt to do so forces the torus to be "self-intersecting" and thus not being a manifold. One way to look at this obstruction is by considering the topology of the space around the tori. Although the knotted and unknotted tori are homeomorphic, their complements in $\mathbb{R}^3$ are not. This motivates us to consider both the notion of an *isotopy*, in which a torus deforms continuously, and the notion of an *ambient isotopy*, in which not only the torus deforms, but the entire $\mathbb{R}^3$ deforms with it.

Definition 1.18. (Isotopy) An *isotopy* connecting two spaces $\mathbb{T} \subseteq \mathbb{R}^d$ and $\mathbb{U} \subseteq \mathbb{R}^d$ is a continuous map $\xi \colon \mathbb{T} \times [0, 1] \to \mathbb{R}^d$ where $\xi(\mathbb{T}, 0) = \mathbb{T}, \xi(\mathbb{T}, 1) = \mathbb{U}$, and for every $t \in [0, 1]$, $\xi(\cdot, t)$ is a homeomorphism between $\mathbb{T}$ and its image $\{\xi(x, t) \colon x \in \mathbb{T}\}$. An *ambient isotopy* connecting $\mathbb{T}$ and $\mathbb{U}$ is a map $\xi \colon$

$\mathbb{R}^d \times [0, 1] \to \mathbb{R}^d$ such that $\xi(\cdot, 0)$ is the identity function on $\mathbb{R}^d$, $\xi(\mathbb{T}, 1) = \mathbb{U}$, and for each $t \in [0, 1]$, $\xi(\cdot, t)$ is a homeomorphism.

For an example, consider the map

$$\xi(x, t) = \frac{1 - (1 - t)\|x\|}{1 - \|x\|} x$$

that sends the open d-ball $\mathbb{B}_o^d$ to itself if $t = 0$, and to the Euclidean space $\mathbb{R}^d$ if $t = 1$. The parameter t plays the role of time, that is, $\xi(\mathbb{B}_o^d, t)$ deforms continuously from a ball at time zero to $\mathbb{R}^d$ at time one. Thus, there is an isotopy between the open d-ball and $\mathbb{R}^d$.

Every ambient isotopy becomes an isotopy if its domain is restricted from $\mathbb{R}^d \times [0, 1]$ to $\mathbb{T} \times [0, 1]$. It is known that if there is an isotopy between two subspaces, then there exists an ambient isotopy between them. Hence, the two notions are equivalent.

There is another notion of similarity among topological spaces that is weaker than homeomorphism, called *homotopy equivalence*. It relates spaces that can be continuously deformed to one another but the transformation may not preserve homeomorphism. For example, a ball can shrink to a point, which is not homeomorphic to it because a bijective function from an infinite point set to a single point cannot exist. However, homotopy preserves some form of connectivity, such as the number of connected components, holes, and/or voids. This is why a coffee cup is homotopy equivalent to a circle, but not to a ball or a point.

To get to homotopy equivalence, we first need the concept of homotopies, which are isotopies without the homeomorphism.

Definition 1.19. (Homotopy) Let $g \colon \mathbb{X} \to \mathbb{U}$ and $h \colon \mathbb{X} \to \mathbb{U}$ be maps. A *homotopy* is a map $H \colon \mathbb{X} \times [0, 1] \to \mathbb{U}$ such that $H(\cdot, 0) = g$ and $H(\cdot, 1) = h$. Two maps are *homotopic* if there is a homotopy connecting them.

For example, let $g \colon \mathbb{B}^3 \to \mathbb{R}^3$ be the identity map on a unit ball and $h \colon \mathbb{B}^3 \to \mathbb{R}^3$ be the map sending every point in the ball to the origin. The fact that g and h are homotopic is demonstrated by the homotopy $H(x, t) = (1 - t) \cdot g(x)$. Observe that $H(\mathbb{B}^3, t)$ deforms continuously a ball at time zero to a point at time one. A key property of a homotopy is that, as H is continuous, at every time t the map $H(\cdot, t)$ remains continuous.

For developing more intuition, consider two maps that are not homotopic. Let $g \colon \mathbb{S}^1 \to \mathbb{S}^1$ be the identity map from the circle to itself, and let $h \colon \mathbb{S}^1 \to \mathbb{S}^1$ map every point on the circle to a single point $p \in \mathbb{S}^1$. Although apparently it seems that we can contract a circle to a point, that view is misleading because

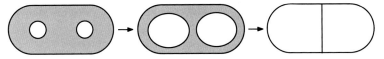

Figure 1.6 All three of the topological spaces are homotopy equivalent, because they are all deformation retracts of the leftmost space.

the map H is required to map every point on the circle at every time to a point on the circle. The contraction of the circle to a point is possible only if we break the continuity, say by cutting or gluing the circle somewhere.

Observe that a homeomorphism relates two topological spaces $\mathbb{T}$ and $\mathbb{U}$ whereas a homotopy or an isotopy (which is a special kind of homotopy) relates two maps, thereby indirectly establishing a relationship between two subspaces $g(\mathbb{X}) \subseteq \mathbb{U}$ and $h(\mathbb{X}) \subseteq \mathbb{U}$. That relationship is not necessarily an equivalent one, but the following is.

Definition 1.20. (Homotopy equivalent) Two topological spaces $\mathbb{T}$ and $\mathbb{U}$ are *homotopy equivalent* if there exist maps $g : \mathbb{T} \to \mathbb{U}$ and $h : \mathbb{U} \to \mathbb{T}$ such that $h \circ g$ is homotopic to the identity map $\iota_\mathbb{T} : \mathbb{T} \to \mathbb{T}$ and $g \circ h$ is homotopic to the identity map $\iota_\mathbb{U} : \mathbb{U} \to \mathbb{U}$.

Homotopy equivalence is indeed an equivalence relation, that is, if A, B and B, C are homotopy equivalent spaces, so are the pair A, C. Homeomorphic spaces necessarily have the same dimension though homotopy equivalent spaces may have different dimensions. To gain more intuition about homotopy equivalent spaces, we show why a 2-ball is homotopy equivalent to a single point p. Consider a map $h : \mathbb{B}^2 \to \{p\}$ and a map $g : \{p\} \to \mathbb{B}^2$ where $g(p)$ is any point q in $\mathbb{B}^2$. Observe that $h \circ g$ is the identity map on $\{p\}$, which is trivially homotopic to itself. In the other direction, $g \circ h : \mathbb{B}^2 \to \mathbb{B}^2$ sends every point in $\mathbb{B}^2$ to q. A homotopy between $g \circ h$ and the identity map $\text{id}_{\mathbb{B}^2}$ is given by the map $H(x, t) = (1 - t)q + tx$.

An useful intuition for understanding the definition of homotopy equivalent spaces can be derived from the fact that two spaces $\mathbb{T}$ and $\mathbb{U}$ are homotopy equivalent if and only if there exists a third space $\mathbb{X}$ so that both $\mathbb{T}$ and $\mathbb{U}$ are *deformation retracts* of $\mathbb{X}$; see Figure 1.6.

Definition 1.21. (Deformation retract) Let $\mathbb{T}$ be a topological space, and let $\mathbb{U} \subset \mathbb{T}$ be a subspace. A *retraction* r of $\mathbb{T}$ to $\mathbb{U}$ is a map from $\mathbb{T}$ to $\mathbb{U}$ such that $r(x) = x$ for every $x \in \mathbb{U}$. The space $\mathbb{U}$ is a *deformation retract* of $\mathbb{T}$ if the identity map on $\mathbb{T}$ can be continuously deformed to a retraction with no motion of the points already in $\mathbb{U}$: specifically, there is a homotopy called *deformation* *retraction* $R: \mathbb{T} \times [0, 1] \rightarrow \mathbb{T}$ such that $R(\cdot, 0)$ is the identity map on $\mathbb{T}$, $R(\cdot, 1)$ is a retraction of $\mathbb{T}$ to $\mathbb{U}$, and $R(x, t) = x$ for every $x \in \mathbb{U}$ and every $t \in [0, 1]$.

Fact 1.1. *If $\mathbb{U}$ is a deformation retract of $\mathbb{T}$, then $\mathbb{T}$ and $\mathbb{U}$ are homotopy equivalent.*

For example, any point on a line segment (open or closed) is a deformation retract of the line segment and is homotopy equivalent to it. The letter M is a deformation retract of the letter W, and also of a 1-ball. Moreover, as we said before, two spaces are homotopy equivalent if they are deformation retractions of a common space. The symbols ∅, ∞, and ∘—∘ (viewed as one-dimensional point sets) are deformation retracts of a double doughnut – a doughnut with two holes. Therefore, they are homotopy equivalent to each other, though none of them is a deformation retract of any of the others because one is not a subspace of the other. They are not homotopy equivalent to A, X, O, ⊕, ⊙, ◎, a ball, nor a coffee cup.

1.4 Manifolds

A manifold is a topological space that is locally connected in a particular way. A 1-manifold has this local connectivity looking like a segment. A 2-manifold (with boundary) has the local connectivity looking like a complete or partial disk. In layman's terms, a 2-manifold has the structure of a piece of paper or rubber sheet, possibly with the boundaries glued together to form a closed surface – a category that includes disks, spheres, tori, and Möbius bands.

Definition 1.22. (Manifold) A topological space M is an *m-manifold*, or simply a *manifold*, if every point $x \in M$ has a neighborhood homeomorphic to $\mathbb{B}_o^m$ or $\mathbb{H}^m$. The *dimension* of M is m.

Every manifold can be partitioned into boundary and interior points. Observe that these words mean very different things for a manifold than they do for a metric space or topological space.

Definition 1.23. (Boundary; Interior) The *interior* Int M of an *m*-manifold M is the set of points in M that have a neighborhood homeomorphic to $\mathbb{B}_o^m$. The *boundary* Bd M of M is the set of points $M \setminus$ Int M. The boundary Bd M, if not empty, consists of the points that have a neighborhood homeomorphic to $\mathbb{H}^m$. If Bd M is the empty set, we say that M is *without boundary*.

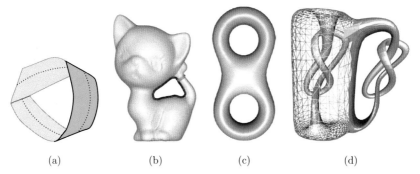

(a)   (b)   (c)   (d)

Figure 1.7 (a) A Möbius band. (b) Removal of the red and green loops opens up the torus into a topological disk. (c) A double torus: every surface without boundary in $\mathbb{R}^3$ resembles a sphere or a conjunction of one or more tori. (d) Double torus knotted.

A single point, a 0-ball, is a 0-manifold without boundary according to this definition. The closed disk $\mathbb{B}^2$ is a 2-manifold whose interior is the open disk $\mathbb{B}^2_o$ and whose boundary is the circle $\mathbb{S}^1$. The open disk $\mathbb{B}^2_o$ is a 2-manifold whose interior is $\mathbb{B}^2_o$ and whose boundary is the empty set. This highlights an important difference between Definitions 1.13 and 1.23 of "boundary": when $\mathbb{B}^2_o$ is viewed as a point set in the space $\mathbb{R}^2$, its boundary is $\mathbb{S}^1$ according to Definition 1.13; but viewed as a manifold, its boundary is empty according to Definition 1.23. The boundary of a manifold is *always* included in the manifold.

The open disk $\mathbb{B}^2_o$, the Euclidean space $\mathbb{R}^2$, the sphere $\mathbb{S}^2$, and the torus are all connected 2-manifolds without boundary. The first two are homeomorphic to each other, but the last two are not. The sphere and the torus in $\mathbb{R}^3$ are compact (bounded and closed with respect to $\mathbb{R}^3$) whereas $\mathbb{B}^2_o$ and $\mathbb{R}^2$ are not.

A d-manifold, $d \geq 2$, can have orientations whose formal definition we skip here. Informally, we say that a 2-manifold M is *non-orientable* if, starting from a point p, one can walk on one side of M and end up on the opposite side of M upon returning to p. Otherwise, M is *orientable*. Spheres and balls are orientable, whereas the *Möbius band* in Figure 1.7(a) is a non-orientable 2-manifold with boundary.

A *surface* is a 2-manifold that is a subspace of $\mathbb{R}^d$. Any compact surface without boundary in $\mathbb{R}^3$ is an orientable 2-manifold. To be non-orientable, a compact surface must have a non-empty boundary (like the Möbius band) or be embedded in a four- or higher-dimensional Euclidean space.

A surface can sometimes be disconnected by removing one or more *loops* (connected 1-manifolds without boundary) from it. The *genus* of an orientable and compact surface without boundary is g if $2g$ is the maximum number of loops that can be removed from the surface without disconnecting it; here the loops are permitted to intersect each other. For example, the sphere has genus zero as every loop cuts it into two disks. The torus has genus one: a circular cut around its neck and a second circular cut around its circumference, illustrated in Figure 1.7(b), allow it to unfold into a topological disk. A third loop would cut it into two pieces. Figure 1.7(c) and (d) each shows a 2-manifold without boundary of genus two. Although a high-genus surface can have a very complex shape, all compact 2-manifolds in $\mathbb{R}^3$ that have the same genus and no boundary are homeomorphic to each other.

1.4.1 Smooth Manifolds

A purely topological manifold has no geometry. But if we embed it in a Euclidean space, it could appear smooth or wrinkled. We now introduce a "geometric" manifold by imposing a differential structure on it. For the rest of this chapter, we focus on only manifolds without boundary.

Consider a map $\phi\colon U \to W$ where U and W are open sets in $\mathbb{R}^k$ and $\mathbb{R}^d$, respectively. The map ϕ has d components, namely $\phi(x) = (\phi_1(x), \phi_2(x), \ldots, \phi_d(x))$, where $x = (x_1, x_2, \ldots, x_k)$ denotes a point in $\mathbb{R}^k$. The *Jacobian* of ϕ at x is the $d \times k$ matrix of the first-order partial derivatives

$$\begin{bmatrix} \dfrac{\partial \phi_1(x)}{\partial x_1} & \cdots & \dfrac{\partial \phi_1(x)}{\partial x_k} \\ \vdots & \ddots & \vdots \\ \dfrac{\partial \phi_d(x)}{\partial x_1} & \cdots & \dfrac{\partial \phi_d(x)}{\partial x_k} \end{bmatrix}.$$

The map ϕ is *regular* if its Jacobian has rank k at every point in U. The map ϕ is C^i-continuous if the i-th-order partial derivatives of ϕ are continuous.

The reader may be familiar with *parametric surfaces*, for which U is a two-dimensional *parameter space* and its image $\phi(U)$ in d-dimensional space is a parametric surface. Unfortunately, a single parametric surface cannot easily represent a manifold with a complicated topology. However, for a manifold to be smooth, it suffices that each point on the manifold has a neighborhood that looks like a smooth parametric surface.

Definition 1.24. (Smooth embedded manifold) For any $i > 0$, an m-manifold M without boundary embedded in $\mathbb{R}^d$ is C^i-*smooth* if for every point $p \in M$, there exists an open set $U_p \subset \mathbb{R}^m$, a neighborhood $W_p \subset \mathbb{R}^d$ of p, and a map $\phi_p\colon U_p \to W_p \cap M$ such that (i) ϕ_p is C^i-continuous, (ii) ϕ_p is a homeomorphism, and (iii) ϕ_p is regular. If $m = 2$, we call M a C^i-*smooth surface*.

The first condition says that each map is continuously differentiable at least *i* times. The second condition requires each map to be bijective, ruling out "wrinkles" where multiple points in *U* map to a single point in *W*. The third condition prohibits any map from having a directional derivative of zero at any point in any direction. The first and third conditions together enforce smoothness, and imply that there is a well-defined tangent *m*-flat at each point in *M*. The three conditions together imply that the maps ϕ_p defined in the neighborhood of each point $p \in M$ overlap smoothly. There are two extremes of smoothness. We say that *M* is C^∞-smooth if for every point $p \in M$, the partial derivatives of ϕ_p of all orders are continuous. On the other hand, *M* is *nonsmooth* if *M* is an *m*-manifold (therefore C^0-smooth) but not C^1-smooth.

1.5 Functions on Smooth Manifolds

In previous sections, we introduced topological spaces, including the special case of (smooth) manifolds. Very often, a space can be equipped with continuous functions defined on it. In this section, we focus on *real-valued* functions of the form $f : X \to \mathbb{R}$ defined on a topological space *X*, also called *scalar functions*; see Figure 1.8(a) for the graph of a function $f : \mathbb{R}^2 \to \mathbb{R}$. Scalar functions appear commonly in practice that describe space/data of interest (e.g., the elevation function defined on the surface of the Earth). We are interested in the topological structures behind scalar functions. In this section, we limit our discussion to nicely behaved scalar functions (called Morse functions) defined on smooth manifolds. Their topological structures are characterized by the so-called critical points which we will introduce below. Later in the book we will also discuss scalar functions on simplicial complex domains, as well as more complex maps defined on a space *X*, for example, a multivariate function $f : X \to \mathbb{R}^d$.

1.5.1 Gradients and Critical Points

In what follows, for simplicity of presentation, we assume that we consider smooth (C^∞-continuous) functions and smooth manifolds embedded in $\mathbb{R}^d$, even though often we only require the functions (resp. manifolds) to be C^2-continuous (resp. C^2-smooth).

To provide intuition, let us start with a smooth scalar function defined on the real line, $f : \mathbb{R} \to \mathbb{R}$; the graph of such a function is shown in Figure 1.8(b). Recall that the *derivative* of a function at a point $x \in \mathbb{R}$ is defined as

$$Df(x) = \frac{d}{dx} f(x) = \lim_{t \to 0} \frac{f(x + t) - f(x)}{t}. \tag{1.1}$$

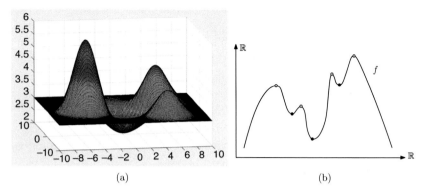

Figure 1.8 (a) The graph of a function $f : \mathbb{R}^2 \to \mathbb{R}$. (b) The graph of a function $f : \mathbb{R} \to \mathbb{R}$ with critical points marked.

The value $Df(x)$ gives the rate of change of the value of f at x. This can be visualized as the slope of the tangent line of the graph of f at $(x, f(x))$. The *critical points* of f are the set of points x such that $Df(x) = 0$. For a function defined on the real line, there are two types of critical points in the generic case: maxima and minima, as marked in Figure 1.8(b).

Now suppose we have a smooth function $f : \mathbb{R}^d \to \mathbb{R}$ defined on $\mathbb{R}^d$. Fix an arbitrary point $x \in \mathbb{R}^d$. As we move a little around x within its local neighborhood, the rate of change of f differs depending on which direction we move. This gives rise to the *directional derivative* $D_v f(x)$ at x in direction (i.e., a unit vector) $v \in \mathbb{S}^{d-1}$, where $\mathbb{S}^{d-1}$ is the unit $(d-1)$-sphere, defined as

$$D_v f(x) = \lim_{t \to 0} \frac{f(x + t \cdot v) - f(x)}{t}. \tag{1.2}$$

The gradient vector of f at $x \in \mathbb{R}^d$ intuitively captures the direction of steepest increase of the function f. More precisely, we have the following.

Definition 1.25. (Gradient for functions on $\mathbb{R}^d$) Given a smooth function $f : \mathbb{R}^d \to \mathbb{R}$, the *gradient vector field* $\nabla f : \mathbb{R}^d \to \mathbb{R}^d$ is defined as follows: for any $x \in \mathbb{R}^d$,

$$\nabla f(x) = \left[\frac{\partial f}{\partial x_1}(x), \ \frac{\partial f}{\partial x_2}(x), \ \ldots, \ \frac{\partial f}{\partial x_d}(x) \right]^{\mathsf{T}}, \tag{1.3}$$

where $(x_1, x_2, \ldots, x_d)$ represents an orthonormal coordinate system for $\mathbb{R}^d$. The vector $\nabla f(x) \in \mathbb{R}^d$ is called the *gradient vector of f at x*. A point $x \in \mathbb{R}^d$ is a *critical point* if $\nabla f(x) = [0\ 0\ \ldots\ 0]^{\mathsf{T}}$; otherwise, x is *regular*.

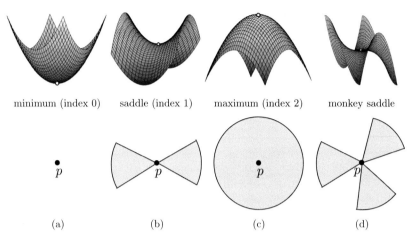

minimum (index 0)   saddle (index 1)   maximum (index 2)   monkey saddle

p p p p

(a)   (b)   (c)   (d)

Figure 1.9 (top) The graph of the function around nondegenerate critical points for a smooth function on $\mathbb{R}^2$, and a degenerate critical point, called "monkey saddle." For example, for an index-0 critical point p, its local neighborhood can be written as $f(x) = f(p) + x_1^2 + x_2^2$, making p a local minimum. (bottom) The local (closed) neighborhood of the corresponding critical point in the domain $\mathbb{R}^2$, where the dark blue colored regions are the portion of the neighborhood of p whose function value is at most $f(p)$.

Observe that for any $v \in \mathbb{R}^d$, the directional derivative satisfies that $D_v f(x) = \langle \nabla f(x), v \rangle$. It then follows that $\nabla f(x) \in \mathbb{R}^d$ is along the unit vector v where $D_v f(x)$ is maximized among the directional derivatives in all unit directions around x; and its magnitude $\|\nabla f(x)\|$ equals the value of this maximum directional derivative. The critical points of f are those points where the directional derivative vanishes in all directions – locally, the rate of change of f is zero no matter which direction one deviates from x. See Figure 1.9 for the three types of critical points, minimum, saddle point, and maximum, for a generic smooth function $f : \mathbb{R}^2 \to \mathbb{R}$.

Finally, we can extend the above definitions of gradients and critical points to a smooth function $f : M \to \mathbb{R}$ defined on a smooth Riemannian m-manifold M. Here, a Riemannian manifold is a manifold equipped with a Riemannian metric, which is a smoothly varying inner product defined on the tangent spaces. This allows the measurements of length so as to define gradient. At a point $x \in M$, denote the tangent space of M at x by TM_x, which is the m-dimensional vector space consisting of all tangent vectors of M at x. For example, TM_x is an m-dimensional linear space $\mathbb{R}^m$ for an m-dimensional manifold M embedded in the Euclidean space $\mathbb{R}^d$, with Riemannian metric (inner product in the tangent space) induced from $\mathbb{R}^d$.

The gradient ∇f is a vector field on M, that is, $\nabla f : M \to TM$ maps every point $x \in M$ to a vector $\nabla f(x) \in TM_x$ in the tangent space of M at x. Similar to the case for a function defined on $\mathbb{R}^d$, the gradient vector field ∇f satisfies that for any $x \in M$ and $v \in TM_x$, $\langle \nabla f(x), v \rangle$ gives rise to the directional derivative $D_v f(x)$ of f in direction v, and $\nabla f(x)$ still specifies the direction of steepest increase of f along all directions in TM_x with its magnitude being the maximum rate of change. More formally, we have the following definition, analogous to Definition 1.25 for the case of a smooth function on $\mathbb{R}^d$.

Definition 1.26. (Gradient vector field; Critical points) Given a smooth function $f : M \to \mathbb{R}$ defined on a smooth m-dimensional Riemannian manifold M, the *gradient vector field* $\nabla f : M \to TM$ is defined as follows: for any $x \in M$, let $(x_1, x_2, \ldots, x_m)$ be a local coordinate system in a neighborhood of x with orthonormal unit vectors x_i, then the gradient at x is

$$\nabla f(x) = \left[\frac{\partial f}{\partial x_1}(x), \ \frac{\partial f}{\partial x_2}(x), \ \ldots, \ \frac{\partial f}{\partial x_m}(x) \right]^{\mathrm{T}}.$$

A point $x \in M$ is *critical* if $\nabla f(x)$ vanishes, in which case $f(x)$ is called a *critical value* for f. Otherwise, x is *regular.*

It follows from the chain rule that the criticality of a point x is independent of the local coordinate system being used.

1.5.2 Morse Functions and Morse Lemma

From the first-order derivatives of a function we can determine critical points. We can learn more about the "type" of the critical points by inspecting the second-order derivatives of f.

Definition 1.27. (Hessian matrix; Nondegenerate critical points) Given a smooth m-manifold M, the *Hessian matrix* of a twice differentiable function $f : M \to \mathbb{R}$ at x is the matrix of second-order partial derivatives,

$$\text{Hessian}(x) = \begin{bmatrix} \dfrac{\partial^2 f}{\partial x_1 \partial x_1}(x) & \dfrac{\partial^2 f}{\partial x_1 \partial x_2}(x) & \cdots & \dfrac{\partial^2 f}{\partial x_1 \partial x_m}(x) \\ \dfrac{\partial^2 f}{\partial x_2 \partial x_1}(x) & \dfrac{\partial^2 f}{\partial x_2 \partial x_2}2(x) & \cdots & \dfrac{\partial^2 f}{\partial x_2 \partial x_m}(x) \\ \vdots & \vdots & \ddots & \vdots \\ \dfrac{\partial^2 f}{\partial x_m \partial x_1}(x) & \dfrac{\partial^2 f}{\partial x_m \partial x_2}2(x) & \cdots & \dfrac{\partial^2 f}{\partial x_m \partial x_m}(x) \end{bmatrix},$$

where $(x_1, x_2, \ldots, x_m)$ is a local coordinate system in a neighborhood of x.

A critical point x of f is *nondegenerate* if its Hessian matrix, Hessian(x), is nonsingular (has nonzero determinant); otherwise, it is a *degenerate critical point*.

For example, consider $f: \mathbb{R}^2 \to \mathbb{R}$ defined by $f(x, y) = x^3 - 3xy^2$. The origin $(0, 0)$ is a degenerate critical point often referred to as a "monkey saddle:" see Figure 1.9(d), where the graph of the function around $(0, 0)$ goes up and down three times (instead of twice as for a nondegenerate saddle shown in Figure 1.9b). It turns out that, as a consequence of the Morse Lemma below, nondegenerate critical points are always isolated whereas the degenerate ones may not be so. A simple example is $f: \mathbb{R}^2 \to \mathbb{R}$ defined by $f(x, y) = x^2$, where all points on the y-axis are degenerate critical points. The local neighborhood of nondegenerate critical points can be completely characterized by the following Morse Lemma.

Proposition 1.2. (Morse Lemma) *Given a smooth function $f: M \to \mathbb{R}$ defined on a smooth m-manifold M, let p be a nondegenerate critical point of f. Then there is a local coordinate system in a neighborhood $U(p)$ of p so that (i) the coordinate of p is $(0, 0, \ldots, 0)$, and (ii) locally for every point $x = (x_1, x_2, \ldots, x_m)$ in neighborhood $U(p)$,*

$$f(x) = f(p) - x_1^2 - \cdots - x_s^2 + x_{s+1}^2 \cdots + x_m^2, \quad for\ some\ s \in [0, m].$$

The number s of minus signs in the above quadratic representation of $f(x)$ is called the index of the critical point p.

For a smooth function $f: M \to \mathbb{R}$ defined on a 2-manifold M, an index-0, index-1, or index-2 (nondegenerate) critical point corresponds to a minimum, a saddle, or a maximum, respectively. For a function defined on an m-manifold, nondegenerate critical points include minima (index 0), maxima (index m), and $m - 1$ types of saddle points.

The behavior of degenerate critical points is more complicated to characterize. Instead, we now introduce a family of "nice" functions, called *Morse functions*, whose critical points cannot be degenerate.

Definition 1.28. (Morse function) A smooth function $f: M \to \mathbb{R}$ defined on a smooth manifold M is a *Morse function* if and only if: (i) none of f's critical points are degenerate; and (ii) the critical points have distinct function values.

Limiting our study only to well-behaved Morse functions is not too restrictive as the Morse functions form an open and dense subset of the space of all smooth functions $C^\infty(M)$ on M. So in this sense, a generic function is a Morse function. On the other hand, it is much cleaner to characterize the topology induced by such a function, which we do now.

1.5.3 Connection to Topology

We now characterize how critical points influence the topology of M induced by the scalar function $f : M \to \mathbb{R}$.

Definition 1.29. (Interval, sublevel, and superlevel sets) Given $f : M \to \mathbb{R}$ and $I \subseteq \mathbb{R}$, the *interval levelset* of f with respect to I is defined as

$$M_I = f^{-1}(I) = \{x \in M \mid f(x) \in I\}.$$

The case for $I = (-\infty, a]$ is also referred to as the *sublevel set* $M_{\leq a} := f^{-1}((-\infty, a])$ of f, while $M_{\geq a} := f^{-1}([a, \infty))$ is called the *superlevel set*; and $f^{-1}(a)$ is called the *levelset* of f at $a \in \mathbb{R}$.

Given $f : M \to \mathbb{R}$, imagine sweeping M with increasing function values of f. It turns out that the topology of the sublevel sets can only change when we sweep through critical values of f. More precisely, we have the following classical result, where a diffeomorphism is a homeomorphism that is smooth in both directions.

Theorem 1.3. (Homotopy type of sublevel sets) *Let $f : M \to \mathbb{R}$ be a smooth function defined on a manifold M. Given $a < b$, suppose the interval levelset $M_{[a,b]} = f^{-1}([a, b])$ is compact and contains no critical points of f. Then $M_{\leq a}$ is diffeomorphic to $M_{\leq b}$.*

Furthermore, $M_{\leq a}$ is a deformation retract of $M_{\leq b}$, and the inclusion map $i : M_{\leq a} \hookrightarrow M_{\leq b}$ is a homotopy equivalence.

As an illustration, consider the example of height function $f : M \to \mathbb{R}$ defined on a vertical torus as shown in Figure 1.10(a). There are four critical points for the height function f, u (minimum), v, w (saddles), and z (maximum). We have that $M_{\leq a}$ is: (i) empty for $a < f(u)$; (ii) homeomorphic to a 2-disk for $f(u) < a < f(v)$; (iii) homeomorphic to a cylinder for $f(v) < a < f(w)$; (iv) homeomorphic to a compact genus-one surface with a circle as boundary for $f(w) < a < f(z)$; and (v) a full torus for $a > f(z)$.

Theorem 1.3 states that the homotopy type of the sublevel set remains the same until it passes a critical point. For Morse functions, we can also characterize the homotopy type of sublevel sets around critical points, captured by *attaching k-cells*.

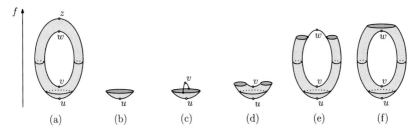

Figure 1.10 (a) The height function defined on a torus with critical points u, v, w, and z. (b)–(f) Passing through an index-k critical point is the same as attaching a k-cell from the homotopy point of view. For example, $M_{\leq a+\varepsilon}$ for $a = f(v)$ (as shown in (d)) is homotopy equivalent to attaching a 1-cell (shown in (c)) to $M_{\leq a-\varepsilon}$ (shown in (b)) for an infinitesimal positive ε.

Specifically, recall that $\mathbb{B}^k$ is the k-dimensional unit Euclidean ball, and its boundary is $\mathbb{S}^{k-1}$, the $(k-1)$-dimensional sphere. Let X be a topological space, and $g\colon \mathbb{S}^{k-1} \to X$ a continuous map. For $k > 0$, *attaching a k-cell to X (w.r.t. g)* is obtained by attaching the k-cell $\mathbb{B}^k$ to X along its boundary as follows: first, take the disjoint union of X and $\mathbb{B}^k$, and next, identify all points $x \in \mathbb{S}^{k-1}$ with $g(x) \in X$. For the special case of $k = 0$, attaching a 0-cell to X is obtained by simply taking the disjoint union of X and a single point.

The following theorem states that, from the homotopy point of view, sweeping past an index-k critical point is equivalent to attaching a k-cell to the sublevel set. See Figure 1.10 for illustrations.

Theorem 1.4. *Given a Morse function $f : M \to \mathbb{R}$ defined on a smooth manifold M, let p be an index-k critical point of f with $\alpha = f(p)$. Assume $f^{-1}([\alpha - \varepsilon, \alpha + \varepsilon])$ is compact for a sufficiently small $\varepsilon > 0$ such that there are no other critical points of f contained in this interval levelset other than p. Then the sublevel set $M_{\leq \alpha+\varepsilon}$ has the same homotopy type as $M_{\leq \alpha-\varepsilon}$ with a k-cell attached to its boundary* Bd $M_{\leq \alpha-\varepsilon}$.

Finally, we state the well-known Morse inequalities, connecting critical points with the so-called Betti numbers of the domain which we will define formally in Section 2.5. In particular, fixing a field coefficient, the i-th Betti number is the rank of the so-called i-th (singular) homology group of a topological space X.

Theorem 1.5. (Morse inequalities) *Let f be a Morse function on a smooth compact d-manifold M. For $0 \leq i \leq d$, let c_i denote the number of critical points of f with index i, and β_i be the i-th Betti number of M. We then have:*

(a) $c_i \geq \beta_i$ *for all* $i \geq 0$; *and* $\sum_{i=0}^{d}(-1)^i c_i = \sum_{i=0}^{d}(-1)^i \beta_i$ *(weak Morse inequality)*;

(b) $c_i - c_{i-1} + c_{i-2} - \cdots \pm c_0 \geq \beta_i - \beta_{i-1} + \beta_{i-2} \cdots \pm \beta_0$ *for all* $i \geq 0$ *(strong Morse inequality)*.

1.6 Notes and Exercises

A good source on point set topology is Munkres [243]. The concepts of various maps and manifolds are well described in Hatcher [186]. The books by Guillemin and Pollack [179] and Milnor [233, 234] are good sources for Morse theory on smooth manifolds and differential topology in general.

Exercises

1. A space is called Hausdorff if every two disjoint point sets have disjoint open sets containing them.
   (a) Give an example of a space that is not Hausdorff.
   (b) Give an example of a space that is Hausdorff.
   (c) Show the above examples on the same ground set $\mathbb{T}$.
2. In every space $\mathbb{T}$, the point sets $\varnothing$ and $\mathbb{T}$ are both closed and open.
   (a) Give an example of a space that has more than two sets that are both closed and open, and list all of those sets.
   (b) Explain the relationship between the idea of connectedness and the number of sets that are both closed and open.
3. A topological space $\mathbb{T}$ is called *path connected* if any two points $x, y \in \mathbb{T}$ can be joined by a path, that is, there exists a continuous map $f : [0, 1] \to \mathbb{T}$ of the segment $[0, 1] \subset \mathbb{R}$ onto $\mathbb{T}$ so that $f(0) = x$ and $f(1) = y$. Prove that a path connected space is also connected but the converse may not be true; however, if $\mathbb{T}$ is finite, then the two notions are equivalent.
4. Prove that for every subset X of a metric space, $\mathrm{Cl\,Cl}\,X = \mathrm{Cl}\,X$. In other words, augmenting a set with its limit points does not give it more limit points.
5. Show that any metric on a finite set induces a discrete topology.
6. Prove that the metric is a continuous function on the Cartesian space $\mathbb{T} \times \mathbb{T}$ of a metric space $\mathbb{T}$.
7. Give an example of a bijective function that is continuous, but its inverse is not. In light of Proposition 1.1, the spaces need to be noncompact.
8. A space is called *normal* if it is Hausdorff and for any two disjoint closed sets X and Y, there are disjoint open sets $U_X \supset X$ and $U_Y \supset Y$. Show that any metric space is normal. Show the same for any compact space.

9. Let $f : \mathbb{T} \to \mathbb{U}$ be a continuous function of a compact space $\mathbb{T}$ into another space $\mathbb{U}$. Prove that the image $f(\mathbb{T})$ is compact.

10. (a) Construct an explicit deformation retraction of $\mathbb{R}^k \setminus \{o\}$ onto $\mathbb{S}^{k-1}$ where o denotes the origin. Also, show $\mathbb{R}^k \cup \{\infty\}$ is homeomorphic to $\mathbb{S}^k$.

   (b) Show that any d-dimensional finite convex polytope is homeomorphic to the d-dimensional unit ball $\mathbb{B}^d$.

11. Deduce that homeomorphism is an equivalence relation. Show that the relation of homotopy among maps is an equivalence relation.

12. Consider the function $f : \mathbb{R}^3 \to \mathbb{R}$ defined as $f(x_1, x_2, x_3) = 3x_1^2 + 3x_2^2 - 9x_3^2$. Show that the origin $(0, 0, 0)$ is a critical point of f. Give the index of this critical point. Let S denote the unit sphere centered at the origin. Show that $f^{(-\infty, 0]} \cap S$ is homotopy equivalent to two points, whereas $f^{[0, \infty)} \cap S$ is homotopy equivalent to $\mathbb{S}^1$, the unit 1-sphere (i.e., circle).

2
Complexes and Homology Groups

This chapter introduces two very basic tools on which topological data analysis (TDA) is built. One is simplicial complexes and the other is homology groups. Data supplied as a discrete set of points do not have an interesting topology. Usually, we construct a scaffold on top of the data which is commonly taken as a simplicial complex. It consists of vertices at the data points, edges connecting them, and triangles, tetrahedra, and their higher-dimensional analogues that establish higher-order connectivity. Section 2.1 formalizes this construction. There are different kinds of simplicial complexes. Some are easier to compute, but take more space. Others are more sparse, but take more time to compute. Section 2.2 presents an important construction called the *nerve* and a complex called the Čech complex which is defined on this construction. This section also presents a commonly used complex in topological data analysis called the Vietoris–Rips complex that interleaves with the Čech complexes in terms of containment. In Section 2.3, we introduce some of the complexes which are sparser in size than the Vietoris–Rips or Čech complexes.

The second topic of this chapter, the homology groups of a simplicial complex, are the essential algebraic structures with which TDA analyzes data. Homology groups of a topological space capture the space of cycles up to those called boundaries that bound "higher-dimensional" subsets. For simplicity, we introduce the concept in the context of simplicial complexes instead of topological spaces. This is called simplicial homology. The essential entities for defining the homology groups are chains, cycles, and boundaries which we cover in Section 2.4. For simplicity and also for relevance in TDA, we define these structures under $\mathbb{Z}_2$-additions.

Section 2.5 defines the simplicial homology group of a simplicial complex as the quotient space of the cycles with respect to the boundaries. Some of the concepts related to homology groups, such as induced homology under a map, singular homology groups for general topological spaces, relative homology groups of a complex with respect to a subcomplex, and the dual concept of homology groups, called cohomology groups are also introduced in this section.

2.1 Simplicial Complex

A complex is a collection of some basic elements that satisfy certain properties. In a simplicial complex, these basic elements are simplices.

Definition 2.1. (Simplex) For $k \geq 0$, a *k-simplex* σ in a Euclidean space $\mathbb{R}^m$ is the convex hull[1] of a set P of $k + 1$ affinely independent points in $\mathbb{R}^m$. In particular, a 0-simplex is a *vertex*, a 1-simplex is an *edge*, a 2-simplex is a *triangle*, and a 3-simplex is a *tetrahedron*. A k-simplex is said to have *dimension* k. For $0 \leq k' \leq k$, a *k'-face* (or simply a *face*) of σ is a k'-simplex that is the convex hull of a non-empty subset of P. Faces of σ come in all dimensions from zero (σ's vertices) to k; and σ is a face of σ. A *proper face* of σ is a simplex that is the convex hull of a proper subset of P; that is, any face except σ. The $(k - 1)$-faces of σ are called *facets* of σ; σ has $k + 1$ facets.

In Figure 2.1(a), triangle *abc* is a 2-simplex which has three vertices as 0-faces and three edges as 1-faces. These are proper faces out of which edges are its facets. Similarly, a tetrahedron has four 0-faces (vertices), six 1-faces (edges), four 2-faces (triangles), and one 3-face (the tetrahedron itself) out of which vertices, edges, and triangles are proper. The triangles are facets.

Definition 2.2. (Geometric simplicial complex) A *geometric simplicial complex K*, also known as a *triangulation*, is a set containing finitely[2] many simplices that satisfies the following two restrictions.

- K contains every face of each simplex in K.
- For any two simplices $\sigma, \tau \in K$, their intersection $\sigma \cap \tau$ is either empty or a face of both σ and τ.

The *dimension k* of K is the maximum dimension of any simplex in K, which is why we also refer to it as a simplicial k-complex.

---

[1] The convex hull of a set of given points $p_0, \ldots, p_k$ in $\mathbb{R}^m$ is the set of all points $x \in \mathbb{R}^m$ that are a convex combination of the given points, that is, $x = \Sigma_{i=0}^{k} \alpha_i p_i$ for $\alpha_i \geq 0$ and $\Sigma \alpha_i = 1$.

[2] Topologists usually define complexes so they have countable cardinality. We restrict complexes to finite cardinality here.

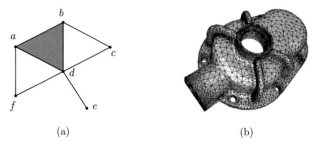

(a)   (b)

Figure 2.1 (a) A simplicial complex with six vertices, eight edges, and one triangle. (b) A simplicial 2-complex triangulating a 2-manifold in $\mathbb{R}^3$.

The above definition of simplicial complexes is very geometric, which is why they are referred to as geometric simplicial complexes. Figure 2.1 shows such a geometric simplicial 2-complex in $\mathbb{R}^2$ (Figure 2.1a) and another in $\mathbb{R}^3$ (Figure 2.1b). There is a parallel notion of simplicial complexes that is devoid of geometry.

Definition 2.3. (Abstract simplex and simplicial complex) A collection K of non-empty subsets of a given set $V(K)$ is an *abstract simplicial complex* if every element $\sigma \in K$ has all of its non-empty subsets $\sigma' \subseteq \sigma$ also in K. Each such element σ with $|\sigma| = k + 1$ is called a *k-simplex* (or simply a *simplex*). Each subset $\sigma' \subseteq \sigma$ with $|\sigma'| = k' + 1$ is called a *k'-face* (or simply a *face*) of σ, and σ with $|\sigma| = k + 1$ is called a *k-coface* (or simply a *coface*) of σ'. Sometimes, σ' is also called a face of σ with *codimension* $k - k'$. Also, a $(k - 1)$-face (resp. $(k + 1)$-coface) of a k-simplex is called its *facet* (resp. *cofacet*). The elements of $V(K)$ are the vertices of K. Each k-simplex in K is said to have dimension k. We also say K is a *simplicial k-complex* if the top dimension of any simplex in K is k.

Remark 2.1. *The collection K can possibly be empty in which case $V(K)$ is empty though a non-empty K cannot have the empty set as one of its elements by definition.*

A geometric simplicial complex K in $\mathbb{R}^m$ is called a *geometric realization* of an abstract simplicial complex K' if and only if there is an embedding $e \colon V(K') \to \mathbb{R}^m$ that takes every k-simplex $\{v_0, v_1, \dots, v_k\}$ in K' to a k-simplex in K that is the convex hull of $e(v_0), e(v_1), \dots, e(v_k)$. For example, the complex drawn in $\mathbb{R}^2$ in Figure 2.1(a) is a geometric realization of the abstract complex with vertices a, b, c, d, e, f, eight 1-simplices $\{a, b\}, \{a, d\}, \{a, f\}, \{b, c\}, \{b, d\}, \{c, d\}, \{d, e\}, \{d, f\}$, and one 2-simplex $\{a, b, d\}$.

Any simplicial k-complex can be geometrically realized in $\mathbb{R}^{2k+1}$ by mapping the vertices generically to the *moment curve* $C(t)$ in $\mathbb{R}^{2k+1}$ given by the parameterization $C(t) = (t, t^2, \ldots, t^{2k+1})$. Also, an abstract simplicial complex K with m vertices can always be geometrically realized in $\mathbb{R}^{m-1}$ as a subcomplex of a geometric $(m-1)$-simplex. To make the realization canonical, we choose the $(m-1)$-simplex to be in $\mathbb{R}^m$ with a vertex v_i having the ith coordinate to be 1 and all other coordinates 0. We define K's underlying space as the underlying space of this canonical geometric realization.

Definition 2.4. (Underlying space) The *underlying space* of an abstract simplicial complex K, denoted $|K|$, is the pointwise union of its simplices in its canonical geometric realization; that is, $|K| = \bigcup_{\sigma \in K} |\sigma|$ where $|\sigma|$ is the restriction of this realization on σ. If K is geometric, its geometric realization can be taken as itself.

Because of the equivalence between geometric and abstract simplicial complexes, we drop the qualifiers "geometric" and "abstract" and call them simply simplicial complexes when it is clear from the context which one we actually mean. Also, sometimes, we denote a simplex $\sigma = \{v_0, v_1, \ldots, v_k\}$ simply as $v_0 v_1 \cdots v_k$.

Definition 2.5. (k-skeleton) For any $k \geq 0$, the *k-skeleton* of a simplicial complex K, denoted by K^k, is the subcomplex formed by all simplices of dimension at most k.

In Figure 2.1, the 1-skeleton of the simplicial complex in Figure 2.1(a) consists of six vertices a, b, c, d, e, f and eight edges adjoining them.

Stars and Links
Given a simplex $\tau \in K$, its *star* in K is the set of simplices that have τ as a face, denoted by $\mathrm{St}(\tau) = \{\sigma \in K \mid \tau \subseteq \sigma\}$ (recall that $\tau \subseteq \sigma$ means that τ is a face of σ). Generally, the star is not closed under the face relation and hence is not a simplicial complex. We can make it so by adding all missing faces. The result is the *closed star*, denoted by

$$\overline{\mathrm{St}}(\tau) = \bigcup_{\sigma \in \mathrm{St}(\tau)} \{\sigma\} \cup \{\sigma' \in K \mid \sigma' \subset \sigma\},$$

which is also the smallest subcomplex that contains the star. The *link of τ* consists of the set of simplices in the closed star that are disjoint from τ, that is,

$$\mathrm{Lk}(\tau) = \{\sigma \in \overline{\mathrm{St}}(\tau) \mid \sigma \cap \tau = \varnothing\}.$$

Intuitively, we can think of the star (resp. the closed star) of a vertex as an open (resp. closed) neighborhood around it, and the link as the boundary of that neighborhood.

In Figure 2.1(a), we have:

- $St(a) = \{\{a\}, \{a, b\}, \{a, d\}, \{a, f\}, \{a, b, d\}\}$, $\overline{St}(a) = St(a) \cup \{\{b\}, \{d\}, \{f\}, \{b, d\}\}$;
- $St(f) = \{\{f\}, \{a, f\}, \{d, f\}\}$, $\overline{St}(f) = St(f) \cup \{\{a\}, \{d\}\}$;
- $St(\{a, b\}) = \{\{a, b\}, \{a, b, d\}\}$, $\overline{St}(\{a, b\}) = St(\{a, b\}) \cup \{\{a\}, \{b\}, \{d\}, \{a, d\}, \{b, d\}\}$;
- $Lk(a) = \{\{b\}, \{d\}, \{f\}, \{b, d\}\}$, $Lk(f) = \{\{a\}, \{d\}\}$, $Lk(\{a, b\}) = \{\{d\}\}$.

Triangulation of a Manifold

Given a simplicial complex K and a manifold M, we say that K is a triangulation of M if the underlying space $|K|$ is homeomorphic to M. Note that if M is a k-manifold, the dimension of K is also k. Furthermore, for any vertex $v \in K$, the underlying space $|St(v)|$ of the star $St(v)$ is homeomorphic to the open k-ball $\mathbb{B}_o^k$ if v maps to an interior point in M and to the k-dimensional half-space $\mathbb{H}^k$ if v maps to a point on the boundary of M. The underlying space $|Lk(v)|$ of the link $Lk(v)$ is homeomorphic to the $(k-1)$-sphere $\mathbb{S}^{k-1}$ if v maps to the interior and to a closed $(k-1)$-ball $\overline{\mathbb{B}_o^{k-1}}$ otherwise.

Simplicial Map

Corresponding to the continuous functions (maps) between topological spaces, we have a notion called simplicial map between simplicial complexes.

Definition 2.6. (Simplicial map) A map $f: K_1 \rightarrow K_2$ is called *simplicial* if for every simplex $\{v_0, \ldots, v_k\} \in K_1$, we have the simplex $\{f(v_0), \ldots, f(v_k)\}$ in K_2.

A map is called a *vertex map* if the domain and codomain of f are only vertex sets $V(K_1)$ and $V(K_2)$, respectively. Every simplicial map is associated with a vertex map. However, a vertex map $f: V(K_1) \rightarrow V(K_2)$ does not necessarily extend to a simplicial map from K_1 to K_2.

Fact 2.1. *Every continuous function $f: |K_1| \rightarrow |K_2|$ can be approximated closely by a simplicial map g on appropriate subdivisions of K_1 and K_2. The approximation being "close" means that, for a point $x \in |K_1|$, there is a simplex in K_2 which contains both $f(x)$ and $h(x)$ in geometric realization.*

There is also a counterpart of homotopic maps in simplicial setting.

Definition 2.7. (Contiguous maps) Two simplicial maps $f_1 \colon K_1 \to K_2$ and $f_2 \colon K_1 \to K_2$ are *contiguous* if for every simplex $\sigma \in K_1$, $f_1(\sigma) \cup f_2(\sigma)$ is a simplex in K_2.

Contiguous maps play an important role in topological analysis. We use a result involving contiguous maps and homology groups. We defer stating it till Section 2.5 where we introduce homology groups.

2.2 Nerves, Čech and Rips Complexes

Recall Definition 1.6 of covers from Chapter 1. A cover of a topological space defines a special simplicial complex called its *nerve*. The nerve plays an important role in bridging topological spaces to complexes which we will see below and also later in Chapter 9. We first define the nerve in general terms which can be specialized to covers easily.

Definition 2.8. (Nerve) Given a finite collection of sets $\mathcal{U} = \{U_\alpha\}_{\alpha \in A}$, we define the *nerve* of the set $\mathcal{U}$ to be the simplicial complex $N(\mathcal{U})$ whose vertex set is the index set A, and where a subset $\{\alpha_0, \alpha_1, \ldots, \alpha_k\} \subseteq A$ spans a k-simplex in $N(\mathcal{U})$ if and only if $U_{\alpha_0} \cap U_{\alpha_1} \cap \cdots \cap U_{\alpha_k} \neq \varnothing$.

Taking $\mathcal{U}$ to be a cover of a topological space in the above definition, one gets a nerve of a cover. Figure 2.2 shows two topological spaces, their covers, and corresponding nerves.

One important result involving nerves is the so-called Nerve Theorem which has different forms that depend on the type of topological spaces and covers. Adapting to our need, we state it for metric spaces (Definition 1.8) which are a special type of topological space as we have observed in Chapter 1.

Theorem 2.1. (Nerve Theorem [45, 300]) *Given a finite cover $\mathcal{U}$ (open or closed) of a metric space M, the underlying space $|N(\mathcal{U})|$ is homotopy equivalent to M if every non-empty intersection $\bigcap_{i=0}^{k} U_{\alpha_i}$ of cover elements is homotopy equivalent to a point, that is, contractible.*

The cover in the top row of Figure 2.2 satisfies the property of the above theorem and its nerve is homotopy equivalent to M whereas the same is not true for the cover shown in the bottom row.

Given a finite subset P for a metric space (M, d), we can build an abstract simplicial complex called a Čech complex with vertices in P using the concept of nerve.

$$M \qquad\qquad \mathcal{U} \qquad\qquad N(\mathcal{U})$$

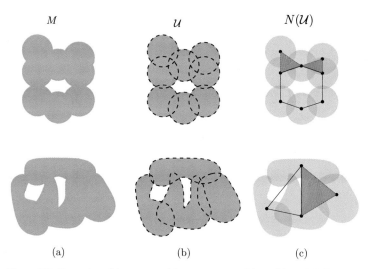

(a)   (b)   (c)

Figure 2.2 Examples of two spaces (a), open covers of them (b), and their nerves
(c). (top) The intersections of covers are contractible. (bottom) The intersections
of covers are not necessarily contractible.

Definition 2.9. (Čech complex) Let (M, d) be a metric space and P be a finite
subset of M. Given a real $r > 0$, the *Čech complex* $\mathbb{C}^r(P)$ is defined to be the
nerve of the set $\{B(p_i, r)\}$ where

$$B(p_i, r) = \{x \in M \mid \mathsf{d}(p_i, x) \le r\}$$

is the geodesic closed ball of radius r centering p_i.

Observe that if M is Euclidean, the balls considered for the Čech complex are necessarily convex and hence their intersections are contractible. By
Theorem 2.1, the Čech complex in this case is homotopy equivalent to the
space of union of the balls. The Čech complex is related to another complex called the Vietoris–Rips complex which is often used in topological data
analysis.

Definition 2.10. (Vietoris–Rips complex) Let (P, d) be a finite metric space.
Given a real $r > 0$, the *Vietoris–Rips* (*Rips* in short) *complex* is the abstract
simplicial complex $\mathbb{VR}^r(P)$ where a simplex $\sigma \in \mathbb{VR}^r(P)$ if and only if
$\mathsf{d}(p, q) \le 2r$ for every pair of vertices of σ.

Notice that the 1-skeleton of $\mathbb{VR}^r(P)$ determines all of its simplices. It is the
completion (in terms of simplices) of its 1-skeleton; see Figure 2.3. Also, we
observe the following fact.

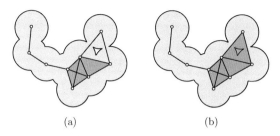

(a)   (b)

Figure 2.3 (a) Čech complex $\mathbb{C}^r(P)$ and (b) Rips complex $\mathbb{VR}^r(P)$.

Fact 2.2. *Let P be a finite subset of a metric space (M, d) where M satisfies the property that, for any real $r > 0$ and two points $p, q \in M$ with $\mathsf{d}(p, q) \leq 2r$, the metric balls $B(p, r)$ and $B(q, r)$ have non-empty intersection. Then, the 1-skeletons of $\mathbb{VR}^r(P)$ and $\mathbb{C}^r(P)$ coincide.*

Notice that if M is Euclidean, it satisfies the condition stated in the above fact and hence for finite point sets in any Euclidean space, the Čech and Rips complexes defined with Euclidean balls share the same 1-skeleton. However, for a general finite metric space (P, d), it may happen that for some $p, q \in P$, one has $\mathsf{d}(p, q) \leq 2r$ and $B(p, r) \cap B(q, r) = \varnothing$.

An easy but important observation is that the Rips and Čech complexes interleave.

Proposition 2.2. *Let P be a finite subset of a metric space (M, d). Then,*

$$\mathbb{C}^r(P) \subseteq \mathbb{VR}^r(P) \subseteq \mathbb{C}^{2r}(P).$$

PROOF. The first inclusion is obvious because if there is a point x in the intersection $\bigcap_{i=1}^{k} B(p_i, r)$, the distances $\mathsf{d}(p_i, p_j)$ for every pair (i, j), $1 \leq i, j \leq k$, are at most $2r$. It follows that for every simplex, $\{p_1, \ldots, p_k\} \in \mathbb{C}^r(P)$ is also in $\mathbb{VR}^r(P)$.

To prove the second inclusion, consider a simplex $\{p_1, \ldots, p_k\} \in \mathbb{VR}^r(P)$. Since by definition of the Rips complex $\mathsf{d}(p_i, p_1) \leq 2r$ for every p_i, $i = 1, \ldots, k$, we have $\bigcap_{i=1}^{k} B(p_i, 2r) \supset p_1 \neq \varnothing$. Then, by definition, $\{p_1, \ldots, p_k\}$ is also a simplex in $\mathbb{C}^{2r}(P)$. $\square$

2.3 Sparse Complexes

The Rips and Čech complexes are often too large to handle in practice. For example, the Rips complex with n points in $\mathbb{R}^d$ can have $\Omega(n^d)$ simplices. In practice, they can become large even in dimension as low as three. Just to give

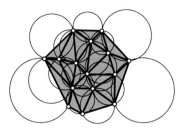

Figure 2.4 Every triangle in a Delaunay complex has an empty open circumdisk.

a sense of the scale of the problem, we note that the Rips or Čech complex built out of a few thousand points often has triangles in the range of millions. There are other complexes that are much sparser in size because of which they may be preferred sometimes for computations.

2.3.1 Delaunay Complex

This is a special complex that can be constructed out of a point set $P \in \mathbb{R}^d$. This complex embeds in $\mathbb{R}^d$ (in the generic setting). Because of its various optimal properties, this complex is used in many applications involving mesh generation, in particular in $\mathbb{R}^2$ and $\mathbb{R}^3$; see [97]. However, computation of Delaunay complexes in high dimensions beyond $\mathbb{R}^3$ can be time-intensive, so it is not yet the preferred choice for applications in dimensions beyond $\mathbb{R}^3$.

Definition 2.11. (Delaunay simplex; Complex) In the context of a finite point set $P \in \mathbb{R}^d$, a k-simplex σ is *Delaunay* if its vertices are in P and there is an open d-ball whose boundary contains its vertices and is *empty* – that is, contains no point in P. Note that any number of points in P can lie on the boundary of this ball. But, for simplicity, we will assume that only the vertices of σ are on the boundary of its empty ball. A *Delaunay complex* of P, denoted Del P, is a (geometric) simplicial complex with vertices in P in which every simplex is Delaunay and |Del P| coincides with the convex hull of P, as illustrated in Figure 2.4.

In $\mathbb{R}^2$, a Delaunay complex of a set of points in general position is made out of Delaunay triangles and all of their lower-dimensional faces. Similarly, in $\mathbb{R}^3$, a Delaunay complex is made out of Delaunay tetrahedra and all of their lower-dimensional faces.

Fact 2.3. *Every nondegenerate point set* (no $d + 2$ points are cospherical) admits a unique Delaunay complex.

Delaunay complexes are dual to the famous Voronoi diagrams defined below.

Definition 2.12. (Voronoi diagram) Given a finite point set $P \subset \mathbb{R}^d$ in generic position, the *Voronoi diagram* Vor (P) of P is the tessellation of the embedding space $\mathbb{R}^d$ into convex cells V_p for every $p \in P$ where

$$V_p = \{x \in \mathbb{R}^d \mid \mathsf{d}(x, p) \leq \mathsf{d}(x, q) \, \forall q \in P\}.$$

A k-face of Vor (P) is the intersection of $(d - k + 1)$ Voronoi cells.

Fact 2.4. *For $P \subset \mathbb{R}^d$, Del (P) is the nerve of the set of Voronoi cells $\{V_p\}_{p \in P}$ which is a closed cover of $\mathbb{R}^d$.*

The above fact actually provides a duality between Delaunay complex and Voronoi diagram. It is expressed by the duality among their faces. Specifically, a Delaunay k-simplex in Del (P) is dual to a Voronoi $(d - k)$-face in Vor (P). The Voronoi diagram dual to the Delaunay complex in Figure 2.4 is shown in Figure 2.5.

The following optimality properties make Delaunay complexes useful for applications.

Fact 2.5. *A triangulation of a point set $P \subset \mathbb{R}^d$ is a geometric simplicial complex whose vertex set is P and whose simplices tessellate the convex hull of P. Among all triangulations of a point set $P \subset \mathbb{R}^d$, Del P achieves the following optimized criteria:*

(a) In $\mathbb{R}^2$, Del P maximizes the minimum angle of triangles in the complex.
(b) In $\mathbb{R}^2$, Del P minimizes the largest circumcircle for triangles in the complex.
(c) For a simplex in Del P, let its min-ball be the smallest ball that contains the simplex in it. In all dimensions, Del P minimizes the largest min-ball.

The 1-skeletons of Delaunay complexes in $\mathbb{R}^2$ are plane graphs and hence Delaunay complexes in $\mathbb{R}^2$ have size $\Theta(n)$ for n points. They can be computed in $\Theta(n \log n)$ time. In $\mathbb{R}^3$, their size grows to $\Theta(n^2)$ and they can be computed in $\Theta(n^2)$ time. In $\mathbb{R}^d$, $d \geq 3$, Delaunay complexes have size $\Theta(n^{\lceil d/2 \rceil})$ and can be computed in optimal time $\Theta(n^{\lceil d/2 \rceil})$ [91].

Alpha Complex
Alpha complexes are subcomplexes of the Delaunay complexes which are parameterized by a real $\alpha \geq 0$. For a given point set P and $\alpha \geq 0$, an alpha

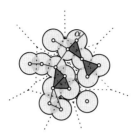

Figure 2.5 The alpha complex of the point set in Figure 2.4 for an α indicated in the figure. The Voronoi diagram of the point set is shown with dotted edges. The triangles and edges in the complex are shown with solid edges which are a subset of the Delaunay complex.

complex consists of all simplices in Del (P) that have a circumscribing ball of radius at most α. It can also be described alternatively as a nerve. For each point $p \in P$, let $B(p, \alpha)$ denote a closed ball of radius α centering p. Consider the closed set D_p^α defined as follows:

$$D_p^\alpha = \{x \in B(p, \alpha) \mid \mathsf{d}(x, p) \le \mathsf{d}(x, q) \, \forall q \in P\}.$$

The alpha complex $\mathrm{Del}^\alpha(P)$ is the nerve of the closed sets $\{D_p^\alpha\}_{p \in P}$. Another interpretation for alpha complex stems from its relation to the Voronoi diagram of the point set P. The alpha complex contains a k-simplex $\sigma = \{p_0, \ldots, p_k\}$ if and only if $\bigcup_{p \in P} B(p, \alpha)$ meets the intersection of Voronoi cells $V_{p_0} \cap V_{p_1} \cap \cdots \cap V_{p_k}$. Figure 2.5 shows an alpha complex for the point set in Figure 2.4 for an α. The Voronoi diagram is shown with the dotted segments.

2.3.2 Witness Complex

The witness complex defined by de Silva and Carlsson [113] sidesteps the size problem by a subsampling strategy. First, we define the witness complex with two point sets, P called the witnesses and Q called the landmarks. The complex is built with vertices in the landmarks where the simplices are defined with a notion of witness from the witness set. Given a point set P equipped with pairwise distances $\mathsf{d} \colon P \times P \to \mathbb{R}$, we can build the witness complex on a finite subsample $Q \subseteq P$.

Definition 2.13. (Weak witness) Let P be a point set with a real-valued function on pairs $\mathsf{d} \colon P \times P \to \mathbb{R}$ and $Q \subseteq P$ be a finite subset. A simplex $\sigma = \{q_1, \ldots, q_k\}$ with $q_i \in Q$ is *weakly witnessed* by $x \in P \setminus Q$ if $\mathsf{d}(q, x) \le \mathsf{d}(p, x)$ for every $q \in \{q_1, \ldots, q_k\}$ and $p \in Q \setminus \{q_1, \ldots, q_k\}$.

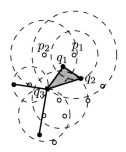

Figure 2.6 A witness complex constructed out of the points in Figure 2.4 where landmarks are the black dots and the witness points are the hollow dots. The witnesses for five edges and the triangle are the centers of the six circles; for example, the triangle $q_1q_2q_3$ and the edge q_1q_3 are weakly witnessed by the points p_1 and p_2, respectively.

We now define the witness complex using the notion of weak witnesses.

Definition 2.14. (Witness complex) Let P and Q be point sets as in Definition 2.13. The *witness complex* $\mathcal{W}(Q, P)$ is defined as the collection of all simplices whose all faces are weakly witnessed by a point in $P \setminus Q$.

Figure 2.6 shows an example of a witness complex. Observe that a simplex which is weakly witnessed may not have all its faces weakly witnessed (Exercise 7). This is why the definition above forces the condition to have a simplicial complex.

When $P = \mathbb{R}^d$ equipped with Euclidean distance and Q is a finite subset of it, we have the notion of strong witness.

Definition 2.15. (Strong witness) Let $Q \subset \mathbb{R}^d$ be a finite set. A simplex $\sigma = \{q_1, \ldots, q_d\}$ with $q_i \in Q$ is *strongly witnessed* by $x \in \mathbb{R}^d$ if $\mathsf{d}(q, x) \leq \mathsf{d}(p, x)$ for every $q \in \{q_1, \ldots, q_d\}$ and $p \in Q \setminus \{q_1, \ldots, q_d\}$ and additionally $\mathsf{d}(q_1, x) = \cdots = \mathsf{d}(q_d, x)$.

When $Q \subset \mathbb{R}^d$ as in the above definition, the following fact holds [112].

Proposition 2.3. *A simplex σ is strongly witnessed if and only if every face $\tau \leq \sigma$ is weakly witnessed.*

Furthermore, when $Q \subset \mathbb{R}^d$, we have some connections of the witness complex to the Delaunay complex. By definition, we know the following.

Fact 2.6. *Let Q be a finite subset of $\mathbb{R}^d$. Then a simplex σ is in the Delaunay triangulation* Del Q *if and only if σ is strongly witnessed by points in $\mathbb{R}^d$.*

By combining the above fact and the observation that every simplex in a witness complex is strongly witnessed, we have the following result which was observed by de Silva [112].

Proposition 2.4. *If P is a finite subset of $\mathbb{R}^d$ and $Q \subseteq P$, then $\mathcal{W}(Q, P) \subseteq$* Del Q.

One important implication of the above observation is that the witness complexes for point samples in a Euclidean space are embedded in that space.

The concept of the witness complex has a parallel in the concept of the restricted Delaunay complex. When the set P in Proposition 2.4 is not necessarily a finite subset, but only a subset X of $\mathbb{R}^d$, and Q is finite, we can relate $\mathcal{W}(Q, P)$ to the *restricted Delaunay complex* Del$|_X Q$ defined as the collection of Delaunay simplices in Del Q whose Voronoi duals have non-empty intersection with X.

Proposition 2.5. *The following hold:*

(a) $\mathcal{W}(Q, \mathbb{R}^d) =$ Del$|_{\mathbb{R}^d} Q :=$ Del Q *[112].*
(b) $\mathcal{W}(Q, M) =$ Del$|_M Q$ *if $M \subseteq \mathbb{R}^d$ is a smooth 1- or 2-manifold [11].*
(c) $\mathcal{W}(Q, P) =$ Del$|_M Q$ *where P and Q are sufficiently dense sample of a 1-manifold M in $\mathbb{R}^2$ and the result does not extend to other cases of submanifolds embedded in Euclidean spaces [178].*

2.3.3 Graph Induced Complex

The witness complex does not capture the topology of a manifold even if the input sample is dense except for smooth curves in the plane. One can modify them with extra structures such as putting weights on the points and changing the metric to weighted distances to tackle this problem as shown in [40]. But this becomes clumsy in terms of the "practicality" of a solution. We study another complex called *graph induced complex* (GIC) [123] which also uses subsampling, but is more powerful in capturing topology and in some case geometry. The advantage of the GIC over the witness complex is that GIC is not necessarily a subcomplex of the Delaunay complex and hence contains a few more simplices which aid topology inference. But, for the same reason, it may not embed in the Euclidean space where its input vertices lie.

In the following definition, the minimization argmin d(p, Q) may be a set instead of a single point in which case the nearest point map ν is set to choose any point in the set.

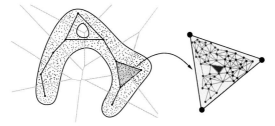

Figure 2.7 A graph induced complex shown with bold vertices, edges, and a shaded triangle on the left. The input graph within the shaded triangle is shown on the right. The 3-clique with three different colors (shown inside the shaded triangle on the right) causes the shaded triangle on the left to be in the graph induced complex.

Definition 2.16. (Graph Induced Complex) Let (P, d) be a metric space where P is a finite set and $G(P)$ be a graph with vertices in P. Let $Q \subseteq P$ and let $\nu \colon P \to Q$ that sets $\nu(p)$ to be any point in argmin $\mathsf{d}(p, Q)$. The *graph induced complex* (GIC) $\mathcal{G}(G(P), Q, \mathsf{d})$ is the simplicial complex containing a k-simplex $\sigma = \{q_1, \ldots, q_{k+1}\}$, $q_i \in Q$, if and only if there exists a $(k + 1)$-clique in $G(P)$ spanned by vertices $\{p_1, \ldots, p_{k+1}\} \subseteq P$ so that $q_i \in \nu(p_i)$ for each $i \in \{1, 2, \ldots, k + 1\}$. To see that it is indeed a simplicial complex, observe that a subset of a clique is also a clique.

Input Graph $G(P)$

The input point set P can be a finite sample of a subset X of a Euclidean space, such as a manifold or a compact subset. In this case, we may consider the input graph $G(P)$ to be the neighborhood graph $G^{\alpha}(P) := (P, E)$ where there is an edge $\{p, q\} \in E$ if and only if $\mathsf{d}(p, q) \leq \alpha$. The intuition is that if P is a sufficiently dense sample of X, then $G^{\alpha}(P)$ captures the local neighborhoods of the points in X. Figure 2.7 shows a graph induced complex for point data in the plane with a neighborhood graph where d is the Euclidean metric. To emphasize the dependence on α we use the notation $\mathcal{G}^{\alpha}(P, Q, \mathsf{d}) := \mathcal{G}(G^{\alpha}(P), Q, \mathsf{d})$.

Subsample Q

Of course, the ability to capture the topology of the sampled space after subsampling with Q depends on the quality of Q. We quantify this quality with a parameter $\delta > 0$.

Definition 2.17. (δ-sample; δ-sparse) A subset $Q \subseteq P$ is called a δ-*sample* of a metric space (P, d), if the following condition holds:

• For all $p \in P$, there exists a $q \in Q$, so that $\mathsf{d}(p, q) \leq \delta$.

Subset Q is called δ-*sparse* if the following condition holds:

- For all $(q, r) \in Q \times Q$ with $q \neq r$, $\mathsf{d}(q, r) \geq \delta$.

The first condition ensures Q to be a good sample of P with respect to the parameter δ and the second condition enforces that the points in Q cannot be too close relative to the distance δ.

Metric d

The metric d assumed in the metric space (P, d) will be of two types in our discussion below: (i) the Euclidean metric denoted d_E; or (ii) the graph metric d_G derived from the input graph $G(P)$ where $\mathsf{d}_G(p, q)$ is the shortest path distance between p and q in the graph $G(P)$ assuming its edges have nonnegative weights such as their Euclidean lengths.

We state two inference results involving GIC below. The first result is about reconstructing a surface from its sample. The other result is about inferring the one-dimensional homology group from a sample. We introduce the homology groups in the next section. The reader can skip this result and come back to it after consulting the relevant definitions later. Also, for details, we refer to [123]. In the following theorem ρ denotes the "reach" of the manifold, an intrinsic feature with respect to which the sampling needs to be dense. We define it more precisely in Definition 6.8 of Chapter 6.

Theorem 2.6. *Let* $M \subset \mathbb{R}^3$ *be a smooth, compact, and connected surface. If* $8\varepsilon \leq \delta \leq \frac{2}{27}\rho$, $\alpha \geq 8\varepsilon$, P *is an* ε-*sample of* (M, d_E), *and* $Q \subseteq P$ *is a* δ-*sparse* δ-*sample of* (P, d_E), *then a triangulation* T *of* M *exists as a subcomplex of* $\mathcal{G}^\alpha(P, Q, \mathsf{d}_E)$ *which can be computed efficiently.*

In the next theorem, d_G is the graph metric where the input graph is $G^\alpha(P)$ for some $\alpha \geq 0$ constructed by the Euclidean metric which is the input P is equipped with.

Theorem 2.7. *Let* P *be an* ε-*sample of an embedded smooth, compact manifold* M *in a Euclidean space with reach* ρ, *and* Q *a* δ-*sample of* (P, d_G). *For* $4\varepsilon \leq \alpha$, $\delta \leq \frac{1}{3}\sqrt{\frac{3}{5}}\rho$, *the map* $h_*: \mathsf{H}_1(\mathbb{VR}^\alpha(P)) \to \mathsf{H}_1(\mathcal{G}^\alpha(P, Q, d_G))$ *is an isomorphism where* $h : \mathbb{VR}^\alpha(P) \to \mathcal{G}^\alpha(P, Q, \mathsf{d}_G)$ *is the simplicial map induced by the nearest point map* $v_{\mathsf{d}_G} : P \to Q$.

Instead of stating other homology inference results precisely, we give some empirical results involving homology groups just to emphasize the advantage of GICs over other complexes in this respect. Again, readers unfamiliar with homology groups can consult the next section.

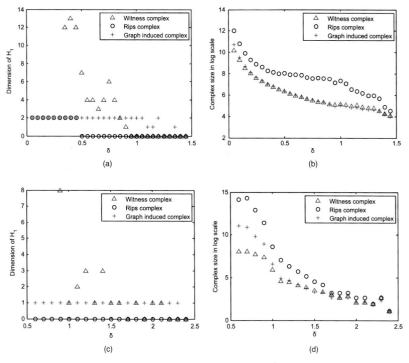

Figure 2.8 Comparison results for the Klein bottle in $\mathbb{R}^4$ (a) and (b) and the primary circle in $\mathbb{R}^{25}$ (c) and (d). The estimated β_1 computed from three complexes are shown in (a) and (c), and their sizes are shown on log scale in (b) and (d). Images are taken from [123].

An Empirical Example

When equipped with appropriate metric, the GIC can decipher the topology from data. It retains the simplicity of the Rips complex as well as the sparsity of the witness complex. It does not build a Rips complex on the subsample and thus is sparser than the Rips complex with the same set of vertices. This fact makes a real difference in practice as experiments show.

Figure 2.8 shows experimental results on two datasets, 40 000 sample points from a Klein bottle in $\mathbb{R}^4$ and 15 000 sample points from the so-called *primary circle* of natural image data considered in $\mathbb{R}^{25}$. The graphs connecting any two points within $\alpha = 0.05$ unit distance for the Klein bottle and $\alpha = 0.6$ unit distance for the primary circle were taken as input for the graph induced complexes. The 2-skeleton of the Rips complexes for these α parameters have 608 200 and 1 329 672 867 simplices, respectively. These sizes are too large to carry out fast computations.

For comparison, we constructed the graph induced complex, a sparsified version of the Rips complex (Section 6.2), and the witness complex on the same subsample determined by a parameter δ. The parameter δ is also used in the graph induced complex and the witness complex. The edges in the Rips complex built on the same *subsample* were of lengths at most $\alpha + 2\delta$. One of the main uses of the sparse complexes in TDA is to infer homology groups (covered in the next section) from samples. To compare GIC with the sparse Rips and witness complexes, we varied δ and observed the rank of the one-dimensional homology group (β_1). As evident from Figure 2.8, the graph induced complex captured β_1 correctly for a significantly wider range of δ (Figure 2.8a and c) while its size remained comparable to that of the witness complex (Figure 2.8b and d). In some cases, the graph induced complex could capture the correct β_1 with remarkably small number of simplices. For example, it had $\beta_1 = 2$ for the Klein bottle when there were 278 simplices for $\delta = 0.7$ and 154 simplices for $\delta = 1.0$. In both cases the Rips and witness complexes had the wrong β_1 while the Rips complex had a much larger size ($\log_e$ scale plot) and the witness complex had comparable size. This illustrates why the graph induced complex can be a better choice than the Rips and witness complexes.

Constructing a GIC

One may wonder how to efficiently construct the graph induced complexes in practice. Experiments show that the following procedure runs quite efficiently in practice. It takes advantage of computing nearest neighbors within a range and, more importantly, computing cliques only in a sparsified graph.

Let the ball $B(q, \delta)$ in metric d be called the δ-cover for the point q. A graph induced complex $\mathcal{G}^\alpha(P, Q, \mathsf{d})$ where Q is a δ-sparse δ-sample can be built easily by identifying δ-covers with a rather standard greedy (farthest point) iterative algorithm. Let $Q_i = \{q_1, \ldots, q_i\}$ be the point set sampled so far from P. We maintain the invariants (i) Q_i is δ-sparse and (ii) every point $p \in P$ that is in the union of δ-covers $\bigcup_{q \in Q_i} B(q, \delta)$ has its closest point $\nu(p) \in \operatorname{argmin}_{q \in Q_i} \mathsf{d}(p, q)$ in Q_i identified. To augment Q_i to $Q_{i+1} = Q_i \cup \{q_{i+1}\}$, we choose a point $q_{i+1} \in P$ that is outside the δ-covers $\bigcup_{q \in Q_i} B(q, \delta)$. Certainly, q_{i+1} is at least δ units away from all points in Q_i thus satisfying the first invariant. For the second invariant, we check every point p in the δ-cover of q_{i+1} and update $\nu(p)$ to include q_{i+1} if its distance to q_{i+1} is smaller than the distance $\mathsf{d}(p, \nu(p))$. At the end, we obtain a sample $Q \subseteq P$ whose δ-covers cover the entire point set P and thus is a δ-sample of (P, d) which is also δ-sparse due to the invariants maintained. Next, we construct the simplices of $\mathcal{G}^\alpha(P, Q, \mathsf{d})$. This needs identifying cliques in $G^\alpha(P)$ that have vertices with different closest points in Q. We delete every edge pp' from $G^\alpha(P)$ where $\nu(p) = \nu(p')$. Then, we determine every clique $\{p_1, \ldots p_k\}$ in the remaining sparsified graph and include the simplex $\{v(p_1), \ldots, v(p_k)\}$ in $\mathcal{G}^\alpha(P, Q, \mathsf{d})$. The main saving here is that many cliques of the original graph are removed before it is processed for clique computation.

Next, we focus on the second topic of this chapter, namely homology groups. They are algebraic structures to quantify topological features in a space. They do not capture all topological aspects of a space in the sense that two spaces with the same homology groups may not be topologically equivalent. However, two spaces that are topologically equivalent must have isomorphic homology groups. It turns out that the homology groups are computationally tractable in many cases, thus making them more attractive in topological data analysis. Before we introduce their definition and variants in Section 2.5, we need the important notions of chains, cycles, and boundaries given in the following section.

2.4 Chains, Cycles, Boundaries

2.4.1 Algebraic Structures

First, we recall briefly the definitions of some standard algebraic structures that are used in the book. For details we refer the reader to any standard book on algebra, for example, [14].

Definition 2.18. (Group; Homomorphism; Isomorphism) A set G together with a binary operation "+" is a *group* if it satisfies the following properties: (i) for every $a, b \in G$, $a + b \in G$; (ii) for every $a, b, c \in G$, $(a + b) + c = a + (b + c)$; (iii) there is an *identity* element denoted 0 in G so that $a + 0 = 0 + a = a$ for every $a \in G$; and (iv) there is an *inverse* $-a \in G$ for every $a \in G$ so that $a + (-a) = 0$. If the operation "+" commutes, that is, $a + b = b + a$ for every $a, b \in G$, then G is called *abelian*. A subset $H \subseteq G$ is a *subgroup* of $(G, +)$ if $(H, +)$ is also a group.

Definition 2.19. (Free abelian group; Basis; Rank; Generator) An abelian group G is called *free* if there is a subset $B \subseteq G$ so that every element of G can be written *uniquely* as a finite sum of elements in B and their inverses disregarding trivial cancellations $a + b = a + c - c + b$. Such a set B is called a *basis* of G and its cardinality is called its *rank*. If the condition of uniqueness is dropped, then B is called a *generator* of G and we also say that B *generates* G.

Definition 2.20. (Coset; Quotient) For a subgroup $H \subseteq G$ and an element $a \in G$, the *left coset* is $aH = \{a + b \mid b \in H\}$ and the *right coset* is

$Ha = \{b + a \mid b \in H\}$. For abelian groups, the left and right cosets are identical and hence are simply called *cosets*. If G is abelian, the *quotient* group of G with a subgroup $H \subseteq G$ is given by $G/H = \{aH \mid a \in G\}$ where the group operation is inherited from G as $aH + bH = (a + b)H$ for every $a, b \in G$.

Definition 2.21. (Homomorphism; Isomorphism; Kernel; Image; Cokernel) A map $h : G \to H$ between two groups $(G, +)$ and $(H, *)$ is called a *homomorphism* if $h(a + b) = h(a) * h(b)$ for every $a, b \in G$. If, in addition, h is bijective, it is called an *isomorphism*. Two groups G and H with an isomorphism are called *isomorphic* and denoted as $G \cong H$. The *kernel*, *image*, and *cokernel* of a homomorphism $h : G \to H$ are defined as subgroups $\ker h = \{a \in G \mid h(a) = 0\}$, $\operatorname{Im} h = \{b \in H \mid \exists a \in G \text{ with } h(a) = b\}$, and the quotient group $\operatorname{coker} h = H/\operatorname{im} h$, respectively.

Definition 2.22. (Ring) A set R equipped with two binary operations, addition "+" and multiplication "·" is called a *ring* if (i) R is an abelian group with the addition, (ii) the multiplication is associative, that is, $(a \cdot b) \cdot c = a \cdot (b \cdot c)$ and is distributive with the addition, that is, $a \cdot (b + c) = a \cdot b + a \cdot c$, for all $a, b, c \in R$, and (iii) there is an identity for the multiplication.

The additive identity of a ring R is usually denoted as 0 whereas the multiplicative identity is denoted as 1. Observe that, by the definition of abelian group, the addition is commutative. However, the multiplication need not be so. When the multiplication is also commutative, R is called a *commutative ring*. A commutative ring in which every nonzero element has a multiplicative inverse is called a *field*.

Definition 2.23. (Module) Given a commutative ring R with multiplicative identity 1, an *R-module* M is an abelian group with an operation $R \times M \to M$ which satisfies the following properties for all $r, r' \in R$ and $x, y \in M$:

- $r \cdot (x + y) = r \cdot x + r \cdot y$;
- $(r + r')x = r \cdot x + r' \cdot x$;
- $1 \cdot x = x$;
- $(r \cdot r') \cdot x = r \cdot (r' \cdot x)$.

Essentially, in an R-module, elements can be added and multiplied with coefficients in R. However, if R is taken as a field **k**, each nonzero element acquires a multiplicative inverse and we get a vector space.

Definition 2.24. (Vector space) An R-module V is called a *vector space* if R is a field. A set of elements $\{g_1, \ldots, g_k\}$ is said to *generate* the vector space

V if every element $a \in V$ can be written as $a = \alpha_1 g_1 + \cdots + \alpha_k g_k$ for some $\alpha_1, \ldots, \alpha_k \in R$. The set $\{g_1, \ldots, g_k\}$ is called a *basis* of V if every $a \in V$ can be written in the above way *uniquely*. All bases of V have the same cardinality which is called the *dimension* of V. We say a set $\{g_1, \ldots, g_m\} \subseteq V$ is *independent* if the equation $\alpha_1 g_1 + \cdots + \alpha_m g_m = 0$ can only be satisfied by setting $\alpha_i = 0$ for $i = 1, \ldots, m$.

Fact 2.7. *A basis of a vector space is a generating set of minimal cardinality and an independent set of maximal cardinality.*

2.4.2 Chains

Let K be a simplicial k-complex with m_p number of p-simplices, $k \leq p \leq 0$. A p-chain c in K is a formal sum of p-simplices added with some coefficients, that is, $c = \sum_{i=1}^{m_p} \alpha_i \sigma_i$ where σ_i are the p-simplices and α_i are the coefficients. Two p-chains $c = \sum \alpha_i \sigma_i$ and $c' = \sum \alpha_i' \sigma_i$ can be added to obtain another p-chain:

$$c + c' = \sum_{i=1}^{m_p} (\alpha_i + \alpha_i') \sigma_i.$$

In general, coefficients can come from a ring R with its associated additions making the chains constituting an R-module. For example, these additions can be integer additions where the coefficients are integers; for example, from two 1-chains (edges) we get

$$(2e_1 + 3e_2 + 5e_3) + (e_1 + 7e_2 + 6e_4) = 3e_1 + 10e_2 + 5e_3 + 6e_4.$$

Notice that while writing a chain, we only write the simplices that have nonzero coefficient in the chain. We follow this convention all along. In our case, we will focus on the cases where the coefficients come from a field $\mathbf{k}$. In particular, we will mostly be interested in $\mathbf{k} = \mathbb{Z}_2$. This means that the coefficients come from the field $\mathbb{Z}_2$ whose elements can only be 0 or 1 with the modulo-2 additions $0 + 0 = 0$, $0 + 1 = 1$, and $1 + 1 = 0$. This gives us $\mathbb{Z}_2$-additions of chains; for example, we have

$$(e_1 + e_3 + e_4) + (e_1 + e_2 + e_3) = e_2 + e_4.$$

Observe that p-chains with $\mathbb{Z}_2$-coefficients can be treated as sets: the chain $e_1 + e_3 + e_4$ is the set $\{e_1, e_3, e_4\}$, and $\mathbb{Z}_2$-addition between two chains is simply the symmetric difference between the corresponding sets.

From now on, unless specified otherwise, we will consider all chain additions to be $\mathbb{Z}_2$-additions. One should keep in mind that one can have parallel

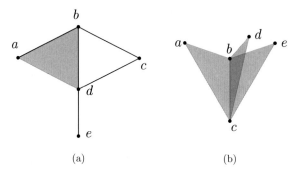

Figure 2.9 Chains, boundaries, cycles.

concepts for coefficients and additions coming from integers, reals, rationals, fields, and other rings. Under $\mathbb{Z}_2$-additions, we have

$$c + c = \sum_{i=1}^{m_p} 0\sigma_i = 0.$$

Below, we show the addition of chains shown in Figure 2.9:

0-chain:	$(\{b\} + \{d\}) + (\{d\} + \{e\})$	$=$	$\{b\} + \{e\}$	(left)
1-chain:	$(\{a, b\} + \{b, d\}) + (\{b, c\} + \{b, d\})$	$=$	$\{a, b\} + \{b, c\}$	(left)
2-chain:	$(\{a, b, c\} + \{b, c, e\}) + (\{b, c, e\})$	$=$	$\{a, b, c\}$	(right)

The p-chains with the $\mathbb{Z}_2$-additions form a group where the identity is the chain $0 = \sum_{i=1}^{m_p} 0\sigma_i$, and the inverse of a chain c is c itself since $c + c = 0$. This group, called the *p-th chain group*, is denoted $\mathsf{C}_p(K)$. We also drop the complex K and use the notation C_p when K is clear from the context. We do the same for other structures that we define afterward.

2.4.3 Boundaries and Cycles

The chain groups at different dimensions are related by a boundary operator. Given a p-simplex $\sigma = \{v_0, \ldots, v_p\}$ (also denoted as $v_0 v_1 \cdots v_p$), let

$$\partial_p \sigma = \sum_{i=0}^{p} \{v_0, \ldots, \hat{v}_i, \ldots, v_p\},$$

where $\hat{v}_i$ indicates that the vertex v_i is omitted. Informally, we can view ∂_p as a map that sends a p-simplex σ to the $(p-1)$-chain that has nonzero coefficients only on σ's $(p-1)$-faces, also referred to as σ's boundary. At this point, it is instructive to note that the boundary of a vertex is empty, that is, $\partial_0 \sigma = \varnothing$.

Extending ∂_p to a p-chain, we obtain a homomorphism $\partial_p: \mathsf{C}_p \to \mathsf{C}_{p-1}$ called the *boundary operator* that produces a $(p-1)$-chain when applied to a p-chain:

$$\partial_p c = \sum_{i=1}^{m_p} \alpha_i (\partial_p \sigma_i) \text{ for a } p\text{-chain } c = \sum_{i=1}^{m_p} \alpha_i \sigma_i \in \mathsf{C}_p.$$

Again, we note the special case of $p = 0$ when we get $\partial_0 c = \varnothing$. The chain group C_{-1} has only one single element which is its identity 0. On the other hand, we also assume that if K is a k-complex, then C_p is 0 for $p > k$.

Consider the complex in Figure 2.9(b). For the 2-chain $abc + bcd$ we get

$$\partial_2(abc + bcd) = (ab + bc + ca) + (bc + cd + db) = ab + ca + cd + db.$$

It means that from the two triangles sharing the edge bc, the boundary operator returns the four boundary edges that are not shared. Similarly, one can check that the boundary of the 2-chains consisting of all three triangles in Figure 2.9(b) contains all seven edges. In particular, the edge bc does not get cancelled because all three (odd) triangles adjoin it:

$$\partial_2(abc + bcd + bce) = ab + bc + ca + be + ce + bd + dc.$$

One important property of the boundary operator is that applying it twice produces an empty chain.

Proposition 2.8. *For $p > 0$ and any p-chain c, $\partial_{p-1} \circ \partial_p(c) = 0$.*

PROOF. Observe that ∂_0 is a zero map by definition. Also, for a k-complex, ∂_p operates on a zero element for $p > k$ by definition. Then, it is sufficient to show that, for $1 \le p \le k$, $\partial_{p-1} \circ \partial_p(\sigma) = 0$ for a p-simplex σ. Observe that $\partial_p \sigma$ is the set of all $(p-1)$-faces of σ and every $(p-2)$-face of σ is contained in exactly two $(p-1)$-faces. Thus, $\partial_{p-1}(\partial_p \sigma) = 0$. □

Extending the boundary operator to the chain groups, we obtain the following sequence of homomorphisms satisfying Proposition 2.8 for a simplicial k-complex; such a sequence is also called a *chain complex*:

$$0 = \mathsf{C}_{k+1} \xrightarrow{\partial_{k+1}} \mathsf{C}_k \xrightarrow{\partial_k} \mathsf{C}_{k-1} \xrightarrow{\partial_{k-1}} \mathsf{C}_{k-2} \cdots \mathsf{C}_1 \xrightarrow{\partial_1} \mathsf{C}_0 \xrightarrow{\partial_0} \mathsf{C}_{-1} = 0. \quad (2.1)$$

Fact 2.8. *We have the following.*

1. For $p \ge -1$, C_p is a vector space because the coefficients are drawn from a field $\mathbb{Z}_2$ – it has a basis so that every element can be expressed uniquely as a sum of the elements in the basis.

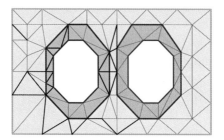

Figure 2.10 Each individual red, blue, and green cycle is not a boundary because they do not bound any 2-chain. However, the sum of the two red cycles, and the sum of the two blue cycles each form a boundary cycle because they bound 2-chains consisting of reddish and bluish triangles, respectively.

2. *There is a basis for* C_p *where every p-simplex forms a basis element because any p-chain is a unique subset of the p-simplices. The dimension of* C_p *is therefore n, the number of p-simplices. When* $p = -1$ *and* $p \geq k+1$, C_p *is trivial with dimension zero. In Figure 2.9(b)* $\{abc, bcd, bce\}$ *is a basis for* C_2 *and so is* $\{abc, (abc + bcd), bce\}$.

Cycle and Boundary Groups

Definition 2.25. (Cycle; Cycle group) A p-chain c is a *p-cycle* if $\partial c = 0$. In words, a chain that has empty boundary is a *cycle*. All p-cycles together form the *p-th cycle group* Z_p under the addition that is used to define the chain groups. In terms of the boundary operator, Z_p is the subgroup of C_p which is sent to the zero of C_{p-1}, that is, $\ker \partial_p = Z_p$.

For example, in Figure 2.9(b), the 1-chain $ab + bc + ca$ is a 1-cycle since

$$\partial_1 (ab + bc + ca) = (a + b) + (b + c) + (c + a) = 0.$$

Also, observe that the above 1-chain is the boundary of the triangle abc. It is no accident that the boundary of a simplex is a cycle. Thanks to Proposition 2.8, the boundary of a p-chain is a $(p - 1)$-cycle. This is a fundamental fact in homology theory.

The set of $(p - 1)$-chains that can be obtained by applying the boundary operator ∂_p on p-chains forms a subgroup of $(p - 1)$-chains, called the *$(p - 1)$-th boundary group* $B_{p-1} = \partial_p(C_p)$; or, in other words, the image of the boundary homomorphism is the boundary group, $B_{p-1} = \text{im } \partial_p$. We have $\partial_{p-1} B_{p-1} = 0$ for $p > 0$ due to Proposition 2.8 and hence $B_{p-1} \subseteq Z_{p-1}$. Figure 2.10 illustrates cycles and boundaries.

Fact 2.9. *For a simplicial k-complex, we have the following.*

(a) $C_0 = Z_0$ *and* $B_k = 0$.
(b) For $p \geq 0$, $B_p \subseteq Z_p \subseteq C_p$.
(c) Like C_p, *both* B_p *and* Z_p *are vector spaces.*

2.5 Homology

The homology groups classify the cycles in a cycle group by putting together those cycles in the same class that differ by a boundary. From a group theoretic point of view, this is done by taking the quotient of the cycle groups with the boundary groups, which is allowed since the boundary group is a subgroup of the cycle group.

Definition 2.26. (Homology group) For $p \geq 0$, the p-th *homology group* is the quotient group $H_p = Z_p/B_p$. Since we use a field, namely $\mathbb{Z}_2$, for coefficients, H_p is a vector space and its dimension is called the p-th Betti number, denoted by β_p:

$$\beta_p := \dim H_p.$$

Every element of H_p is obtained by adding a p-cycle $c \in Z_p$ to the entire boundary group, $c + B_p$, which is a coset of B_p in Z_p. All cycles constructed by adding an element of B_p to c form the class $[c]$, referred to as the *homology class* of c. Two cycles c and c' in the same homology class are called *homologous*, which also means $[c] = [c']$. By definition, $[c] = [c']$ if and only if $c \in c' + B_p$, and under $\mathbb{Z}_2$ coefficients, this also means that $c + c' \in B_p$. For example, in Figure 2.10, the outer cycle c_5 is homologous to the sum $c_2 + c_4$ because they together bound the 2-chain consisting of all triangles. Also, observe that the group operation for H_p is defined by $[c]+[c'] = [c+c']$.

Example 2.1. *Consider the boundary complex K of a tetrahedron which consists of four triangles, six edges, and four vertices. Consider the 0-skeleton K^0 of K which consists of four vertices only (see Figure 2.11a). All four vertices whose classes coincide with them are necessary to generate $H_0(K^0)$. Therefore, these four vertices form a basis of $H_0(K^0)$. However, one can verify that $H_0(K^1)$ for the 1-skeleton K^1 is generated by any one of the four vertices because all four vertices belong to the same class when we consider K^1. This exemplifies the fact that the rank of $H_0(K)$ captures the number of connected components in a complex K.*

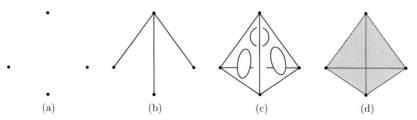

Figure 2.11 Complex K of a tetrahedron: (a) vertices, (b) spanning tree of the 1-skeleton, (c) 1-skeleton, and (d) 2-skeleton of K.

The 1-skeleton K^1 of the tetrahedron is a graph with four vertices and six edges. Consider a spanning tree with any vertex and the three edges adjoining it as in Figure 2.11(b). There is no 1-cycle in this configuration. However, each of the other three edges creates a new 1-cycle which is not a boundary because there is no triangle in K^1. These three cycles c_1, c_2, and c_3 as indicated in Figure 2.11(c) form their own classes in $H_1(K^1)$. Observe that the 1-cycle at the base can be written as a combination of the other three and thus all classes in $H_1(K^1)$ can be generated by only three classes $[c_1]$, $[c_2]$, and $[c_3]$ and no fewer. Hence, these three classes form a basis of $H_1(K^1)$. To develop more intuition, consider a simplicial surface M without boundary embedded in $\mathbb{R}^3$. If the surface has genus g, that is, g tunnels and handles in the complement space, then $H_1(M)$ has dimension 2g (Exercise 4).

The 2-chain of the sum of four triangles in K (Figure 2.11d) makes a 2-cycle c because its boundary is 0. Since K does not have any 3-simplex (the tetrahedron is not part of the complex), this 2-cycle cannot be added to any 2-boundary other than 0 to form its class. Therefore, the homology class of c is c itself, $[c] = \{c\}$. There is no other 2-cycle in K. Therefore, $H_2(K)$ is generated by $[c]$ alone. Its dimension is only one. If the tetrahedron is included in the complex, c becomes a boundary element, and hence $[c] = [0]$. In that case, $H_2(K) = 0$. Intuitively, one may think of $H_2(K)$ as capturing the voids in a complex K embedded in $\mathbb{R}^3$. (Convince yourself that $H_1(K) = 0$ no matter whether the tetrahedron belongs to K or not.)

Fact 2.10. *For $p \geq 0$, we have the following.*

(a) H_p is a vector space (when defined over $\mathbb{Z}_2$).

(b) H_p may not be a vector space when defined over $\mathbb{Z}$, the integer coefficients. In this case, there could be torsion subgroups.

(c) The Betti number, $\beta_p = \dim H_p$. is given by $\beta_p = \dim Z_p - \dim B_p$.

(d) There are exactly 2^{β_p} homology classes in H_p when defined with $\mathbb{Z}_2$ coefficients.

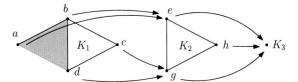

Figure 2.12 Induced homology by simplicial map. Simplicial map f obtained by the vertex map $a \to e, b \to e, c \to g, d \to g$ induces a map at the homology level $f_* : \mathsf{H}_1(K_1) \to \mathsf{H}_1(K_2)$ which takes the only nontrivial class created by the empty triangle abc to zero even though $\mathsf{H}_1(K_1) \cong \mathsf{H}_1(K_2)$. Another simplicial map $K_2 \to K_3$ destroys the single homology class born by the empty triangle egh in K_2.

2.5.1 Induced Homology

Continuous functions from a topological space to another topological space take cycles to cycles and boundaries to boundaries. Therefore, they induce a map in their homology groups as well. Here we will restrict ourselves only to simplicial complexes and simplicial maps that are the counterpart of continuous maps between topological spaces. Simplicial maps between simplicial complexes take cycles to cycles and boundaries to boundaries with the following definition.

Definition 2.27. (Chain map) Let $f : K_1 \to K_2$ be a simplicial map. The *chain map* $f_\# : \mathsf{C}_p(K_1) \to \mathsf{C}_p(K_2)$ corresponding to f is defined as follows. If $c = \sum \alpha_i \sigma_i$ is a p-chain, then $f_\#(c) = \sum \alpha_i \tau_i$ where

$$\tau_i = \begin{cases} f(\sigma_i), & \text{if } f(\sigma_i) \text{ is a } p\text{-simplex in } K_2, \\ 0, & \text{otherwise.} \end{cases}$$

For example, in Figure 2.12, the 1-cycle $bc + cd + db$ in K_1 is mapped to the 1-chain $eg + eg = 0$ by the chain map $f_\#$.

Proposition 2.9. *Let $f : K_1 \to K_2$ be a simplicial map. Let $\partial_p^{K_1}$ and $\partial_p^{K_2}$ denote the boundary homomorphisms in dimension $p \geq 0$. Then, the induced chain maps commute with the boundary homomorphisms, that is, $f_\# \circ \partial_p^{K_1} = \partial_p^{K_2} \circ f_\#$.*

The statement in the above proposition can also be represented with the following diagram, which we say *commutes* since starting from the top left corner, one reaches the same chain at the lower right corner using both paths – first going right and then down, or first going down and then right (see Definition 3.15 in the next chapter).

$$C_p(K_1) \xrightarrow{\;f_\#\;} C_p(K_2) \qquad\qquad (2.2)$$

$$\downarrow \partial_p^{K_1} \qquad\qquad\quad \downarrow \partial_p^{K_2}$$

$$C_{p-1}(K_1) \xrightarrow{\;f_\#\;} C_{p-1}(K_2)$$

For example, in Figure 2.12, we have $f_\#(c = ab + bd + da) = 0$ and $\partial_p^{K_1}(c) = 0$. Therefore, $\partial_p^{K_2}(f_\#(c)) = \partial_p^{K_2}(0) = 0 = f_\#(0) = f_\#(\partial_p^{K_1}(c))$.

Since $B_p(K_1) \subseteq Z_p(K_1)$, we have that $f_\#(B_p(K_1)) \subseteq f_\#(Z_p(K_1))$. Thus, the induced map in the quotient space, namely,

$$f_*(Z_p(K_1)/B_p(K_1)) := f_\#(Z_p(K_1))/f_\#(B_p(K_1))$$

is well defined. Furthermore, by the commutativity of the diagram (2.2), $f_\#(Z_p(K_1)) \subseteq Z_p(K_2)$ and $f_\#(B_p(K_1)) \subseteq B_p(K_2)$, which gives an induced homomorphism in the homology groups:

$$f_* : Z_p(K_1)/B_p(K_1) \to Z_p(K_2)/B_p(K_2) \quad \text{or equivalently} \quad f_* : H_p(K_1) \to H_p(K_2)$$

A homology class $[c] = c + B_p$ from K_1 is mapped to the homology class $f_\#(c) + f_\#(B_p)$ from K_2 by f_*. In Figure 2.12, we have $B_1 = \{0, ab+bd+da\}$. Then, for $c = bd + dc + cb$, we have $f_*([c]) = \{f_\#(c), f_\#(c) + f_\#(ab + bd + da)\} = \{0, 0\} = [0]$.

Now we can state a result relating contiguous maps (Definition 2.7) and homology groups that we promised in Section 2.1.

Fact 2.11. *For two contiguous maps $f_1 : K_1 \to K_2$ and $f_2 : K_1 \to K_2$, the induced maps $f_{1*} : H_p(K_1) \to H_p(K_2)$ and $f_{2*} : H_p(K_1) \to H_p(K_2)$ are equal.*

2.5.2 Relative Homology

As the name suggests, we can define a homology group of a complex relative to a subcomplex. Let K_0 be a subcomplex of K. By definition, the chain group $C_p(K_0)$ is a subgroup of $C_p(K)$. Therefore, the quotient group $C_p(K)/C_p(K_0)$ is well defined which is called a *relative chain group* and is denoted $C_p(K, K_0)$. It is an abelian group whose elements are the cosets $[c_p] = c_p + C_p(K_0)$ for every chain $c_p \in C_p(K)$.

The boundary operator $\partial_p : C_p(K) \to C_{p-1}(K)$ extends to the relative chain groups in a natural way:

$$\partial_p^{K, K_0} : C_p(K, K_0) \to C_{p-1}(K, K_0), \quad [c_p] \mapsto [\partial_p c_p].$$

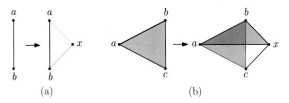

Figure 2.13 Illustration for relative homology: the subcomplex K_0 consists of (a) vertices a and b, and (b) vertices a, b, c, and the edge ab; the coned complex K^* is indicated with a coning from a dummy vertex x.

One may verify that $\partial_{p-1}^{K,K_0} \circ \partial_p^{K,K_0} = 0$ as before. Therefore, we can define

$$\mathsf{Z}_p(K, K_0) = \ker \partial_p^{K,K_0}, \text{ the } p\text{-th relative cycle group,}$$

$$\mathsf{B}_p(K, K_0) = \operatorname{Im} \partial_{p+1}^{K,K_0}, \text{ the } p\text{-th relative boundary group,}$$

$$\mathsf{H}_p(K, K_0) = \mathsf{Z}_p(K, K_0)/\mathsf{B}_p(K, K_0), \text{ the } p\text{-th relative homology group.}$$

The relative homology $\mathsf{H}_p(K, K_0)$ is related to a coned complex K^*. A coned complex K^* of a simplicial complex K with respect to the pair (K, K_0) is a simplicial complex which has all simplices from K and every coned simplex $\sigma \cup \{x\}$ from an additional vertex x to every simplex $\sigma \in K_0$. Figure 2.13 shows the coned complexes on the right in each panel. The following fact is useful to build an intuition about relative homology groups.

Fact 2.12. *One has* $\mathsf{H}_p(K, K_0) \cong \mathsf{H}_p(K^*)$ *for all* $p > 0$ *and* $\beta_0(\mathsf{H}_0(K, K_0))$ $= \beta_0(\mathsf{H}_0(K^*)) - 1$.

For example, consider K to be an edge $\{a, b, ab\}$ with $K_0 = \{a, b\}$ as in Figure 2.13(a). The 1-chain ab is a relative 1-cycle because $\partial_1(ab) = a + b \in \mathsf{C}_0(K_0)$ and hence $\partial_1^{K,K_0}([ab])$ is 0 in $\mathsf{C}_0(K, K_0)$. This is indicated by the presence of the loop in the coned space.

Now, consider K to be a triangle $\{a, b, c, ab, ac, bc, abc\}$ with $K_0 = \{a, b, c, ab\}$ as in Figure 2.13(b). The 1-chains bc and ac are both relative 1-cycles because $\partial_1(bc) = b + c \in \mathsf{C}_0(K_0)$ and hence $\partial_1^{K,K_0}([bc])$ is 0 in $\mathsf{C}_0(K, K_0)$; similarly, $\partial_1^{K,K_0}([ac]) = 0$. The 1-chain ab is of course a relative 1-cycle because it is already 0 as a relative chain. Therefore, the relative 1-cycle group $\mathsf{Z}_1(K, K_0)$ has a basis $\{[bc], [ac]\}$. The relative 1-boundary group $\mathsf{B}_1(K, K_0)$ is given by $\partial_2^{K,K_0}(abc) = [ab] + [bc] + [ac] = [bc] + [ac]$. The relative homology group $\mathsf{H}_1(K, K_0)$ has one nontrivial class, namely the class of either $[bc]$ or $[ac]$ but not both because $[bc] + [ac]$ is a relative boundary.

2.5.3 Singular Homology

So far we have considered only simplicial homology which is defined on a simplicial complex without any assumption of a particular topology. Now, we extend this definition to topological spaces. Let X be a topological space. We bring the notion of simplices in the context of X by considering maps from the standard d-simplices to X. A standard p-simplex Δ^p is defined by the convex hull of $p + 1$ points $\left\{ (x_1, \ldots, x_i, \ldots, x_{p+1}) \mid x_i = 1 \text{ and } x_j = 0 \text{ for } j \neq i \right\}_{i=1,\ldots,p+1}$ in $\mathbb{R}^{p+1}$.

Definition 2.28. (Singular simplex) A *singular p-simplex* for a topological space X is defined as a map $\sigma : \Delta^p \to X$.

Notice that the map σ need not be injective and thus Δ^p may be "squashed" arbitrarily in its image. Nevertheless, we can still have a notion of the chains, boundaries, and cycles which are the main ingredients for defining a homology group called the *singular homology* of X.

The boundary of a p-simplex σ is given by $\partial\sigma = \tau_0 + \tau_2 + \cdots + \tau_p$ where $\tau_i : (\partial\Delta^p)_i \to X$ is the restriction of the map σ on the i-th facet $(\partial\Delta^p)_i$ of Δ^p.

A p-chain is a sum of singular p-simplices with coefficients from integers, reals, or some appropriate rings. As before, under our assumption of $\mathbb{Z}_2$ coefficients, a singular p-chain is given by $\sum_i \alpha_i \sigma_i$ where $\alpha_i = 0$ or 1. The boundary of a singular p-chain is defined the same way as we did for simplicial chains, the only difference being that we have to accommodate for infinite chains:

$$\partial(c_p = \sigma_1 + \sigma_2 + \cdots) = \partial\sigma_1 + \partial\sigma_2 + \cdots$$

We get the usual chain complex with $\partial_p \circ \partial_{p-1} = 0$ for all $p > 0$,

$$\cdots \overset{\partial_{p+1}}{\to} \mathsf{C}_p \overset{\partial_p}{\to} \mathsf{C}_{p-1} \overset{\partial_{p-1}}{\to} \cdots,$$

and can define the cycle and boundary groups as $\mathsf{Z}_p = \ker \partial_p$ and $\mathsf{B}_p = \operatorname{im} \partial_{p+1}$. We have the singular homology defined as the quotient group $\mathsf{H}_p = \mathsf{Z}_p / \mathsf{B}_p$.

A useful fact is that singular and simplicial homology coincide when both are well defined.

Theorem 2.10. *Let X be a topological space with a triangulation K, that is, the underlying space $|K|$ is homeomorphic to X. Then $\mathsf{H}_p(K) \cong \mathsf{H}_p(X)$ for any $p \geq 0$.*

Note that the above theorem also implies that different triangulations of the same topological space give rise to isomorphic simplicial homology.

2.5.4 Cohomology

There is a dual concept to homology called cohomology. Although cohomology can be defined with coefficients in rings as in the case of homology groups, we will mainly focus on defining it over a field, thus becoming a vector space.

A vector space V defined with a field $\mathbf{k}$ admits a dual vector space V^* whose elements are linear functions $\phi \colon V \to \mathbf{k}$. These linear functions themselves can be added and multiplied over $\mathbf{k}$ forming the dual vector space V^*. The homology group $\mathsf{H}_p(K)$ as we defined in Definition 2.26 over the field $\mathbb{Z}_2$ is a vector space and hence admits a dual vector space which is usually denoted as $\mathrm{Hom}(\mathsf{H}_p(K), \mathbb{Z}_2)$. The p-th cohomology group denoted $\mathsf{H}^p(K)$ is not equal to this dual space, though over the coefficient field $\mathbb{Z}_2$, one has that $\mathsf{H}^p(K)$ is isomorphic to $\mathrm{Hom}(\mathsf{H}_p(K), \mathbb{Z}_2)$ and $\mathsf{H}^p(K)$ is also defined with spaces of linear maps.

Cochains, Coboundaries, and Cocycles

A p-cochain is a homomorphism $\phi \colon \mathsf{C}_p \to \mathbb{Z}_2$ from the chain group to the coefficient ring over which C_p is defined which is $\mathbb{Z}_2$ here. In this case, a p-cochain ϕ is given by its evaluation $\phi(\sigma)$ (0 or 1) on every p-simplex σ in K, or more precisely, a p-chain $c = \sum_{i=1}^{m_p} \alpha_i \sigma_i$ gets a value

$$\phi(c) = \alpha_1 \phi(\sigma_1) + \alpha_2 \phi(\sigma_2) + \cdots + \alpha_{m_p} \phi(\sigma_{m_p}).$$

Also, verify that $\phi(c + c') = \phi(c) + \phi(c')$ satisfying the property of group homomorphism. For a chain c, the particular cochain that assigns 1 to a simplex if and only if it has a nonzero coefficient in c is called its dual cochain c^*. The p-cochains form a cochain group C^p dual to C_p where the addition is defined by $(\phi + \phi')(c) = \phi(c) + \phi'(c)$ by taking $\mathbb{Z}_2$-addition on the right. We can also define a scalar multiplication $(\alpha\phi)(c) = \alpha\phi(c)$ by using the $\mathbb{Z}_2$-multiplication. This makes C^p a vector space.

Similar to boundaries of chains, we have the notion of coboundaries of cochains $\delta_p \colon \mathsf{C}^p \to \mathsf{C}^{p+1}$. Specifically, for a p-cochain ϕ, its $(p + 1)$-coboundary is given by the homomorphism $\delta\phi \colon \mathsf{C}^{p+1} \to \mathbb{Z}_2$ defined as $\delta\phi(c) = \phi(\partial c)$ for any $(p + 1)$-chain c. Therefore, the coboundary operator δ takes a p-cochain and produces a $(p + 1)$-cochain giving the sequence for a simplicial k-complex:

$$0 = \mathsf{C}^{-1} \xrightarrow{\delta_{-1}} \mathsf{C}^0 \xrightarrow{\delta_0} \mathsf{C}^1 \xrightarrow{\delta_1} \cdots \xrightarrow{\delta_{k-1}} \mathsf{C}^k \xrightarrow{\delta_k} \mathsf{C}^{k+1} = 0.$$

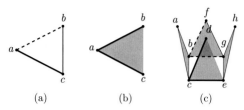

(a)   (b)   (c)

Figure 2.14 Illustration for cohomology. (a) and (c) The 1-cochain with support
on the solid thick edges is a 1-cocycle which is not a 1-coboundary, so it constitutes a nontrivial class in H^1. The 1-cochain with support on dashed edges
constitutes a cohomologous class. (b) The 1-cochain with support on the solid
thick edges is a 1-cocycle which is also a 1-coboundary and hence belongs to a
trivial class.

The set of p-coboundaries forms the coboundary group (vector space) B^p
where the group addition and scalar multiplication are given by the same in
C^p.

Now we come to cocycles, the dual notion to cycles. A p-cochain ϕ is
called a *p-cocycle* if its coboundary $\delta\phi$ is a zero homomorphism. The set of
p-cocycles form a group Z^p (a vector space) where again the addition and
multiplication are induced by the same in C^p.

Similar to the boundary operator ∂, the coboundary operator δ satisfies the
following property.

Fact 2.13. *For $p > 0$, $\delta_p \circ \delta_{p-1} = 0$ which implies $\mathsf{B}^p \subseteq \mathsf{Z}^p$.*

Definition 2.29. (Cohomology group) Since B^p is a subgroup of Z^p, the quotient group $\mathsf{H}^p = \mathsf{Z}^p/\mathsf{B}^p$ is well defined which is called the *p-th cohomology
group*.

Example 2.2. *Consider the three complexes in Figure 2.14. In the following
discussion, for convenience, we refer to the p-simplices on which c^p evaluates to 1 as the* support *of c^p. The 1-cochain ϕ with the support on the
edge ac is a cocycle because $\delta_1\phi = 0$ as there is no triangle and hence no
nonzero 2-cochain. It is also not a coboundary because there is no 0-cochain
ϕ' (assignment of 0 and 1 on vertices) so that*

$$\delta_0\phi'(ac) = \phi'(a+c) = 1 = \phi(ac),$$
$$\delta_0\phi'(ab) = \phi'(a+b) = 0 = \phi(ab),$$
$$\delta_0\phi'(bc) = \phi'(b+c) = 0 = \phi(bc).$$

*The 1-cochain ϕ with support on edges ab and ac in Figure 2.14(b) is a 1-
cocycle because $\delta_1\phi(abc) = \phi(ab+ac+bc) = 0$. Notice that now a cochain*

with support only on one edge ac cannot be a cocycle because of the presence of the triangle abc. The 1-cochain ϕ is also a 1-coboundary because a 0-cochain with assignment of 1 on the vertex a produces ϕ as a coboundary.

Similarly, verify that the 1-cochain ϕ with support on edges cd and ce in Figure 2.14(c) is a cocycle but not a coboundary. Thus, the class $[\phi]$ is nontrivial in one-dimensional cohomology H^1. Any other nontrivial class is cohomologous to it. For example, the class $[\phi']$ where ϕ' has support on edges bf and bg is cohomologous to $[\phi]$. This follows from the fact that $[\phi] + [\phi'] = [\phi + \phi'] = [0]$ because $\phi + \phi'$ is a 1-coboundary obtained by assigning 1 to vertices a, b, and c.

Similar to the homology groups, a simplicial map $f: K_1 \to K_2$ also induces a homomorphism f^* between the two cohomology groups, but in the *opposite* direction. To see this, consider the chain map $f_\#$ induced by f (Definition 2.27). Then, a cochain map $f^\# : \mathsf{C}^p(K_2) \to \mathsf{C}^p(K_1)$ is defined as $f^\#(\phi)(c) = \phi(f_\#(c))$. The cochain map $f^\#$ in turn defines the induced homomorphism between the respective cohomology groups. We will use the following fact in Section 4.2.1.

Fact 2.14. *A simplicial map $f: K_1 \to K_2$ induces a homomorphism $f^*: \mathsf{H}^p(K_2) \to \mathsf{H}^p(K_1)$ for every $p \geq 0$.*

2.6 Notes and Exercises

Simplicial complexes are a fundamental structure in algebraic topology. A good source for the subject is Munkres [242].

The concept of nerve is credited to Aleksandroff [7]. The Nerve Theorem has different versions. It holds for open covers for topological spaces with some mild conditions [300]. Borsuk proved it for closed covers again with some conditions on the space and covers [45]. The assumptions of both are satisfied by metric spaces and finite covers with which we state the theorem in Section 2.2. A version of the theorem is also credited to Leray [220].

Čech and Vietoris–Rips complexes have turned out to be very effective data structures in topological data analysis. Čech complexes were introduced to define Čech homology. Leonid Vietoris [293] introduced the Vietoris complex to extend the homology theory from simplicial complexes to metric spaces. Later, Eliyahu Rips used it in hyperbolic group theory [176]. Jean-Claude Hausmann named it as Vietoris–Rips complex and showed that it is homotopy equivalent to a compact Riemannian manifold when the vertex set spans all points of the manifold and the parameter to build it is sufficiently small [187].

This result was further improved by Latschev [218] who showed that the homotopy equivalence holds even when the vertex set is finite.

Delaunay complexes are a very well known and useful data structure for various geometric applications in two and three dimensions. They enjoy various optimal properties. For example, for a given point set $P \subset \mathbb{R}^2$, among all simplicial complexes linearly embedded in $\mathbb{R}^2$ with vertex set P, the Delaunay complex maximizes the minimum angle over all triangles as stated in Fact 2.5. Many such properties and algorithms for computing Delaunay complexes are described in the books by Edelsbrunner [148] and Cheng et al. [96]. The alpha complex was proposed in [151] and further developed in [153]. The first author of this book can attest to the historic fact that the development of the persistence algorithm was motivated by the study of alpha complexes and their Betti numbers. The book by Edelsbrunner and Harer [149] confirms this. Witness complexes were proposed by de Silva and Carlsson [113] in an attempt to build a sparser complex out of a dense point sample. The graph induced complex is also another such construction proposed in [123].

Homology groups and their associated concepts are main algebraic tools used in topological data analysis. Many associated structures and results about them exist in algebraic topology. We only cover the main necessary concepts that are used in this book and leave others. Interested readers can familiarize themselves with these omitted topics by reading Munkres [242], Hatcher [186], or Ghrist [170], among many other excellent sources.

Exercises

1. Suppose we have a collection of sets $\mathcal{U} = \{U_\alpha\}_{\alpha \in A}$ where there exists an element $U \in \mathcal{U}$ that contains all other elements in $\mathcal{U}$. Show that the nerve complex $N(\mathcal{U})$ is contractible to a point.

2. Given a parameter α and a set of points $P \subset \mathbb{R}^d$, show that the alpha complex $\mathrm{Del}^\alpha(P)$ is contained in the intersection of Delaunay complex and Čech complex at scale α; that is, $\mathrm{Del}^\alpha(P) \subseteq \mathrm{Del}(P) \cap \mathbb{C}^\alpha(P)$.

3. Let K be the simplicial complex of a tetrahedron. Write a basis for the chain groups C_1, C_2, boundary groups B_1, B_2, and cycle groups Z_1, Z_2. Write the boundary matrix representing the boundary operator ∂_2 with rows and columns representing bases of C_1 and C_2, respectively.

4. Let K be a triangulation of an orientable surface without boundary that has genus g. Prove that $\beta_1(K) = 2g$.

5. Let K be a triangulation of a two-dimensional sphere $\mathbb{S}^2$. Now remove h number of vertex-disjoint triangles from K, and let the resulting simplicial complex be K'. Describe the Betti numbers of K', and justify your answer.

6. We state the Nerve Theorem (Theorem 2.1) for covers where either all cover elements are closed or all cover elements are open. Show that the theorem does not hold if we mix open and closed elements in the cover.

7. Give an example where a simplex which is weakly witnessed may not have all its faces weakly witnessed. Show that
   (a) $\mathcal{W}(Q, P') \subseteq \mathcal{W}(Q, P)$ for $P' \subseteq P$,
   (b) $\mathcal{W}(Q', P)$ may not be a subcomplex of $\mathcal{W}(Q, P)$ where $Q' \subseteq Q$.

8. Consider Definition 2.16 for graph induced complex. Let $\mathbb{VR}(G)$ be the clique complex given by the input graph $G(P)$. Assume that the map $\nu \colon P \to 2^Q$ sends every point to a singleton under input metric d. Then, $\nu \colon P \to \nu(P)$ is a well-defined vertex map. Prove that the vertex map $\nu : P \to Q$ extends to a simplicial map $\bar{\nu} : \mathbb{VR}(G) \to \mathcal{G}(G(P), Q, \mathsf{d})$. Also, show that every simplicial complex $K(Q)$ with the vertex set Q for which $\bar{\nu} \colon \mathbb{VR}(G) \to K(Q)$ becomes simplicial must contain $\mathcal{G}(G(P), Q, \mathsf{d})$.

9. Prove Proposition 2.9.

10. Consider a complex $K = \{a, b, c, ab, bc, ca, abc\}$. Enumerate all elements in the 1-chain, 1-cycle, and 1-boundary groups defined on K under $\mathbb{Z}_2$ coefficient. Do the same for cochains, cocycles, and coboundaries.

11. Show an example for each of the following:
    (a) a chain that is a cycle but its dual cochain is not a cocycle;
    (b) a chain that is a cycle and its dual cochain is a cocycle;
    (c) a chain that is a boundary and its dual cochain is not a coboundary;
    (d) a chain that is a boundary and its dual cochain is a coboundary.

12. Prove that $\partial_{p-1} \circ \partial_p = 0$ for relative chain groups and also $\delta_p \circ \delta_{p-1} = 0$ for cochain groups.

3

Topological Persistence

Suppose we have point cloud data P sampled from a 3D model. A quantified summary of the topological features of the model that can be computed from this sampled representation helps in further processing such as shape analysis in geometric modeling. Persistent homology offers this avenue, as Figure 3.1 illustrates. For further explanation, consider P sampled from a curve in $\mathbb{R}^2$ as in Figure 3.3 later. Our goal is to get the information that the sampled space had two loops, one bigger and more prominent than the other. The notion of persistence captures this information. Consider the distance function $r : \mathbb{R}^2 \to \mathbb{R}$ defined over $\mathbb{R}^2$ where $r(x)$ equals $\mathsf{d}(x, P)$, that is, the minimum distance of x to the points in P. Now let us look at the sublevel sets of r, that is, $r^{-1}[-\infty, a]$ for some $a \in \mathbb{R}^+ \cup \{0\}$. These sublevel sets are the union of closed balls of radius a centering the points. We can observe from Figure 3.3 that if we increase a starting from zero, we come across different holes surrounded by the union of these balls which ultimately get filled up at different times. However, the two holes corresponding to the original two loops persist longer than the others. We can abstract out this observation by looking at how long a feature (homological class) survives when we scan over the increasing sublevel sets. This weeds out the "false" features (noise) from the true ones. The notion of persistent homology formalizes and discretizes this idea: *It takes a function defined on a topological space (simplicial complex) and quantifies the changes in homology classes as the sublevel sets (subcomplexes) grow with increasing value of the function.*

There are two predominant scenarios where persistence appears, though in slightly different contexts. One is when the function is defined on a topological space which requires considering singular homology groups of the sublevel sets. The other is when the function is defined on a simplicial complex and the sequence of sublevel sets is implicitly given by a nested sequence of subcomplexes called a *filtration*. This involves simplicial homology. Section 3.1

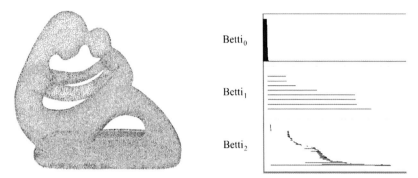

Figure 3.1 Persistence barcodes computed from point cloud data. The barcodes
on the right show a single long bar for H_0 (top), signifying one connected component, eight long bars for H_1 (middle), signifying eight fundamental classes two for
each of the four "through holes," and a single long bar for H_2 (bottom), signifying
the connected closed surface. This picture taken from [134].

introduces persistence in both of these contexts, though we focus mainly on the
simplicial setting which is availed most commonly for computational purposes.

The birth and death of homological classes give rise to intervals during
which a class remains alive. These intervals together called a *barcode* summarize the topological persistence of a filtration; see, for example, Figure 3.1.
An equivalent notion called *persistence diagrams* plots the intervals as points
in the extended plane $\bar{\mathbb{R}}^2 := (\mathbb{R} \cup \{\pm\infty\})^2$; specifically, the birth and death
constitute the x- and y-coordinates of a point. The stability of the persistence
diagrams against the perturbation of the functions that generate the filtrations
is an important result. It makes topological persistence robust against noise.
When filtrations are given without any explicit mention of a function, we
can still talk about the stability of the persistence diagrams with respect to
the so-called *interleaving distance* between the induced *persistence modules*.
Sections 3.2 and 3.4 are devoted to these concepts.

The algorithms that compute the persistence diagram from a given filtration
are presented in Section 3.3. First, we introduce an algorithm assuming that
the input is presented combinatorially with simplices added one at a time in a
filtration. The algorithm pairs simplices, one creating and the other destroying
an interval. Then, this pairing is translated into matrix operations assuming
that the input is a boundary matrix representing the filtration. A more efficient
version of the algorithm is obtained by some simple but effective modification.

Finally, we consider the case of a piecewise-linear function (PL-function)
on a simplicial complex and derive a filtration out of it from which the actual
persistence of the input PL-function can be computed. This is presented in
Section 3.5.

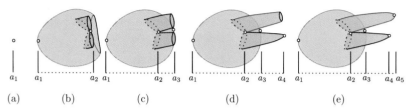

Figure 3.2 Persistence of a function on a topological space that has five critical values: (a) $\mathbb{T}_{a_1}$, only a new class in H_0 is created; (b) $\mathbb{T}_{a_2}$, two new independent classes in H_1 are created; (c) $\mathbb{T}_{a_3}$, one of the two classes in H_1 dies; (d) $\mathbb{T}_{a_4}$, the single remaining class in H_1 dies; and (e) $\mathbb{T}_{a_5}$, a new class in H_2 is created.

3.1 Filtrations and Persistence

At the core of topological persistence is the notion of filtrations which can arise in the context of topological spaces or simplicial complexes.

3.1.1 Space Filtration

Consider a real-valued function $f : \mathbb{T} \to \mathbb{R}$ defined on a topological space $\mathbb{T}$. Let $\mathbb{T}_a = f^{-1}(-\infty, a]$ denote the sublevel set for the function value a. Certainly, we have inclusions:

$$\mathbb{T}_a \subseteq \mathbb{T}_b \text{ for } a \leq b.$$

Now consider a sequence of reals $a_1 \leq a_2 \leq \cdots \leq a_n$ which are often chosen to be critical values where the homology group of the sublevel sets change as illustrated in Figure 3.2. Considering the sublevel sets at these values and a dummy value $a_0 = -\infty$ with $\mathbb{T}_{a_0} = \varnothing$, we obtain a nested sequence of subspaces of $\mathbb{T}$ connected by inclusions which gives a *filtration* $\mathcal{F}_f$:

$$\mathcal{F}_f : \varnothing = \mathbb{T}_{a_0} \hookrightarrow \mathbb{T}_{a_1} \hookrightarrow \mathbb{T}_{a_2} \hookrightarrow \cdots \hookrightarrow \mathbb{T}_{a_n}. \qquad (3.1)$$

Figure 3.2 shows an example of the inclusions of the sublevel sets. The inclusions in a filtration induce linear maps in the singular homology groups of the subspaces involved. So, if $\iota : \mathbb{T}_{a_i} \to \mathbb{T}_{a_j}$, $i \leq j$, denotes the inclusion map $x \mapsto x$, we have an induced homomorphism

$$h_p^{i,j} = \iota_* : H_p(\mathbb{T}_{a_i}) \to H_p(\mathbb{T}_{a_j}) \qquad (3.2)$$

for all $p \geq 0$ and $0 \leq i \leq j \leq n$. Therefore, we have a sequence of homomorphisms induced by inclusions forming what we call a *homology module*:

$$0 = H_p(\mathbb{T}_{a_0}) \to H_p(\mathbb{T}_{a_1}) \to H_p(\mathbb{T}_{a_2}) \to \cdots \to H_p(\mathbb{T}_{a_n}).$$

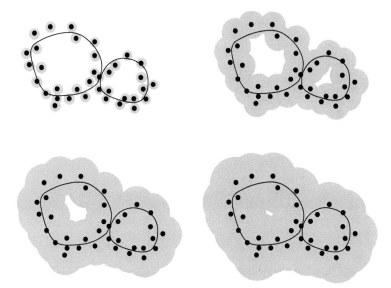

Figure 3.3 Noisy sample of a curve with two loops and the growing sublevel sets of the distance function to the sample points. The larger loop appearing as the bigger hole in the complement of the union of balls persists longer than the same for the smaller loop while other spurious holes persist even shorter.

It is worthwhile to mention that writing a group to be 0 means that it is a trivial group containing only the identity element 0. The homomorphism $h_p^{i,j}$ sends the homology classes of the sublevel set $\mathbb{T}_{a_i}$ to those of the sublevel set of $\mathbb{T}_{a_j}$. Some of these classes may die (become trivial) while the others survive. The image Im $h_p^{i,j}$ contains this information.

The inclusions of sublevel sets give rise to persistence also in the context of point clouds, a common input form in data analysis.

Point Cloud

For a point set P in a metric space (M, d), define the distance function $f : M \to \mathbb{R}$, $x \mapsto \mathsf{d}(x, p)$, where $p \in \operatorname{argmin}_{q \in P} \mathsf{d}(x, q)$. Observe that the sublevel sets $f^{-1}(-\infty, a]$ are the union of closed metric balls of radius a centering points in P. Now we have exactly the same setting as we described for general topological spaces above where $\mathbb{T}$ is replaced with M and sublevel sets $\mathbb{T}_a$ by the union of metric balls that grows with increasing value of a. Figure 3.3 illustrates an example where M is the Euclidean plane $\mathbb{R}^2$.

3.1.2 Simplicial Filtrations and Persistence

Persistence on topological spaces involves computing singular homology groups for sublevel sets. Computationally, this is cumbersome. So, we take

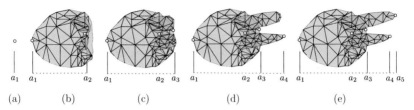

Figure 3.4 Persistence of the piecewise-linear version of the function on a triangulation of the topological space considered in Figure 3.2.

refuge in the discrete analogue of the topological persistence. This involves two important adaptations: first, the topological space is replaced with a simplicial complex; second, singular homology groups are replaced with simplicial homology groups. This means that the topological space $\mathbb{T}$ considered before is replaced with one of its triangulations, as Figure 3.4 illustrates. For point cloud data, the union of balls can be replaced by their nerve, the Čech complex or its cousin Vietoris–Rips complex introduced in Section 2.2. Figure 3.5 illustrates this conversion for example in Figure 3.3. Of course, these replacements need to preserve the original persistence in some sense, which is addressed in general by the notion of stability introduced in Section 3.4.

The nested sequence of topological spaces that arise with growing sublevel sets translates into a nested sequence of simplicial complexes in the discrete analogue. This brings in the concept of filtration of simplicial complexes that allows defining the persistence using simplicial homology groups.

Definition 3.1. (Simplicial filtration) A *filtration* $\mathcal{F} = \mathcal{F}(K)$ of a simplicial complex K is a nested sequence of its subcomplexes

$$\mathcal{F} \colon \varnothing = K_0 \subseteq K_1 \subseteq \cdots \subseteq K_n = K$$

which is also written with inclusion maps as

$$\mathcal{F} \colon \varnothing = K_0 \hookrightarrow K_1 \hookrightarrow \cdots \hookrightarrow K_n = K.$$

$\mathcal{F}$ is called *simplex-wise* if either $K_i \setminus K_{i-1}$ is empty or a single simplex for every $i \in [1, n]$. Notice that the possibility of difference being empty allows two consecutive complexes to be the same.

Simplicial filtrations can appear in various contexts.

Simplex-wise Monotone Function
Consider a simplicial complex K and a (simplex-wise) function $f \colon K \to \mathbb{R}$ on it. We call the function f *simplex-wise monotone* if for every $\sigma' \subseteq \sigma$, we

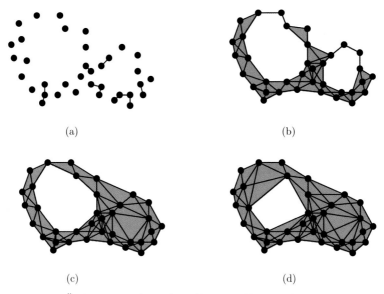

(a)　　　　　　　　　　　(b)

(c)　　　　　　　　　　　(d)

Figure 3.5 Čech complex of the union of balls considered in Figure 3.3. Homology classes in H_1 are being born and die as the union grows. The two most prominent holes appear as the two most persistent homology classes in H_1. Other classes appear and disappear quickly with relatively much shorter persistence.

have $f(\sigma') \leq f(\sigma)$. This property ensures that the sublevel sets $f^{-1}(-\infty, a]$ are subcomplexes of K for every $a \in \mathbb{R}$. Denoting $K_i = f^{-1}(-\infty, a_i]$ and a dummy value $a_0 = -\infty$, we get a filtration:

$$\varnothing = K_0 \hookrightarrow K_1 \hookrightarrow \cdots \hookrightarrow K_n = K.$$

Vertex Function
A *vertex function* $f \colon V(K) \to \mathbb{R}$ is defined on the vertex set $V(K)$ of the complex K. We can construct a filtration $\mathcal{F}$ from such a function.

Lower/Upper Stars
Recall that in Section 2.1 we have defined the star and link of a vertex $v \in K$ which intuitively captures the concept of local neighborhood of v in K. We infuse the information about a vertex function f into these structures. First, we fix a total order on vertices $V = \{v_1, \ldots, v_n\}$ of K so that their f-values are in nondecreasing order, that is, $f(v_1) \leq f(v_2) \leq \cdots \leq f(v_n)$. The *lower star* of a vertex $v \in V$, denoted by $\mathrm{Lst}(v)$, is the set of simplices in $\mathrm{St}(v)$ whose vertices except v appear before v in this order. The *closed lower star* $\overline{\mathrm{Lst}}(v)$ is the closure of $\mathrm{Lst}(v)$, that is, it consists of simplices in $\mathrm{Lst}(v)$ and their faces. The *lower link* $\mathrm{Llk}(v)$ is the set of simplices in $\overline{\mathrm{Lst}}(v)$ disjoint from v. Symmetrically, we can define the *upper star* $\mathrm{Ust}(v)$, *closed upper star* $\overline{\mathrm{Ust}}(v)$,

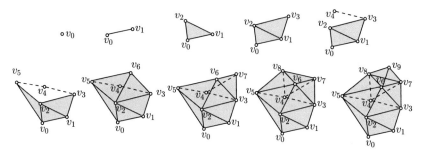

Figure 3.6 The sequence shows a *lower-star* filtration of K induced by a vertex function which is a 'height function' that records the vertical height of a vertex increasing from bottom to top here.

and *upper link* $\mathrm{Ulk}(v)$, spanned by vertices in the star of v which appear after v in the chosen order.

One gets a filtration using the lower stars of the vertices: $K_{f(v_i)}$ in the following filtration denotes all simplices in K spanned by vertices in $\{v_1, \ldots, v_i\}$. Let v_0 denote a dummy vertex with $f(v_0) = -\infty$:

$$\varnothing = K_{f(v_0)} \subseteq K_{f(v_1)} \subseteq K_{f(v_2)} \subseteq \cdots \subseteq K_{f(v_n)} = K.$$

Observe that the $K_{f(v_i)} \setminus K_{f(v_{i-1})} = \mathrm{Lst}(v_i)$ for $i \in [1, n]$ in the above filtration, that is, each time we add the lower star of the next vertex in the filtration. This filtration called the *lower-star filtration* for f is studied in Section 3.5 in more detail. Figure 3.6 shows a lower-star filtration. A lower-star filtration can be made simplex-wise by adding the simplices in a lower star in any order that puts a simplex after all of its faces.

Alternatively, we may consider the vertices in non-increasing order of f-values and obtain an upper-star filtration. For this we take $K^{f(v_i)}$ to be all simplices spanned by vertices in $\{v_i, v_{i+1}, \ldots, v_n\}$. Assuming a dummy vertex v_{n+1} with $f(v_{n+1}) = \infty$, one gets a filtration:

$$\varnothing = K^{f(v_{n+1})} \subseteq K^{f(v_n)} \subseteq K^{f(v_{n-1})} \subseteq \cdots \subseteq K^{f(v_1)} = K.$$

Observe that the $K^{f(v_i)} \setminus K^{f(v_{i+1})} = \mathrm{Ust}(v_i)$ for $i \in [1, n]$ in the above filtration, that is, each time we add the upper star of the next vertex in the filtration. This filtration called the *upper-star filtration* for f is in some sense a symmetric version of the lower-star filtration though they may provide different persistence pairs. An upper-star filtration can also be made simplex-wise by adding the simplices in an upper star in any order that puts a simplex after all of its faces. In this book, by default, we will assume that the function values along a filtration are nondecreasing. This means that we consider only lower filtrations by default.

Vertex functions are closely related to the so-called PL-functions. A vertex function $f \colon K \to \mathbb{R}$ defines a PL-function on the underlying space $|K|$ of K which is obtained by linearly interpolating f over all simplices. On the other hand, the restriction of a PL-function to vertices trivially provides a vertex function.

Definition 3.2. (PL-functions) Given a simplicial complex K, a *piecewise-linear (PL) function* $f \colon |K| \to \mathbb{R}$ is defined to be the linear extension of a vertex function $f_V \colon V(K) \to \mathbb{R}$ defined on vertices $V(K)$ of K so that for every point $x \in |K|$, $\bar{f}(x) = \sum_{i=1}^{k+1} \alpha_i f_V(v_i)$ where $\sigma = \{v_1, \ldots, v_{k+1}\}$ is the unique lowest-dimensional simplex of dimension $k \geq 0$ containing x and $\alpha_1, \ldots, \alpha_{k+1}$ are the barycentric coordinates of x in σ.[1]

Fact 3.1. *We have the following.*

(a) A PL-function $f \colon |K| \to \mathbb{R}$ naturally provides a vertex function $f_V \colon V(K) \to \mathbb{R}$.

(b) A simplex-wise lower-star filtration for f is also a filtration for the simplex-wise monotonic function $\bar{f} \colon K \to \mathbb{R}$ where $\bar{f}(\sigma) = \max_{v \in \sigma} f(v)$.

(c) Similarly, a simplex-wise upper-star filtration for f is also a filtration for the simplex-wise monotonic function $\bar{f}(\sigma) = \max_{v \in \sigma}(-f(v))$.

Observe that a given vertex function $f_V \colon K \to \mathbb{R}$ induces a PL-function $f \colon |K| \to \mathbb{R}$ whose persistence on the topological space $|K|$ can be defined by taking sublevel sets at critical values (see Definition 3.23 for critical points in the PL case) and then applying Definition 3.4. The relation of this persistence to the persistence of the lower-star filtration of K induced by f_V is studied in Section 3.5.2. Indeed, the persistence of f can be read from the persistence of the lower-star filtration of f_V.

Finally, we note that any simplicial filtration $\mathcal{F}$ can naturally be induced by a function. We introduce this association to unify the definition of persistence pairing later in Definition 3.7.

Definition 3.3. (Filtration function) If a simplicial filtration $\mathcal{F}$ is obtained from a simplex-wise monotone function or a vertex function f, then $\mathcal{F}$ is induced by f. Conversely, if $\mathcal{F}$ is given without any explicit input function, we say $\mathcal{F}$ is induced by the simplex-wise monotone function f where every simplex $\sigma \in (K_i \setminus K_{i-1})$ for $K_i \neq K_{i+1}$ is given the value $f(\sigma) = i$.

---

[1] Unique numbers $\alpha_1, \ldots, \alpha_{k+1}$ for which $x = \sum_{i=1}^{k+1} \alpha_i v_i$ with $\sum \alpha_i = 1$ and $\alpha_i \geq 0$ for all i are called barycentric coordinates of x in σ.

Naturally, every simplicial filtration gives rise to a sequence of homomorphisms $h_p^{i,j}$ as in Eq. (3.2) induced by inclusions again forming a homology module:

$$0 = \mathsf{H}_p(K_0) \to \mathsf{H}_p(K_1) \to \cdots \to \mathsf{H}_p(K_i) \xrightarrow{\;h_p^{i,j}\;} \mathsf{H}_p(K_j) \to \cdots \to \mathsf{H}_p(K_n) = \mathsf{H}_p(K).$$

3.2 Persistence

In both cases of space and simplicial filtration $\mathcal{F}$, we arrive at a homology module:

$$\mathsf{H}_p\mathcal{F} : 0 = \mathsf{H}_p(X_0) \to \mathsf{H}_p(X_1) \to \cdots \to \mathsf{H}_p(X_i) \xrightarrow{\;h_p^{i,j}\;} \mathsf{H}_p(X_j) \to \cdots \to \mathsf{H}_p(X_n) = \mathsf{H}_p(X),$$

$$(3.3)$$

where $X_i = \mathbb{T}_{a_i}$ if $\mathcal{F}$ is a space filtration of a topological space $X = \mathbb{T}$ or $X_i = K_i$ if $\mathcal{F}$ is a simplicial filtration of a simplicial complex $X = K$. Persistent homology groups for a homology module are algebraic structures capturing the survival of the homology classes through this sequence. In general, we will call homology modules *persistence modules* in Section 3.4 recognizing that we can replace homology groups with vector spaces.

Definition 3.4. (Persistent Betti number) The p-th persistent homology groups are the images of the homomorphisms: $\mathsf{H}_p^{i,j} = \mathrm{im}\, h_p^{i,j}$, for $0 \leq i \leq j \leq n$. The p-th *persistent Betti numbers* are the dimensions $\beta_p^{i,j} = \dim \mathsf{H}_p^{i,j}$ of the vector spaces $\mathsf{H}_p^{i,j}$.

The p-th persistent homology groups contain the important information of when a homology class is born or when it dies. The issue of birth and death of a class becomes more subtle because when a new class is born, many other classes that are the sum of this new class and any other existing class are also born. Similarly, when a class ceases to exist, many other classes may cease to exist along with it. Therefore, we need a mechanism to pair births and deaths canonically. Figure 3.7 illustrates the birth and death of a class, though the pairing of birth and death events is more complicated as stated in Fact 3.3.

Observe that the nontrivial elements of p-th persistent homology groups $\mathsf{H}_p^{i,j}$ consist of classes that survive from X_i to X_j, that is, the classes which do not get "quotiented out" by the boundaries in X_j. So, one can observe the following.

Fact 3.2. *One has* $\mathsf{H}_p^{i,j} = \mathsf{Z}_p(X_i)/(\mathsf{B}_p(X_j) \cap \mathsf{Z}_p(X_i))$ *and* $\beta_p^{i,j} = \dim \mathsf{H}_p^{ij}$.

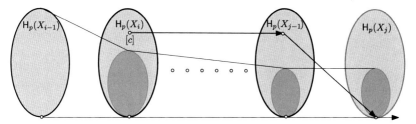

Figure 3.7 A simplistic view of birth and death of classes. A class $[c]$ is born at X_i since it is not in the image of $H_p(X_{i-1})$. It dies entering X_j since this is the first time its image becomes trivial.

Notice that $Z_p(X_i)$ is a subgroup of $Z_p(X_j)$ because $X_i \subseteq X_j$ and hence the above quotient is well defined. We now formally state when a class is born or dies.

Definition 3.5. (Birth and death) A nontrivial p-th homology class $\xi \in H_p(X_a)$ is born at X_i, $i \leq a$, if $\xi \in H_p^{i,a}$ but $\xi \notin H_p^{i-1,a}$. Similarly, a nontrivial p-th homology class $\xi \in H_p(X_a)$ dies entering X_j, $a < j$, if $h_p^{a,j-1}(\xi)$ is not zero (nontrivial) but $h_p^{a,j}(\xi) = 0$.

Observe that not all classes that are born at X_i necessarily die entering some X_j though more than one such may do so.

Fact 3.3. *Let $[c] \in H_p(X_{j-1})$ be a p-th homology class that dies entering X_j. Then, it is born at X_i if and only if there exists a sequence $i_1 \leq i_2 \leq \cdots \leq i_k = i$ for some $k \geq 1$ so that (i) $0 \neq [c_{i_\ell}] \in H_p(X_{j-1})$ is born at X_{i_ℓ} for every $\ell \in \{1, \ldots, k\}$ and (ii) $[c] = [c_{i_1}] + \cdots + [c_{i_k}]$.*

One may interpret the above fact as follows. When a class dies, it may be thought of as a merger of several classes among which the youngest one $[c_{i_k}]$ determines the birth point. This viewpoint is particularly helpful while pairing simplices in the persistence algorithm PAIRPERSISTENCE presented later.

Notice that each X_i, $i = 0, \ldots, n$, is associated with a value of the function f that induces $\mathcal{F}$. For a space filtration, we say $f(X_i) = a_i$ where $X_i = \mathbb{T}_{a_i}$. For a simplicial filtration, we say $f(X_i) = a_i$ where $a_i = f(\sigma)$ for any $\sigma \in X_i$ when the filtration function (Definition 3.3) is simplex-wise monotone. When it is a vertex function f, then we extend f to a simplex-wise monotone function as stated in Fact 3.1.

3.2.1 Persistence Diagram

Fact 3.3 provides a qualitative characterization of the pairing of births and deaths of classes. Now we give a quantitative characterization which helps to draw a visual representation of this pairing called a *persistence diagram*; see Figure 3.8(a). Consider the extended plane $\bar{\mathbb{R}}^2 := (\mathbb{R} \cup \{\pm\infty\})^2$ on which we represent the birth at a_i paired with the death at a_j as a point (a_i, a_j). This pairing uses a *persistence pairing function* $\mu_p^{i,j}$ defined below. Strictly positive values of this function correspond to multiplicities of points in the persistence diagram (Definition 3.8). In what follows, to account for classes that never die, we extend the induced module in Eq. (3.3) on the right end by assuming that $\mathsf{H}_p(X_{n+1}) = 0$.

Definition 3.6. For $0 < i < j \le n + 1$, define

$$\mu_p^{i,j} = \left(\beta_p^{i,j-1} - \beta_p^{i,j} \right) - \left(\beta_p^{i-1,j-1} - \beta_p^{i-1,j} \right). \tag{3.4}$$

The first difference on the right-hand side counts the number of independent classes that are born at or before X_i and die entering X_j. The second difference counts the number of independent classes that are born at or before X_{i-1} and die entering X_j. The difference between the two differences thus counts the number of independent classes that are born at X_i and die entering X_j. When $j = n + 1$, $\mu_p^{i,n+1}$ counts the number of independent classes that are born at X_i and die entering X_{n+1}. They remain alive till the end in the original filtration without extension, or we say that they never die. To emphasize that classes which exist in X_n actually never die, we equate $n + 1$ with ∞ and take $a_{n+1} = a_\infty = \infty$. Observe that, with this assumption, we have $\beta^{i,n+1} = \beta^{i,\infty} = 0$ for every $i \le n$.

Remark 3.1. *The p-th homology classes in $\mathsf{H}_p(X_{j-1})$ that get born at X_i and die entering X_j may not form a vector space. Hence, we cannot talk about its dimension. In fact, the definition of $\mu_p^{i,j}$, in some sense, compensates for this limitation. This definition involves alternating sums of dimensions (β_{ij}) of vector spaces. The dimensions appearing with negative signs lead to this anomaly. However, one can express $\mu_p^{i,j}$ as the dimension of a vector space which is a quotient of a subspace; see [18] for details.*

Definition 3.7. (Class persistence) For $\mu_p^{i,j} \ne 0$, the *persistence* Pers $([c])$ of a class $[c]$ that is born at X_i and dies at X_j is defined as Pers $([c]) = a_j - a_i$. When $j = n + 1 = \infty$, Pers $([c])$ equals $a_{n+1} - a_i = \infty$.

Notice that values a_i can be the index i when no explicit function is given (Definition 3.3). In that case, the persistence of a class is sometimes referred to as *index persistence* which is $j - i$.

Definition 3.8. (Persistence diagram) The *persistence diagram* $\mathrm{Dgm}_p(\mathcal{F}_f)$ (also written $\mathrm{Dgm}_p f$) of a filtration $\mathcal{F}_f$ induced by a function f is obtained by drawing a point (a_i, a_j) with nonzero multiplicity $\mu_p^{i,j}$, $i < j$, on the extended plane $\bar{\mathbb{R}}^2 := (\mathbb{R} \cup \{\pm\infty\})^2$ where the points on the diagonal $\Delta : \{(x,x)\}$ are added with infinite multiplicity.

The addition of the diagonal is a technical necessity for results that we will see afterward.

A class born at a_i and never dying is represented as a point $(a_i, a_{n+1}) = (a_i, \infty)$ (point v in Figure 3.8) – we call such points in the persistence diagram *essential persistent points*, and their corresponding homology classes are called *essential homology classes*. Classes may have the same coordinates because they may be born and die at the same time. This happens only when we allow mutiple homology classes to be created or destroyed at the same function value or filtration point. In general, this also opens up the possibility of creating infinitely many birth–death pairs even if the filtration is finite. To avoid such pathological cases, we always assume that the linear maps in the homology modules have finite rank, a condition known as q-tameness in the literature [80].

There is also an alternative representation of persistence called *barcode* where each birth–death pair (a_i, a_j) is represented by a line segment $[a_i, a_j)$ called a *bar* which is open on the right. The open end signifies that the class dying entering X_j does not exist in X_j. Points at infinity such as (a_i, ∞) are represented with a ray $[a_i, \infty)$ giving an *infinite bar*. See Figure 3.8(b). Figure 3.9 shows typical persistence diagrams and barcodes (ignoring the types of endpoints) for $p = 0, 1$.

Fact 3.4. *We have the following.*

(a) *If a class has persistence s, then the point representing it will be at a Euclidean distance $s/\sqrt{2}$ from the diagonal Δ (distance between t, $\bar{t}$ and r, $\bar{r}$ in Figure 3.8).*

(b) *For sublevel set filtrations, all points (a_i, a_j) representing a class have $a_i \leq a_j$, so they lie on or above the diagonal.*

(c) *If m_i denotes the multiplicity of an essential point (a_i, ∞) in $\mathrm{Dgm}_p(\mathcal{F})$, where $\mathcal{F}$ is a filtration of $X = X_n$, one has $\sum_i m_i = \dim \mathsf{H}_p(X)$, the p-th Betti number of X.*

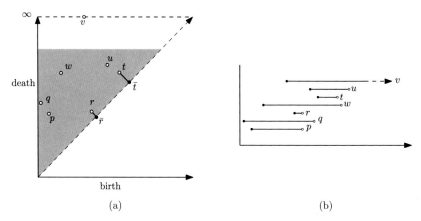

(a)   (b)

Figure 3.8 (a) A persistence diagram with nondiagonal points only in the positive quadrant, and (b) corresponding barcode.

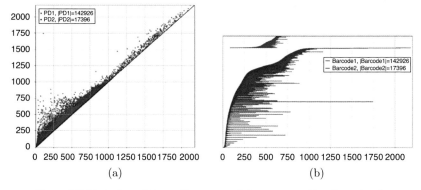

(a)   (b)

Figure 3.9 Typical persistence diagrams and the corresponding barcodes for image data; red and blue correspond to zeroth and first persistence diagrams, respectively. The bars are sorted in increasing order of their birth time from bottom to top.

Here is one important fact relating persistent Betti numbers and persistence diagrams.

Theorem 3.1. *For every pair of indices* $0 \leq k \leq \ell \leq n$ *and every p, the p-th persistent Betti number satisfies* $\beta_p^{k,\ell} = \sum_{i \leq k} \sum_{j > \ell} \mu_p^{i,j}$.

Observe that $\beta_p^{k,\ell}$ is the number of points in the upper left quadrant of the corner (a_k, a_ℓ). A class that is born at X_i and dies entering X_j is counted for $\beta_p^{k,\ell}$ if and only if $i \leq k$ and $j > \ell$. The quadrant is therefore closed on the right and open at the bottom.

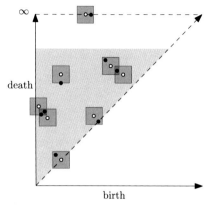

Figure 3.10 Two persistence diagrams and their bottleneck distance, which is half of the side lengths of the squares representing bijections.

Stability of Persistence Diagrams

A persistence diagram $\mathrm{Dgm}_p(\mathcal{F}_f)$, as a set of points in the extended plane $\bar{\mathbb{R}}^2$, summarizes certain topological information of a simplicial complex (space) in relation to the function f that induces the filtration $\mathcal{F}_f$. However, this is not useful in practice unless we can be certain that a slight change in f does not change this diagram dramatically. In practice f is seldom measured accurately, and if its persistence diagram can be approximated from a slightly perturbed version, it becomes useful. Fortunately, persistence diagrams are stable. To formulate this stability, we need a notion of distance between persistence diagrams.

Let $\mathrm{Dgm}_p(\mathcal{F}_f)$ and $\mathrm{Dgm}_p(\mathcal{F}_g)$ be two persistence diagrams for two functions f and g. We want to consider bijections between points from $\mathrm{Dgm}_p(\mathcal{F}_f)$ and $\mathrm{Dgm}_p(\mathcal{F}_g)$. However, they may have different cardinality for off-diagonal points. Recall that persistence diagrams include the points on the diagonal Δ each with infinite multiplicity. This addition allows us to borrow points from the diagonal when necessary to define the bijections. Note that we are considering only filtrations of finite complexes which also make each homology group finite.

Definition 3.9. (Bottleneck distance) Let $\Pi = \{\pi : \mathrm{Dgm}_p(\mathcal{F}_f) \rightarrow \mathrm{Dgm}_p(\mathcal{F}_g)\}$ denote the set of all bijections. Consider the distance between two points $x = (x_1, x_2)$ and $y = (y_1, y_2)$ in L_∞-norm $\|x - y\|_\infty = \max\{|x_1 - x_2|, |y_1 - y_2|\}$ with the assumption that $\infty - \infty = 0$. The *bottleneck distance* between the two diagrams (see Figure 3.10) is

$$\mathrm{d}_b(\mathrm{Dgm}_p(\mathcal{F}_f), \mathrm{Dgm}_p(\mathcal{F}_g)) = \inf_{\pi \in \Pi} \sup_{x \in \mathrm{Dgm}_p(\mathcal{F}_f)} \|x - \pi(x)\|_\infty.$$

Fact 3.5. *The bottleneck distance* d_b *is a metric on the space of persistence diagrams. Clearly,* $\mathsf{d}_b(X, Y) = 0$ *if and only if* $X = Y$. *Moreover,* $\mathsf{d}_b(X, Y) = \mathsf{d}_b(Y, X)$ *and* $\mathsf{d}_b(X, Y) \leq \mathsf{d}_b(X, Z) + \mathsf{d}_b(Z, Y)$.

There is a caveat for the above fact. If d_b is taken as a distance on the space of homology modules $\mathsf{H}_p\mathcal{F}$ instead of the persistence diagrams $\mathrm{Dgm}_p(\mathcal{F})$ they generate, that is, if we define $\mathsf{d}_b(\mathsf{H}_p\mathcal{F}_f, \mathsf{H}_p\mathcal{F}_g) := \mathsf{d}_b(\mathrm{Dgm}_p(\mathcal{F}_f), \mathrm{Dgm}_p(\mathcal{F}_g))$, then it may not be a metric. The first axiom for a metric becomes false if the homology modules are allowed to have classes created and destroyed at the same function values. These classes of zero persistence generate points on the diagonal Δ in the diagram. Since points on the diagonal have infinite multiplicity, two modules differing in the number of such classes of zero persistence may have diagrams with zero bottleneck distance. If we allow such cases, d_b becomes a pseudometric on the space of homology modules, meaning that it satisfies all the axioms of a metric except the first one.

The following theorems originally proved in [101] and further detailed in [149] quantify the notion of the stability of the persistence diagram. There are two versions: one involves simplicial filtrations and the other involves space filtrations. For two functions, $f, g : X \to \mathbb{R}$, the infinity norm is defined as $\|f - g\|_\infty := \sup_{x \in X} |f(x) - g(x)|$.

Theorem 3.2. (Stability for simplicial filtrations) *Let* $f, g : K \to \mathbb{R}$ *be two simplex-wise monotone functions giving rise to two simplicial filtrations* $\mathcal{F}_f$ *and* $\mathcal{F}_g$. *Then, for every* $p \geq 0$,

$$\mathsf{d}_b(\mathrm{Dgm}_p(\mathcal{F}_f), \mathrm{Dgm}_p(\mathcal{F}_g)) \leq \|f - g\|_\infty.$$

For the second version of the stability theorem, we require that the functions referred to in the theorem are "nice" in the sense that they are tame. A function $f : X \to \mathbb{R}$ is *tame* if the homology groups of its sublevel sets have finite rank and these ranks change only at finitely many values called *critical*.

Theorem 3.3. (Stability for space filtrations) *Let* X *be a triangulable space and* $f, g : X \to \mathbb{R}$ *be two tame functions giving rise to two space filtrations* $\mathcal{F}_f$ *and* $\mathcal{F}_g$ *where the values for sublevel sets include critical values. Then, for every* $p \geq 0$,

$$\mathsf{d}_b(\mathrm{Dgm}_p(\mathcal{F}_f), \mathrm{Dgm}_p(\mathcal{F}_g)) \leq \|f - g\|_\infty.$$

There is another distance called q-Wasserstein distance with which persistence diagrams are also compared.

Definition 3.10. (Wasserstein distance) Let Π be the set of bijections as defined in Definition 3.9. For any $p \geq 0$, $q \geq 1$, the *q-Wasserstein distance* is defined as

$$\mathsf{d}_{W,q}(\mathrm{Dgm}_p(\mathcal{F}_f), \mathrm{Dgm}_p(\mathcal{F}_g)) = \inf_{\pi \in \Pi} \left[\sum_{x \in \mathrm{Dgm}_p(\mathcal{F}_f)} \left(\|x - \pi(x)\|_q \right)^q \right]^{1/q}.$$

The distance $\mathsf{d}_{W,q}$ also is a metric on the space of persistence diagrams just like the bottleneck distance. It also enjoys a stability property, though it is not as strong as in Theorem 3.3.

Fact 3.6. *Let $f, g : X \to \mathbb{R}$ be two Lipschitz functions defined on a triangulable compact metric space X. Then, there exist constants C and k depending on X and the Lipschitz constants of f and g so that for every $p \geq 0$ and $q \geq k$,*

$$\mathsf{d}_{W,q}(\mathrm{Dgm}_p(\mathcal{F}_f), \mathrm{Dgm}_p(\mathcal{F}_g)) \leq C \cdot \|f - g\|_\infty^{1-k/q}.$$

The above result was improved recently [278] by considering the L^q-distance between functions defined on a common domain X:

$$\|f - g\|_q = \left(\sum_{x \in X} |f(x) - g(x)|^q \right)^{1/q}.$$

Theorem 3.4. (Stability for Wasserstein distance) *Let $f, g : K \to \mathbb{R}$ be two simplex-wise monotone functions on a simplicial complex K. Then, one has*

$$\mathsf{d}_{W,q}(\mathrm{Dgm}_p(\mathcal{F}_f), \mathrm{Dgm}_p(\mathcal{F}_g)) \leq \|f - g\|_q.$$

Bottleneck distances can be computed using perfect matchings in bipartite graphs. Computing Wasserstein distance becomes more difficult. It can be computed using an algorithm for minimum weight perfect matching in weighted bipartite graphs. We leave it as an exercise question (Exercise 5).

Computing Bottleneck Distances

Let A and B be the nondiagonal points in two persistence diagrams $\mathrm{Dgm}_p(\mathcal{F}_f)$ and $\mathrm{Dgm}_p(\mathcal{F}_g)$, respectively. For a point $a \in A$, let $\bar{a}$ denote the nearest point of a on the diagonal. Define $\bar{b}$ for every point $b \in B$ similarly. Let $\bar{A} = \{\bar{a}\}$ and $\bar{B} = \{\bar{b}\}$. Let $\tilde{A} = A \cup \bar{B}$ and $\tilde{B} = B \cup \bar{A}$. We want to bijectively match points in $\tilde{A}$ and $\tilde{B}$. Let $\Pi = \{\pi\}$ denote such a matching. It follows from the definition that

$$d_b(\mathrm{Dgm}_p(\mathcal{F}_f), \mathrm{Dgm}_p(\mathcal{F}_g)) = \min_{\pi \in \Pi} \sup_{a \in \tilde{A}, \pi(a) \in \tilde{B}} \|a - \pi(a)\|_\infty.$$

Then, the bottleneck distance we want to compute must be L_∞ distance $\max\{|x_a - x_b|, |y_a - y_b|\}$ for two points $a \in \tilde{A}$ and $b \in \tilde{B}$. We do a binary search on all such possible $O(n^2)$ distances where $|\tilde{A}| = |\tilde{B}| = n$. Let $\delta_0, \delta_1, \ldots, \delta_{n'}$ be the sorted sequence of these distances in a nondecreasing order.

Given a $\delta = \delta_i \geq 0$ where i is the median of the index in the binary search interval $[\ell, u]$, we construct a bipartite graph $G = (\tilde{A} \cup \tilde{B}, E)$ where an edge $e = (a, b)_{\{a \in \tilde{A}, b \in \tilde{B}\}}$ is in E if and only if either both $a \in \bar{A}$ and $b \in \bar{B}$ (weight$(e) = 0$) or $\|a - b\|_\infty \leq \delta$ (weight$(e) = \|a - b\|_\infty$). A complete matching in G is a set of n edges so that every vertex in $\tilde{A}$ and $\tilde{B}$ is incident to exactly one edge in the set. To determine if G has a complete matching, one can use an $O(n^{2.5})$ algorithm of Hopcroft and Karp [198] for complete matching in a bipartite graph. However, exploiting the geometric embedding of the points in the persistence diagrams, we can apply an $O(n^{1.5})$ time algorithm of Efrat et al. [154] for the purpose. If such an algorithm affirms that a complete matching exists, we do the following: If $\ell = u$ we output δ, otherwise we set $u = i$ and repeat. If no matching exists, we set $\ell = i$ and repeat. Observe that matching has to exist for some value of δ, in particular for $\delta_{n'}$ and thus the binary search always succeeds. Algorithm 1: BOTTLENECK lays out the pseudocode for this matching. The algorithm runs in $O(n^{1.5} \log n)$ time accounting for the $O(\log n)$ probes for binary search each applying an $O(n^{1.5})$ time matching algorithm. However, to achieve this complexity, we have to avoid sorting $n' = O(n^2)$ values taking $O(n^2 \log n)$ time. Again, using the geometric embedding of the points, one can perform the binary probes without incurring the cost for sorting. For details and an efficient implementation of this algorithm, see [209].

3.3 Persistence Algorithm

For computational purposes, we focus on simplicial filtrations because it is not always easy to compute the singular homology of topological spaces. We present algorithms that, given a simplicial filtration, compute its persistence diagram. For this, it is sufficient to compute every pair of simplices that ensue the birth and death of a homology class. First, we describe a combinatorial algorithm originally proposed in [152] and later present a version of it in terms

---

Algorithm 1 BOTTLENECK(Dgm$_p(\mathcal{F}_f)$, Dgm$_p(\mathcal{F}_g)$)

Input:
  Two persistent diagrams Dgm$_p(\mathcal{F}_f)$, Dgm$_p(\mathcal{F}_g)$
Output:
  Bottleneck distance d$_b$(Dgm$_p(\mathcal{F}_f)$, Dgm$_p(\mathcal{F}_g)$)

1: Compute sorted distances $\delta_0 \leq \delta_1 \leq \cdots \leq \delta_{n'}$ from Dgm$_p(\mathcal{F}_f)$ and Dgm$_p(\mathcal{F}_g)$
2: $\ell := 0; u := n'$
3: **while** $\ell < u$ **do**
4:    $i := \lfloor (u + \ell)/2 \rfloor; \delta := \delta_i$
5:    Compute graph $G = (\tilde{A} \cup \tilde{B}, E)$ where for all $e \in E$, weight(e) $\leq \delta$
6:    **if** there exists complete matching in G **then**
7:      $u := i$
8:    **else**
9:      $\ell := i$
10:    **end if**
11: **end while**
12: Output δ

---

of matrix reductions. We assume that the input is a *simplex-wise filtration* that begins with an empty complex

$$\varnothing = K_0 \hookrightarrow K_1 \hookrightarrow K_2 \hookrightarrow \cdots \hookrightarrow K_n = K,$$

where $K_j \setminus K_{j-1} = \sigma_j$ is a single simplex for each $j \in [1, n]$.

Remark 3.2. *The assumption of simplex-wise filtration does not pose any limitation because any filtration can be expanded into a simplex-wise filtration. For this, put all simplices in the difference of two consecutive complexes in the given filtration in any order only ensuring that all faces of a simplex appear before it in the expanded filtration. The persistence diagram of the original filtration can be read from the diagram of this expanded simplex-wise filtration by considering the original filtration function values associated with the simplices.*

Observe that a simplex-wise filtration necessarily renders the persistence pairing function $\mu_p^{i,j}$ to assume a value of at most 1 due to the following fact.

Fact 3.7. *When a p-simplex $\sigma_j = K_j \setminus K_{j-1}$ is added, exactly one of the following two possibilities occurs:*

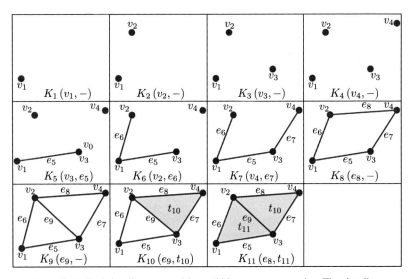

Figure 3.11 Red simplices are positive and blue ones are negative. The simplices are indexed to coincide with their order in the filtration. The $(\cdot, \cdot)$ in each subcomplex $K_i(\cdot, \cdot)$ shows the pairing between the positive and the negative. The second component missing in the parentheses shows the introduction of a positive simplex.

(a) A nonboundary p-cycle c along with its classes $[c] + h$ for any class $h \in$ $\mathsf{H}_p(K_{j-1})$ are born (created). In this case we call σ_j a positive simplex (also called a creator).

(b) An existing $(p-1)$-cycle c along with its class $[c]$ dies (is destroyed). In this case we call σ_j a negative simplex (also called a destructor).

To elaborate upon the above two changes, consider the example depicted in Figure 3.11. When one moves from K_7 to K_8, a nonboundary loop which is a 1-cycle $(e_5 + e_6 + e_7 + e_8)$ is created after adding edge e_8. Strictly speaking, a positive p-simplex σ_j may create more than one p-cycle. Only one of them can be taken as independent and the others become its linear combinations with the existing ones in K_{j-1}. From K_8 to K_9, the introduction of edge e_9 creates two nonboundary loops $(e_5 + e_6 + e_9)$ and $(e_7 + e_8 + e_9)$. But either one of them is the linear combination of the other one with the existing loop $(e_5+e_6+e_7+e_8)$. Notice that there is no canonical way to choose an independent one. However, the creation of a loop is reflected in the increase of the rank of H_1. In other words, in general, the Betti number β_p increases by 1 for a positive simplex. For a negative simplex, we get the opposite effect. In this case β_{p-1} decreases by 1, signifying the death of a cycle. However, unlike positive simplices, the destroyed cycle is determined uniquely up to homology, which is the equivalent class carried by the boundary of σ_j. For example, in Figure 3.11, the loop $(e_7 + e_8 + e_9)$ gets destroyed by triangle t_{10} when we go from K_9 to K_{10}.

Pairing

We already saw that destruction of a class is uniquely paired with the creation of a class through the "youngest first" rule; see the discussion after Fact 3.3. By Fact 3.7, this means that each negative simplex is paired uniquely with a positive simplex. The goal of the persistence algorithm is to find out these pairs.

Consider the birth and death of the classes by addition of simplices into a filtration. When a p-simplex σ_j is added, we explore whether it destroys the class $[c]$ of its boundary $c = \partial\sigma_j$ if it is not a boundary already. The cycle c was created when the youngest $(p-1)$-simplex in it, say σ_i, was added. Note that a simplex is younger if it comes later in the filtration. If σ_i, a positive $(p-1)$-simplex, has already been paired with a p-simplex σ'_j, then a class also created by σ_i was destroyed when σ'_j appeared. We can get the $(p-1)$-cycle representing this destroyed class and add it to $\partial\sigma_j$. The addition provides a cycle that existed before σ_i. We update c to be this new cycle and look for the youngest $(p-1)$-simplex σ_i in c and continue the process till we find one that is unpaired, or the cycle c becomes empty. In the latter case, we discover that $c = \partial\sigma_j$ was a boundary cycle already and thus σ_j creates a new class in $\mathsf{H}_p(K_j)$. In the other case, we discover that σ_j is a negative p-simplex which destroys a class created by σ_i. We pair σ_j with σ_i. Indeed, one can show that the above algorithm produces the persistence pairs according to Definition 3.11 below, that is, their function values lead to the persistence diagram (Definition 3.8). We give a proof for a matrix version of the algorithm later (Theorem 3.6).

Definition 3.11. (Persistence pairs) Given a simplex-wise filtration $\mathcal{F}: K_0 \hookrightarrow K_1 \hookrightarrow \cdots \hookrightarrow K_n$, for $0 < i < j \leq n$, we say a p-simplex $\sigma_i = K_i \setminus K_{i-1}$ and a $(p+1)$-simplex $\sigma_j = K_j \setminus K_{j-1}$ form a *persistence pair* (σ_i, σ_j) if and only if $\mu_p^{i,j} > 0$.

The full algorithm is presented in Algorithm 2: PAIRPERSISTENCE, which takes as input a sequence of simplices $\sigma_1, \sigma_2, \ldots, \sigma_n$ ordered according to the filtration of a complex whose persistence diagram is to be computed. It assumes that the complex is represented combinatorially with adjacency structures among its simplices.

Let us again consider the example in Figure 3.11 and see how the algorithm PAIRPERSISTENCE works. From K_7 to K_8, e_8 is added. Its boundary is $c = (v_2 + v_4)$. The vertex v_4 is the youngest positive vertex in c but it is paired with e_7 in K_7. Thus, c is updated to $(v_3 + v_4 + v_4 + v_2) = (v_3 + v_2)$. The vertex

---

Algorithm 2 PAIRPERSISTENCE($\sigma_1, \sigma_2, \cdots, \sigma_n$)

Input:
An ordered sequence of simplices forming a filtration of a complex
Output:
Determine if a simplex is "positive" or "negative" and generate persistence
pairs

1: **for** $j = 1$ to n **do**
2:     $c := \partial_p \sigma_j$
3:     σ_i is the youngest positive $(p-1)$-simplex in c
4:     **while** σ_i is paired and c is not empty **do**
5:        Let c' be the cycle destroyed by the simplex paired with σ_i *
         computed previously in step 10 *\
6:        $c := c' + c$ * this addition may cancel simplices *\
7:        Update σ_i to be the youngest positive $(p-1)$-simplex in c
8:     **end while**
9:     **if** c is not empty **then**
10:       σ_j is a negative p-simplex; generate pair (σ_i, σ_j); associate c with σ_j
        as destroyed
11:     **else**
12:       σ_j is a positive p-simplex * σ_j may get paired later *\
13:     **end if**
14: **end for**

---

v_3 becomes the youngest positive one but it is paired with e_5. So, c is updated
to $(v_1 + v_2)$. The vertex v_2 becomes the youngest positive one but it is paired
with e_6. So, c is updated to be empty. Hence e_8 is a positive edge. Now we
examine the addition of the triangle t_{11} from K_{10} to K_{11}. The boundary of t_{11}
is $c = (e_5 + e_6 + e_9)$. The youngest positive edge e_9 is paired with t_{10}. Thus,
c is updated by adding the cycle destroyed by t_{10} to $(e_5 + e_6 + e_7 + e_8)$. Since
e_8 is the youngest positive edge that is not yet paired, t_{11} finds e_8 as its paired
positive edge. Observe that we finally obtain a loop that is destroyed by adding
the negative triangle. For example, we obtain the loop $(e_5 + e_6 + e_7 + e_8)$ by
adding t_{11}.

3.3.1 Matrix Reduction Algorithm

There is a version of the algorithm PAIRPERSISTENCE that uses only matrix
operations. First notice the following:

- The boundary operator $\partial_p : \mathsf{C}_p \to \mathsf{C}_{p-1}$ can be represented by a boundary matrix D_p where the columns correspond to the p-simplices and rows correspond to $(p-1)$-simplices.
- It represents the transformation of a basis of C_p given by the set of p-simplices to a basis of C_{p-1} given by the set of $(p-1)$-simplices:

$$D_p[i, j] = \begin{cases} 1 & \text{if } \sigma_i \in \partial_p \sigma_j, \\ 0 & \text{otherwise.} \end{cases}$$

- One can combine all boundary matrices into a single matrix D that represents all linear maps $\bigoplus_p \partial_p = \bigoplus_p (\mathsf{C}_p \to \mathsf{C}_{p-1})$, that is, transformation of a basis of all chain groups together to a basis of itself, but with a shift to a one lower dimension:

$$D[i, j] = \begin{cases} 1 & \text{if } \sigma_i \in \partial_* \sigma_j, \\ 0 & \text{otherwise.} \end{cases}$$

Definition 3.12. (Filtered boundary matrix) Let $\mathcal{F} : \varnothing = K_0 \hookrightarrow K_1 \hookrightarrow \cdots \hookrightarrow K_n = K$ be a filtration induced by an ordering of simplices $(\sigma_1, \sigma_2, \ldots, \sigma_n)$ in K. Let D denote the boundary matrix for simplices in K that respects the ordering of the simplices in the filtration, that is, the simplex σ_i in the filtration occupies column i and row i in D. We call D the *filtered boundary matrix* for $\mathcal{F}$.

Given any matrix A, let $\text{row}_A[i]$ and $\text{col}_A[j]$ denote the ith row and jth column of A, respectively. We abuse notation slightly to let $\text{col}_A[j]$ denote also the chain $\{\sigma_i \mid A[i, j] = 1\}$, which is the collection of simplices corresponding to 1s in the column $\text{col}_A[j]$.

Definition 3.13. (Reduced matrix) Let $\text{low}_A[j]$ denote the row index of the last 1 in the jth column of A, which we call the *low-row index* of the column j. It is undefined for empty columns (marked with -1 in Algorithm 3). The matrix A is *reduced* (or is in *reduced form*) if $\text{low}_A[j] \neq \text{low}_A[j']$ for any $j \neq j'$; that is, no two columns share the same low-row indices.

Fact 3.8. *Given a matrix A in reduced form, we have that the set of nonzero columns in A are all linearly independent over $\mathbb{Z}_2$.*

We define a matrix A over $\mathbb{Z}_2$ to be *upper triangular* if all of its diagonal elements are 1, and there is no entry $A[i, j] = 1$ with $i > j$. We will compute a reduced matrix from a given boundary matrix by left-to-right column additions. A series of such column additions is equivalent to multiplying the boundary matrix on the right with an upper triangular matrix.

Now, we state a result saying that if a reduced form is obtained via only left-to-right column additions, then for each column, the low-row index is unique in the sense that it does not depend on how the reduced form is obtained. Using this result we show that persistence pairing of simplices can be obtained from these low-row indices. Given an $n_1 \times n_2$ matrix A, let $A_{[a,b]}^{[c,d]}$, $a \leq b$ and $c \leq d$, denote the submatrix formed by rows a to b, and columns from c to d. In cases when $b = n_2$ and $c = 1$, we also write it as $A_a^d := A_{a,n_2}^{1,d}$ for simplicity. For any $1 \leq i < j \leq n$, define the quantity $r_A(i, j)$ as follows:

$$r_A(i, j) = \text{rank}\,(A_i^j) - \text{rank}\,(A_{i+1}^j) + \text{rank}\,(A_{i+1}^{j-1}) - \text{rank}\,(A_i^{j-1}). \quad (3.5)$$

Proposition 3.5. (Pairing uniqueness [105]) *Let $R = DV$, where R is in reduced form and V is upper triangular. Then for any $1 \leq j \leq n$, $\text{low}_R[j] = i$ if and only if $r_D(i, j) = 1$.*

Next, we show that a pairing based on low-row indices indeed provides the persistence pairs according to Definition 3.11.

Theorem 3.6. *Let D be the $m \times m$ filtered boundary matrix for a filtration $\mathcal{F}$ (Definition 3.12). Let $R = DV$, where R is in reduced form and V is upper triangular. Then, the simplices σ_i and σ_j in $\mathcal{F}$ form a persistence pair if and only if $\text{low}_R[j] = i$.*

PROOF. First, it is easy to verify that $r_R(i, j) = r_D(i, j)$ for any $1 \leq i < j \leq n$ (in particular, rank $(R_a^d) = $ rank (D_a^d) as the effect of V is to add columns of D to columns on the right only). Combining this and Proposition 3.5, we only need to show that there is a persistence pair (σ_i, σ_j) (i.e., $\mu_p^{i,j} = 1$) if and only if $r_R(i, j) = 1$.

Next, we observe that due to the uniqueness of the entry $\text{low}_R[j]$ (Proposition 3.5), if we prove the theorem for a specific reduced matrix $R' = DV'$, then it holds for any reduced form $R = DV$. In what follows, we assume that the reduced form $R = DV$ is obtained by Algorithm 3: MATPERSISTENCE(D). For this specific reduction algorithm, it is easy to see that if a simplex σ_j is of dimension p, then all columns ever added to the j-th column correspond to simplices of dimension p. In particular, let D_p denote the matrix obtained by setting all columns in D corresponding to simplices of dimension $\neq p$ to be all zero; hence all nonzero columns in D_p represent the p-th boundary operator $\partial_p : C_p(K) \to C_{p-1}(K)$. Define R_p similarly. Then observe that the algorithm MATPERSISTENCE simply reduces each matrix D_p independently, for all dimensions p, and the reduced form for D_p is R_p.

In what follows, we assume that the dimension of simplex σ_j (corresponding to the j-th column of D) is p; and for simplicity, set $\tilde{R} := R_p$. We leave the proof of the following claim as an exercise (Exercise 5).

Proposition 3.7. *Let the dimension for σ_j be p and construct $\tilde{R} = R_p$ as described above. For any $1 \leq i < j$, we have that $r_R(i, j) = r_{\tilde{R}}(i, j)$.*

To this end, let Z_{p-1}^k and B_{p-1}^k denote the $(p-1)$-th cycle group and the $(p-1)$-th boundary group for K_k, respectively. Consider the persistence pairing function for $1 \leq i < j \leq n$:

$$\mu_{p-1}^{i,j} = \left(\beta_{p-1}^{i,j-1} - \beta_{p-1}^{i,j} \right) - \left(\beta_{p-1}^{i-1,j-1} - \beta_{p-1}^{i-1,j} \right). \tag{3.6}$$

On the other hand, note that for any $1 \leq a < b \leq n$,

$$\beta_{p-1}^{a,b} := \text{rank}\,(H_{p-1}^{a,b}) = \text{rank}\left(\frac{Z_{p-1}^a}{Z_{p-1}^a \cap B_{p-1}^b} \right) = \text{rank}\,(Z_{p-1}^a) - \text{rank}\,(Z_{p-1}^a \cap B_{p-1}^b). \tag{3.7}$$

Let $\Gamma_a^b := \{\text{col}_{\tilde{R}}[k] \mid k \in [1, b] \text{ and } 1 \leq \text{low}_{\tilde{R}}[k] \leq a\}$. Using the facts that all nonzero columns in $\tilde{R}$ with index at most b form a basis for B_{p-1}^b, and that each low-row index for every nonzero column is unique, one can show that rank $\left(Z_{p-1}^a \cap B_{p-1}^b \right) = |\Gamma_a^b|$. Now consider the set of all nonzero columns in $\tilde{R}$ with index at most b that are not in Γ_a^b, denoted by $\widehat{\Gamma}_a^b$. Note that $|\widehat{\Gamma}_a^b| = $ rank $\left(\tilde{R}_{a+1}^b \right) = $ rank $\left(B_{p-1}^b \right) - |\Gamma_a^b|$; hence

$$\text{rank}\left(Z_{p-1}^a \cap B_{p-1}^b \right) = |\Gamma_a^b| = \text{rank}\left(B_{p-1}^b \right) - |\widehat{\Gamma}_a^b| = \text{rank}\left(\tilde{R}_1^b \right) - \text{rank}\left(\tilde{R}_{a+1}^b \right).$$

Combining the above with Proposition 3.7, Eq. (3.6) and Eq. (3.7), we thus have that

$$
\begin{aligned}
\mu_{p-1}^{i,j} &= \text{rank}\left(Z_{p-1}^i \right) - \text{rank}\left(Z_{p-1}^i \cap B_{p-1}^{j-1} \right) - \left(\text{rank}\left(Z_{p-1}^i \right) - \text{rank}\left(Z_{p-1}^i \cap B_{p-1}^j \right) \right) \\
&\quad - \left(\text{rank}\left(Z_{p-1}^{i-1} \right) - \text{rank}\left(Z_{p-1}^{i-1} \cap B_{p-1}^{j-1} \right) \right) + \text{rank}\left(Z_{p-1}^{i-1} \right) - \text{rank}\left(Z_{p-1}^{i-1} \cap B_{p-1}^j \right) \\
&= \text{rank}\left(Z_{p-1}^i \cap B_{p-1}^j \right) - \text{rank}\left(Z_{p-1}^i \cap B_{p-1}^{j-1} \right) + \text{rank}\left(Z_{p-1}^{i-1} \cap B_{p-1}^{j-1} \right) - \text{rank}\left(Z_{p-1}^{i-1} \cap B_{p-1}^j \right) \\
&= -\text{rank}\left(\tilde{R}_{i+1}^j \right) + \text{rank}\left(\tilde{R}_{i+1}^{j-1} \right) - \text{rank}\left(\tilde{R}_i^{j-1} \right) + \text{rank}\left(\tilde{R}_i^j \right) = r_{\tilde{R}}(i, j) = r_R(i, j) = r_D(i, j).
\end{aligned}
$$

By Proposition 3.5, the theorem then follows. $\square$

Matrix Reduction Algorithm

Notice that there are possibly many R and V for a fixed D forming a *reduced-form decomposition*. Theorem 3.6 implies that the persistence pairing is independent of the particular contents of R and V as long as R is reduced and V is upper triangular. If we reduce a given filtered boundary matrix D to a reduced form R only with left-to-right column additions, indeed then we obtain

---

Algorithm 3 MATPERSISTENCE(D)

Input:
Boundary matrix D of a complex with columns and rows ordered by a given filtration
Output:
Reduced matrix with each column j either being empty or having a unique $\text{low}_D[j]$ entity

1: **for** $j = 1 \to |\text{col}_D|$ **do**
2:     **while** there exists $j' < j$ such that $\text{low}_D[j'] == \text{low}_D[j]$ **and** $\text{low}_D[j] \neq -1$ **do**
3:         $\text{col}_D[j] := \text{col}_D[j] + \text{col}_D[j']$
4:     **end while**
5:     **if** $\text{low}_D[j] \neq -1$ **then**
6:         $i := \text{low}_D[j]$ \∗ generate pair (σ_i, σ_j) ∗\
7:     **end if**
8: **end for**

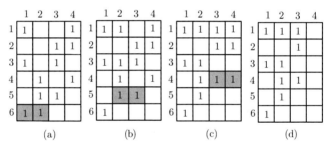

(a)   (b)   (c)   (d)

Figure 3.12 Matrix reduction for a 6×4 matrix D. The low of columns are shaded to point out the conflicts: (a) $\text{low}_D[1]$ conflicts with $\text{low}_D[2]$ and $\text{col}_D[1]$ is added to $\text{col}_D[2]$; (b) $\text{low}_D[2]$ conflicts with $\text{low}_D[3]$ and $\text{col}_D[2]$ is added to $\text{col}_D[3]$; (c) $\text{low}_D[3]$ conflicts with $\text{low}_D[4]$; and (d) the addition of $\text{col}_D[3]$ to $\text{col}_D[4]$ zeroes out the entire column $\text{col}_D[4]$.

$R = DV$ as required. With this principle, Algorithm 3: MATPERSISTENCE is designed to compute the persistence pairs of simplices. See Figure 3.12 for an example of the execution of the algorithm. We process the columns of D from left to right which correspond to the order in which they appear in the filtration. The row indices also follow the same order top down (thus "lower" refers to a larger index, which also means that a simplex is "younger" in the filtration). We assume that $|\text{col}_D|$ denotes the number of columns in D. Suppose we have processed all columns up to $j - 1$ and now are going to process the column j.

We check if the row $\text{low}_D[j]$ contains any other lowest 1 for any column j' to the left of j, that is, $j' < j$. If so, we add $\text{col}_D[j']$ to $\text{col}_D[j]$. This decreases $\text{low}_D[j]$. We continue this process until either we turn all entries in $\text{col}_D[j]$ to be 0, or settle on $\text{low}_D[j]$ that does not conflict with any other $\text{low}_D[j']$ to its left. In the latter case, σ_j is a negative p-simplex that pairs with the positive $(p-1)$-simplex $\sigma_{\text{low}_D[j]}$. In the algorithm MATPERSISTENCE, we assume that when a column j is zeroed out completely, $\text{low}_D[j]$ returns -1.

To compute the persistence diagram $\text{Dgm}(\mathcal{F}_f)$ for a filtration $\mathcal{F}_f$, we first run MATPERSISTENCE on the filtered boundary matrix D representing $\mathcal{F}_f$. Every computed persistence pair (σ_i, σ_j) gives a finite bar $[f(\sigma_j), f(\sigma_i))$ or a point with finite coordinates $(f(\sigma_i), f(\sigma_j))$ in $\text{Dgm}(\mathcal{F}_f)$. Every simplex σ_i that remains unpaired provides an infinite bar $[f(\sigma_i), \infty)$ or a point $(f(\sigma_i), \infty)$ at infinity in $\text{Dgm}(\mathcal{F}_f)$. Observe that not every positive p-simplex σ_i (column i is zeroed out) gives a point at infinity in $\text{Dgm}_p(\mathcal{F}_f)$, the only ones that do are the ones that are not paired with a $(p+1)$-simplex whose column is processed afterward. A simple fact about unpaired simplices follows from Fact 3.4.

Fact 3.9. *The number of unpaired p-simplices in a simplex-wise filtration of a simplicial complex K equals its p-th Betti number $\beta_p(K)$.*

We have already mentioned that the input boundary matrix D should respect the filtration order, that is, the row and column indices of D correspond to the indices of the simplices in the input filtration. Observe that we can consider a slightly different filtration without changing the persistence pairs. We can arrange all of the p-simplices for any $p \geq 0$ together in the filtration without changing their relative orders as follows, where σ_j^i denotes the j-th i-simplex among all i-simplices in the original filtration:

$$\left(\sigma_1^0, \sigma_2^0, \ldots, \sigma_{n_0}^0\right), \ldots, \left(\sigma_1^p, \sigma_2^p, \ldots, \sigma_{n_p}^p\right), \ldots, \left(\sigma_1^d, \sigma_2^d, \ldots, \sigma_{n_d}^d\right). \quad (3.8)$$

This means that the columns and rows of p-simplices in D become adjacent though retaining their relative ordering from the original matrix. Observe that, by this rearrangement, all columns that are added to a column j in the original D still remain to the left of j in their newly assigned indices. In other words, processing the rearranged matrix D can be thought of as processing each individual p-boundary matrix $D_p = [\partial_p]$ separately where the column and row indices respect the relative orders of p- and $(p-1)$-simplices in the original filtration.

Complexity of MATPERSISTENCE
Let the filtration $\mathcal{F}$ based on which the boundary matrix D is constructed insert n simplices. This means that D has at most n rows and columns. Then, the

outer **for** loop is executed at most $O(n)$ times. Within this **for** loop, steps 5–7 take only $O(1)$ time. The complexity is indeed determined by the **while** loop (steps 2–4). We argue that this loop iterates at most $O(n)$ times. This follows from the fact that each column addition in step 3 decreases $low_D[j]$ by at least one, and over the entire algorithm it cannot decrease by more than the length of the column, which is $O(n)$. Each column addition in step 3 takes at most $O(n)$ time, giving a total time of $O(n^2)$ for the **while** loop. Accounting for the outer **for** loop, we get a complexity of $O(n^3)$ for MATPERSISTENCE.

One can implement the above matrix reduction algorithm with a more efficient data structure noting that most of the entries in the input matrix D are empty. A linked list representing the nonzero entries in the columns of D is space-wise more efficient. Edelsbrunner and Harer [149] present a clever implementation of MATPERSISTENCE using such a sparse matrix representation. For every column j, the algorithm executes $O(j - i)$ column additions of $O(j - i)$ length each incurring a cost $O((j - i)^2)$, where $i = 1$ if σ_j is positive and is the index of the simplex σ_i with which it pairs if σ_j is negative. Therefore, the total time complexity becomes $O(\sum_{j \in [1,n]} (j - i)^2)$. Here, we assume that the dimension of the complex K is a constant.

It is worth noting that essentially the matrix reduction algorithm is a version of the classical Gaussian elimination method with a given column order and a specific choice of row pivots. In this respect, persistence of a given filtration can be computed by the PLU factorization of a matrix for which Bunch and Hopcroft [57] give an $O(M(n))$ time algorithm, where $M(n)$ is the time to multiply two $n \times n$ matrices. It is known that $M(n) = O(n^\omega)$ where $\omega \in [2, 2.373)$ is called the exponent for matrix multiplication.

3.3.2 Efficient Implementation

The matrix reduction algorithm considers a column from left to right and reduces it by left-to-right additions. As we have observed, every addition to a column with index j pushes $low_D[j]$ upward. In the case that σ_j is a positive simplex, the entire column is zeroed out. In general, positive simplices incur more cost than negative ones because $low_D[\cdot]$ needs to be pushed all the way up for zeroing out the entire column. However, they do not participate in any future left-to-right column additions. Therefore, if it is known beforehand that the simplex σ_j will be a positive simplex, then the costly step of zeroing out the column j can be avoided.

Chen and Kerber [94] observed the following simple fact. If we process the input filtration backward in dimension, that is, process the boundary matrices D_p, $p = 1, \ldots, d$, in decreasing order of dimensions, then a persistence pair

---

Algorithm 4 CLEARPERSISTENCE($D_1, D_2, \ldots, D_d$)

Input:
    Boundary matrices ordered by dimension of the boundary operators with columns ordered by filtration

Output:
    Reduced matrices with each column for a negative simplex having a unique low entry

1: MATPERSISTENCE(D_d)
2: **for** $i = (d-1) \to 1$ **do**
3:     **for** $j = 1 \to |\mathrm{col}_{D_i}|$ **do**
4:       **if** σ_j is not paired while processing D_{i+1} **then**
5:         * column j is not processed if σ_j is already paired *\
6:         **while** there exists $j' < j$ such that $\mathrm{low}_D[j] \neq -1$ **and** $\mathrm{low}_{D_i}[j'] == \mathrm{low}_{D_i}[j]$ **do**
7:           $\mathrm{col}_{D_i}[j] := \mathrm{col}_{D_i}[j] + \mathrm{col}_{D_i}[j']$
8:         **end while**
9:         **if** $\mathrm{low}_D[j] \neq -1$ **then**
10:           $k := \mathrm{low}_{D_i}[j]$ * generate pair (σ_k, σ_j) *\
11:         **end if**
12:       **end if**
13:     **end for**
14: **end for**

(σ^{p-1}, σ^p) is detected from D_p before processing the column for σ^{p-1} in D_{p-1}. Fortunately, we already know that σ^{p-1} has to be a positive simplex because it cannot pair with a negative simplex σ^p otherwise. So, we can simply ignore the column of σ^{p-1} while processing D_{p-1}. We call it *clearing* out column $p - 1$. In practice, this saves a considerable amount of computation in cases where a lot of positive simplices occur such as in Rips filtrations. Algorithm 4: CLEARPERSISTENCE implements this idea.

We cannot take advantage of the clearing for the last dimension in the filtration. If d is the highest dimension of the simplices in the input filtration, the matrix D_d has to be processed for all columns because the pairings for the positive d-simplices are not available.

If the number of d-simplices is large compared to the number of simplices of lower dimensions, the incurred cost of processing their columns can still be high. For example, in a Rips filtration restricted to a certain dimension d, the number of d-simplices becomes usually much larger than the number of, say,

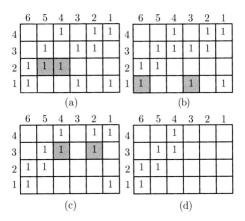

Figure 3.13 Matrix reduction with the twisted matrix D^* of the matrix D in Figure 3.12, which is first transposed and then its rows and columns are reversed in order. The conflicts in $\mathrm{low}_D[\cdot]$ are resolved to obtain the intermediate matrices shown in (a) through (d); the last transformation from (c) to (d) assumes the completion of all conflict resolutions from columns 3 through 1. Observe that every column–row pair corresponds to a row–column pair in the original matrix. Also, all columns that are zeroed out here correspond to all rows in the original that did not get paired with any column, meaning that they are either a negative simplex, or a positive simplex not paired with any other.

1-simplices. In those cases, the clearing can be more cost-effective if it can be applied forward.

In this respect, the following observation becomes helpful. Let D_p^* denote the anti-transpose of the matrix D_p, defined by the transpose of D_p with the columns and rows being ordered in reverse. This means that if D_p has row and column indices $1, \ldots, m$ and $1, \ldots, n$, respectively, then $D_p^*(i, j) = D_p(n + 1 - j, m + 1 - i)$. We call it the twisted matrix of D_p. Figure 3.13 shows the twisted matrix D^* of the matrix D in Figure 3.12 where the rows and columns are marked with the indices of the original matrix. The following proposition guarantees that we can compute the persistence pairs in D_p from the matrix D_p^*.

Proposition 3.8. *Pair (σ^{p-1}, σ^p) is a persistence pair computed from D_p if and only if (σ^p, σ^{p-1}) is computed as a pair from D_p^* by* MATPERSISTENCE(D_p^*).

PROOF. Let the indices of σ^{p-1} and σ^p in D_p be i and j, respectively. Then, by Theorem 3.6, one has $\mathrm{low}_R[j] = i$ where R is the reduced matrix obtained from D_p by left-to-right column additions. Consider bottom-to-top row additions in D_p each of which takes a row and adds it to a row above it. Similar to $\mathrm{low}_A[j]$ for a matrix A, let $\mathrm{lft}_A[i]$ denote the column index of the leftmost 1 in the row i of A. Call A left reduced if every nonzero row i has a unique $lft_A[i]$. In the rest of the proof, for simplicity, we use the row and column indices of D_p also for D_p^*, that is, by an index pair (j, i) in D_p^* we actually mean the pair $(n + 1 - j, m + 1 - i)$.

First, observe that each bottom-to-top row addition in D_p is equivalent to a left-to-right column addition in D_p^*. Also, a reduced matrix by left-to-right column additions in D_p^* corresponds to a left reduced matrix obtained by corresponding bottom-to-top row additions in D_p. So, if S denotes the reduced matrix obtained from D_p^* by left-to-right column additions and L denotes the left reduced matrix obtained from D_p by bottom-to-top row additions, then $low_S[i] = j$ if and only if $lft_L[i] = j$. Furthermore, MAT-PERSISTENCE(D_p^*) computes the pair (j, i) (hence (σ^p, σ^{p-1})) if and only if $low_S[i] = j$.

Therefore, to prove the proposition, it is sufficient to argue that $low_R[j] = i$ if and only if $lft_L[i] = j$. By Proposition 3.5, $low_R[j] = i$ if and only if $r_{D_p}(i, j)$ as defined in Eq. (3.5) equals 1. Therefore, it is sufficient to show that $lft_L[i] = j$ if and only if $r_{D_p}(i, j) = 1$.

The above claim can be proved in exactly the same way as Proposition 3.5 is proved in [105] while replacing the role of $low_R[j]$ with $lft_L[i]$. Observe that bottom-to-top row additions do not change the rank of the lower left minors. Hence, $r_{D_p} = r_L$. Therefore, it is sufficient to show that $lft_L[i] = j$ if and only if $r_L(i, j) = 1$. Assume first that $lft_L[i] = j$. The rows of L_i^j (see the definitions above Eq. (3.5)) are linearly independent and hence rank $(L_i^j) -$ rank $(L_{i+1}^j) = 1$. Now delete the last column in L_i^j which leaves the top row having only zeros. This implies that rank $(L_i^{j-1}) -$ rank $(L_{i+1}^{j-1}) = 0$. This gives $r_L(i, j) = 1$ as needed. Next, assume that $lft_L[i] \neq j$. Consider L_i^j and L_i^{j-1}. If $lft_L[i] > j$, the top row in both matrices is zero. Therefore, rank $(L_i^j) -$ rank $(L_{i+1}^j) = 0$ and also rank $(L_i^{j-1}) =$ rank (L_{i+1}^{j-1}), giving $r_L(i, j) = 0$ as required. If $lft_L[i] < j$, the top row in both matrices is nonzero giving rank $(L_i^j) -$ rank $(L_{i+1}^j) = 1$ and rank $(L_i^{j-1}) -$ rank $(L_{i+1}^{j-1}) = 1$ giving again $r_L(i, j) = 0$ as required. $\qquad\square$

To apply clearing we process D_{p+1}^* after D_p^* by calling CLEARPERSISTENCE($D_d^*, \ldots, D_2^*, D_1^*$) because if we get a pair (σ^{p+1}, σ^p) while processing D_p^*, we already know that σ^{p+1} is a negative simplex and its column in D_{p+1}^* cannot contain a defined low entry. This means that the column of σ^{p+1} in D_{p+1}^* can be zeroed out and hence can be ignored. Now, the only boundary matrix that needs to be processed without any clearing is D_1^*. So, depending on whether D_d or D_1 is large, one can choose to process the filtration in increasing or decreasing dimensions, respectively.

3.4 Persistence Modules

We have seen in Section 3.2.1 that persistence diagrams are stable with respect to the perturbation of the function that defines the filtration on a given simplicial complex or a space. This requires the domain of the function to be fixed. The result depends on the observation that perturbations in the filtrations are bounded by the perturbations in the function which in turn also results in bounded perturbations at the homology level. A natural follow-up is to derive a bound of the perturbations of the persistence diagrams directly in terms of the perturbations at the homology level. Toward this goal, we now define a generalized notion of homology modules called *persistence modules* and a distance among them called the *interleaving distance*.

Recall that a filtration gives rise to a homology module which is a sequence of homology groups connected by homomorphisms that are induced by inclusions defining the filtration. These homology groups when defined over a field (e.g., $\mathbb{Z}_2$) can be thought of as vector spaces connected by linear maps. Persistence modules extend homology modules by taking vector spaces in place of homology groups and linear maps in place of inclusion induced homomorphisms.

We make one more extension. So far, the sequences in a filtration and homology modules have been indexed over a finite subset of natural numbers. It turns out that we can enlarge the index set to be any subposet A of $\mathbb{R}$. In the following definition, persistence modules and their interleaving distance are defined over the poset $(A, \leq)$.

Definition 3.14. (Persistence module) A *persistence module* over a poset $A \subseteq \mathbb{R}$ is any collection $\mathbb{V} = \left\{ V_a \right\}_{a \in A}$ of vector spaces V_a together with linear maps $v_{a,a'} : V_a \to V_{a'}$ so that $v_{a,a} = \mathrm{id}$ and $v_{a',a''} \circ v_{a,a'} = v_{a,a''}$ for all $a, a', a'' \in A$ where $a \leq a' \leq a''$. Sometimes we write $\mathbb{V} = \left\{ V_a \xrightarrow{v_{a,a'}} V_{a'} \right\}_{a \leq a'}$ to denote this collection with the maps.

Remark 3.3. *A persistence module defined over a subposet A of $\mathbb{R}$ can be "extended" in a module over $\mathbb{R}$. For this, for any $a < a' \in A$ where the open interval (a, a') is not in A and for any $a \leq b < b' < a'$, assume that $v_{b,b'}$ is an isomorphism and $\lim_{a \to -\infty} V_a = 0$ if it is not given.*

Our goal is to define a distance between two persistence modules with respect to which we would bound the distance between their persistence diagrams. Given two persistence modules defined over $\mathbb{R}$, we define a distance between them by identifying maps between constituent vector spaces of the modules.

We will come across a structural property involving maps called *commutative diagrams* quite often in this and the following chapters.

Definition 3.15. (Commutative diagram) A *commutative diagram* is a collection of maps $A_i \xrightarrow{f_i} B_i$ where any two compositions of maps beginning and ending in the same sets result in equal maps. Formally, whenever we have two sequences in the collection of the form

$$A = U_1 \xrightarrow{f_1} U_2 \xrightarrow{f_2} \cdots \xrightarrow{f_m} U_{m+1} = B,$$
$$A = V_1 \xrightarrow{g_1} V_2 \xrightarrow{g_2} \cdots \xrightarrow{g_n} V_{n+1} = B,$$

we have $f_m \circ \cdots \circ f_1 = g_n \circ \cdots \circ g_1$. Commutative diagrams are usually formed by commutative triangles and squares.

Definition 3.16. (ε-interleaving) Let $\mathbb{U}$ and $\mathbb{V}$ be two persistence modules over the index set $\mathbb{R}$. We say $\mathbb{U}$ and $\mathbb{V}$ are ε-*interleaved* if there exist two families of maps $\varphi_a : U_a \to V_{a+\varepsilon}$ and $\psi_a : V_a \to U_{a+\varepsilon}$ satisfying the following two conditions:

- $v_{a+\varepsilon,a'+\varepsilon} \circ \varphi_a = \varphi_{a'} \circ u_{a,a'}$ and $u_{a+\varepsilon,a'+\varepsilon} \circ \psi_a = \psi_{a'} \circ v_{a,a'}$ (rectangular commutativity);
- $\psi_{a+\varepsilon} \circ \varphi_a = u_{a,a+2\varepsilon}$ and $\varphi_{a+\varepsilon} \circ \psi_a = v_{a,a+2\varepsilon}$ (triangular commutativity).

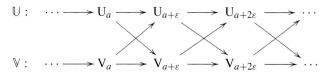

Some of the relevant maps for interleaving between two modules are shown above, whereas the two parallelograms and the two triangles below depict the rectangular and the triangular commutativities, respectively.

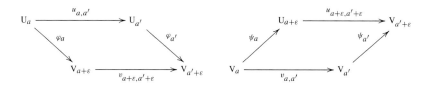

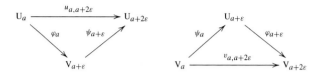

Definition 3.17. (Interleaving distance) Given two persistence modules $\mathbb{U}$ and $\mathbb{V}$, their *interleaving distance* is defined as

$$\mathsf{d}_I(\mathbb{U}, \mathbb{V}) = \inf\{\varepsilon \mid \mathbb{U} \text{ and } \mathbb{V} \text{ are } \varepsilon\text{-interleaved}\}.$$

Observe that, when $\varepsilon = 0$, Definition 3.16 implies that the maps $\varphi_a : \mathrm{U}_a \to \mathrm{V}_a$ and $\psi_a : \mathrm{V}_a \to \mathrm{U}_a$ are isomorphisms. In that case, we get the following diagrams where each vertical map is an isomorphism and each square commutes. We get two isomorphic persistence modules.

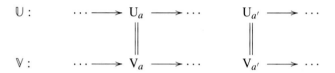

Definition 3.18. (Isomorphic persistence modules) We say two persistence modules $\mathbb{U}$ and $\mathbb{V}$ indexed over an index set $A \subseteq \mathbb{R}$ are *isomorphic* if the following two conditions hold (illustrated by the diagram above):

- $\mathrm{U}_a \cong \mathrm{V}_a$ for every $a \in A$, and
- for every $x \in \mathrm{U}_a$, if x is mapped to $y \in \mathrm{V}_a$ by the isomorphism, then $u_{a,a'}(x) \in \mathrm{U}_{a'}$ is mapped to $v_{a,a'}(y) \in \mathrm{V}_{a'}$ also by the isomorphism.

Fact 3.10. *If two persistence modules arising from two filtrations $\mathcal{F}_f$ and $\mathcal{F}_g$ are isomorphic, the persistence diagrams* $\mathrm{Dgm}_p(\mathcal{F}_f)$ *and* $\mathrm{Dgm}_p(\mathcal{F}_g)$ *are identical.*

We have seen earlier that filtrations give rise to homology modules and hence persistence modules. Just like the persistence modules, we can define an interleaving distance between two filtrations too. In the following definition, $\iota_{a,a'}$ denotes the inclusion map from X_a to $X_{a'}$ and also from Y_a to $Y_{a'}$ for $a' \geq a$. For simplicial filtrations, we need contiguity of simplicial maps to assert equality of maps at the homology level, whereas for space filtrations, we need homotopy of continuous maps to assert equality at the homology level. These maps are between filtrations and not internal maps within the filtrations which are still inclusions. In the next chapter, we go from inclusions to simplicial maps as internal maps (see Definition 4.2).

Definition 3.19. (ε-interleaving) We say two simplicial (resp. space) filtrations $\mathcal{X}$ and $\mathcal{Y}$ defined over $\mathbb{R}$ are *ε-interleaved* if there exist two families of simplicial (resp. continuous) maps $\varphi_a : X_a \to Y_{a+\varepsilon}$ and $\psi_a : Y_a \to X_{a+\varepsilon}$ satisfying the following two conditions:

- $\varphi_a \circ \iota_{a+\varepsilon,a'+\varepsilon}$ is contiguous (homotopic) to $\varphi_{a'} \circ \iota_{a,a'}$ and $\iota_{a+\varepsilon,a'+\varepsilon} \circ \psi_a$ is contiguous (homotopic) to $\psi_{a'} \circ \iota_{a,a'}$ (rectangular commutativity);
- $\psi_{a+\varepsilon} \circ \varphi_a$ is contiguous (homotopic) to $\iota_{a,a+2\varepsilon}$ and $\varphi_{a+\varepsilon} \circ \psi_a$ is contiguous (homotopic) to $\iota_{a,a+2\varepsilon}$ (triangular commutativity).

Similar to the persistence modules, we can define the interleaving distance between two filtrations:

$$\mathsf{d}_I(\mathcal{X}, \mathcal{Y}) = \inf\{\varepsilon \mid \mathcal{X} \text{ and } \mathcal{Y} \text{ are } \varepsilon\text{-interleaved}\}.$$

Two ε-interleaved filtrations give rise to ε-interleaved persistence modules at the homology level. Since contiguous simplicial (resp. homotopic continuous) maps become equal at the homology level, we obtain the following inequality.

Proposition 3.9. *One has* $\mathsf{d}_I(\mathsf{H}_p\mathcal{X}, \mathsf{H}_p\mathcal{Y}) \leq \mathsf{d}_I(\mathcal{X}, \mathcal{Y})$ *for any* $p \geq 0$ *where* $\mathsf{H}_p\mathcal{X}$ *and* $\mathsf{H}_p\mathcal{Y}$ *denote the persistence modules of* $\mathcal{X}$ *and* $\mathcal{Y}$, *respectively, at the homology level.*

Now we relate the interleaving distance between two persistence modules and the persistence diagrams they define. For this, we consider a special type of persistence module called an *interval module*. Below, we use the standard convention that an open end of an interval is denoted with round parentheses "(" or ")" and a closed end of an interval with square brackets "[" or "]".

Definition 3.20. (Interval module) Given an index set $A \subseteq \mathbb{R}$ and a pair of indices $b, d \in A$, $b \leq d$, four types of *interval modules* denoted $\mathbb{I}[b, d)$, $\mathbb{I}(b, d]$, $\mathbb{I}[b, d]$, and $\mathbb{I}(b, d)$, respectively are special persistence modules defined as follows.

- closed–open: $\mathbb{I}[b, d) : \{V_a \xrightarrow{v_{a,a'}} V_{a'}\}_{a,a' \in A}$ where (i) $V_a = \mathbb{Z}_2$ for all $a \in [b, d)$ and $V_a = 0$ otherwise, and (ii) $v_{a,a'}$ is the identity map for $b \leq a \leq a' < d$ and the zero map otherwise.

- open–closed: $\mathbb{I}(b, d]$: $\{V_a \xrightarrow{v_{a,a'}} V_{a'}\}_{a,a' \in A}$ where (i) $V_a = \mathbb{Z}_2$ for all $a \in (b, d]$ and $V_a = 0$ otherwise, and (ii) $v_{a,a'}$ is the identity map for $b < a \le a' \le d$ and the zero map otherwise.

- closed–closed: $\mathbb{I}[b, d]$: $\{V_a \xrightarrow{v_{a,a'}} V_{a'}\}_{a,a' \in A}$ where (i) $V_a = \mathbb{Z}_2$ for all $a \in [b, d]$ and $V_a = 0$ otherwise, and (ii) $v_{a,a'}$ is the identity map for $b \le a \le a' \le d$ and the zero map otherwise.

- open–open: $\mathbb{I}(b, d)$: $\{V_a \xrightarrow{v_{a,a'}} V_{a'}\}_{a,a' \in A}$ where (i) $V_a = \mathbb{Z}_2$ for all $a \in (b, d)$ and $V_a = 0$ otherwise, and (ii) $v_{a,a'}$ is the identity map for $b < a \le a' < d$ and the zero map otherwise.

In general, we denote the four types of interval modules as $\mathbb{I}\langle b, d\rangle$ being oblivious to the particular type. The two endpoints b and d signify the birth and the death points of the interval in analogy to the bars we have seen for persistence diagrams. This is why sometimes we also write $\mathbb{I}\langle b, d\rangle = \langle b, d\rangle$. Gabriel [163] showed that a persistence module decomposes uniquely into interval modules when the index set is finite. This condition can be relaxed further as stated in the proposition below. A persistence module $\mathbb{U}$ for which each of the vector spaces U_a, $a \in A \subseteq \mathbb{R}$, has finite dimension is called a *pointwise finite-dimensional* (p.f.d. in short) persistence module. A persistence module for which the connecting linear maps have finite rank is called *q-tame*. The results below are part of a more general concept called *quiver theory*.

Proposition 3.10. *The following hold for persistence modules:*

(a) Any persistence module over a finite index set decomposes uniquely into closed–closed interval modules, that is, $\mathbb{U} \cong \bigoplus_{j \in J} \mathbb{I}[b_j, d_j]$ [163].

(b) Any p.f.d. persistence module decomposes uniquely into interval modules, that is, $\mathbb{U} \cong \bigoplus_{j \in J} \mathbb{I}\langle b_j, d_j\rangle$ [110, 298].

(c) Any q-tame persistence module decomposes uniquely into interval modules [80].

The birth and death points of the interval modules that a given persistence module $\mathbb{U}$ decomposes into (Proposition 3.10) can be plotted as points in $\mathbb{R}^2$. This defines a persistence diagram $\mathrm{Dgm}\,\mathbb{U}$ for a persistence module $\mathbb{U}$. We aim to relate the interleaving distance between persistence modules and the bottleneck distance between their persistence diagrams thus defined.

Definition 3.21. (Persistence diagram for persistence module) Let $\mathbb{U} \cong \bigoplus_j \langle b_j, d_j\rangle$ be the interval decomposition of a given persistence module $\mathbb{U}$

(Proposition 3.10). The collection of points $\{(b_j, d_j)\}$ with proper multiplicity and the points on the diagonal $\Delta : \{(x, x)\}$ with infinite multiplicity constitute the *persistence diagram* Dgm $\mathbb{U}$ of the persistence module $\mathbb{U}$.

For the index set $A = \mathbb{R}$, Chazal et al. [77] showed that the bottleneck distance between two persistence diagrams of p.f.d. modules is bounded from above by their interleaving distance. The result also holds for q-tame modules. It is proved in [23, 221] that the two distances are indeed equal.

Theorem 3.11. *Given two q-tame persistence modules defined over the totally ordered index set $\mathbb{R}$, $\mathsf{d}_I(\mathbb{U}, \mathbb{V}) = \mathsf{d}_b(\text{Dgm } \mathbb{U}, \text{Dgm } \mathbb{V})$.*

Remark 3.4. *The isometry theorem stated for the index set $\mathbb{R}$ does not apply directly to the persistence modules that are not defined over the index set $\mathbb{R}$. In this case, to define the interleaving distance, we can extend the module to be indexed over $\mathbb{R}$ as described in Remark 3.3. For example, consider a persistence module $\mathsf{H}_p \mathcal{F}$ obtained from a filtration $\mathcal{F}$ defined on a finite index set A or when $A = \mathbb{Z}$. Observe that all interval modules for $\mathsf{H}_p \mathcal{F}$ (without extension) are of closed–closed type $[b, d]$ for some $b, d \in A$. This brings out a subtlety. Intervals of the form $[b, d]$ where $b = d$ are mapped to the diagonal Δ in the persistence diagram. These points get ignored while computing the bottleneck distance as both diagrams have diagonal points with infinite multiplicity. In fact, the isometry theorem (Theorem 3.11) does not hold if this is not taken care of. To address the issue, for persistence modules $\mathsf{H}_p \mathcal{F}$ generated by a finite filtration $\mathcal{F}$, we map each interval $[b, d]$ in the decomposition of $\mathsf{H}_p \mathcal{F}$ to a point $(b, d + 1)$ in $\text{Dgm}_p(\mathcal{F})$ (Definition 3.8). This aligns with the observation that, after extension over the index set $\mathbb{R}$, the interval $[b, d]$ indeed stretches to $[b, d + 1)$.*

3.5 Persistence for PL-Functions

Given a PL-function $f : |K| \to \mathbb{R}$ on a simplicial complex K (Definition 3.2), we can produce a simplicial filtration of K as well as a space filtration of the topological space $|K|$. In this section, we study the relation between the persistent homology of the two, noting that the former involves simplicial homology and the latter singular homology. Observe that the PL framework allows us to inspect the topological space $|K|$ through the lens of the function f, and is useful in practice as one can describe different properties of K by designing an appropriate (descriptor) function f.

In Section 3.5.1, we describe the critical points of such functions. The restriction of f on the vertex set of K is a vertex function $f_V : V(K) \to \mathbb{R}$ which naturally induces a simplicial filtration (the lower-star filtration). In Section 3.5.2, we relate the space filtration of the PL-function $f : |K| \to \mathbb{R}$ with the simplicial filtration induced by f_V, which in turn allows us to apply the output of the persistence algorithm run on $\mathcal{F}_{f_V}$ to the space filtration. Finally, in Section 3.5.3, we present a simple algorithm to compute the zeroth persistence diagram induced by a PL-function (thus also a vertex function).

3.5.1 PL-Functions and Critical Points

In Section 1.5, we discussed smooth functions defined on smooth manifolds. However, often the domain is a piecewise-linear domain such as a simplicial complex, and a natural family of functions defined on a simplicial complex is a PL-function as introduced in Definition 3.2; in particular, recall that a PL-function $f : |K| \to \mathbb{R}$ is determined by its restriction on vertices $f|_{V(K)} : V(K) \to \mathbb{R}$ and linearly extending it within each simplex $\sigma \in K$. From now on, we will simplify notation and use f to denote the vertex function $f|_{V(K)}$ as well; that is, we write the vertex function also as $f : V(K) \to \mathbb{R}$.

PL-Critical Points

For a Morse function f defined on a smooth d-manifold M, the Morse Lemma (see Proposition 1.2) suggests that the index of a critical point p is completely determined by its local neighborhood within the sublevel set $M_{\leq f(p)}$. For PL-functions, this is captured by lower star and lower link. We define the PL-critical points for PL-functions using homology groups. However, as the neighborhood of a point is not necessarily a topological ball, we now need to consider both the lower and upper links. In this context, it is more convenient to use the p-th reduced Betti number $\tilde{\beta}_p(X)$ of a space/complex X.

Definition 3.22. (Reduced Betti number) One has $\tilde{\beta}_p(X) = \beta_p(X)$ for $p > 0$. For $p = 0$, one has $\tilde{\beta}_0(X) = \beta_0(X) - 1$ and $\tilde{\beta}_{-1}(X) = 0$ if X is not empty; otherwise, $\tilde{\beta}_0(X) = 0$ and $\tilde{\beta}_{-1}(X) = 1$.

Definition 3.23. (PL-critical points) Given a PL-function $f : |K| \to \mathbb{R}$, we say that a vertex $v \in K$ is a *regular* vertex or point if $\tilde{\beta}_p(\mathrm{Llk}(v)) = 0$ and $\tilde{\beta}_p(\mathrm{Ulk}(v)) = 0$ for any $p \geq -1$. Otherwise, it is a *PL-critical* (or simply *critical*) vertex or point. Furthermore, we say that v has *lower-link-index* p if $\tilde{\beta}_{p-1}(\mathrm{Llk}(v)) > 0$. Similarly v has *upper-link-index* p if $\tilde{\beta}_{p-1}(\mathrm{Ulk}(v)) > 0$.

The function value of a critical point is a *critical value for* f.

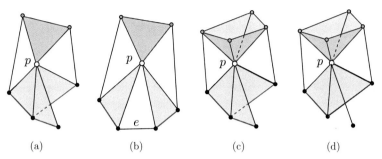

(a)　　　　　(b)　　　　　(c)　　　　　(d)

Figure 3.14 The point p is a regular point in (a). The point p is PL-critical in (b), (c), and (d). Light-blue shaded triangles are in the lower star, light-pink shaded ones are in the upper star, while light-yellow shaded ones are in neither. In (b), note that edge e is not in Llk(p); here p has lower-link-index 1 as $\tilde{\beta}_0(\text{Llk}(p)) = 1$. In (c), the point p has upper-link-index 2. In (d), the point p has lower-link-index 1 and upper-link-index 2.

Discussions of PL-Critical Points

Some examples of PL-critical points are given in Figure 3.14. As mentioned above, in the smooth case for a Morse function defined on an m-manifold M, the type of a nondegenerate critical point v is completely defined by its local neighborhood lower than $f(v)$ (as the portion higher than $f(v)$ is its complement with respect to an m-ball). This is no longer the case for the PL case, as we see in Figure 3.14. We also note that a PL-critical point could have multiple lower-link-indices and upper-link-indices. Nevertheless, as we will see later (e.g., Theorem 3.13), these PL-critical points are related to the change of homology groups within the sublevel sets or superlevel sets, somewhat analogous to the smooth setting.

We note that other concepts of "critical values" exist in the literature. In particular, the concept of *homological critical values* is introduced in [101] for a function $f : M \to \mathbb{R}$ defined on a topological space M. In particular, $\alpha \in \mathbb{R}$ is a *homological critical value* if there exists some $p \geq 0$ such that for all sufficiently small $\varepsilon > 0$, the homomorphism $\mathsf{H}_p(f^{-1}(-\infty, \alpha - \varepsilon]) \to \mathsf{H}_p(f^{-1}(-\infty, \alpha + \varepsilon])$ induced by inclusion is **not** an isomorphism. It can be shown that for a PL-function, any PL-critical point with a nonzero lower-link-index is homological critical. In other words, we can think of our definition of PL-critical points as providing an explicit characterization for the local neighborhood of points giving rise to "critical values" for the PL-setting – this now allows us to identify critical points using only the star/link of a point.

The homological critical value of [101] is not symmetric with respect to the role of sub- versus superlevel sets for general spaces. Indeed, one could also define a symmetric version using superlevel sets. The point in Figure 3.14(c) does not give rise to a homological critical value with respect to the sublevel sets, but does so with respect to the superlevel sets. A more general (and symmetric) concept of critical values is introduced in [63], which we formally define later (Definition 4.14) in Chapter 4 when we describe the more general zigzag persistence modules.

Two Choices of "Sublevel Sets"

Consider a PL-function $f : |K| \to \mathbb{R}$. Its sublevel set at a is given by

$$|K|_a := \{x \in |K| \mid f(x) \leq a\},$$

which gives rise to a space filtration over $|K|$ as a increases. Let us call it a *space sublevel set*.

On the other hand, given $a \in \mathbb{R}$, we can also consider the subcomplex K_a spanned by all vertices from K whose function value is at most a; that is,

$$K_a := \{\{u_0, \ldots, u_d\} \in K \mid f(u_i) \leq a\}.$$

We refer to K_a as the *simplicial sublevel set* with respect to $f : |K| \to \mathbb{R}$ (or w.r.t. the vertex function $f|_{V(K)} : V(K) \to \mathbb{R}$). Assume vertices $v_1, \ldots, v_n \in V(K)$ are ordered so that $f(v_1) \leq f(v_2) \leq \cdots \leq f(v_n)$. It is easy to see that $K_a = K_{f(v_i)}$ if $a \in [f(v_i), f(v_{i+1}))$. Note that this is also the sublevel set for the simplex-wise monotonic function $\bar{f}$ introduced in Fact 3.1. These two "types" of sublevel sets relate to each other via the following result.

Theorem 3.12. *Given a PL-function $f : |K| \to \mathbb{R}$, for any $a \in \mathbb{R}$, the space and simplicial sublevel sets have isomorphic homology groups; that is, $\mathsf{H}_*(K_a) \cong \mathsf{H}_*(|K|_a)$.*

Furthermore, the following diagram commutes, where the horizontal homomorphisms are induced by natural inclusions.

$$
\begin{array}{ccc}
\mathsf{H}_*(K_a) & \longrightarrow & \mathsf{H}_*(K_b) \\
\downarrow{\scriptstyle\cong} & & \downarrow{\scriptstyle\cong} \\
\mathsf{H}_*(|K|_a) & \longrightarrow & \mathsf{H}_*(|K|_b)
\end{array}
$$

PROOF. If $a < f(v_1)$, $K_a = \varnothing$ and $|K|_a = \varnothing$. If $a \geq f(v_n)$, $K_a = K$ and $|K|_a = |K|$. Thus the theorem holds in both cases. Now assume $a \in [f(v_i), f(v_{i+1}))$ for some $i \in [1, n)$. In this case, $K_a = K_{f(v_i)} = \bigcup_{j \leq i} \overline{\mathrm{Lst}}(v_j)$. It follows that $|K_a| \subseteq |K|_a$. We now show that there is a continuous map $\mu : |K|_a \times [0, 1] \to |K|_a$ that will continuously deform the identity map on $|K|_a$ to a retraction from $|K|_a$ to $|K_a|$. In other words, μ is a deformation retraction from $|K|_a$ onto $|K_a|$; thus $|K_a| \hookrightarrow |K|_a$ induces an isomorphism at the homology level. This will then establish the first part of the theorem.

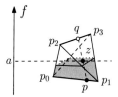

Figure 3.15 Consider the simplex $\sigma = \{p_0, p_1, p_2, p_3\}$, where $\tau_I = \{p_0, p_1\}$ and $\tau_O = \{p_2, p_3\}$. The shaded region equals $|\sigma| \cap |K|_a$. This shaded region is the union of a set of segments $\overline{pz}$ which are disjoint in their interior. The map μ deformation retracts the segment $\overline{pz}$ to the point $p \in |\tau_I| \subseteq |K_a|$.

For any point $x \in |K_a|$, we set $\mu(t, x) = x$ for any $t \in [0, 1]$. Now the set of points in $A := |K|_a \setminus |K_a|$ form a set of "partial simplices": In particular, since f is a PL-function, there is a set C of simplices in K such that $A = \bigcup_{\sigma \in C} interior(\sigma) \cap |K|_a$, where $interior(\sigma)$ denotes the set of points in $|\sigma|$ that are not in any proper face of σ. We construct the map μ on A by constructing its restriction to each simplex $\sigma \in C$.

Specifically, consider $\sigma = \{p_0, \dots, p_d\} \in C$. Let $\tau_I = \{p_0, \dots, p_s\}$ be the maximal face of σ contained in $|K|_a$, and $\tau_O = \{p_{s+1}, \dots, p_d\}$ is then the face outside $|K|_a$ spanned by vertices of σ not in $|K|_a$; see Figure 3.15. On the other hand, we can write the underlying space $|\sigma|$ as $|\sigma| = \bigcup_{p \in |\tau_I|} \bigcup_{q \in |\tau_O|} \overline{pq}$, where $\overline{pq}$ denotes the convex combination of p and q (line segment from p to q). Furthermore, $\overline{pq} \cap |K|_a = \overline{pz}$ with $f(z) = a$ as f is a PL-function. For any point $x \in \overline{pz}$, we simply set $\mu(t, x) = (1 - t)x + tp$. This map is well defined as all lines $\overline{pq}$, $p \in |\tau_I|$ and $q \in |\tau_O|$, are disjoint in their interior. Since f is piecewise-linear on σ, the map μ as constructed is continuous. Also, $\mu(0, \cdot)$ is the identity on $|K|_a$, and $\mu(1, \cdot) \colon |K|_a \to |K_a|$ is a retraction map. Thus μ is a deformation retraction and, by Fact 1.1, $|K|_a$ and $|K_a|$ are homotopy equivalent, implying that $\mathsf{H}_*(|K|_a) \cong \mathsf{H}_*(|K_a|)$. The first part of the theorem then follows.

Furthermore, given that μ is a deformation retraction, the natural inclusion $|K_a| \subseteq |K|_a$ induces an isomorphism at the homology level. The second part of the theorem follows from this, combined with the naturality of the isomorphism $\mathsf{H}_*(K_a) \cong \mathsf{H}_*(|K_a|)$. □

We note that we can also inspect the superlevel sets for the underlying space $|K|$ and for the simplicial setting in a symmetric manner. A result analogous to the above theorem also holds for the superlevel sets.

Relation to PL-Critical Points

Similar to critical points for smooth functions, the homology group of the sublevel sets can only change at the PL-critical points. For simplicity, in what follows we set $K_i := K_{f(v_i)}$, for any $i \in [1, n]$. Observe that for any $a \in \mathbb{R}$, if complex K_a is non-empty, then it equals K_i for some i; in particular, $K_a = K_i$ where $a \in [f(v_i), f(v_{i+1}))$.

Theorem 3.13. (PL-critical points) *Let $f : |K| \to \mathbb{R}$ be a PL-function defined on a simplicial complex K. For any index $r \in [2, n]$ and dimension $p \geq 0$, the inclusion $K_{r-1} \hookrightarrow K_r$ induces an isomorphism $\mathsf{H}_p(K_{r-1}) \cong \mathsf{H}_p(K_r)$ unless v_r is a PL-critical point of lower-link-index p or $p + 1$.*

A symmetric statement for the superlevel sets and PL-critical points of nonzero upper-link-index also holds.

PROOF. Let $A = \overline{\mathrm{Lst}}(v_r)$ be the closed lower star of v_r, and $B = K_{r-1}$. Set $U = A \cup B$ and $V = A \cap B$; it is easy to see that $U = K_r$, while $V = \mathrm{Llk}(v_r)$. Furthermore, by the definition of lower stars and lower links over a simplicial complex, $A = \overline{\mathrm{Lst}}(v_r)$ equals the coning of v_r and $\mathrm{Llk}(v_r)$. It follows that A has trivial reduced homology for all dimensions. Now consider the following (Mayer–Vietoris) exact sequence:

$$\cdots \longrightarrow \tilde{\mathsf{H}}_p(V) \longrightarrow \tilde{\mathsf{H}}_p(A) \oplus \tilde{\mathsf{H}}_p(B) \xrightarrow{\phi} \tilde{\mathsf{H}}_p(U) \longrightarrow \tilde{\mathsf{H}}_{p-1}(V) \longrightarrow \cdots \quad (3.9)$$

Assume that v_r is neither a lower-link-index-$(p + 1)$ PL-critical point, nor an index-p one. Since v_r is not lower-link-index-$(p + 1)$ PL-critical, $\tilde{\mathsf{H}}_p(V)$ is trivial. Thus by exactness of the sequence, the homomorphism ϕ must be injective. Similarly, as v_r is not lower-link-index-p PL-critical, $\tilde{\mathsf{H}}_{p-1}(V)$ is trivial; thus ϕ must be surjective. Hence ϕ is an isomorphism. Furthermore, note that $\tilde{\mathsf{H}}_p(A) \oplus \tilde{\mathsf{H}}_p(B) = 0 \oplus \tilde{\mathsf{H}}_p(B)$ as A has trivial homology. It then follows that $\mathsf{H}_p(K_{r-1}) \cong \mathsf{H}_p(K_r)$ induced by the inclusion map $K_{r-1} \hookrightarrow K_r$. The claim then follows.   □

Corollary 3.14. *Given a PL-function $f : |K| \to \mathbb{R}$ defined on a finite simplicial complex K, let $[a, b] \subset \mathbb{R}$ be such that it does not contain any PL-critical value for f.*

(a) *Then the inclusion map $K_a \hookrightarrow K_b$ induces an isomorphism between the simplicial homology groups, that is, $\mathsf{H}_p(K_a) \cong \mathsf{H}_p(K_b)$ for any dimension $p \geq 0$.*

(b) *This also implies that $|K|_a \hookrightarrow |K|_b$ induces an isomorphism between the singular homology groups, that is, $\mathsf{H}_p(|K|_a) \cong \mathsf{H}_p(|K|_b)$ for any dimension $p \geq 0$.*

Again, a version of the above corollary also holds for superlevel sets.

3.5.2 Lower-Star Filtration and Its Persistent Homology

Let $f : |K| \to \mathbb{R}$ be a PL-function. Recall that $K_i := K_{f(v_i)}$ where $v_1, \dots, v_n$ are ordered in nondecreasing values of f. Setting $a_i = f(v_i)$, we write $|K|_{a_i} = |K|_{f(v_i)}$. The two different types of sublevel sets give rise to two sequences of growing spaces:

lower-star simplicial filtration $\mathcal{F}_f : \varnothing \hookrightarrow K_1 \hookrightarrow K_2 \hookrightarrow \cdots \hookrightarrow K_{n-1} \hookrightarrow K_n = K;$ (3.10)

sublevel set space filtration $\widehat{\mathcal{F}}_f : \varnothing \hookrightarrow |K|_{a_1} \hookrightarrow |K|_{a_2} \hookrightarrow \cdots \hookrightarrow |K|_{a_{n-1}} \hookrightarrow |K|_{a_n} = |K|.$

(3.11)

As $K_i := \bigcup_{j \le i} \mathrm{Lst}(v_j)$ is the union of the lower stars of $v_1, \dots, v_i$, we call the filtration in Eq. (3.10) the *lower-star filtration for* f; see also Section 3.1.2 and Figure 3.6. The two homology modules $\mathsf{H}_p \mathcal{F}_f$ and $\mathsf{H}_p \widehat{\mathcal{F}}_f$ can be shown to be isomorphic due to Theorem 3.12, and thus they produce identical persistence diagrams (Fact 3.10).

Corollary 3.15. *The homology module* $\mathsf{H}_p \mathcal{F}_f$ *is isomorphic to the homology module* $\mathsf{H}_p \widehat{\mathcal{F}}_f$ *for every* $p \ge 0$. *This implies that these two persistence modules have the same persistence diagrams.*

Intuitively, the lower-star filtration of the simplicial complex K can be thought of as the discrete version of the sublevel set filtration of the space $|K|$ with respect to the PL-function f. By Corollary 3.15, the lower-star simplicial filtration $\mathcal{F}_f$ and the sublevel set space filtration $\widehat{\mathcal{F}}_f$ have identical persistence diagrams. We refer to this common persistence diagram as *the persistence diagram of the PL-function* f, denoted by $\mathrm{Dgm} f$.

For a space filtration induced by a Morse function defined on a Riemannian manifold, the birth and death coordinates of the points in the persistence diagrams correspond to critical values of this Morse function. A similar result holds for the PL case. In particular, one can prove, using Corollary 3.14, that, for a PL-function f, the persistence pairings for $\mathcal{F}_f$ occur only between PL-critical points. That is expressed as follows.

Fact 3.11. *Given a PL-function* $f : |K| \to \mathbb{R}$ *and its associated filtration* $\mathcal{F}_f$, *let* $\mu_{f,p}^{i,j}$ *denote the corresponding* p-th *persistence pairing function with respect to* $\mathcal{F}_f$. *If* $\mu_{f,p}^{i,j} \ne 0$, *then vertices* v_i *and* v_j *must be PL-critical.*

However, not all PL-critical points necessarily appear in persistence pairings with respect to the lower-star filtration $\mathcal{F}_f$.

Computing Persistence Diagram Induced by $\mathcal{F}_f$ and $\widehat{\mathcal{F}}_f$

By Corollary 3.15, we only need to describe how to compute the persistence diagram for the lower-star filtration $\mathcal{F}_f$. We will do so via Algorithm 3:

MATPERSISTENCE from Section 3.3. However, recall that algorithm MAT-PERSISTENCE works on simplex-wise filtrations (Definition 3.1). The algorithm either pairs each simplex with another simplex, producing a *persistence pairing*, or leaves it unpaired, producing an *essential persistent point* $(b, \infty) \in$ the persistence diagram. To compute $\mathrm{Dgm} f$, we first expand $\mathcal{F}_f$ into a simplex-wise filtration $\mathcal{F}_s$ induced by a total ordering of all m simplices in K,

$$\sigma_1, \ldots, \sigma_{I_1}, \sigma_{I_1+1}, \ldots, \sigma_{I_2}, \sigma_{I_2+1}, \ldots, \sigma_{I_j}, \sigma_{I_j+1}, \ldots, \sigma_{I_{j+1}}, \sigma_{I_{j+1}+1}, \ldots, \sigma_{I_{n-1}}, \sigma_{I_{n-1}+1}, \ldots, \sigma_{I_n=m},$$
$$(3.12)$$

so that following two conditions hold:

- $\mathrm{Lst}(v_j) = \{\sigma_{I_{j-1}+1}, \ldots, \sigma_{I_j}\}$, for any $j \in [1, n]$ (here $I_0 = 0$);
- for any simplex σ, its faces appear earlier than in the total ordering of simplices.

With this total ordering of simplices, the induced simplex-wise filtration becomes

$$\mathcal{F}_s : \quad L_1 \hookrightarrow L_2 \hookrightarrow \cdots \hookrightarrow L_m, \quad \text{where } L_i := \{\sigma_j \mid j \leq i\} \text{ and thus } \sigma_i = L_i \setminus L_{i-1}. \quad (3.13)$$

Note that $K_i = L_{I_i}$; thus $\mathcal{F}_f$ is a subsequence of the simplex-wise filtration $\mathcal{F}_s$. The construction of $\mathcal{F}_s$ from $\mathcal{F}_f$ is not necessarily unique. We can simply choose $\sigma_{I_{j-1}+1}, \ldots, \sigma_{I_j}$ to be the set of simplices in $\mathrm{Lst}(v_j)$ sorted by their dimension. We now construct the map $\pi : [0, m] \to [0, n]$ as $\pi(j) = k$ if $j \in [I_{k-1} + 1, I_k]$; that is, $\pi(j) = k$ means that simplex σ_j is in the lower star of vertex v_k.

We run the persistence algorithm MATPERSISTENCE on the simplex-wise filtration $\mathcal{F}_s$. Let $\mu_{s,p}^{i,j}$ denote the persistence pairing function with respect to $\mathcal{F}_s$. Many of the pairings are between two simplices within the same lower star of a vertex and are not interesting. Instead, we aim to compute the persistence diagram $\mathrm{Dgm} f$ for the filtration $\mathcal{F}_f$, which captures only the nonlocal pairings where the birth and death are from different K_i. The following theorem specifies how we can compute the persistence diagram $\mathrm{Dgm} f$ for the filtration $\mathcal{F}_f$ from the output of the persistence algorithm with the simplex-wise filtration $\mathcal{F}_s$ as input.

Theorem 3.16. (Computation of $\mathrm{Dgm} f$ in the PL case) *Given a PL-function $f : |K| \to \mathbb{R}$, let $\mu_{s,p}^{i,j}$ denote the p-dimensional persistence pairing function with respect to the simplex-wise filtration $\mathcal{F}_s$ as described above. We can compute the* persistence pairing function $\mu_{f,p}^{i,j}$ *with respect to $\mathcal{F}_f$ as follows:*

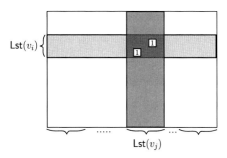

Figure 3.16 Persistence pairing function $\mu_{f,p}^{i,j} = 2$.

$$\mu_{f,p}^{i,j} := \sum_{b \in (I_{i-1}, I_i], d \in (I_{j-1}, I_j]} \mu_{s,p}^{b,d} \text{ for any } i < j \le n; \text{ and } \mu_{f,p}^{i,\infty} := \sum_{b \in (I_{i-1}, I_i]} \mu_{s,p}^{b,\infty} \text{ for any } i \le n.$$

If $\mu_{f,p}^{i,j} \ne 0$, we refer to (v_i, v_j) as a persistence pair with respect to f and we add the corresponding persistent point $(f(v_i), f(v_j))$, with multiplicity $\mu_{f,p}^{i,j} \ne 0$, to the persistence diagram Dgm f. The persistence of this pair (v_i, v_j) is $|f(v_i) - f(v_j)|$.

Remark 3.5. *As an example, see Figure 3.16, which shows the reduced matrix after running Algorithm 3:* MATPERSISTENCE *on the filtered boundary matrix D for $\mathcal{F}_s$ where "1" indicates the lowest "1" in the shaded columns. Only columns corresponding to p-simplices are shown. We have $\mu_{f,p}^{i,j} = 2$. One can have an alternative view of the persistence pairs given by $\mu_{f,p}^{i,j}$ as follows: For each persistence index pair $(i, j) \in \text{Dgm}(\mathcal{F}_s)$ (i.e., $\mu_p^{i,j} > 0$ w.r.t. $\mathcal{F}_s$), one has a persistence pair $(v_{\pi(i)}, v_{\pi(j)})$ for $\mathcal{F}_f$ if and only if $\pi(i) \ne \pi(j)$. In other words, all local pairs $(i, j) \in \text{Dgm}(\mathcal{F}_s)$ with $\pi(i) = \pi(j)$ signifying that σ_i and σ_j are from the lower star of the same vertex are ignored for the persistence diagram Dgm$(\mathcal{F}_f)$.*

PROOF OF THEOREM 3.16. Recall that $\mu_{f,p}^{i,j}$ and $\mu_{s,p}^{i,j}$ are the persistence pairing functions induced by the filtration $\mathcal{F}_f$ and $\mathcal{F}_s$, respectively. Similarly, we use $\beta_{f,p}^{i,j}$ and $\beta_{s,p}^{i,j}$ to denote the persistent Betti numbers induced by filtrations $\mathcal{F}_f$ and $\mathcal{F}_s$, respectively. In what follows, we will just prove that, for any dimension $p \ge 0$ and $i, j \in [1, n]$, we have that $\mu_{f,p}^{i,j}$ can be computed as stated in the theorem. The case when $j = \infty$ can be handled in a similar manner and is left as an exercise.

For any $i \in [1, n]$, let I_i be as defined in Eq. (3.12). Given the relation of $\mathcal{F}_f$ and $\mathcal{F}_s$, it follows that for any $i', j' \in [1, n]$, we have that $K_{i'} = L_{I_{i'}}$, $K_{j'} = L_{I_{j'}}$, and thus $\beta_{f,p}^{i',j'} = \beta_{s,p}^{I_{i'}, I_{j'}}$ as $\mathcal{F}_f$ is a subsequence of $\mathcal{F}_s$.

Now fix the dimension $p \geq 0$, and for simplicity omit p from all subscripts. Given any $i, j \in [1, n]$, we have that

$$\mu_f^{i,j} = (\beta_f^{i,j-1} - \beta_f^{i,j}) - (\beta_f^{i-1,j-1} - \beta_f^{i-1,j})$$

$$= (\beta_s^{I_i,I_{j-1}} - \beta_s^{I_i,I_j}) - (\beta_s^{I_{i-1},I_{j-1}} - \beta_s^{I_{i-1},I_j}) = \mu_s^{I_i,I_j}.$$

Hence we aim to show that $\mu_s^{I_i,I_j} = \sum_{b \in (I_{i-1}, I_i], \, d \in (I_{j-1}, I_j]} \mu_s^{b,d}$ which then proves the theorem. To this end, note that by Theorem 3.1, we have the following:

$$\beta_s^{I_i,I_{j-1}} - \beta_s^{I_i,I_j} = \sum_{b \leq I_i, \, d > I_{j-1}} \mu_s^{b,d} - \sum_{b \leq I_i, \, d > I_j} \mu_s^{b,d} = \sum_{b \leq I_i, \, d \in (I_{j-1}, I_j]} \mu_s^{b,d};$$

$$\beta_s^{I_{i-1},I_{j-1}} - \beta_s^{I_{i-1},I_j} = \sum_{b \leq I_{i-1}, \, d > I_{j-1}} \mu_s^{b,d} - \sum_{b \leq I_{i-1}, \, d > I_j} \mu_s^{b,d} = \sum_{b \leq I_{i-1}, \, d \in (I_{j-1}, I_j]} \mu_s^{b,d};$$

$$\Rightarrow \mu_s^{I_i,I_j} = \sum_{b \leq I_i, \, d \in (I_{j-1}, I_j]} \mu_s^{b,d} - \sum_{b \leq I_{i-1}, \, d \in (I_{j-1}, I_j]} \mu_s^{b,d} = \sum_{b \in (I_{i-1}, I_i], \, d \in (I_{j-1}, I_j]} \mu_s^{b,d}.$$

The theorem then follows.   □

An implication of the above result is that any simplex-wise filtration $\mathcal{F}_s$ obtained from the lower-star filtration $\mathcal{F}_f$ produces the same pairing between critical points and the same persistence diagram.

3.5.3 Persistence Algorithm for Zeroth Persistent Homology

The best known running time for a general persistence algorithm is $O(n^\omega)$, where n is the total number of simplices in the filtration. However, the zeroth persistent homology (zeroth persistence diagram $\text{Dgm}_0 f$) for a PL-function $f : |K| \to \mathbb{R}$ can be computed efficiently in $O(n \log n + m\alpha(n))$ time, where n and m are the number of vertices and edges in K, respectively.

Indeed, first observe that we only need the 1-skeleton of K to compute $\text{Dgm}_0 f$. So, in what follows, assume that K contains only vertices V and edges E. Assume that all vertices in V are sorted in nondecreasing order of their f-values. As before, let K_i be the union of lower stars of all vertices v_j where $j \leq i$. Since we are only interested in the zeroth homology, we only need to track the zeroth homology group of K_i, which essentially embodies the information about connected components.

Assume we are at vertex v_j. Consider $\text{Lst}(v_j)$. There are three cases.

Case 1: $\text{Lst}(v_j) = \{v_j\}$. Then v_j starts a new connected component in K_j. Hence v_j is a creator.

Case 2: All edges in $\text{Lst}(v_j)$ connect to vertices from the same connected component C in K_{j-1}. In this case, the component C grows in the sense that it now includes also vertex v_j and its incident edges in the lower star. However, $\mathsf{H}_0(K_{j-1})$ and $\mathsf{H}_0(K_j)$ are isomorphic where $K_{j-1} \subseteq K_j$ induces an isomorphism.

Case 3: Edges in $\mathrm{Lst}(v_j)$ link to two or more components, say $C_1, \ldots, C_r$, in K_{j-1}. In this case, after the addition of $\mathrm{Lst}(v_j)$, all $C_1, \ldots, C_r$ are merged into a single component,

$$C' = C_1 \cup C_2 \cup \cdots \cup C_r \cup \mathrm{Lst}\, v_j.$$

Hence inclusion $K_{j-1} \hookrightarrow K_j$ induces a surjective homomorphism $\xi \colon \mathsf{H}_0(K_{j-1}) \to \mathsf{H}_0(K_j)$ and $\beta_0(K_j) = \beta_0(K_{j-1}) - (r-1)$. That is, we can consider that $r-1$ number of components are destroyed; only one stays on as C'.

Proposition 3.17. *Suppose case 3 happens where edges in* $\mathrm{Lst}(v_j)$ *merge components* $C_1, \ldots, C_r$ *in* K_{j-1}. *Let* v_{k_i} *be the global minimum of component* C_i *for* $i \in [1, r]$. *Assume without loss of generality that* $f(v_{k_1}) \leq f(v_{k_2}) \leq \cdots \leq f(v_{k_r})$. *Then the node* v_j *participates in exactly* $r-1$ *number of persistence pairings* $(v_{k_2}, v_j), \ldots, (v_{k_r}, v_j)$ *for the zero-dimensional persistent diagram* $\mathrm{Dgm}_0 f$, *corresponding to points* $(f(v_{k_2}), f(v_j)), \ldots, (f(v_{k_r}), f(v_j))$ *in* $\mathrm{Dgm}_0 f$.

Intuitively, when case 3 happens, consider the set of 0-cycles $c_2 = v_{k_2} + v_{k_1}, c_3 = v_{k_3} + v_{k_1}, \ldots, c_r = v_{k_r} + v_{k_1}$. On the one hand, it is easy to see that their corresponding homology classes $[c_i]$ are independent within $\mathsf{H}_0(K_{j-1})$. Furthermore, each c_i is created upon entering K_{k_i} for each $i \in [1, r]$. On the other hand, the homology classes $[c_2], \ldots, [c_r]$ become trivial in $\mathsf{H}_0(K_j)$ (thus they are destroyed upon entering K_j). Hence $\mu_0^{k_i, j} > 0$ for $i \in [2, r]$, corresponding to persistence pairings $(v_{k_2}, v_j), \ldots, (v_{k_r}, v_j)$. Furthermore, consider any 0-cycle $c_1 = v_{k_1} + c$ where c is a 0-chain from K_{k_1-1}. The class $[c_1]$ is created at K_{k_1} yet remains nontrivial at K_j. Hence there is no persistence pairing (v_{k_1}, v_j).

Based on Proposition 3.17 we can compute the persistence pairings for the zero-dimensional persistent homology without the matrix reduction algorithm. We only need to maintain connected components information for each K_i, and potentially merge multiple components. We would also need to be able to query the membership of a given vertex u in the components of the current sublevel set. Such operations can be implemented by a standard union–find data structure.

Specifically, a union–find data structure is a standard data structure that maintains dynamic disjoint sets [108]. Given a set of elements U called the

---

Algorithm 5 ZEROPERDG($K = (V, E), f$)

Input:
  K: a 1-complex with a vertex function f on it
Output:
  Vertex pairs generating $\text{Dgm}_0(f)$ for the PL-function given by f

1: Sort vertices in V so that $f(v_1) \leq f(v_2) \leq \cdots \leq f(v_n)$
2: **for** $j = 1 \to n$ **do**
3:   CREATESET(v_j)
4:   $flag := 0$
5:   **for** each $(v_k, v_j) \in \text{Lst}(v_j)$ **do**
6:     **if** ($flag == 0$) **then**
7:       UNION(v_k, v_j)
8:       $flag := 1$
9:     **else**
10:       **if** FINDSET(v_k) $\neq$ FINDSET(v_j) **then**
11:         Set $\ell_1 = \text{REPSET}(v_k)$ and $\ell_2 = \text{REPSET}(v_j)$
12:         UNION(v_k, v_j)
13:         Output pairing ($\text{argmax}\{f(\ell_1), f(\ell_2)\}, v_j$)
14:       **end if**
15:     **end if**
16:   **end for**
17: **end for**
18: **for** each disjoint set C **do**
19:   Output pairing (REPSET(C), ∞)
20: **end for**

---

universe, this data structure typically supports the following three operations to maintain a set $\mathcal{S}$ of disjoint subsets of U, where each subset also maintains a representative element: (1) MAKESET(x) which creates a new set $\{x\}$ and adds it to $\mathcal{S}$; (2) FINDSET(x) returns the representative of the set from $\mathcal{S}$ containing x; and (3) UNION(x, y) merges the sets from $\mathcal{S}$ containing x and y, respectively, into a single one if they are different.

We now present Algorithm 5: ZEROPERDG. Here the universe U is the set of all vertices V of K. Note that each vertex v is also associated with its function value $f(v)$. In this algorithm, we assume that the *representative* of a set C is the minimum in it, that is, the vertex with the smallest f-value, and the query REPSET(v) returns the representative of the set containing vertex v. We assume that this query takes the same time as FINDSET(v). Given a disjoint set C, we also use REPSET(C) to represent the representative (minimum) of this set. One can view a disjoint set C in the collection S as the maximal set of elements sharing the same representative.

Let n and m denote the number of vertices and edges in K, respectively. Sorting all vertices in V takes $O(n \log n)$ time. There are $O(n + m)$ number of CREATESET, FINDSET, UNION, and REPSET operations. By using the standard union–find data structure, the total time for all these operations is $(n + m)\alpha(n)$ where $\alpha(n)$ is the inverse Ackermann's function that grows extremely slowly with n [108]. Hence the total time complexity of algorithm ZEROPERDG is $O(n \log n + m\alpha(n))$.

Note that lines 18–20 of algorithm ZEROPERDG inspect all disjoint sets after processing all vertices and their lower stars; each such disjoint set corresponds to a connected component in K. Hence each of them generates an essential pair in the zeroth persistence diagram.

Theorem 3.18. *Given a PL-function* $f : |K| \to \mathbb{R}$, *the zero-dimensional persistence diagram* $\mathrm{Dgm}_0 f$ *for the lower-star filtration of* f *can be computed by the algorithm* ZEROPERDG *in* $O(n \log n + m\alpha(n))$ *time, where* n *and* m *are the number of vertices and edges in* K, *respectively.*

Connection to Minimum Spanning Tree

If we view the 1-skeleton of K as a graph $G = (V, E)$, then ZEROPERDG(K, f) essentially computes the minimum spanning forest of G with the following edge weights: for every edge $e = (u, v)$, we set its weight $w(e) = \max\{f(u), f(v)\}$. Then, we can get the persistence pairs output of ZEROPERDG by running the well-known Kruskal's algorithm on the weighted graph G. When we come across an edge $e = (u, v)$ that joins two disjoint components in this algorithm, we determine the two minimum vertices ℓ_1, ℓ_2 in these two components and pair e with the one among ℓ_1, ℓ_2 that has the larger f-value. After generating all such vertex–edge pairs (u, e), we convert them to vertex–vertex pairs (u, v) where $e \in \mathrm{Lst}(v)$. We throw away any pair of the form (u, u) because they signify local pairs.

Graph Filtration

The algorithm ZEROPERDG can be easily adapted to compute persistence for a given filtration of a graph. In this case, we process the vertices and edges in their order in the filtration and maintain connected components using the union–find data structure as in ZEROPERDG. For each edge $e = (u, v)$, we check if it connects two disconnected components represented by vertices ℓ_1 and ℓ_2 (line 11) and if so, e is paired with the younger vertex between ℓ_1 and ℓ_2 (line 13). We output all vertex–edge pairs thus computed. The vertices and edges that remain unpaired provide the infinite bars in the zeroth and first persistence diagrams. The algorithm runs in $O(n\alpha(n))$ time if the graph has n vertices and edges in total. The $O(n \log n)$ term in the complexity is eliminated because sorting of the vertices is implicitly given by the input filtration.

3.6 Notes and Exercises

The concept of topological persistence came to the fore in early 2000 with the paper by Edelsbrunner, Letscher, and Zomorodian [152] though the concept was proposed in a rudimentary form (e.g., zero-dimensional homology) in other papers by Frosini [162] and Robins [266]. The persistence algorithm as described in this chapter was presented in [152] which has become the cornerstone of topological data analysis. The original algorithm was described without any matrix reduction, which first appeared in [105]. Since then various versions of the algorithm have been presented. We have already seen that persistence for filtrations of simplicial 1-complexes (graphs) with n simplices can be computed in $O(n\alpha(n))$ time. Persistence for filtrations of simplicial 2-manifolds also can be computed in an $O(n\alpha(n))$ time algorithm by essentially reducing the problem to computing persistence on a dual graph. In general, for any constant $d \geq 1$, the persistence pairs between d- and $(d-1)$-simplices of a simplicial d-manifold can be computed in $O(n\alpha(n))$ time by considering the dual graph. If the manifold has a boundary, then one has to consider a "dummy" vertex that connects to every dual vertex of a d-simplex adjoining a *boundary* $(d-1)$-simplex.

For efficient implementation, clearing and compression strategies as described in Section 3.3.2 were presented by Chen and Kerber [94]. We have given a proof based on matrix reduction that the same persistence pairs can be computed by considering the anti-transpose of the boundary matrix. This is termed the *cohomology algorithm* first introduced in [114]. The name is justified by the fact that, considering cohomology groups and the resulting persistence module that reverses the arrows (Fact 2.14), we obtain the same barcode. The anti-transpose of the boundary matrix indeed represents the coboundary matrix filtered reversely. These tricks are further used by Bauer for processing Rips filtration efficiently in the Ripser software [19]; see also [304]. Boissonnat et al. [41, 42] have suggested a technique to reduce the size of a given filtration using the strong collapse of Barmak and Minian [17]. The collapse on the complex can be efficiently achieved through only simple manipulations of the boundary matrix.

The concept of bottleneck distance for persistence diagrams was first proposed by Cohen-Steiner et al. [101] who also showed the stability of such diagrams in terms of bottleneck distances with respect to the infinity norm of the difference between functions generating them. This result was extended to Wasserstein distance, though in a weaker form in [103] which got improved recently [278]. The more general concept of interleaving distance between persistence modules and the stability of persistence diagrams with respect to them was presented by Chazal et al. [77]. The fact that the bottleneck distance between persistence diagrams is not only bounded from above by the interleaving distance but is indeed equal to it was shown by Lesnick [221] which was further studied by Bauer and Lesnick [23] later. Also, see [54] for more generalization at algebraic level.

The use of the reduced Betti numbers for the lower link of a vertex to quantify its criticality was originally introduced in [150] for a PL-function defined on a triangulation of a d-manifold. Our PL-criticality considers both the lower link and upper link for more general simplicial complexes. As far as we know, the relations between such PL-critical points and homology groups of sublevel sets for the PL-setting have not been stated explicitly elsewhere in the literature. The concept of homological critical values was first introduced in [101], and the more general concept of "levelset critical values" (and levelset tame functions) was originally introduced in [63].

The idea of using union–find data structure to compute the zeroth persistent homology group was already introduced in the original persistence algorithm paper [152]. In this chapter, we present a modification for the PL-function setting.

Exercises

1. Let K be a p-complex with every $(p-1)$-simplex incident to exactly two p-simplices. Let M be a boundary matrix of the boundary operator ∂_p for K. We run a different version of the persistence algorithm on M. We scan its columns from left to right as before, but we add the current column to its right to resolve conflict, that is, for each $i = 1, \ldots, n$ in this order if there exists $j > i$ so that $\text{low}_M[i] = \text{low}_M[j]$, then add $\text{col}_M[i]$ to $\text{col}_M[j]$. Show the following.
   (a) There can be at most one such j.
   (b) At termination, every column of M is either empty or has a unique low entry.
   (c) The algorithm outputs in $O(n^2)$ time the same $\text{low}_M[i]$ as the original persistence algorithm returns on M.
2. For a given matrix with binary entries, a valid column operation is one that adds a column to its right ($\mathbb{Z}_2$-addition). Similarly, define a valid row

operation as the one that adds a row to another one above it. Show that there exists a set of valid column and row operations that leave every row and column either empty or with a single nonzero entry.

3. Let D be a boundary matrix of a simplicial complex K. Modify the algorithm MATPERSISTENCE to compute a set of p-cycles, $p \geq 0$, whose classes form a basis of $H_p(K)$ (Hint: Consider interpreting the role of the matrix V in the decomposition $R = DV$ of the reduced matrix R.)

4. Prove Theorem 3.1.

5. Give a polynomial-time algorithm for computing $d_{W,q}$.

6. Prove Proposition 3.7.

7. Let m_q and β_q be the number of q-simplices and qth Betti number of a simplicial complex of dimension p. Using pairing in persistence, show that

$$m_p - m_{p-1} + \cdots + \pm m_0 = \beta_p - \beta_{p-1} + \cdots \pm \beta_0.$$

8. Let $\mathcal{F}$ be a filtration where every p-simplex appears only after all $(p-1)$-simplices like in Eq. (3.8). Let $\mathcal{F}'$ be a modified filtration of $\mathcal{F}$ as follows. For every $p \geq 0$, all p-simplices in $\mathcal{F}$ are ordered in nondecreasing order of their persistence values in $\mathcal{F}'$ assuming that unpaired p-simplices have persistence value ∞. Show that the persistence pairing remains the same for $\mathcal{F}$ and $\mathcal{F}'$.

9. Let $\mathcal{F}$ be a simplex-wise filtration $\mathcal{F}$ of complex K induced by the sequence of simplices: $\sigma_1, \ldots, \sigma_N$. Let $\mathcal{F}'$ be a modification of $\mathcal{F}$ where only two consecutive simplices σ_k and σ_{k+1} swap their order, that is, $\mathcal{F}'$ is induced by the sequence

$$\sigma_1, \ldots, \sigma_{k-1}, \sigma_{k+1}, \sigma_k, \sigma_{k+2}, \ldots, \sigma_N.$$

Describe the relation between their corresponding persistence diagrams $\mathrm{Dgm}(\mathcal{F})$ and $\mathrm{Dgm}(\mathcal{F}')$.

10. Give an example of a piecewise-linear function $f: |K| \to \mathbb{R}$ where a vertex v_i is a PL-critical point, but $H_*(K_{i-1}) \cong H_*(K_i)$ as induced by inclusion.

11. Let $f: V(K) \to \mathbb{R}$ be a vertex function defined on the vertex set $V(K)$ of complex K. Consider $g = h \circ f + a$, where $h: \mathbb{R} \to \mathbb{R}$ is a monotone function and $a \in \mathbb{R}$ is a real value. Consider the lower-star filtrations $\mathcal{F}_f$ and $\mathcal{F}_g$ induced by induced PL-functions $f, g: |K| \to \mathbb{R}$ as in Eq. (3.10). Describe the relation between their corresponding persistence diagrams $\mathrm{Dgm}(\mathcal{F}_f)$ and $\mathrm{Dgm}(\mathcal{F}_g)$.

12. Consider two PL-functions $f, g: |K| \to \mathbb{R}$ on K induced by vertex functions $f, g: V(K) \to \mathbb{R}$, respectively. Suppose $\|f - g\|_\infty = \delta$, where

$\|f - g\|_\infty = \max_{v \in V} |f(v) - g(v)|$. Consider the persistence modules $\mathbb{P}_f$ and $\mathbb{P}_g$ induced by the lower-star filtration for f and g respectively.

(a) Show that $\mathsf{d}_I(\mathbb{P}_f, \mathbb{P}_g) \leq \delta$.

(b) Give an example of K, f, and g so that $\mathsf{d}_I(\mathbb{P}_f, \mathbb{P}_g) < \delta$.

13. For a PL-function $f : |K| \to \mathbb{R}$, we know how to produce a simplex-wise filtration $\mathcal{F}$ so that the barcode for f can be read from the barcode for $\mathcal{F}$. Design an algorithm to do the reverse, that is, given a filtration $\mathcal{F}$ on a complex K, produce a filtration $\mathcal{G}$ of a simplicial complex K' so that $\mathcal{G}$ is indeed a simplex-wise filtration of a PL function $g : |K'| \to \mathbb{R}$ where bars for $\mathcal{F}$ can be obtained from those for $\mathcal{G}$. (Hint: Use barycentric subdivision of K.)

14. Prove Proposition 3.9.

15. Consider the two persistence modules $\mathbb{U}$ and $\mathbb{V}$ as shown below and a sequence of linear maps $f_i : U_i \to V_i$ so that all squares commute.

$$\mathbb{U}: \quad U_1 \longrightarrow U_2 \longrightarrow U_3 \longrightarrow \cdots \longrightarrow U_m$$
$$\quad \quad \downarrow f_1 \quad \quad \downarrow f_2 \quad \quad \downarrow f_3 \quad \quad \quad \quad \quad \downarrow f_m$$
$$\mathbb{V}: \quad V_1 \longrightarrow V_2 \longrightarrow V_3 \longrightarrow \cdots \longrightarrow V_m$$

Consider the sequences

$$\ker \mathcal{F}: \{\ker f_i \subseteq U_i \to \ker f_{i+1} \subseteq U_{i+1}\},$$

where the maps are induced from the module $\mathbb{U}$. Prove that $\ker \mathcal{F}$ is a persistence module. Show the same for the sequences

$$\operatorname{im} \mathcal{F}: \{\operatorname{im} f_i \subseteq V_i \to \operatorname{im} f_{i+1} \subseteq V_{i+1}\}$$

and

$$\operatorname{coker} \mathcal{F} : \{\operatorname{coker} f_i = V_i / \operatorname{im} f_i \to \operatorname{coker} f_{i+1} = V_{i+1} / \operatorname{im} f_{i+1}\}.$$

4

General Persistence

We have considered filtrations so far for defining persistence and its stability. In a filtration, the connecting maps between consecutive spaces or complexes are inclusions. Assuming a discrete subset of reals, $A: a_0 \leq a_1 \leq \cdots \leq a_n$, as an index set, we write a filtration as

$$\mathcal{F}: X_{a_0} \hookrightarrow X_{a_1} \hookrightarrow \cdots \hookrightarrow X_{a_n}.$$

A more generalized scenario occurs when the inclusions are replaced with continuous maps for space filtrations and simplicial maps for simplicial filtrations, that is, $x_{ij}: X_{a_i} \to X_{a_j}$. In that case, we call the sequence a space and a simplicial *tower*, respectively:

$$\mathcal{X}: X_{a_0} \xrightarrow{x_{01}} X_{a_1} \xrightarrow{x_{12}} \cdots \xrightarrow{x_{(n-1)n}} X_{a_n}. \tag{4.1}$$

Considering the homology group of each space (resp. complex) in the sequence, we obtain a sequence of vector spaces connected with linear maps, which we have seen before. Specifically, we obtain the following *tower* of vector spaces:

$$\mathsf{H}_p\mathcal{X}: \mathsf{H}_p(X_{a_0}) \xrightarrow{x_{01*}} \mathsf{H}_p(X_{a_1}) \xrightarrow{x_{12*}} \cdots \xrightarrow{x_{(n-1)n*}} \mathsf{H}_p(X_{a_n}).$$

In the above sequence, each linear map x_{ij*} is the homomorphism induced by the map x_{ij}. We have already seen that the persistent homology of such a sequence of vector spaces and linear maps are well defined. However, since the linear maps here are not induced by inclusions, the original persistence algorithm as described in the previous chapter does not work. In Section 4.2, we describe a new algorithm to compute the persistence diagram of simplicial towers. Next, we generalize a filtration by allowing the inclusion maps to be directed either way giving rise to what is called a *zigzag* filtration:

$$\mathcal{F}: X_{a_0} \leftrightarrow X_{a_1} \leftrightarrow \cdots \leftrightarrow X_{a_n}. \tag{4.2}$$

Here each bidirectional arrow "↔" is either a forward or a backward inclusion map. In Section 4.3, we present an algorithm to compute the persistence of a zigzag filtration. A juxtaposition of a zigzag filtration with a tower provides a further generalization referred to as a *zigzag tower*. Section 4.4 presents an approach for computing the persistence of such a tower.

Before presenting the algorithms, we generalize the notion of stability for towers. We have seen such a notion in Section 3.4 for persistence modules arising out of filtrations. Here, we adapt it to a tower.

4.1 Stability of Towers

Just like in the previous chapter, we define the stability with respect to the perturbation of the towers themselves, forgetting the functions that generate them. This requires a definition of the distance between towers at simplicial (space) levels and homology levels.

It turns out that it is convenient and sometimes appropriate if the objects (spaces, simplicial complexes, or vector spaces) in a tower are indexed over the positive real axis instead of a discrete subset of it. This, in turn, requires one to spell out the connecting map between every pair of objects.

Definition 4.1. (Tower) A *tower* indexed in an ordered set $A \subseteq \mathbb{R}$ is any collection $\mathsf{T} = \left\{ T_a \right\}_{a \in A}$ of objects T_a, $a \in A$, together with maps $t_{a,a'} : T_a \to T_{a'}$ so that $t_{a,a} = \mathrm{id}$ and $t_{a',a''} \circ t_{a,a'} = t_{a,a''}$ for all $a \leq a' \leq a''$. Sometimes we write $\mathsf{T} = \left\{ T_a \xrightarrow{t_{a,a'}} T_{a'} \right\}_{a \leq a'}$ to denote the collection with the maps. We say that the tower T has *resolution r* if $a \geq r$ for every $a \in A$.

When T is a collection of topological spaces connected with continuous maps, we call it a *space tower*. When it is a collection of simplicial complexes connected with simplicial maps, we call it a *simplicial tower*, and when it is a collection of *vector spaces* connected with linear maps, we call it a *vector space tower*.

Remark 4.1. *As we have already seen, in practice, it may happen that a tower needs to be defined over a discrete set or more generally an index set A that is only a subposet of $\mathbb{R}$. In such a case, one can "embed" A into $\mathbb{R}$ and convert the input to a tower according to Definition 4.1 by assuming that for any $a < a' \in A$ with $(a, a') \notin A$ and for any $a \leq b < b' < a'$, $t_{b,b'}$ is an isomorphism.*

Definition 4.2. (Interleaving of simplicial (space) towers) Let $\mathcal{X} = \left\{ X_a \xrightarrow{x_{a,a'}} X_{a'} \right\}_{a \leq a'}$ and $\mathcal{Y} = \left\{ Y_a \xrightarrow{y_{a,a'}} Y_{a'} \right\}_{a \leq a'}$ be two towers of simplicial complexes (resp. spaces) indexed in $\mathbb{R}$. For any real $\varepsilon \geq 0$, we say that they are ε-*interleaved* if for every a one can find simplicial maps (resp. continuous maps) $\varphi_a \colon X_a \to Y_{a+\varepsilon}$ and $\psi_a \colon Y_a \to X_{a+\varepsilon}$ so that:

(i) for all $a \in \mathbb{R}$, $\psi_{a+\varepsilon} \circ \varphi_a$ and $x_{a,a+2\varepsilon}$ are contiguous (resp. homotopic), and

(ii) for all $a \in \mathbb{R}$, $\varphi_{a+\varepsilon} \circ \psi_a$ and $y_{a,a+2\varepsilon}$ are contiguous (resp. homotopic),

(iii) for all $a' \geq a$, $\varphi_{a'} \circ x_{a,a'}$ and $y_{a+\varepsilon,a'+\varepsilon} \circ \varphi_a$ are contiguous (resp. homotopic), and

(iv) for all $a' \geq a$, $x_{a+\varepsilon,a'+\varepsilon} \circ \psi_a$ and $\psi_{a'} \circ y_{a,a'}$ are contiguous (resp. homotopic).

If no such finite ε exists, we say the two towers are ∞-*interleaved*.

These four conditions are summarized by requiring that the four diagrams below commute up to contiguity (resp. homotopy):

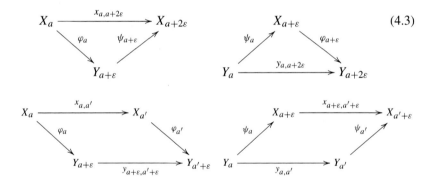

(4.3)

If we replace the operator "$+$" by the multiplication "$\cdot$" with respect to the indices in the above definition, then we say that $\mathcal{X}$ and $\mathcal{Y}$ are *multiplicatively ε-interleaved*. By interleaving we will mean additive interleaving by default and use the term multiplicative interleaving where necessary to signify that the shift is multiplicative rather than additive.

Definition 4.3. (Interleaving distance between simplicial (space) towers) The *interleaving distance* between two simplicial (space) towers $\mathcal{X}$ and $\mathcal{Y}$ is

$$\mathsf{d}_I(\mathcal{X}, \mathcal{Y}) = \inf_{\varepsilon}\{\mathcal{X} \text{ and } \mathcal{Y} \text{ are } \varepsilon\text{-interleaved}\}.$$

Similar to the simplicial (space) towers, we can define interleaving of vector space towers. But, in that case, we replace contiguity (homotopy) with equality in the four conditions in Definition 4.2.

Definition 4.4. (Interleaving of vector space towers) Let $\mathbb{U} = \left\{ U_a \stackrel{u_{a,a'}}{\longrightarrow} U_{a'} \right\}_{a \leq a'}$ and $\mathbb{V} = \left\{ V_a \stackrel{v_{a,a'}}{\longrightarrow} V_{a'} \right\}_{a \leq a'}$ be two vector space towers indexed in $\mathbb{R}$. For any real $\varepsilon \geq 0$, we say that they are ε-*interleaved* if for each $a \in \mathbb{R}$ one can find linear maps $\varphi_a : U_a \to V_{a+\varepsilon}$ and $\psi_a : V_a \to U_{a+\varepsilon}$ so that:

- for all $a \in \mathbb{R}$, $\psi_{a+\varepsilon} \circ \varphi_a = u_{a,a+2\varepsilon}$,
- for all $a \in \mathbb{R}$, $\varphi_{a+\varepsilon} \circ \psi_a = v_{a,a+2\varepsilon}$.
- for all $a' \geq a$, $\varphi_{a'} \circ u_{a,a'} = v_{a+\varepsilon,a'+\varepsilon} \circ \varphi_a$,
- for all $a' \geq a$, $u_{a+\varepsilon,a'+\varepsilon} \circ \psi_a = \psi_{a'} \circ v_{a,a'}$.

If no such finite ε exists, we say the two towers are ∞-*interleaved*.

Analogous to the simplicial (space) towers, if we replace the operator "+" by the multiplication "·" in the above definition, then we say that $\mathbb{U}$ and $\mathbb{V}$ are *multiplicatively ε-interleaved*.

Definition 4.5. (Interleaving distance between vector space towers) The *interleaving distance* between two towers of vector spaces $\mathbb{U}$ and $\mathbb{V}$ is

$$\mathsf{d}_I(\mathbb{U}, \mathbb{V}) = \inf_{\varepsilon}\{\mathbb{U} \text{ and } \mathbb{V} \text{ are } \varepsilon\text{-interleaved}\}.$$

Suppose that we have two simplicial (space) towers $\mathfrak{X} = \left\{ X_a \stackrel{x_{a,a'}}{\to} X_{a'} \right\}$ and $\mathfrak{Y} = \left\{ Y_a \stackrel{y_{a,a'}}{\to} Y_{a'} \right\}$. Consider the two vector space towers, also called homology towers, obtained by taking the homology groups of the complexes (spaces), that is,

$$\mathbb{V}_X = \{\mathsf{H}_p(X_a) \stackrel{x_{(a,a')*}}{\to} \mathsf{H}_p(X_{a'})\} \quad \text{and} \quad \mathbb{V}_Y = \{\mathsf{H}_p(Y_a) \stackrel{y_{(a,a')*}}{\to} \mathsf{H}_p(Y_{a'})\}.$$

The following should be obvious because simplicial (resp. continuous) maps become linear maps and contiguous (resp. homotopic) maps become equal at the homology level.

Proposition 4.1. *One has* $\mathsf{d}_I(\mathbb{V}_X, \mathbb{V}_Y) \leq \mathsf{d}_I(\mathfrak{X}, \mathfrak{Y})$.

One can recognize that the vector space tower is a persistence module defined in Section 3.4. Therefore, we can use Definition 3.21 to define the persistence diagram Dgm $\mathbb{V}$ of the tower $\mathbb{V}$. Recall that d_b denotes the bottleneck distance between persistence diagrams. The *isometry theorem* as stated in Theorem 3.11 also holds for towers that are *q-tame* (or simply *tame*), that is, towers with all linear maps having finite rank.

Theorem 4.2. *For any two tame vector space towers* $\mathbb{U}$ *and* $\mathbb{V}$, *we have* $\mathsf{d}_b(\mathrm{Dgm}(\mathbb{U}), \mathrm{Dgm}(\mathbb{V})) = \mathsf{d}_I(\mathbb{U}, \mathbb{V})$.

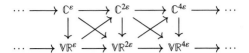

Figure 4.1 Čech and Rips complexes interleave multiplicatively.

Combining Proposition 4.1 and Theorem 4.2, we obtain the following result.

Theorem 4.3. *Let* $\mathcal{X}$ *and* $\mathcal{Y}$ *be two simplicial (space) towers and* $\mathbb{V}_X$ *and* $\mathbb{V}_Y$ *be their homology towers, respectively, that are tame. Then,* $d_b(\mathrm{Dgm}(\mathbb{V}_X), \mathrm{Dgm}(\mathbb{V}_Y)) \leq d_I(\mathcal{X}, \mathcal{Y})$.

We want to apply the above result to translate the multiplicative interleaving distances into a bottleneck distance of the persistence diagrams. For that we need to consider log scale. Given a persistence diagram Dgm for a tower with a positive resolution, we denote its *log-scaled version* $\mathrm{Dgm}_{\log}$ to be the diagram consisting of the set of nondiagonal points $\{(\log_2 x, \log_2 y) \mid (x, y) \in \mathrm{Dgm}\}$ along with the usual diagonal points. In log scale, a multiplicative interleaving turns into an additive interleaving by which the following corollary is deduced immediately from Theorem 4.3.

Corollary 4.4. *Let* $\mathcal{X}$ *and* $\mathcal{Y}$ *be two simplicial (space) towers with a positive resolution that are multiplicatively c-interleaved and* $\mathbb{V}_X$ *and* $\mathbb{V}_Y$ *be their homology towers, respectively, that are tame. Then,*

$$d_b(\mathrm{Dgm}_{\log}(\mathbb{V}_X), \mathrm{Dgm}_{\log}(\mathbb{V}_Y)) \leq \log c.$$

Interleaving Between Čech and Rips Filtrations
We show an example where we can use the stability result in Corollary 4.4. Let $P \subseteq M$ be a finite subset of a metric space (M, d). Consider the Rips and Čech filtrations:

$$\mathcal{R} \colon \{\mathbb{VR}^{\varepsilon}(P) \hookrightarrow \mathbb{VR}^{\varepsilon'}(P)\}_{0 < \varepsilon \leq \varepsilon'} \quad \text{and} \quad \mathcal{C} \colon \{\mathbb{C}^{\varepsilon}(P) \hookrightarrow \mathbb{C}^{\varepsilon'}(P)\}_{0 < \varepsilon \leq \varepsilon'}.$$

From Proposition 2.2, we know that the following inclusions hold:

$$\cdots \subseteq \mathbb{C}^{\varepsilon}(P) \subseteq \mathbb{VR}^{\varepsilon}(P) \subseteq \mathbb{C}^{2\varepsilon}(P) \subseteq \mathbb{VR}^{2\varepsilon}(P) \subseteq \mathbb{C}^{4\varepsilon}(P) \subseteq \mathbb{VR}^{4\varepsilon}(P) \subseteq \cdots.$$

Figure 4.1 illustrates that the Čech and Rips complexes are multiplicatively 2-interleaved. Then, according to Corollary 4.4, the persistence diagrams $\mathrm{Dgm}_{\log}\mathcal{C}$ and $\mathrm{Dgm}_{\log}\mathcal{R}$ have bottleneck distance of $\log 2 = 1$.

4.2 Computing Persistence of Simplicial Towers

In this section, we present an algorithm for computing the persistence of a simplicial tower. Consider a simplicial tower $\mathcal{K}$: $K_0 \xrightarrow{f_0} K_1 \xrightarrow{f_1} K_2 \xrightarrow{f_2} \cdots \xrightarrow{f_{n-1}} K_n$ and the map $f_{ij}: K_i \to K_j$ where $f_{ij} = f_{j-1} \circ \cdots \circ f_{i+1} \circ f_i$. To compute the persistent homology for a simplicial filtration, the persistence algorithm in the previous chapter essentially maintains a consistent basis by computing the image $f_{ij_*}(B_i)$ of a basis B_i of $H_*(K_i)$. As the algorithm moves through an inclusion in the filtration, the homology basis elements get created (birth) or are destroyed (death). Here, for towers, instead of a consistent homology basis, we maintain a consistent cohomology basis. We need to be aware that, for cohomology, the induced maps from $f_{ij}: K_i \to K_j$ are reversed, that is, $f_{ij}^*: H^p(K_i) \leftarrow H^p(K_j)$; refer to Section 2.5.4. So, if B^i is a cohomology basis of $H^p(K_i)$ maintained by the algorithm, it computes implicitly the pre-image $f_{ij}^{*-1}(B^i)$. Dually, this implicitly maintains a consistent homology basis and thus captures all information about persistent homology as well.

4.2.1 Annotations

We maintain a consistent cohomology basis using a notion called annotations [60] which are binary vectors assigned to simplices. These annotations are updated as we go forward through the sequence in the given tower. This implicitly maintains a cohomology basis in the reverse direction where the birth and death of cohomology classes coincide with the death and birth, respectively of homology classes.

Definition 4.6. (Annotation) Given a simplicial complex K, let $K(p)$ denote the set of p-simplices in K. An *annotation* for $K(p)$ is an assignment $a : K(p) \to \mathbb{Z}_2^g$ of a binary vector $a_\sigma = a(\sigma)$ of length g to each p-simplex $\sigma \in K$. The binary vector a_σ is called the *annotation* for σ. Each entry "0" or "1" of a_σ is called its *element*. Annotations for simplices provide an annotation for every p-chain c_p: $a_{c_p} = \sum_{\sigma \in c_p} a_\sigma$.

An annotation $a : K(p) \to \mathbb{Z}_2^g$ is *valid* if the following two conditions are satisfied:

- $g = \operatorname{rank} H_p(K)$, and
- two p-cycles z_1 and z_2 have $a_{z_1} = a_{z_2}$ if and only if their homology classes are identical, that is, $[z_1] = [z_2]$.

Proposition 4.5. *The following two statements are equivalent.*

(a) An annotation $\mathsf{a} : K(p) \to \mathbb{Z}_2^g$ *is valid.*

(b) The cochains $\{\phi_i\}_{i=1,\ldots,g}$ *given by* $\phi_i(\sigma) = \mathsf{a}_\sigma[i]$ *for every* $\sigma \in K(p)$
are cocycles whose cohomology classes $\{[\phi_i]\}, i = 1, \ldots, g,$ *constitute a*
basis of $\mathsf{H}^p(K)$.

In light of the above result, an annotation is simply one way to represent
a cohomology basis. However, by representing the corresponding basis as an
explicit vector associated with each simplex, it localizes the basis to each simplex. As a result, we can update the cohomology basis locally by changing
the annotations locally (see Proposition 4.8). This point of view also helps to
reveal how we can process elementary collapses, which are neither inclusions
nor deletions, by transferring annotations (see Proposition 4.9).

4.2.2 Algorithm

Consider the persistence module $\mathsf{H}_p\mathcal{K}$ induced by a simplicial tower
$\mathcal{K} : \{K_i \xrightarrow{f_i} K_{i+1}\}$ where every f_i is a so-called elementary simplicial map
which we will introduce shortly:

$$\mathsf{H}_p\mathcal{K} : \ \mathsf{H}_p(K_0) \xrightarrow{f_{0*}} \mathsf{H}_p(K_1) \xrightarrow{f_{1*}} \mathsf{H}_p(K_2) \xrightarrow{f_{2*}} \cdots \xrightarrow{f_{n-1*}} \mathsf{H}_p(K_n).$$

Instead of tracking a consistent homology basis for the module $\mathsf{H}_p\mathcal{K}$, we track
a cohomology basis in the module $\mathsf{H}^p\mathcal{K}$ where the homomorphisms are in
reverse direction:

$$\mathsf{H}^p\mathcal{K} : \ \mathsf{H}^p(K_0) \xleftarrow{f_0^*} \mathsf{H}^p(K_1) \xleftarrow{f_1^*} \mathsf{H}^p(K_2) \xleftarrow{f_2^*} \cdots \xleftarrow{f_{n-1}^*} \mathsf{H}^p(K_n).$$

As we move from left to right in the above sequence, the annotations *implicitly*
maintain a cohomology basis whose elements are also *timestamped* to signify when a basis element is born or dies. We keep in mind that the *birth* and
death of a cohomology basis element coincides with the *death* and *birth* of a
homology basis element because the two modules run in opposite directions.

To jump start the algorithm, we need annotations for simplices in K_0 at
the beginning whose nonzero elements are timestamped with 0. This can be
achieved by considering an arbitrary filtration of K_0 and then applying the
generic algorithm as we describe for inclusions in Section 4.2.3. The first
vertex in this filtration gets the annotation of [1].

Before describing the algorithm, we observe a simple fact that simplicial
maps can be decomposed into elementary maps which let us design simpler
atomic steps for the algorithm.

Definition 4.7. (Elementary simplicial maps) A simplicial map $f : K \to K'$ is called *elementary* if it is of one of the following two types:

- Simplicial map f is injective, and K' has at most one more simplex than K. In this case, f is called an *elementary inclusion*.
- Simplicial map f is not injective but is surjective, and the vertex map f_V is injective everywhere except on a pair $\{u, v\} \subseteq V(K)$. In this case, f is called an *elementary collapse*. An elementary collapse maps a pair of vertices into a single vertex, and is injective on every other vertex.

We observe that any simplicial map is a composition of elementary simplicial maps.

Proposition 4.6. *If $f : K \to K'$ is a simplicial map, then there are elementary simplicial maps f_i where*

$$K = K_0 \xrightarrow{f_0} K_1 \xrightarrow{f_1} K_2 \xrightarrow{f_2} \cdots \xrightarrow{f_{n-2}} K_{n-1} \xrightarrow{f_{n-1}} K_n = K' \quad \text{so that} \quad f = f_{n-1} \circ f_{n-2} \circ \cdots \circ f_0.$$

In view of Proposition 4.6, it is sufficient to show how one can design the persistence algorithm for an elementary simplicial map. At this point, we make a change in Definition 4.7 of elementary simplicial maps that eases further discussions. We let f_V be the identity (which is an injective map) everywhere except possibly on a pair of vertices $\{u, v\} \subseteq V(K)$ for which f_V maps to one of these two vertices, say u, in K'. This change can be implemented by renaming the vertices in K' that are mapped onto injectively.

4.2.3 Elementary Inclusion

Consider an elementary inclusion $K_i \hookrightarrow K_{i+1}$. Assume that K_i has a valid annotation. We describe how we obtain a valid annotation for K_{i+1} from that of K_i after inserting the p-simplex $\sigma = K_{i+1} \setminus K_i$. We compute the annotation $\mathsf{a}_{\partial\sigma}$ for the boundary $\partial\sigma$ in K_i and take actions as follows which ultimately lead to computing the persistence diagram.

Case (i): If $\mathsf{a}_{\partial\sigma}$ is a zero vector, the class $[\partial\sigma]$ is trivial in $\mathsf{H}_{p-1}(K_i)$. This means that σ creates a p-cycle in K_{i+1} and by duality a p-cocycle is killed while going left from K_{i+1} to K_i. In this case we augment the annotations for all p-simplices by one element with a timestamp $i + 1$, that is, the annotation $[b_1, b_2, \ldots, b_g]$ for every p-simplex τ is updated to $[b_1, b_2, \ldots, b_g, b_{g+1}]$ with b_{g+1} being timestamped $i + 1$. We set $b_{g+1} = 0$ for $\tau \neq \sigma$ and $b_{g+1} = 1$ for $\tau = \sigma$. The element b_i of a_σ is set to zero for $1 \leq i \leq g$. Other annotations for other simplices remain unchanged. See Figure 4.2(a).

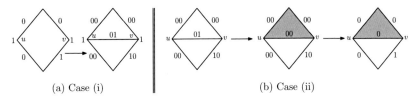

(a) Case (i)   (b) Case (ii)

Figure 4.2 Case (i) of inclusion: the boundary $\partial uv = u + v$ of the edge uv has annotation $1 + 1 = 0$. After its addition, every edge gains an element in its annotation which is 0 for all except the edge uv. Case (ii) of inclusion: the boundary of the top triangle has annotation 01. It is added to the annotation of uv which is the only edge having the second element 1. Consequently the second element is zeroed out for every edge, and is then deleted.

Case (ii): If $a_{\partial\sigma}$ is not a zero vector, the class of the $(p - 1)$-cycle $\partial\sigma$ is nontrivial in $H_{p-1}(K_i)$. Therefore, σ kills the class of this $(p - 1)$-cycle and a corresponding class of $(p - 1)$-cocycles is born in the reverse direction. We simulate it by forcing $a_{\partial\sigma}$ to be zero which affects other annotations as well. Let $i_1 < i_2 < \cdots < i_k$ be the set of indices in nondecreasing order so that $b_{i_1}, b_{i_2}, \ldots, b_{i_k}$ are all of the nonzero elements in $a_{\partial\sigma} = [b_1, b_2, \ldots, b_{i_k}, \ldots, b_g]$. Recall that ϕ_j denotes the $(p - 1)$-cocycle given by its evaluation $\phi_j(\sigma') = a_{\sigma'}[j]$ for every $(p - 1)$-simplex $\sigma' \in K_i$ (Proposition 4.5). With this notation, the cocycle $\phi = \phi_{i_1} + \phi_{i_2} + \cdots + \phi_{i_k}$ is born after deleting σ in the reverse direction. This cocycle does not exist after time i_k in the reverse direction. In other words, the cohomology class $[\phi]$ which is born leaving the time $i + 1$ is killed at time i_k. This pairing matches that of the standard persistence algorithm where the youngest basis element is chosen to be paired among all those whose combination is killed. We add the vector $a_{\partial\sigma}$ to the annotation of every $(p-1)$-simplex whose i_k-th element is nonzero. This zeroes out the i_k-th element of the annotation for every $(p - 1)$-simplex and at the same time updates other elements so that a valid annotation according to Proposition 4.5 is maintained. We simply delete the i_k-th element from the annotation for every $(p - 1)$-simplex. See Figure 4.2(b). We further set the annotation a_σ for σ to be a zero vector of length s, where s is the length of the annotation vector of every p-simplex at this point.

Notice that determining if we have case (i) or (ii) can be done easily in $O(pg)$ time by checking the annotation of $\partial\sigma$. Indeed, this is achieved because the annotation already localizes the cohomology basis to each individual simplex.

Before going to the next case of elementary collapse, here we present Algorithm 6: ANNOT for computing the annotations for all simplices in a given simplicial complex using the steps of elementary inclusions. The algorithm proceeds in the order of increasing dimension because it needs to have the annotations of $(p-1)$-simplices before dealing with p-simplices. It starts with vertices whose annotations are readily computable. In the following algorithm K^p denotes the p-skeleton of the input simplicial d-complex K.

Algorithm 6 ANNOT(K)

Input:
  K: input complex
Output:
  Annotation for every simplex in K

1: Let $m := |K^0|$
2: For every vertex $v_i \in K^0$, assign an m-vector $\mathsf{a}(v_i)$ where $\mathsf{a}(v_i)[j] = 1$ if and only if $j = i$
3: **for** $p = 1 \to d$ **do**
4:   **for all** simplex $\sigma \in K^p$ **do**
5:     Let annotation of every p-simplex be a vector of length g so far
6:     **if** $\mathsf{a}(\partial\sigma) \neq 0$ **then**
7:       Assign $\mathsf{a}(\sigma)$ to be a zero vector of size g
8:       Pick any nonzero entry b_u in $\mathsf{a}(\partial\sigma)$
9:       Add $\mathsf{a}(\partial\sigma)$ to every $(p-1)$-simplex σ' so that $\mathsf{a}(\sigma')[u] = 1$
10:       Delete the u-th entry from annotation of every $(p-1)$-simplex
11:     **else**
12:       Extend $\mathsf{a}(\tau)$ for every p-simplex τ so far added by appending a 0 bit
13:       Create vector $\mathsf{a}(\sigma)$ of length $g + 1$ with only the last bit being 1
14:     **end if**
15:   **end for**
16: **end for**

4.2.4 Elementary Collapse

The case for handling collapse is more interesting. It has three distinct steps: (i) elementary inclusions to satisfy the so-called link condition; (ii) local annotation transfer to prepare for the collapse; and (iii) collapse of the simplices with updated annotations. We explain each of these steps now.

The elementary inclusions that may precede the final collapse are motivated by a result that connects collapses with the change in cohomology. Consider an elementary collapse $K_i \xrightarrow{f_i} K_{i+1}$ where the vertex pair (u, v) collapses to

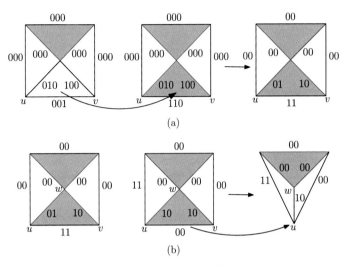

Figure 4.3 Annotation updates for elementary collapse: inclusion of a triangle so as to satisfy the link condition (a), annotation transfer and actual collapse (b); annotation 11 of the vanishing edge uv is added to all edges (cofacets) adjoining u.

u. The following link condition, introduced in [121] and later used to preserve homotopy [12], becomes relevant.

Definition 4.8. (Link condition) A vertex pair (u, v) in a simplicial complex K_i satisfies the *link condition* if the edge $uv \in K_i$ and $\operatorname{Lk} u \cap \operatorname{Lk} v = \operatorname{Lk} uv$. An elementary collapse $f_i \colon K_i \to K_{i+1}$ satisfies the link condition if the vertex pair on which f_i is not injective satisfies the link condition.

Proposition 4.7. [12] *If an elementary collapse $f_i \colon K_i \to K_{i+1}$ satisfies the link condition, then the underlying spaces $|K_i|$ and $|K_{i+1}|$ remain homotopy equivalent. Hence, the induced homomorphisms $f_{i*} \colon \mathsf{H}_p(K_i) \to \mathsf{H}_p(K_{i+1})$ and $f_i^* \colon \mathsf{H}^p(K_i) \leftarrow \mathsf{H}^p(K_{i+1})$ are isomorphisms.*

If an elementary collapse satisfies the link condition, we can perform the collapse knowing that the cohomology does not change. Otherwise, we know that the cohomology is affected by the collapse and it should be reflected in our updates for annotations. The diagram below provides a precise means to carry out the change in cohomology. Let S be the minimal set of simplices ordered in nondecreasing order of their dimensions whose addition to K_i makes (u, v) satisfy the link condition. One can describe a construction of S recursively as follows. In dimension one, if the edge (u, v) is missing, it is added to S.

Recursively assume that S has all of the necessary p-simplices. Then, all missing $(p + 1)$-simplices adjoining the edge (a, b) whose boundary is already present are added to S. For each simplex $\sigma \in S$, we modify the annotations of every simplex which we would have done if σ were to be inserted. Thereafter, we carry out the rest of the elementary collapse. In essence, implicitly, we obtain an intermediate complex $\hat{K}_i = K_i \cup S$ where the diagram below commutes. Here, f_i' is induced by the same vertex map that induces f_i, and j is an inclusion. This means that the persistence of f_i is identical to that of $f_i' \circ j$ which justifies our action of elementary inclusions followed by the actual collapses.

$$\begin{array}{ccc} K_i & \xrightarrow{\ f_i\ } & K_{i+1} \\ & \searrow{\scriptstyle j} & \uparrow{\scriptstyle f_i'} \\ & & \hat{K}_i \end{array}$$

We remark that this is the only place where we may insert implicitly a simplex σ in the current approach. The number of such σ is usually much smaller than the number of simplices that one may need for a *coning* strategy detailed in Section 4.4 to process simplicial towers.

After constructing $\hat{K}_i$ with annotations, we transfer annotations to prepare for the collapse. This step locally changes the annotations for simplices containing the vertices u and/or v. The following definition facilitates the description.

Definition 4.9. (Vanishing; Mirror simplices) For the elementary collapse $f_i' \colon \hat{K}_i \to K_{i+1}$, a simplex $\sigma \in K_i$ is called *vanishing* if the cardinality of $f_i'(\sigma)$ is one less than that of σ. Two simplices σ and σ' are called *mirror* partners if one contains u and the other v, and share the rest of the vertices. In Figure 4.3(b), the vanishing simplices are $\{uv, uvw\}$ and the mirror partners are $\{u, v\}$ and $\{uw, vw\}$.

In an elementary collapse that sends (u, v) to u, all vanishing simplices need to be deleted, and all simplices containing v need to be pulled to corresponding ones containing the vertex u (which are their mirror partners). We update the annotations in such a way that the annotations of all vanishing simplices become zero, and those of both mirror partners become the same. Once this is achieved, the collapse is implemented by simply deleting the vanishing simplices and replacing v with u in all simplices containing v (effectively this identifies mirror partners) without changing their annotations. The following proposition provides the justification behind the specific update operations that we perform.

Proposition 4.8. *Let K be a simplicial complex and* $\mathsf{a}\colon K(p) \to \mathbb{Z}_2^g$ *be a valid annotation. Let* $\sigma \in K(p)$ *be any p-simplex and τ any of its $(p-1)$-faces. Then, adding* a_σ *to the annotation for all cofacets of τ including σ produces a valid annotation for $K(p)$. Furthermore, the cohomology basis corresponding to the annotations (Proposition 4.5) remains unchanged by this modification.*

Consider now the elementary collapse $f_i'\colon \hat{K}_i \to K_{i+1}$ that sends (u, v) to u. We update the annotations for simplices in $\hat{K}_i$ as follows. First, note that the vanishing simplices are exactly those simplices containing the edge $\{u, v\}$. For every p-simplex containing $\{u, v\}$, that is, a vanishing simplex, exactly two of its $(p-1)$-faces are mirror simplices, and all other remaining $(p-1)$-faces are vanishing simplices. Let σ be a vanishing p-simplex and τ be its $(p-1)$-face that is a mirror simplex containing u. We add a_σ to the annotations for all cofacets (cofaces of codimension one) of τ including σ. This implements the *annotation transfer* for σ. By Proposition 4.8, the new annotation generated by this process corresponds to the old cohomology basis for $\hat{K}_i$. This new annotation has a_σ as zero since $\mathsf{a}_\sigma + \mathsf{a}_\sigma = 0$. See Figure 4.3(b). We perform the above operation for each vanishing simplex. It turns out that by using the relations of vanishing simplices and mirror simplices, each mirror simplex eventually acquires an identical annotation to that of its partner. Specifically, we have the following observation.

Proposition 4.9. *After all possible annotation transfers involved in a collapse, (i) each vanishing simplex has a zero annotation and (ii) each mirror simplex τ has the same annotation as its mirror partner simplex τ'.*

Subsequent to the annotation transfer, the annotation of $\hat{K}_i$ fits for actual collapse since each pair of mirror simplices which are collapsed to a single simplex get the identical annotation and the vanishing simplex acquires the zero annotation. Furthermore, Proposition 4.8 tells us that the cohomology basis does not change by annotation transfer which aligns with the fact that $f_i'^*\colon \mathsf{H}^p(K_{i+1}) \to \mathsf{H}^p(\hat{K}_i)$ is indeed an isomorphism. Accordingly, no timestamp changes after the annotation transfer and the actual collapse. Propositions 5.2 and 5.3 in [122] provide formal statements justifying the algorithm for annotation updates.

The persistence diagram of a given simplicial tower $\mathcal{K}$ can be retrieved easily from the annotation algorithm. Each time during an elementary operation either we add a new element into the annotation of all p-simplices for some $p \geq 0$ or we delete an element from the annotations of all of them. During the deletion, we add the point (bar) (a, b) into $\mathrm{Dgm}_p\mathcal{K}$ where b is the current time of deletion (death) and a is the timestamp of the element when it was added (birth).

4.3 Persistence for Zigzag Filtration

Now we consider another generalization of filtration where all inclusions are not necessarily in the forward direction. The possibility of backward inclusions allows simplices to be deleted as we move forward. So, essentially, we allow both insertions and deletions, making it possible for the complex to grow and shrink as we move forward with the filtration. It is not obvious *a priori* that the resulting persistence module admits barcodes as in the original filtration where all inclusions are in the forward direction. The existence of such barcodes is essential for defining persistence pairs and designing an algorithm to compute them. We are assured by quiver theory [163] that such barcodes also exist for zigzag filtration with both forward and backward insertions. We aim to compute them.

Specifically, a *zigzag filtration* $\mathcal{F}$ of a complex K (space $\mathbb{T}$) is a zigzag diagram of the form

$$\mathcal{F}\colon X_0 \leftrightarrow X_1 \leftrightarrow \cdots \leftrightarrow X_{n-1} \leftrightarrow X_n, \tag{4.4}$$

where for each i, $X_i = K_i \subseteq K$ for a simplicial filtration and $X_i = \mathbb{T}_i \subseteq \mathbb{T}$ for a space filtration, and $X_i \leftrightarrow X_{i+1}$ is either a *forward* inclusion $X_i \hookrightarrow X_{i+1}$ or a *backward* inclusion $X_i \hookleftarrow X_{i+1}$. Figure 4.4 illustrates a simplicial zigzag filtration and its barcode. Observe that reverse arrows can be interpreted

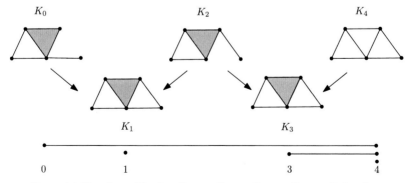

Figure 4.4 The zigzag filtration $K_0 \hookrightarrow K_1 \hookleftarrow K_2 \hookrightarrow K_3 \hookleftarrow K_4$ has four intervals (bars) for one-dimensional homology H_1, namely $[0, 4]$, $[1, 1]$, $[3, 4]$, and $[4, 4]$.

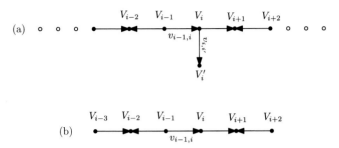

Figure 4.5 A representation of a quiver (a) and a representation of an A_n-type quiver (b).

as simplex deletions. For any $j \in [0, n]$, we let $\mathcal{F}^j$ denote the *prefix* of $\mathcal{F}$ consisting of the complexes (spaces) $X_0, \ldots, X_j$.

For $p \geq 0$, considering the p-th homology groups with coefficient in a field $\mathbf{k}$ (which is $\mathbb{Z}_2$ here), we obtain a sequence of vector spaces connected by forward or backward linear maps, called a *zigzag persistence module*:

$$\mathsf{H}_p\mathcal{F}: \mathsf{H}_p(X_0) \xleftrightarrow{\varphi_0} \mathsf{H}_p(X_1) \xleftrightarrow{\varphi_1} \cdots \xleftrightarrow{\varphi_{n-2}} \mathsf{H}_p(X_{n-1}) \xleftrightarrow{\varphi_{n-1}} \mathsf{H}_p(X_n), \quad (4.5)$$

where the map $\varphi_i : \mathsf{H}_p(X_i) \leftrightarrow \mathsf{H}_p(X_{i+1})$ can be either forward or backward and is induced by the inclusion.

In the non-zigzag case, when index set for $\mathsf{H}_p\mathcal{F}$ is finite, Proposition 3.10 says that $\mathsf{H}_p\mathcal{F}$ is a direct sum of interval modules. In the zigzag case, a similar statement holds due to quiver theory [163].

Definition 4.10. (Quiver) A *quiver* $Q = (N, E)$ is a directed graph which can be finite or infinite. A representation $\mathbb{V}(Q)$ of Q is an assignment of a vector space V_i to every node $N_i \in N$ and a linear map $v_{ij} : V_i \to V_j$ for every directed edge $(N_i, N_j) \in E$. Figure 4.5 illustrates representations of two quivers.

A zigzag persistence module is a special type of quiver representation where the graph is finite and linear shaped, also known as A_n-type (see Figure 4.5b), where every node has at most two directed edges incident to it. Such a quiver representation has an interval decomposition, though we need to define the intervals afresh to take into account the fact that arrows can be bidirectional.

Definition 4.11. (Interval module) An *interval module* $\mathfrak{I}_{[b,d]}$ also called an *interval* or a *bar* over an index set $0, 1, \ldots, n$ with field $\mathbf{k}$ is a sequence of vector spaces

$$\mathfrak{I}_{[b,d]}: I_0 \leftrightarrow I_1 \leftrightarrow \cdots \leftrightarrow I_n,$$

where $I_k = \mathbf{k}$ for $b \le k \le d$ and 0 otherwise, with the maps $\mathbf{k} \leftarrow \mathbf{k}$ and $\mathbf{k} \to \mathbf{k}$ being identities.

Remark 4.2. *Notice that unlike the bars that we defined in Chapter 3 for nonzigzag filtration, here the bars are closed on both ends. However, we will see that we can designate them to be of four types similar to what we have seen for the persistence modules for non-zigzag persistence.*

Theorem 4.10. [13, 163, 267] *Every quiver representation $\mathbb{V}(Q)$ for an A_n-type quiver Q has an interval decomposition, that is, $\mathbb{V}(Q) \cong \bigoplus_i \mathcal{I}_{[b_i, d_i]}$. Furthermore, this decomposition is unique up to isomorphism and permutation of the intervals.*

The underlying graph of a zigzag persistence module as shown in Eq. (4.5) is of A_n type. Hence, we have the decomposition $\mathsf{H}_p \mathcal{F} \cong \bigoplus_i \mathcal{I}_{[b_i, d_i]}$ that provides the barcode for zigzag persistence. Notice that Theorem 4.10 does not require the vector spaces to be finite-dimensional. Hence, we still have a valid decomposition even if the vector spaces in the zigzag persistence module are not finite-dimensional. However, for finite computation, we will assume that our zigzag persistence module is finite both in terms of the index set and also in terms of the dimension of the vector spaces.

Recall from Section 3.2.1 that each bar (interval) in a barcode (interval decomposition) corresponds to a point in the persistence diagram $\mathrm{Dgm}_p(\mathcal{F})$ and thus we also say that the bar belongs to the diagram. Sometimes, we also abuse the notation $[b, d]$ to denote both an interval in the index set and an interval module in a p-th zigzag persistence module.

Types of Bars
A bar $[b, d]$ for a zigzag persistence module $\mathsf{H}_p \mathcal{F}$ can be of four types depending on the direction of the arrow between X_{b-1} and X_b and the arrow between X_d and X_{d+1} in $\mathcal{F}$. They are defined as follows.

- *closed–closed* $[b, d]$: $X_{b-1} \hookrightarrow X_b \cdots X_d \hookleftarrow X_{d+1}$: either $b = 0$ or the inclusion $X_{b-1} \hookrightarrow X_b$ is a forward arrow; and $d < n$ with the inclusion $X_d \hookleftarrow X_{d+1}$ being a backward arrow.
- *closed–open* $[b, d]$: $X_{b-1} \hookrightarrow X_b \cdots X_d \hookrightarrow X_{d+1}$: either $b = 0$ or the inclusion $X_{b-1} \hookrightarrow X_b$ is a forward arrow; and either $d = n$ or the inclusion $X_d \hookrightarrow X_{d+1}$ is a forward arrow.
- *open–closed* $[b, d]$: $X_{b-1} \hookleftarrow X_b \cdots X_d \hookleftarrow X_{d+1}$: $b > 0$ and the inclusion $X_{b-1} \hookleftarrow X_b$ is a backward arrow; and $d < n$ with the inclusion $X_d \hookleftarrow X_{d+1}$ being a backward arrow.

- *open–open* $[b, d]$: $X_{b-1} \leftarrowtail X_b \cdots X_d \hookrightarrow X_{d+1}$: $b > 0$ and the inclusion $X_{b-1} \leftarrowtail X_b$ is a backward arrow; and either $d = n$ or the inclusion $X_d \hookrightarrow X_{d+1}$ is a forward arrow.

With the four types of bars, when we compute the bottleneck distance between persistence diagrams for two zigzag persistence modules, we consider matching between bars of similar types. That is, $\mathsf{d}_b(\mathrm{Dgm}_p(\mathcal{F}_1), \mathrm{Dgm}_p(\mathcal{F}_2))$ is computed with the understanding that only similar types of bars are compared while matching the bars and the points on the diagonal are assumed to have any type. We face a difficulty in defining an interleaving distance between zigzag modules because of the zigzag nature of the arrows. However, one can define such an interleaving distance by mapping the module to a two-parameter persistence module. See the notes in Chapter 12 for more details.

4.3.1 Approach

We briefly describe an overview of our approach for computing zigzag persistent intervals for a simplicial zigzag filtration:

$$\mathcal{F}: \varnothing = K_0 \leftrightarrow K_1 \leftrightarrow \cdots \leftrightarrow K_{n-1} \leftrightarrow K_n. \tag{4.6}$$

We assume that the filtration is simplex-wise, which means that K_i and K_{i+1} differ by only one simplex σ_i, and also begins with the empty complex. We have seen similar conditions before for the non-zigzag case in Section 3.1.2. This is not a serious restriction because we can expand an inclusion of a set of simplices to a series of inclusions by a single simplex while using any order that puts a simplex after all its faces and we can always pad an empty complex at the beginning with the first inclusion being forward.

The method we describe is derived from maintaining a consistent basis with a set of representative cycles over the intervals as we define now. These cycles generate an interval module in a straightforward way by associating a cycle to a homology class at each position.

Definition 4.12. (Representative cycles) Let $p \geq 0$, $\mathcal{F}: K_0 \leftrightarrow \cdots \leftrightarrow K_n$ be a zigzag filtration, and $\mathsf{H}_p \mathcal{F} = \{\mathsf{H}_p(K_i) \overset{\phi_i}{\leftrightarrow} \mathsf{H}_p(K_{i+1})\}_{i=0,\ldots,n-1}$ be the corresponding zigzag persistence module. Let $[b, d]$ be an interval in $\mathrm{Dgm}_p(\mathcal{F})$. A set of *representative p-cycles* for $[b, d]$ is an indexed set of p-cycles $\{c_i \subseteq K_i \mid i \in [b, d]\}$ so that we have the following.

- For $b > 0$, $[c_b]$ is not in the image of φ_{b-1} if $K_{b-1} \leftrightarrow K_b$ is a forward inclusion, or $[c_b]$ is the nonzero class mapped to 0 by φ_{b-1} otherwise.

- For $d < n$, $[c_d]$ is not in the image of φ_d if $K_d \leftrightarrow K_{d+1}$ is a backward inclusion, or $[c_d]$ is the nonzero class mapped to 0 by φ_d otherwise.
- For each $i \in [b, d-1]$, $[c_i] \leftrightarrow [c_{i+1}]$ by φ_i, that is, either $[c_i] \mapsto [c_{i+1}]$ or $[c_i] \leftarrow\!\shortmid [c_{i+1}]$ by φ_i.

The interval module *induced by* the representative p-cycles is a zigzag persistence module $\mathcal{I} : I_0 \leftrightarrow I_1 \leftrightarrow \cdots \leftrightarrow I_n$ such that I_i equals the one-dimensional vector space generated by $[c_i] \in \mathsf{H}_p(K_i)$ for $i \in [b, d]$ and equals 0 otherwise.

The following theorem justifies the definition of representative cycles, which says that representative cycles always produce an interval decomposition of a zigzag module and vice versa

Theorem 4.11. *Let $p \geq 0$, $\mathcal{F} : K_0 \leftrightarrow \cdots \leftrightarrow K_n$ be a zigzag filtration with $\mathsf{H}_p(K_0) = 0$ and A be an index set. One has that $\mathsf{H}_p\mathcal{F}$ is **equal to** (not merely isomorphic to) a direct sum of interval submodules $\bigoplus_{\alpha \in A} \mathcal{I}_{[b_\alpha, d_\alpha]}$ if and only if for each $\alpha \in A$, $\mathcal{I}_{[b_\alpha, d_\alpha]}$ is an interval module induced by a set of representative p-cycles for $[b_\alpha, d_\alpha]$ where $\mathrm{Dgm}_p(\mathcal{F}) = \big\{ [b_\alpha, d_\alpha] \,|\, \alpha \in A \big\}$.*

We now present an abstract algorithm based on an approach in [230] which helps us design a concrete algorithm later. Given a filtration $\mathcal{F}: \varnothing = K_0 \leftrightarrow \cdots \leftrightarrow K_n$ starting with an empty complex, first let $\mathrm{Dgm}_p(\mathcal{F}^0) = \varnothing$. The algorithm then iterates for $i \leftarrow 0, \ldots, n-1$. At the beginning of the i-th iteration, inductively assume that the intervals and their representative cycles for $\mathsf{H}_p\mathcal{F}^i$ have already been computed. The aim of the i-th iteration is to compute these for $\mathsf{H}_p\mathcal{F}^{i+1}$. Let $\mathrm{Dgm}_p(\mathcal{F}^i) = \big\{ [b_\alpha, d_\alpha] \,|\, \alpha \in A^i \big\}$ be indexed by a set A^i, and let $\big\{ c_k^\alpha \subseteq K_k \,|\, k \in [b_\alpha, d_\alpha] \big\}$ be a set of representative p-cycles for each $[b_\alpha, d_\alpha]$. For ease of presentation, we also let $c_k^\alpha = 0$ for each $\alpha \in A^i$ and each $k \in [0, i]$ not in $[b_\alpha, d_\alpha]$. We call intervals of $\mathrm{Dgm}_p(\mathcal{F}^i)$ ending with i as *surviving intervals* at index i. Each nonsurviving interval of $\mathrm{Dgm}_p(\mathcal{F}^i)$ is directly included in $\mathrm{Dgm}_p(\mathcal{F}^{i+1})$ and its representative cycles stay the same. For surviving intervals of $\mathrm{Dgm}_p(\mathcal{F}^i)$, the i-th iteration proceeds with the following cases determined by the types of the linear maps $\varphi_i : \mathsf{H}_p(K_i) \leftrightarrow \mathsf{H}_p(K_{i+1})$.

φ_i **is isomorphic:** In this case, no intervals are created or cease to persist. For each surviving interval $[b_\alpha, d_\alpha]$ in $\mathrm{Dgm}_p(\mathcal{F}^i)$, $[b_\alpha, d_\alpha]$ now corresponds to an interval $[b_\alpha, i+1]$ in $\mathrm{Dgm}_p(\mathcal{F}^{i+1})$. The representative cycles for $[b_\alpha, i+1]$ are set by the following rule.

Trivial setting rule of representative cycles: For each j with $b_\alpha \leq j \leq i$, the representative cycle for $[b_\alpha, i+1]$ at index j stays the same. The representative cycle for $[b_\alpha, i+1]$ at $i+1$ is set to a $c_{i+1}^\alpha \subseteq K_{i+1}$ such that $[c_i^\alpha] \leftrightarrow [c_{i+1}^\alpha]$ by φ_i.

φ_i **points forward and is injective:** A new interval $[i+1, i+1]$ is added to $\mathrm{Dgm}_p(\mathcal{F}^{i+1})$ and its representative cycle at $i+1$ is set to a p-cycle in K_{i+1} containing σ_i. All surviving intervals of $\mathrm{Dgm}_p(\mathcal{F}^i)$ persist to index $i+1$ and their representative cycles are set by the trivial setting rule.

φ_i **points backward and is surjective:** A new interval $[i+1, i+1]$ is added to $\mathrm{Dgm}_p(\mathcal{F}^{i+1})$ and its representative cycle at $i+1$ is set to a p-cycle homologous to $\partial(\sigma_i)$ in K_{i+1}. All surviving intervals of $\mathrm{Dgm}_p(\mathcal{F}^i)$ persist to index $i+1$ and their representative cycles are set by the trivial setting rule.

φ_i **points forward and is surjective:** A surviving interval of $\mathrm{Dgm}_p(\mathcal{F}^i)$ does not persist to $i+1$. Let $\mathcal{B}^i \subseteq A^i$ consist of indices of all surviving intervals. We have that $\{[c_i^\alpha] \mid \alpha \in \mathcal{B}^i\}$ forms a basis of $\mathsf{H}_p(K_i)$. Suppose that $\varphi_i([c_i^{\alpha_1}] + \cdots + [c_i^{\alpha_\ell}]) = 0$, where $\alpha_1, \ldots, \alpha_\ell \in \mathcal{B}^i$. We can rearrange the indices such that $b_{\alpha_1} < b_{\alpha_2} < \cdots < b_{\alpha_\ell}$ and $\alpha_1 < \alpha_2 < \cdots < \alpha_\ell$. Let λ be α_1 if the arrow $\circ_{b_\alpha - 1} \leftrightarrow \circ_{b_\alpha}$ points backward for every $\alpha \in \{\alpha_1, \ldots, \alpha_\ell\}$ and otherwise be the largest $\alpha \in \{\alpha_1, \ldots, \alpha_\ell\}$ such that $\circ_{b_\alpha - 1} \leftrightarrow \circ_{b_\alpha}$ points forward. Then, $[b_\lambda, i]$ forms an interval of $\mathrm{Dgm}_p(\mathcal{F}^{i+1})$. For each $k \in [b_\lambda, i]$, let $z_k = c_k^{\alpha_1} + \cdots + c_k^{\alpha_\ell}$; then, $\{z_k \mid k \in [b_\lambda, i]\}$ is a set of representative cycles for $[b_\lambda, i]$. All the other surviving intervals of $\mathrm{Dgm}_p(\mathcal{F}^i)$ persist to $i+1$ and their representative cycles are set by the trivial setting rule.

φ_i **points backward and is injective:** A surviving interval of $\mathrm{Dgm}_p(\mathcal{F}^i)$ does not persist to $i+1$. Let $\mathcal{B}^i \subseteq A^i$ consist of indices of all surviving intervals, and let $c_i^{\alpha_1}, \ldots, c_i^{\alpha_\ell}$ be the cycles in $\{c_i^\alpha \mid \alpha \in \mathcal{B}^i\}$ containing σ_i. We can rearrange the indices such that $b_{\alpha_1} < b_{\alpha_2} < \cdots < b_{\alpha_\ell}$ and $\alpha_1 < \alpha_2 < \cdots < \alpha_\ell$. Let λ be α_1 if the arrow $\circ_{b_\alpha - 1} \leftrightarrow \circ_{b_\alpha}$ points forward for every $\alpha \in \{\alpha_1, \ldots, \alpha_\ell\}$ and otherwise be the largest $\alpha \in \{\alpha_1, \ldots, \alpha_\ell\}$ such that $\circ_{b_\alpha - 1} \leftrightarrow \circ_{b_\alpha}$ points backward. Then, $[b_\lambda, i]$ forms an interval of $\mathrm{Dgm}_p(\mathcal{F}^{i+1})$ and the representative cycles for $[b_\lambda, i]$ stay the same. For each $\alpha \in \{\alpha_1, \ldots, \alpha_\ell\}$ not equal to λ, let $z_k = c_k^\alpha + c_k^\lambda$ for each k such that $b_\alpha \leq k \leq i$, and let $z_{i+1} = z_i$; then, $\{z_k \mid k \in [b_\alpha, i+1]\}$ is a set of representative cycles for $[b_\alpha, i+1]$. For the other surviving intervals, the setting of representative cycles follows the trivial setting rule.

Remark 4.3. *Note that in the above algorithm, there is no canonical choice for the representative classes. However, all choices produce the same intervals.*

4.3.2 Zigzag Persistence Algorithm

We now present a concrete version of our approach which runs in cubic time. In this algorithm, given a zigzag filtration $\mathcal{F} : \varnothing = K_0 \leftrightarrow K_1 \leftrightarrow \cdots \leftrightarrow K_n$, the main loop iterates for $i \leftarrow 0, \ldots, n-1$ so that the i-th iteration takes care of the changes from K_i to K_{i+1}. A unique integral id less than n is assigned to each simplex in K_i and id$[\sigma]$ is used to record the id of a simplex σ. Note that the id of a simplex is subject to change during the execution. For each dimension p, a *cycle matrix* $\mathbf{Z}^p$ and a *chain matrix* $\mathbf{C}^{p+1}$ with entries in $\mathbb{Z}_2$ are maintained. The number of columns of $\mathbf{Z}^p$ and $\mathbf{C}^{p+1}$ equals rank $\mathsf{Z}_{\mathsf{p}}(\mathsf{K}_i)$ and the number of rows of $\mathbf{Z}^p$ and $\mathbf{C}^{p+1}$ equals n. We will see that certain columns j of $\mathbf{C}^{p+1}$ maintain a $(p + 1)$-chain whose boundary is in column j of $\mathbf{Z}^p$. Each column of $\mathbf{Z}^p$ and $\mathbf{C}^p$ represents a p-chain in K_i such that for each simplex $\sigma \in K_i$, σ belongs to the p-chain if and only if the bit with index id$[\sigma]$ in the column equals 1. For convenience, we make no distinction between a column of the matrix $\mathbf{Z}^p$ or $\mathbf{C}^p$ and the chain it represents. We use $\mathbf{Z}^p[j]$ to denote the j-th column of $\mathbf{Z}^p$ (columns of $\mathbf{C}^p$ are denoted similarly). For each column $\mathbf{Z}^p[j]$, a *birth timestamp* $b^p[j]$ is maintained. This timestamp is usually nonnegative, but can possibly be negative one (-1). We will see that this special negative value is assigned only to indicate that the column represents a boundary cycle. Moreover, we let the *pivot* of $\mathbf{Z}^p[j]$ be the largest index whose corresponding bit equals 1 in $\mathbf{Z}^p[j]$ and denote it as pivot($\mathbf{Z}^p[j]$). At the start of the i-th iteration, for each p, the following properties for the matrices are preserved:

1. The columns of $\mathbf{Z}^p$ form a basis of $\mathsf{Z}_p(K_i)$ and have distinct pivots.
2. The columns of $\mathbf{Z}^p$ with negative birth timestamps form a basis of $\mathsf{B}_p(K_i)$. Moreover, for each column $\mathbf{Z}^p[j]$ of $\mathbf{Z}^p$ with a negative birth timestamp, one has that $\mathbf{Z}^p[j] = \partial(\mathbf{C}^{p+1}[j])$.
3. For columns of $\mathbf{Z}^p$ with nonnegative birth timestamps, their birth timestamps bijectively map to the starting indices of the intervals of $\mathrm{Dgm}_p(\mathcal{F}^i)$ ending with i. Moreover, for each column $\mathbf{Z}^p[j]$ of $\mathbf{Z}^p$ such that $b^p[j]$ is nonnegative, one has that $\mathbf{Z}^p[j]$ is a representative cycle at index i for the interval $[b^p[j], i]$.

The above properties indicate that a column $\mathbf{Z}^p[j]$ of $\mathbf{Z}^p$ is a boundary if $b^p[j] < 0$ and is not a boundary otherwise. Furthermore, we have that columns of $\mathbf{Z}^p$ with nonnegative birth timestamps represent a homology basis for $\mathsf{H}_p(K_i)$ at the start of the i-th iteration.

Zigzag Algorithm

For each $i \leftarrow 0, \ldots, n - 1$, the algorithm does the following:

Case φ_i is forward: From K_i to K_{i+1}, a p-simplex σ_i is added and the id of σ_i is set as $\mathrm{id}[\sigma_i] = i$. Since the columns of $\mathbf{Z}^{p-1}$ form a basis of $\mathbf{Z}_{p-1}(K_i)$ and have distinct pivots, $\partial(\sigma_i)$ can be represented as a sum of the columns of $\mathbf{Z}^{p-1}$ by a reduction algorithm. Suppose that $\partial(\sigma_i) = \sum_{\alpha \in I} \mathbf{Z}^{p-1}[\alpha]$ where I is a set of column indices of $\mathbf{Z}^{p-1}$. The algorithm then checks the timestamp of $\mathbf{Z}^{p-1}[\alpha]$ for each $\alpha \in I$ to see whether all of them are boundaries. After this, it is known whether or not $\partial(\sigma_i)$ is a boundary in K_i. An interval in dimension p gets born if $\partial(\sigma_i)$ is a boundary in K_i and an interval in dimension $p - 1$ dies otherwise.

- Birth: Append a new column $\sigma_i + \sum_{\alpha \in I} \mathbf{C}^p[\alpha]$ with birth timestamp $i + 1$ to $\mathbf{Z}^p$.
- Death: Let J consist of indices in I whose corresponding columns in $\mathbf{Z}^{p-1}$ have nonnegative birth timestamps. If $\varphi_{b^{p-1}[\alpha]-1}$ points backward for all $\alpha \in J$, let λ be the smallest index in J; otherwise, let λ be the largest α in J such that $\varphi_{b^{p-1}[\alpha]-1}$ points forward. Then, do the following:
    1. Output the $(p-1)$-th interval $\left[b^{p-1}[\lambda], i\right]$.
    2. Set $\mathbf{Z}^{p-1}[\lambda] = \partial(\sigma_i)$, $\mathbf{C}^p[\lambda] = \sigma_i$, and $b^{p-1}[\lambda] = -1$.

Since the pivot of the column $\partial(\sigma_i)$ may conflict with that of another column in $\mathbf{Z}^{p-1}$, we perform steps 1–13 described next to keep the pivots distinct. The total order $\preceq_b$ used in step 10 of the algorithm and later is defined as follows.

Definition 4.13. Let $I \subseteq \big\{1, \ldots, n - 1\big\}$ be a set of indices. For $i, j \in I$, $i \preceq_b j$ in the total order if and only if one of the following holds:

- $i = j$;
- $i < j$ and the function φ_{j-1} points forward;
- $j < i$ and the function φ_{i-1} points backward.

1. **while** there are two columns $\mathbf{Z}^{p-1}[\alpha]$, and $\mathbf{Z}^{p-1}[\beta]$ with the same pivot
   do
2.   **if** $b^{p-1}[\alpha] < 0$ **and** $b^{p-1}[\beta] < 0$ **then**
3. $\mathbf{Z}^{p-1}[\alpha] \leftarrow \mathbf{Z}^{p-1}[\alpha] + \mathbf{Z}^{p-1}[\beta]$
4. $\mathbf{C}^p[\alpha] \leftarrow \mathbf{C}^p[\alpha] + \mathbf{C}^p[\beta]$
5.   **if** $b^{p-1}[\alpha] < 0$ **and** $b^{p-1}[\beta] \geq 0$ **then**
6. $\mathbf{Z}^{p-1}[\beta] \leftarrow \mathbf{Z}^{p-1}[\alpha] + \mathbf{Z}^{p-1}[\beta]$
7.   **if** $b^{p-1}[\alpha] \geq 0$ **and** $b^{p-1}[\beta] < 0$ **then**
8. $\mathbf{Z}^{p-1}[\alpha] \leftarrow \mathbf{Z}^{p-1}[\alpha] + \mathbf{Z}^{p-1}[\beta]$

9.   **if** $b^{p-1}[\alpha] \geq 0$ **and** $b^{p-1}[\beta] \geq 0$ **then**

10.   **if** $b^{p-1}[\alpha] \preceq_b b^{p-1}[\beta]$ **then**

11. $\mathbf{Z}^{p-1}[\beta] \leftarrow \mathbf{Z}^{p-1}[\alpha] + \mathbf{Z}^{p-1}[\beta]$

12.   **else**

13. $\mathbf{Z}^{p-1}[\alpha] \leftarrow \mathbf{Z}^{p-1}[\alpha] + \mathbf{Z}^{p-1}[\beta]$

Case φ_i is backward: From K_i to K_{i+1}, a p-simplex σ_i is deleted. If there is a column in $\mathbf{Z}^p$ containing σ_i, then there are some p-cycles missing going from K_i to K_{i+1} and an interval in dimension p dies. Otherwise, an interval in dimension $p - 1$ gets born.

- Birth: First, the boundaries in $\mathbf{Z}^{p-1}$ need to be updated so that they form a basis of $\mathsf{B}_{p-1}(K_{i+1})$:

    1. **while** there are two columns $\mathbf{Z}^{p-1}[\alpha]$ and $\mathbf{Z}^{p-1}[\beta]$ with negative birth timestamps such that $\mathbf{C}^p[\alpha]$ and $\mathbf{C}^p[\beta]$ contain σ_i **do**

    2.   **if** $\mathrm{pivot}(\mathbf{Z}^{p-1}[\alpha]) > \mathrm{pivot}(\mathbf{Z}^{p-1}[\beta])$ **then**

    3. $\mathbf{Z}^{p-1}[\alpha] \leftarrow \mathbf{Z}^{p-1}[\alpha] + \mathbf{Z}^{p-1}[\beta]$

    4. $\mathbf{C}^p[\alpha] \leftarrow \mathbf{C}^p[\alpha] + \mathbf{C}^p[\beta]$

    5.   **else**

    6. $\mathbf{Z}^{p-1}[\beta] \leftarrow \mathbf{Z}^{p-1}[\alpha] + \mathbf{Z}^{p-1}[\beta]$

    7. $\mathbf{C}^p[\beta] \leftarrow \mathbf{C}^p[\alpha] + \mathbf{C}^p[\beta]$

    Then, let $\mathbf{Z}^{p-1}[\alpha]$ be the only column with negative birth timestamp in $\mathbf{Z}^{p-1}$ such that $\mathbf{C}^p[\alpha]$ contains σ_i; set $b^{p-1}[\alpha] = i + 1$. Note that $\mathbf{Z}^{p-1}[\alpha]$ is homologous to $\partial(\sigma_i)$ in K_{i+1}, and the pivots are automatically distinct.

- Death: First, update $\mathbf{C}^p$ so that no columns of $\mathbf{C}^p$ contain σ_i:

    1. Let $\mathbf{Z}^p[\alpha]$ be a column of $\mathbf{Z}^p$ containing σ_i.

    2. For each column[1]; $\mathbf{C}^p[\beta]$ of $\mathbf{C}^p$ containing σ_i, set $\mathbf{C}^p[\beta] = \mathbf{C}^p[\beta] + \mathbf{Z}^p[\alpha]$.

    Then, remove σ_i from $\mathbf{Z}^p$:

    1. $\alpha_1, \ldots, \alpha_k \leftarrow$ indices of all columns of $\mathbf{Z}^p$ containing σ_i

    2. sort $\alpha_1, \ldots, \alpha_k$ so that $b^p[\alpha_1] \preceq_b \cdots \preceq_b b^p[\alpha_k]$

    3. $z \leftarrow \mathbf{Z}^p[\alpha_1]$

    4. **for** $\alpha \leftarrow \alpha_2, \ldots, \alpha_k$ **do**

    5.   **if** $\mathrm{pivot}(\mathbf{Z}^p[\alpha]) > \mathrm{pivot}(z)$ **then**

    6. $\mathbf{Z}^p[\alpha] \leftarrow \mathbf{Z}^p[\alpha] + z$

    7.   **else**

    8. $\texttt{temp} \leftarrow \mathbf{Z}^p[\alpha]$

---

[1] Note here we only iterate over columns $\mathbf{C}^p[\beta]$ for which $\mathbf{Z}^{p-1}[\beta]$ is a boundary.

9. $\mathbf{Z}^p[\alpha] \leftarrow \mathbf{Z}^p[\alpha] + z$

10. $z \leftarrow \texttt{temp}$

11. output the p-th interval $\left[b^p[\alpha_1], i\right]$

12. delete the column $\mathbf{Z}^p[\alpha_1]$ from $\mathbf{Z}^p$ and delete $b^p[\alpha_1]$ from b^p

At the end of the algorithm, for each p and each column $\mathbf{Z}^p[\alpha]$ of $\mathbf{Z}^p$ with nonnegative birth timestamp, output the p-th interval $\left[b^p[\alpha], n\right]$. Notice that while spewing out the bars, the algorithm can easily output the types of the bars by looking at the relevant arrows as described before.

4.4 Persistence for Zigzag Towers

So far, we have considered computing persistence for towers where maps are all in the forward direction though they may not be inclusions and of zigzag filtrations where maps may be in both forward and backward directions but cannot be other than inclusions. In this section, we consider the zigzag towers that combine both, that is, maps are simplicial (not necessarily inclusions) and may point in both the forward and backward directions:

$$\mathcal{K}\colon K_0 \xleftrightarrow{f_0} K_1 \xleftrightarrow{f_1} K_2 \xleftrightarrow{f_2} \cdots \xleftrightarrow{f_{n-1}} K_n. \tag{4.7}$$

Recall that by Proposition 4.6 each map $f_i\colon K_i \to K_{i+1}$ can be decomposed into elementary inclusions and elementary collapses. So, without loss of generality, we assume that every f_i is either an elementary inclusion or an elementary collapse.

First, we propose a simulation of an elementary collapse with a coning strategy that only requires additions of simplices.

Let $f\colon K \to K'$ be an elementary collapse. Assume that the induced vertex map collapses vertices $u, v \in K$ to $u \in K'$, and is identity on other vertices. For a subcomplex $X \subseteq K$, define the cone $u * X$ to be the complex $\bigcup_{\sigma \in X} \{\overline{\sigma \cup \{u\}}\}$. Consider the augmented complex

$$\hat{K} := K \cup (u * \overline{\mathrm{St}\, v}).$$

In other words, for every simplex $\{u_0, \ldots, u_d\} \in \overline{\mathrm{St}\, v}$ of K, we add the simplex $\{u_0, \ldots, u_d\} \cup \{u\}$ to $\hat{K}$ if it is not already in. See Figure 4.6. Notice that K' is a subcomplex of $\hat{K}$ in this example, which we observe is true in general.

Claim 4.1. *We claim that* $K' \subseteq \hat{K}$.

Now consider the inclusions $\iota\colon K \hookrightarrow \hat{K}$ and $\iota'\colon K' \hookrightarrow \hat{K}$. These inclusions along with the elementary collapse constitute a diagram in Figure 4.6 which

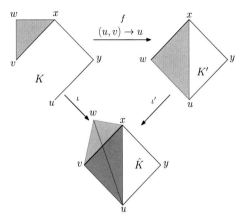

Figure 4.6 Elementary collapse $(u, v) \rightarrow u$: the cone $u * \overline{\mathrm{St}\, v}$ adds edges uw, uv, ux, triangles uwx, uvx, uvw, and the tetrahedron $uvwx$.

does not necessarily commute. Nevertheless, it commutes at the homology level which is precisely stated below.

Proposition 4.12. *In the zigzag module* $\mathsf{H}_p(K) \xrightarrow{\iota_*} \mathsf{H}_p(\hat{K}) \xleftarrow{\iota'_*} \mathsf{H}_p(K')$ *induced by inclusions* ι *and* ι', *the linear map* ι'^* *is an isomorphism and* $f_* : \mathsf{H}_p(K) \rightarrow \mathsf{H}_p(K')$ *equals* $(\iota'_*)^{-1} \circ \iota_*$.

PROOF. We use the notion of contiguous maps which induces equal maps at the homology level. Recall that two maps $f_1 : K_1 \rightarrow K_2$ and $f_2 : K_1 \rightarrow K_2$ are contiguous if for every simplex $\sigma \in K_1$, $f_1(\sigma) \cup f_2(\sigma)$ is a simplex in K_2. We observe that the simplicial maps $\iota' \circ f$ and ι are contiguous and ι' induces an isomorphism at the homology level, that is, $\iota'_* : \mathsf{H}_p(K') \rightarrow \mathsf{H}_p(\hat{K})$ is an isomorphism.

Since ι is contiguous to $\iota' \circ f$, we have $\iota_* = (\iota' \circ f)_* = \iota'_* \circ f_*$. Since ι'_* is an isomorphism, $(\iota'_*)^{-1}$ exists and is an isomorphism. It then follows that $f_* = (\iota'_*)^{-1} \circ \iota_*$. $\qquad\square$

Proposition 4.12 allows us to simulate the persistence of a simplicial tower with only inclusion-induced homomorphisms, which, in turn, allows us to consider a simplicial zigzag filtration. More specifically, the simplicial tower in Eq. (4.7) generates the zigzag persistence module by induced homomorphisms f_{i*}:

$$\mathsf{H}_p(K_0) \xleftrightarrow{f_{0*}} \mathsf{H}_p(K_1) \xleftrightarrow{f_{1*}} \mathsf{H}_p(K_2) \xleftrightarrow{f_{2*}} \cdots \xleftrightarrow{f_{n-1*}} \mathsf{H}_p(K_n). \quad (4.8)$$

With our observation that every map f_{i*} can be simulated with an inclusion induced map, our goal is to replace the original simplicial tower in Eq. (4.7)

$$H_p(K_i) \xrightarrow{f_{i*}} H_p(K_{i+1}) \xleftarrow{\ =\ } H_p(K_{i+1}) \qquad\quad H_p(K_i) \xrightarrow{\ =\ } H_p(K_i) \xleftarrow{f_{i*}} H_p(K_{i+1})$$

$$\downarrow{=} \qquad\qquad \downarrow{\simeq} \qquad\qquad \downarrow{=} \qquad\qquad\qquad \downarrow{=} \qquad\qquad \downarrow{\simeq} \qquad\qquad \downarrow{=}$$

$$H_p(K_i) \xrightarrow{\iota_{i*}} H_p(\hat{K}_i) \xleftarrow{\simeq} H_p(K_{i+1}) \qquad\quad H_p(K_i) \xrightarrow{\simeq} H_p(\hat{K}_{i+1}) \xleftarrow{\iota_{i*}} H_p(K_{i+1})$$

Figure 4.7 The modules in the top row induced from an elementary collapse are isomorphic to the modules in the bottom row induced by inclusions.

$$H_p(K_0) \xrightarrow{g_0} H_p(S_0) \xleftarrow{h_0} H_p(K_1) \xrightarrow{g_1} H_p(S_1) \xleftarrow{h_1} H_p(K_2) \xrightarrow{g_2} \cdots \longleftarrow H_p(K_n)$$

$$\downarrow{=} \qquad\quad \downarrow{\simeq} \qquad\quad \downarrow{=} \qquad\quad \downarrow{\simeq} \qquad\quad \downarrow{=} \qquad\qquad\quad \downarrow{=}$$

$$H_p(K_0) \longrightarrow H_p(T_0) \longleftarrow H_p(K_1) \longrightarrow H_p(T_1) \longleftarrow H_p(K_2) \longrightarrow \cdots \longleftarrow H_p(K_n)$$

Figure 4.8 The modules in Eq. (4.9) and (4.10) are isomorphic.

with a zigzag filtration so that we can take advantage of the algorithm in Section 4.3. In view of Proposition 4.12, the two diagrams shown in Figure 4.7 commute: the one on the left corresponds to a forward collapse $f_i : K_i \to K_{i+1}$ and the other on the right corresponds to a backward collapse $f_i : K_i \leftarrow K_{i+1}$.

Observe that, if f_i is an inclusion instead of a collapse, we can still construct similar commuting diagrams. In that case, we simply take $\hat{K}_i = K_{i+1}$ when f_i is a forward inclusion and take $\hat{K}_{i+1} = K_i$ when f_i is a backward inclusion.

Now, we can expand each f_{i*} of the persistence module in Eq. (4.8) by juxtaposing it with an equality as in the top modules shown in Figure 4.7. Then, this expanded module becomes isomorphic to the modules induced by inclusions at the bottom of the commuting diagrams.

In general, we first consider the expansion of the module in Eq. (4.8) to the following module in Eq. (4.9) where $S_i = K_{i+1}$, $g_i = f_i$, and h_i is equality when f_i is forward, and $S_i = K_i$, g_i is equality, and $h_i = f_i$ when f_i is backward:

$$H_p(K_0) \xrightarrow{g_0} H_p(S_0) \xleftarrow{h_0} H_p(K_1) \xrightarrow{g_1} H_p(S_1) \xleftarrow{h_1} H_p(K_2) \xrightarrow{g_2} \cdots \xleftarrow{h_{n-1}} H_p(K_n). \quad (4.9)$$

Using Figure 4.7, a module isomorphic to the module in Eq. (4.9) can be constructed as given in Eq. (4.10) where $T_i = \hat{K}_i$ when f_i is forward and $T_i = \hat{K}_{i+1}$ when f_i is backward; and all maps are induced by inclusions:

$$H_p(K_0) \longrightarrow H_p(T_0) \longleftarrow H_p(K_1) \longrightarrow H_p(T_1) \longleftarrow H_p(K_2) \longrightarrow \cdots \longleftarrow H_p(K_n). \quad (4.10)$$

The two persistence modules in Eq. (4.9) and Eq. (4.10) are isomorphic because all vertical maps in Figure 4.8 are isomorphisms and all squares commute (Figure 4.7).

In view of the module in Eq. (4.10), we convert the tower $\mathcal{K}$ in Eq. (4.7) to the zigzag filtration below where $T_i = \hat{K}_i$ when f_i is forward and $T_i = \hat{K}_{i+1}$ when f_i is backward:

$$H_p(K_0) \xrightarrow{f_{0*}} H_p(K_1) \longleftarrow H_p(K_1) \xrightarrow{=} H_p(K_1) \xleftarrow{f_{1*}} H_p(K_2) \xrightarrow{f_{2*}} \cdots \longleftarrow H_p(K_n)$$

$$\downarrow{=} \qquad \downarrow{\simeq} \qquad \downarrow{=} \qquad \downarrow{\simeq} \qquad \downarrow{=} \qquad \qquad \downarrow{=}$$

$$H_p(K_0) \xrightarrow{i_{0*}} H_p(\hat{K}_0) \xrightarrow{\simeq} H_p(K_1) \xrightarrow{\simeq} H_p(\hat{K}_2) \xleftarrow{i_{1*}} H_p(K_2) \xrightarrow{i_{2*}} \cdots \longleftarrow H_p(K_n)$$

Figure 4.9 Commuting diagram for the module in Eq. (4.13) and its isomorphic module.

$$\mathcal{F} : K_0 \hookrightarrow T_0 \hookleftarrow K_1 \hookrightarrow T_1 \hookleftarrow K_2 \hookrightarrow \cdots \hookleftarrow K_n. \tag{4.11}$$

The zigzag filtration above is simplex-wise but does not begin with an empty complex. We can expand K_0 simplex-wise to convert the filtration to a simplex-wise filtration that begins with an empty complex. Then, we can apply the zigzag algorithm in Section 4.3.2 to compute the barcode.

Theorem 4.13. *The persistence diagram of* $\mathcal{K}$ *can be derived from that of the filtration* $\mathcal{F}$.

Example 4.1. *Consider the tower in Eq. (4.12), where each map is an elementary collapse, and the persistence module induced by it in Eq. (4.13). This module can be expanded, and its isomorphic module is shown at the bottom of the commuting diagram in Figure 4.9.*

$$K_0 \xrightarrow{f_0} K_1 \xleftarrow{f_1} K_2 \xrightarrow{f_2} \cdots \xrightarrow{f_{n-1}} K_n \tag{4.12}$$

$$H_p(K_0) \xrightarrow{f_{0*}} H_p(K_1) \xleftarrow{f_{1*}} H_p(K_2) \xrightarrow{f_{2*}} \cdots \xrightarrow{f_{n-1*}} H_p(K_n) \tag{4.13}$$

We obtain the following zigzag filtration that corresponds to the module at the bottom of the diagram in Figure 4.9. Hence, we can compute the barcode for the input tower in Eq. (4.12) from this zigzag filtration.

$$K_0 \hookrightarrow \hat{K}_0 \hookleftarrow K_1 \hookrightarrow \hat{K}_2 \hookleftarrow K_2 \hookrightarrow \cdots \hookleftarrow K_n \tag{4.14}$$

Remark 4.4. *Notice that, when* f_i *is an inclusion, we can eliminate introducing the middle column in Figure 4.8 which will translate into eliminating some of the inclusions in the sequence in Eq. (4.11). We introduced these extraneous inclusions just to make the expanded module generic in the sense that its inclusions reverse the directions alternately.*

4.5 Levelset Zigzag Persistence

Now, we consider a special type of zigzag persistence stemming from a function over a topological space. In standard persistence, growing sublevel sets of the function constitute the filtration over which the persistence is defined. In levelset zigzag persistence, we replace the sublevel sets with *levelsets* and *interval sets* and the maps going from the levelsets to the adjacent interval sets give rise to a zigzag filtration. To produce a zigzag filtration corresponding to a levelset persistence, we consider a PL-function on the underlying space of a simplicial complex and then convert a zigzag sequence of subspaces (levelsets and interval sets) into subcomplexes. This is similar to what we did while considering the standard persistence for a PL-function in Sections 3.1 and 3.5.

Before we focus on a PL-function, let us consider a more general real-valued continuous function $f : X \to \mathbb{R}$ on a topological space X. We need a restriction on f that keeps all homology groups being considered to be finite. For a real value $s \in \mathbb{R}$ and an interval $I \subseteq \mathbb{R}$, we denote the *levelset* $f^{-1}(s)$ by $X_{=s}$ and the *interval set* $f^{-1}(I)$ by X_I.

Definition 4.14. (Critical; Regular value) An open interval $I \subseteq \mathbb{R}$ is called a *regular interval* if there exist a topological space Y and a homeomorphism $\Phi : Y \times I \to X_I$ so that $f \circ \Phi$ is the projection onto I and Φ extends to a continuous function $\bar{\Phi} : Y \times \bar{I} \to X_{\bar{I}}$ where $\bar{I}$ is the closure of I. We assume that f is of *Morse type* [63] meaning that each levelset $X_{=s}$ has finitely generated homology groups and there are finitely many values called *critical* $a_0 = -\infty < a_1 < \cdots < a_n < a_{n+1} = +\infty$, so that each interval (a_i, a_{i+1}) is a maximal interval that is regular. A value $s \in (a_i, a_{i+1})$ is then called a *regular value*.

The original construction [63] of levelset zigzag persistence picks regular values $s_0, s_1, \ldots, s_n$ so that each $s_i \in (a_i, a_{i+1})$. Then, the *levelset zigzag filtration* of f is defined as follows:

$$X_{[s_0,s_1]} \leftrightarrow \cdots \hookrightarrow X_{[s_{i-1},s_i]} \leftrightarrow X_{=s_i} \hookrightarrow X_{[s_i,s_{i+1}]} \leftrightarrow \cdots \hookrightarrow X_{[s_{n-1},s_n]}.$$

This construction relies on a choice of regular values and there is no canonical choice. As we work on simplicial complexes, different regular values can result in different complexes in the filtration. Therefore, we adopt the following alternative definition of a levelset zigzag filtration $\mathcal{X}$, which does not rely on a choice of regular values:

$$\mathcal{X} : X_{(a_0,a_2)} \leftrightarrow \cdots \hookrightarrow X_{(a_{i-1},a_{i+1})} \leftrightarrow X_{(a_i,a_{i+1})} \hookrightarrow X_{(a_i,a_{i+2})} \leftrightarrow \cdots \hookrightarrow X_{(a_{n-1},a_{n+1})}. \quad (4.15)$$

The space of the type $X_{(a_{i-1},a_{i+1})}$ contains a critical value a_i and hence is called a *critical space*. For a similar reason, a space of the type $X_{(a_i,a_{i+1})}$ is called a *regular space* which does not contain any critical value. Considering the homology groups of the spaces, we get the zigzag persistence module:

$$\mathsf{H}_p \mathcal{X} : \mathsf{H}_p(X_{(a_0,a_2)}) \leftarrow \cdots \to \mathsf{H}_p(X_{(a_{i-1},a_{i+1})}) \leftarrow \mathsf{H}_p(X_{(a_i,a_{i+1})}) \to \mathsf{H}_p(X_{(a_i,a_{i+2})}) \leftarrow \cdots \to \mathsf{H}_p(X_{(a_{n-1},a_{n+1})}).$$

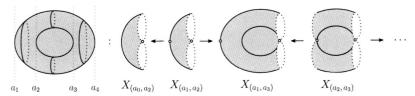

a_1 a_2 a_3 a_4 $X_{(a_0,a_2)}$ $X_{(a_1,a_2)}$ $X_{(a_1,a_3)}$ $X_{(a_2,a_3)}$

Figure 4.10 A torus with four critical values. The real-valued function is the height function over the horizontal line. The first several subspaces in the levelset zigzag diagram are given and the remaining ones are symmetric. An empty dot indicates that the point is not included.

Note that $X_{(a_i,a_{i+1})}$ deformation retracts to $X_{=s_i}$ and $X_{(a_{i-1},a_{i+1})}$ deformation retracts to $X_{[s_{i-1},s_i]}$, so the zigzag modules induced by the two diagrams are isomorphic, that is, equivalent at the persistent homology level. See Figure 4.10 for an example of a levelset zigzag filtration.

Generation of Barcode for Levelset Zigzag
The interval decomposition of the module $H_p\mathcal{X}$ gives the barcode for the zigzag persistence. However, the endpoints of the bars may belong to the index of either a critical or a regular space. If it belongs to a critical space $X_{(a_{i-1},a_{i+1})}$, we map it to the critical value a_i. Otherwise, if it belongs to a regular space $X_{(a_i,a_{i+1})}$, we map it to the regular value s_i. After this conversion, still the bars do not end solely in critical values. We modify the endpoints further. In keeping with the understanding that even the levelset homology classes do not change in the regular spaces, we convert an endpoint s_i to an adjacent critical value and make the bar (interval module) open at that critical value. Precisely we modify the bars as (i) $[a_i, a_j] \Leftrightarrow [a_i, a_j]$, (ii) $[a_i, s_j] \Leftrightarrow [a_i, a_{j+1})$, (iii) $[s_i, a_j] \Leftrightarrow (a_i, a_j]$, or (iv) $[s_i, s_j] \Leftrightarrow (a_i, a_{j+1})$. As in the case of standard zigzag filtration, the intervals in (i)–(iv) are referred to as *closed–closed*, *closed–open*, *open–closed*, and *open–open* bars, respectively. Our goal is to compute these four types of bars for a PL-function where the space X is the underlying space of a simplicial complex K.

4.5.1 Simplicial Levelset Zigzag Filtration

We now turn to a *simplicial version* of the construction we just described. For a given complex K, let $X = |K|$ and $f : X \to \mathbb{R}$ be a PL-function defined by interpolating values on the vertices of K (Definition 3.2). We also assume f to be *generic*, that is, no two vertices of K have the same function value.

We know that f can have critical values only at K's vertices (Section 3.5.1). We call these vertices *critical* and call other vertices *regular*. Let $v_1, \ldots, v_n$ be all the critical vertices of f with values $a_1 < \cdots < a_n$, and let $a_0 = -\infty$ and

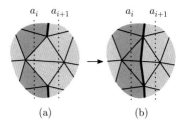

Figure 4.11 Simplicial zigzag filtration is made equivalent to space filtration by subdivision.

$a_{n+1} = +\infty$ be two additional critical values. For two critical values $a_i < a_j$, let $X_{(i,j)} := X_{(a_i,a_j)}$ and $K_{(i,j)}$ be the complex $\{\sigma \in K \mid \forall v \in \sigma, f(v) \in (a_i, a_j)\}$. Then, the *space* $\mathcal{X}$ and *simplicial levelset zigzag filtration* $\mathcal{K}$ of f are defined, respectively, as

$$\mathcal{X}\colon X_{(0,2)} \hookleftarrow \cdots \hookrightarrow X_{(i-1,i+1)} \hookleftarrow X_{(i,i+1)} \hookrightarrow X_{(i,i+2)} \hookleftarrow \cdots \hookrightarrow X_{(n-1,n+1)} \quad (4.16)$$

and

$$\mathcal{K}\colon K_{(0,2)} \hookleftarrow \cdots \hookrightarrow K_{(i-1,i+1)} \hookleftarrow K_{(i,i+1)} \hookrightarrow K_{(i,i+2)} \hookleftarrow \cdots \hookrightarrow K_{(n-1,n+1)}. \quad (4.17)$$

A complex of the form $K_{(i,i+1)}$ in the filtration is called a *regular complex* and a complex of the form $K_{(i,i+2)}$ is called a *critical complex*. Note that while we can expect the space and simplicial levelset zigzag filtrations for a finely tessellated complex to be equivalent, this is not always the case. For example, in Figure 4.11, let K' be the complex in Figure 4.11(a); $|K'_{(i,i+1)}|$ (thick edges) is not homotopy equivalent to $|K'|_{(i,i+1)}$, and hence the simplicial levelset zigzag filtration is not equivalent to the space one. We observe that the non-equivalence is caused by the two central triangles which contain more than one critical value. A subdivision of the two central triangles in the complex K'' in Figure 4.11(b), where no triangles contain more than one critical value, renders $|K''|_{(i,i+1)}$ deformation retracting to $|K''_{(i,i+1)}|$. Based on the above observation, we formulate the following property, which guarantees that the module of the simplicial levelset zigzag filtration remains isomorphic to that of the space one.

Definition 4.15. A complex K is called *compatible with the levelsets* of a PL-function $f\colon |K| \to \mathbb{R}$ if for every simplex σ of K and its convex hull $|\sigma|$, function values of points in $|\sigma|$ contain at most one critical value of f.

Given a PL-function f on a complex K, one can make K compatible with the levelsets of f by subdividing K with barycentric subdivisions (see, e.g., [102]).

Proposition 4.14. *Let K be compatible with the levelsets of f, and let $X =$ $|K|$; one has that $X_{(a_i, a_j)}$ deformation retracts to $\left| K_{(i,j)} \right|$ for any two critical values $a_i < a_j$. Therefore, the zigzag modules induced by the space and the simplicial levelset zigzag filtrations are isomorphic.*

Our goal is to compute the four types of bars for the zigzag filtration X from its simplicial version $\mathcal{K}$. For this, we make $\mathcal{K}$ simplex-wise and call it $\mathcal{F}$. First, $\mathcal{F}$ starts and ends with the same original complexes in $\mathcal{K}$. Second, whenever an inclusion in $\mathcal{K}$ is expanded so that one simplex is added at a time, the addition follows the order of the simplices' function values. Formally, for the inclusion $K_{(i,i+1)} \hookrightarrow K_{(i,i+2)}$ in $\mathcal{K}$, let $u_1 = v_{i+1}, u_2, \ldots, u_k$ be all the vertices with function values in $[a_{i+1}, a_{i+2})$ such that $f(u_1) < f(u_2) < \cdots < f(u_k)$; then, the lower stars of $u_1, \ldots, u_k$ are added in sequence by $\mathcal{F}$. Note that for each $u_j \in \{u_1, \ldots, u_k\}$, we do not restrict how simplices in the lower star of u_j are added. For the inclusion $K_{(i-1,i+1)} \hookleftarrow K_{(i,i+1)}$ in $\mathcal{K}$, everything is reversed, that is, vertices are ordered in decreasing function values and upper stars are added. With this expansion, the zigzag filtration $\mathcal{K}$ in Eq. (4.17) is converted to a filtration $\mathcal{F}$ shown below where a dashed arrow indicates insertions of one or more simplices and a solid arrow indicates a single simplex insertion. In particular, we indicate that the backward inclusion $K_{(i-1,i+1)} \leftarrow\!\!- \!\!\rightarrow K_{(i,i+1)}$ is expanded into a simplex-wise filtration.

$$\mathcal{F}: \quad \cdots \dashrightarrow K_{(i-1,i+1)} \hookleftarrow \cdots \hookleftarrow K_{\ell-1} \hookleftarrow K_\ell \hookleftarrow \cdots \hookleftarrow K_{(i,i+1)} \dashrightarrow K_{(i,i+2)} \dashleftarrow \cdots$$
$$(4.18)$$

After expanding all forward and backward inclusions to make them simplex-wise, we obtain a zigzag filtration whose complexes can be indexed by $0, 1, \ldots, n$ as we assume next.

4.5.2 Barcode for Levelset Zigzag Filtration

One can compute the barcode for the zigzag filtration $\mathcal{F}$ in Eq. (4.18) that is derived from the original zigzag filtration $\mathcal{K}$ in Eq. (4.17). There is one technicality that we need to take care of. To apply the algorithm in Section 4.3.2, we need the input zigzag filtration to begin with an empty complex. The filtration $\mathcal{F}$ as constructed from expanding $\mathcal{K}$ has the first complex $K_{(0,2)}$ that is non-empty. So, as before, we expand $K_{(0,2)}$ simplex-wise and begin $\mathcal{F}$ with an empty complex. We assume below that this is the case for $\mathcal{F}$.

The bars in the barcode for $\mathcal{F}$ do not necessarily coincide with the four types of bars for $\mathcal{K}$ with endpoints only in critical values. However, we can read the bars for $\mathcal{K}$ from the bars of $\mathcal{F}$.

First, assume that $\mathcal{F}$ is indexed as

$$\mathcal{F}: \varnothing = K_0 \leftrightarrow K_1 \leftrightarrow \cdots \leftrightarrow K_{n-1} \leftrightarrow K_n.$$

This means that a complex K_j, $j > 0$, is one of the four categories: (i) it is a complex in the expansion of the backward inclusion $K_{(i-1,i+1)} \leftarrow\!\!\!-\!\!\!\rightarrow K_{(i,i+1)}$; (ii) it is a complex in the expansion of the forward inclusion $K_{(i,i+1)} \hookrightarrow\!\!\!\rightarrow$ $K_{(i,i+2)}$; (iii) it is a regular complex $K_{(i,i+1)}$ for some $i > 0$; or (iv) it is a critical complex $K_{(i-1,i+1)}$ for some $i > 0$. The types of complexes where the endpoints of a bar $[b, d]$ for $\mathcal{F}$ are located determine the bars for $\mathcal{K}$ and hence $\mathcal{X}$ which can be of four types: *closed–closed* $[a_i, a_j]$, *closed–open* $[a_i, a_j)$, *open–closed* $(a_i, a_j]$, and *open–open* (a_i, a_j).

Let $[b, d]$ be a bar for $\mathcal{F}$. If both K_b and K_d appear in the expansion of a forward inclusion $K_{(i,i+1)} \hookrightarrow\!\!\!\rightarrow K_{(i,i+2)}$, we ignore the bar because it is an artificial bar created due to expanding the filtration $\mathcal{K}$ into the filtration $\mathcal{F}$. Similarly, we ignore the bar if both K_b and K_d appear in the expansion of a backward inclusion $K_{(i-1,i+1)} \leftarrow\!\!\!-\!\!\!\rightarrow K_{(i,i+1)}$. We explain other cases below.

(Case 1.) K_b is either a regular complex $K_{(i,i+1)}$ or in the expansion of $K_{(i-1,i+1)} \leftarrow\!\!\!-\!\!\!\rightarrow K_{(i,i+1)}$: the complex K_b is a subcomplex of the critical complex $K_{(i-1,i+1)}$ which stands for the critical value a_i. So, the end b is mapped to a_i and made open because the class for the bar $[b, d]$ does not exist in $K_{(i-1,i+1)}$.

(Case 2.) K_b is either the critical complex $K_{(i,i+2)}$ or in the expansion of $K_{(i,i+1)} \hookrightarrow\!\!\!\rightarrow K_{(i,i+2)}$: the complex is a subcomplex of the critical complex $K_{(i,i+2)}$ which stands for the critical value a_{i+1}. So, the end b is mapped to a_{i+1} and is closed because the class for $[b, d]$ is alive in $K_{(i,i+2)}$.

(Case 3.) K_d is the critical complex $K_{(i-1,i+1)}$ or is in the expansion of the backward inclusion $K_{(i-1,i+1)} \leftarrow\!\!\!-\!\!\!\rightarrow K_{(i,i+1)}$: the complex is a subcomplex of the critical complex $K_{(i-1,i+1)}$ which stands for the critical value a_i. So, the end d is mapped to a_i and made closed because the class for the bar $[b, d]$ exists in $K_{(i-1,i+1)}$.

(Case 4.) K_d is either the regular complex $K_{(i,i+1)}$ or in the expansion of $K_{(i,i+1)} \hookrightarrow\!\!\!\rightarrow K_{(i,i+2)}$: the complex is a subcomplex of the critical complex $K_{(i,i+2)}$ which stands for the critical value a_{i+1}. So, the end d is mapped to a_{i+1} and is open because the class for $[b, d]$ is not alive in $K_{(i,i+2)}$.

4.5.3 Correspondence to Sublevel Set Persistence

Standard persistence as we have seen already is defined by considering the sublevel sets of f, that is, $X_{[0,i]} = f^{-1}[s_0, s_i] = f^{-1}(-\infty, s_i]$ where $s_i \in (a_i, a_{i+1})$ is a regular value. We get the following sublevel set diagram:

$$\mathcal{X}: X_{[0,0]} \to X_{[0,1]} \to \cdots \to X_{[0,n]}.$$

Then, considering f to be a PL-function on $X = |K|$, we have already seen in Section 3.5 that $\mathcal{X}$ can be converted to a simplicial filtration $\mathcal{K}$ shown below where $K_{[0,i]} = \{\sigma \in K \mid f(\sigma) \le a_i\}$. This filtration can further be converted into a simplex-wise filtration which can be used for computing $\mathrm{Dgm}_p(\mathcal{K})$ for $p \ge 0$.

$$\mathcal{K}: K_{[0,0]} \to K_{[0,1]} \to K_{[0,2]} \to \cdots \to K_{[0,n]}$$

The bars for this case have the form $[a_i, a_j)$ where a_j can be $a_{n+1} = \infty$. Each such bar is closed at the left endpoint because the homology class being born exists at $K_{[0,i]}$. However, it is open at the right endpoint because it does not exist at $K_{[0,j]}$.

One can see that there are two types of bars in the sublevel set persistence, one of the type $[a_i, a_j)$, $j \le n$, which is bounded on the right, and the other of the type $[a_i, \infty) = [a_i, a_{n+1})$ which is unbounded on the right. The unbounded bars are the infinite bars we introduced in Section 3.2.1. They correspond to the essential homology classes since $\mathsf{H}_p(K) \cong \bigoplus_i [a_i, \infty)$. The work of [59, 63] imply that both types of barcodes of the standard persistence can be recovered from those of the levelset zigzag persistence as the theorem below states.

Theorem 4.15. *Let $\mathcal{K}$ and $\mathcal{K}'$ denote the filtrations for the sublevel sets and levelsets respectively induced by a continuous function f on a topological space with critical values $a_0, a_1, \ldots, a_{n+1}$ where $a_0 = -\infty$ and $a_{n+1} = \infty$. For every $p \ge 0$,*

(a) $[a_i, a_j)$, $j \ne n + 1$, is a bar for $\mathrm{Dgm}_p(\mathcal{K})$ if and only if it is so for $\mathrm{Dgm}_p(\mathcal{K}')$;

(b) $[a_i, a_{n+1})$ is a bar for $\mathrm{Dgm}_p(\mathcal{K})$ if and only if either $[a_i, a_j]$ is a closed–closed bar for $\mathrm{Dgm}_p(\mathcal{K}')$ for some $a_j > a_i$, or (a_j, a_i) is an open–open bar for $\mathrm{Dgm}_{p-1}(\mathcal{K}')$ for some $a_j < a_i$.

4.5.4 Correspondence to Extended Persistence

There is another persistence considered in the literature under the name *extended persistence* [103], and it turns out that there is a correspondence between extended persistence and levelset zigzag persistence. For a real-valued function $f : X \to \mathbb{R}$, let $X_{[0,i]}$ denote the sublevel set $f^{-1}[s_0, s_i]$ as before and $X_{[i,n]}$ denote the superlevel set $f^{-1}[s_i, s_n]$. Then, a persistence module that considers the sublevel set filtration first and then juxtaposes it with a filtration of quotient spaces of X as shown below gives the notion of extended persistence:

$$\mathfrak{X} : X_{[0,0]} \hookrightarrow \cdots \hookrightarrow X_{[0,n]} \hookrightarrow (X_{[0,n]}, X_{[n,n]}) \hookrightarrow \cdots \hookrightarrow (X_{[0,n]}, X_{[0,n]}).$$

Observe that each inclusion map between two quotient spaces induces a linear map in their relative homology groups. One can read that the above sequence arises by first growing the space to the full space $X_{[0,n]}$ with sublevel sets and then shrinking it by quotienting with the superlevel sets. Again, taking $f : X \to \mathbb{R}$ as a PL-function on $X = |K|$, we get the simplicial extended filtration where $K_{[0,i]} = \{\sigma \in K \mid f(\sigma) \le a_i\}$ and $K_{[i,n]} = \{\sigma \in K \mid f(\sigma) \ge a_i\}$:

$$\mathcal{E} : K_{[0,0]} \hookrightarrow \cdots \hookrightarrow K_{[0,n]} \hookrightarrow (K_{[0,n]}, K_{[n,n]}) \hookrightarrow \cdots \hookrightarrow (K_{[0,n]}, K_{[0,n]}).$$

The decomposition of the persistence module $\mathsf{H}_p \mathcal{E}$ arising out of $\mathcal{E}$ provides the bars in $\mathrm{Dgm}_p(\mathcal{E})$. For the first part of the sequence, the endpoints of the bars are designated with respective function values a_i as before. For the second part, the birth or death point of a bar is designated as a_{n+i} if its class either is born in $(K_{[0,n]}, K_{[i,n]})$ or dies entering into $(K_{[0,n]}, K_{[i,n]})$, respectively for $0 \le i \le n$. We leave the proof of the following theorem as an exercise; see also [63].

Theorem 4.16. *Let $\mathfrak{K}$ and $\mathcal{E}$ denote the simplicial levelset zigzag filtration and the extended filtration of a PL-function $f : |K| \to \mathbb{R}$. Then, for every $p \ge 0$,*

(a) *$[a_i, a_j)$ is a bar for $\mathrm{Dgm}_p(\mathfrak{K})$ if and only if it is a bar for $\mathrm{Dgm}_p(\mathcal{E})$*
(b) *$(a_i, a_j]$ is a bar for $\mathrm{Dgm}_p(\mathfrak{K})$ if and only if $[a_{n+j}, a_{n+i})$ is a bar for $\mathrm{Dgm}_{p+1}(\mathcal{E})$*
(c) *$[a_i, a_j]$ is a bar for $\mathrm{Dgm}_p(\mathfrak{K})$ if and only if $[a_i, a_{n+j})$ is a bar for $\mathrm{Dgm}_p(\mathcal{E})$*
(d) *(a_i, a_j) is a bar for $\mathrm{Dgm}_p(\mathfrak{K})$ if and only if $[a_j, a_{n+i})$ is a bar for $\mathrm{Dgm}_{p+1}(\mathcal{E})$.*

Clearly, for two persistence modules $\mathsf{H}_p \mathcal{E}$ and $\mathsf{H}_p \mathcal{E}'$ arising out of two extended filtrations $\mathcal{E}$ and $\mathcal{E}'$, the stability of persistence diagrams holds, that is, $\mathsf{d}_b(\mathrm{Dgm}_p \mathcal{E}, \mathrm{Dgm}_p \mathcal{E}') = \mathsf{d}_I(\mathsf{H}_p \mathcal{E}, \mathsf{H}_p \mathcal{E}')$ (Theorem 3.11).

4.6 Notes and Exercises

Computation of persistent homology induced by simplicial towers generalizing filtrations was considered in the context of topological data analysis (TDA) by Dey, Fan, and Wang [121]. They gave two approaches to compute persistence diagrams for such towers, one by converting a tower to a zigzag filtration, which we described in Section 4.4 and the other by considering annotations in combination with the link conditions allowing edge collapses without altering homotopy types which is described in Section 4.2.1. The first approach apparently increases the size of the filtration, which motivated the second approach. Kerber and Schreiber showed that indeed the first approach can be leveraged to produce filtrations instead of zigzag filtrations and without blowing up sizes [211].

The concept of zigzag modules obtained from a zigzag filtration by taking the homology groups and linear maps induced by inclusions is closely related to quiver theory due to Gabriel [163] which was brought to the attention of the TDA community by Carlsson and de Silva [62]. They were the first to propose the concept of zigzag persistence and its computation [62]. They observed that any zigzag module can be decomposed into a set of other zigzag modules where the forward nonzero maps are only injective and the backward nonzero maps are only surjective. Although they did not compute this decomposition, they used its existence to design an algorithm for computing the interval decomposition of a given zigzag module. Later, with Morozov, they used these concepts to present an $O(n^3)$ algorithm for computing the persistence of a simplex-wise zigzag filtration with n arrows [63]. Milosavljević et al. [235] improved the algorithm for any zigzag filtration with n arrows to have a time complexity of $O(n^\omega + n^2 \log^2 n)$, where $\omega \in [2, 2.373)$ is the exponent for matrix multiplication. Maria and Oudot [229] presented a different algorithm where they showed how a filtration of the last complex in the prefix of a zigzag filtration can help in computing the persistence incrementally. The algorithm in this chapter draws upon these approaches though is presented quite differently. Indeed, adaptation of the presented approach on graphs led to recent near-linear-time algorithms for zigzag persistence on graphs [126].

Given a real-valued function $f : X \to \mathbb{R}$ on a topological space X, the levelsets at the critical and intermediate values give rise to a levelset zigzag filtration as shown in Section 4.5. Carlsson, de Silva, and Morozov [63] introduced this setup and observed the decomposition of the zigzag module into interval modules with open or closed ends. The four types of bars arising out of this zigzag module give more information than the standard sublevel set persistence which only outputs closed–open and infinite bars. It was observed

in [59] that the open–open and closed–closed bars indeed capture the infinite
bars of the sublevel set persistence with an appropriate dimension shift. Theorem 4.15 summarizes this connection. The extended persistence originally
proposed for surfaces [5] and later extended for filtrations [103] also computes
all four types of bars, but they are described differently using the persistence
diagrams rather than open and closed ends.

Exercises

1. Show that the inequality in Proposition 4.1 cannot be improved to equality
   by giving a counterexample.
2. Prove Proposition 4.5.
3. Prove Proposition 4.6.
4. Prove Proposition 4.7.
5. Prove Proposition 4.8.
6. For computing the persistence of a simplicial tower, we checked the
   link condition in all dimensions. Argue that it is sufficient to check the
   condition only for three relevant dimensions.
7. Let K be a triangulated 2-manifold of genus g without boundary. Consider
   the following tasks:

   (a) Compute the genus g by the formula $2 - 2g = \#vertices - \#edges +$
       $\#triangles$.
   (b) Compute a spanning tree T of the 1-skeleton of K, and a spanning
       tree T^* of the dual graph none of whose edges are dual to any edge
       in T.
   (c) Annotate the edges in T with a zero vector of length g, and index the
       edges not in T and whose duals are not in T^* as $e_1, \ldots, e_{2g}$. Annotate
       e_i with a vector that has the ith entry 1 and all other entries 0.
   (d) Propagate systematically the annotation to the rest of the edges.
   (e) Complete the above approach with a proof of correctness into an algo-
       rithm that computes the annotation for edges in $O(gn)$ time if K has
       n simplices.

8. Do we get the same barcode if we run the zigzag persistence algorithm
   given in Section 4.3.1 and the standard persistence algorithm on a non-
   zigzag filtration? If so, prove it. If not, show the difference and suggest
   a modification to the zigzag persistence algorithm so that both outputs
   become the same.
9. Suppose that a persistence module $\{V_i \xrightarrow{f_i} V_{i+1}\}$ is presented with the
   linear maps f_i as matrices whose columns and rows are fixed bases of V_i

and V_{i+1}, respectively. Design an algorithm to compute the barcode for the input module. Do the same when the input module is a zigzag tower.

10. (See [127].) We have seen that for graphs a near-linear-time algorithm exists for computing non-zigzag persistence. Design a near-linear-time algorithm for computing zigzag persistence for graphs.

11. Consider a PL-function $f : |K| \to \mathbb{R}$.

    (a) Design an algorithm to compute the barcode of $-f$ from a levelset zigzag filtration of f.

    (b) Show that f and $-f$ produce the same closed–closed and open–open bars for the levelset zigzag filtration.

    (c) In general, given a zigzag filtration $\mathcal{F}$, consider the filtration $\mathcal{F}' = -\mathcal{F}$ in the opposite direction from right to left. What is the relation between the barcodes of these two filtrations?

12. We computed the persistence of zigzag towers by first converting it into a zigzag filtration and then using the algorithm in Section 4.3 to compute the bars. Design an algorithm that skips the intermediate conversion to a filtration.

13. Design an algorithm for computing the extended persistence from a given PL-function on an input simplicial complex.

14. (See [63].) Prove Theorem 4.16.

5

Generators and Optimality

So far we have focused mainly on the rank of the homology groups. However, the homology generators, that is, the cycles whose classes constitute the elements of the homology groups, carry information about the space. Computing just *some* generating cycles (cycle basis) typically can be done by the standard algorithms for computing homology groups such as the persistence algorithms. In practice, however, we may sometimes be interested in generating cycles that have some optimal property; see Figure 5.1.

In particular, if the space has a metric associated with it, one may associate a measure with the cycles that can differentiate them in terms of their "size." For example, if K is a simplicial complex embedded in $\mathbb{R}^d$, the measure of a 1-cycle can be its length. Then, we can ask to compute a set of 1-cycles whose classes generate $H_1(K)$ and has minimum total length among all such sets of cycles. Typically, the locality of these cycles captures interesting geometric features of the space $|K|$. Some applications may benefit from computing such cycles respecting geometry. For example, in computer graphics often a surface is cut along a set of cycles to make it flat for parameterization. The classes of

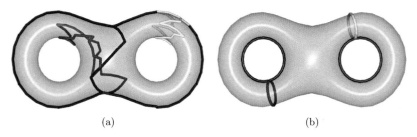

(a)   (b)

Figure 5.1 The double torus has first homology group of rank four, meaning that classes of four representative cycles generate H_1: (a) a non-optimal cycle basis, and (b) an optimal cycle basis.

148

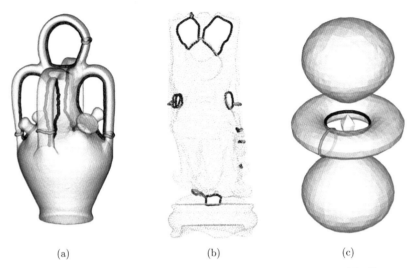

(a)   (b)   (c)

Figure 5.2 Computed shortest basis cycles (a) on a triangular mesh of Botijo, a well-known surface model in computer graphics, (b) on a point cloud data sampling the surface of Buddha, another well-known surface model in computer graphics, and (c) on an isosurface generated from a volume data in visualization.

these cycles constitute a basis of the first homology group. In general, a shortest (optimal) cycle basis is desired because it produces good parameterization for graphic rendering. Figure 5.2 shows examples of such cycles for three kinds of input where a shortest (optimal) cycle basis has been computed with an algorithm that we describe in this chapter. The algorithm works for simplicial complexes, though we can apply it on point cloud data as well after computing an appropriate complex such as Čech or Rips complex on top of the input points.

It turns out that, for $p > 1$, the problem of computing an optimal homology basis for the p-th homology group H_p is NP-hard [93]. However, the problem is polynomial-time solvable for $p = 1$ [135]. A greedy algorithm which was originally devised for computing an optimal H_1-basis for surfaces [156] extends to general simplicial complexes as described in Section 5.1.

There is another case of optimality, namely the *localization* of homology classes. In this problem, given a p-cycle c, we want to compute an optimal p-cycle c^* in the same homology class of c, that is, $[c] = [c^*]$. This problem is NP-hard even for $p = 1$ [73]. Interestingly, there are some special cases for which an integer program formulated for the problem can be solved with a linear program [125]. This is the topic of Section 5.2.

The two versions mentioned above do not consider the persistence framework. We may ask what are the optimal cycles for persistent homology classes.

Toward formulating the problem precisely, we define a persistent cycle for a given bar in the barcode of a filtration. This is a cycle whose class is created at the birth point and becomes a boundary at the death point of the bar. Among all persistent cycles for a given bar, we want to compute an optimal one. The problem in general is NP-hard, but one can devise polynomial-time algorithms for some special cases such as filtrations of what we call weak pseudomanifolds [128]. Section 5.3 describes these algorithms.

5.1 Optimal Generators/Basis

We now formalize the definition for optimal cycles whose classes generate the homology group. Strictly speaking these cycles should not be called generators because it is their classes which generate the group. We take the liberty to call the cycles themselves as the generators.

Definition 5.1. (Weight of cycles) Let $w \colon K_p \to \mathbb{R}_{\geq 0}$ be a nonnegative weight function defined on the set of p-simplices in a simplicial complex K. We extend w to the cycle space Z_p by defining $w(c) = \sum_i \alpha_i w(\sigma_i)$ where $c = \sum_i \alpha_i \sigma_i$ for $\alpha_i \in \mathbb{Z}_2$. For a set of cycles $\mathcal{C} = \{c_1, \ldots, c_g \mid c_i \in Z_p(K)\}$ define its *weight* $w(\mathcal{C}) = \sum_{i=1}^{g} w(c_i)$.

Definition 5.2. (Optimal generator) A set of cycles $\mathcal{C} = \{c_1, c_2, \ldots, c_g \mid c_i \in Z_p(K)\}$ is an $H_p(K)$-generator if the classes $\{[c_i] \mid i = 1, \ldots, g\}$ generate $H_p(K)$. An $H_p(K)$-*generator* is *optimal* if there is no other generator $\mathcal{C}'$ with $w(\mathcal{C}') < w(\mathcal{C})$.

Observe that an optimal generator may not have minimal number of cycles whose classes generate the homology group because we allow zero weights and hence an optimal generator may contain extra cycles with zero weights. This prompts us to define the following.

Definition 5.3. (Optimal basis) An $H_p(K)$-generator $\mathcal{C} = \{c_1, c_2, \ldots, c_g \mid c_i \in Z_p(K)\}$ is an $H_p(K)$-cycle basis or $H_p(K)$-basis in short if $g = \dim H_p(K)$. The classes of cycles in such a cycle basis constitute a basis for $H_p(K)$. An optimal $H_p(K)$-generator that is also an $H_p(K)$-basis is called an *optimal $H_p(K)$-basis*.

We observe that optimal $H_p(K)$-generators with positively weighted cycles are necessarily cycle bases. Notice that to generate $H_p(K)$, the number of cycles in any $H_p(K)$-generator has to be at least $\beta_p(K) = \dim H_p(K)$. On the other hand, an optimal $H_p(K)$-generator with positively weighted cycles cannot have more than β_p cycles because such a generator must contain a cycle whose class is a linear combination of the classes of other cycles in the generator. Thus, omission of this cycle still generates $H_p(K)$ while decreasing the weight of the generator. For one dimension, similar reasoning can also be applied to conclude that each cycle in an $H_1(K)$-cycle basis necessarily contains a *simple* cycle which together form a cycle basis (Exercise 1). A 1-cycle is simple if it has a single connected component (viewed as a graph) and every vertex has exactly two incident edges.

Fact 5.1. *We have the following.*

(a) *An optimal $H_p(K)$-generator with positively weighted cycles is an optimal $H_p(K)$-basis.*
(b) *Every cycle c_i in an $H_1(K)$-basis has a simple cycle $c_i' \subseteq c_i$ so that $\{c_i'\}_i$ form an $H_1(K)$-basis.*

We now focus on computing an optimal $H_p(K)$-basis, also known as the optimal homology basis problem or OHBP in short. One may observe that Definition 5.3 formulates OHBP as a *weighted ℓ^1-optimization* of representatives of bases. This allows for different types of optimality to be achieved by choosing different weights. For example, assume that the simplicial complex K of dimension p or greater is embedded in $\mathbb{R}^d$, where $d \geq p+1$. Let the Euclidean p-dimensional volume of p-simplices be their weights. This specializes OHBP to the *Euclidean ℓ^1-optimization* problem. The resulting optimal $H_p(K)$-basis has the smallest p-dimensional volume amongst all such bases. If the weights are taken to be unity, the resulting optimal solution has the smallest *number* of p-simplices amongst all $H_p(K)$-bases.

5.1.1 Greedy Algorithm for Optimal $H_p(K)$-Basis

Consider the following greedy algorithm in which we first sort the input cycles in nondecreasing order of their weights, and then choose a cycle following this order if its class is independent of the classes of the cycles chosen before.

The greedy algorithm Algorithm 7: GREEDYBASIS is motivated by Proposition 5.1. The specific implementation of line 4 (on independence test) will be given in Section 5.1.2.

Proposition 5.1. *Suppose that $\mathcal{C}$, the input to the algorithm GREEDYBASIS, contains an optimal $H_p(K)$-basis. Then, the output of GREEDYBASIS is an optimal $H_p(K)$-basis.*

Algorithm 7 GREEDYBASIS($\mathcal{C}$)

Input:
   A set of p-cycles $\mathcal{C}$ in a complex
Output:
   A maximal set of cycles from $\mathcal{C}$ whose classes are independent and total
weight is minimum

1: Sort the cycles from $\mathcal{C}$ in nondecreasing order of their weights; that is,
   $\mathcal{C} = \{c_1, \ldots, c_n\}$ implies $w(c_i) \leq w(c_j)$ for $i \leq j$
2: Let $B := \{c_1\}$
3: **for** $i = 2$ **to** n **do**
4:    **if** $[c_i]$ is independent with respect to B **then**
5:       $B := B \cup \{c_i\}$
6:    **end if**
7: **end for**
8: **if** $[c_1]$ is trivial (boundary), **then**
9:    output $B \setminus \{c_1\}$
10: **else**
11:    output B

PROOF. Let $\mathcal{C}$ contain an optimal $\mathsf{H}_p(K)$-basis $\mathcal{C}^* = \{c_1^*, \ldots, c_g^*\}$ sorted according to their appearance in the ordered sequence of $\mathcal{C} = \{c_1, \ldots, c_n\}$. Let $\mathcal{C}' = \{c_1', \ldots, c_{g'}'\}$ be the output of GREEDYBASIS again sorted according to the appearance of the cycles in $\mathcal{C}$. By Definition 5.3, g, the cardinality of $\mathcal{C}^*$, is the dimension of $\mathsf{H}_p(K)$ and hence $g' \leq g$ because $g + 1$ or more classes cannot be independent in $\mathsf{H}_p(K)$.

   Among all optimal $\mathsf{H}_p(K)$-bases that $\mathcal{C}$ contains, take $\mathcal{C}^*$ to be lexicographically smallest, that is, there is no other sorted $\tilde{\mathcal{C}}^* = \{\tilde{c}_1^*, \ldots, \tilde{c}_g^*\}$ so that there exists a $j \geq 1$ where $\tilde{c}_1^* = c_1^*, \ldots, \tilde{c}_{j-1}^* = c_{j-1}^*$ and $\tilde{c}_j^* = c_k$ and $c_j^* = c_\ell$ with $k < \ell$.

   First, we show that $\mathcal{C}'$ is a prefix of $\mathcal{C}^*$. If not, there is a least index $j \geq 1$ so that $c_j^* \neq c_j'$. Since the classes of the cycles in $\mathcal{C}^*$ form a basis for $\mathsf{H}_p(K)$, and $\mathcal{C}'$ cannot contain any trivial cycle (ensured by step 8), the class $[c_j']$ can be written as a linear combination of the classes of the cycles in $\mathcal{C}^*$. Consider the class $[c_k^*]$ in this linear combination with the largest index k. It is not possible that c_k^* appears before c_j' in the order. This is because then $[c_j']$ will be a linear combination of the classes of the cycles appearing before c_j' in $\mathcal{C}'$, which is impossible by the construction of $\mathcal{C}'$. So, assume that c_k^* appears after c_j'. Then, consider the sorted sequence of cycles $\tilde{\mathcal{C}}^*$ constructed by replacing c_k^* in $\mathcal{C}^*$

with c'_j. First, notice that $\tilde{\mathcal{C}}^*$ is lexicographically smaller than $\mathcal{C}^*$ and it is also an $H_p(K)$-basis contradicting the fact that $\mathcal{C}^*$ is the lexicographically smallest optimal cycle basis. The fact that $\tilde{\mathcal{C}}^*$ is an $H_p(K)$-cycle basis follows from the observation that $[c'_j]$ is independent of the classes of the cycles in $\mathcal{C}^* \setminus \{c^*_k\}$ because $[c'_j]$ is a linear combination of the classes that necessarily include $[c^*_k]$.

Now, to complete the proof, we note that $g' = g$. If not, then $g' < g$ and $\mathcal{C}'$ is a prefix of $\mathcal{C}^*$. But then one can add $c^*_{g'+1}$ from $\mathcal{C}^*$ to $\mathcal{C}'$ where $[c^*_{g'+1}]$ is independent of all classes of the cycles already in $\mathcal{C}'$. This suggests that the algorithm GREEDYBASIS cannot stop without enlarging $\mathcal{C}'$.   □

The above proposition suggests that GREEDYBASIS can compute an optimal cycle basis if its input set $\mathcal{C}$ contains one. We show next that such an input (i.e., a set of 1-cycles containing an optimal $H_1(K)$-basis) can be computed for $H_1(K)$ in $O(n^2 \log n)$ time where the 2-skeleton of K has n simplices.

Specifically, given a simplicial complex K, notice that $H_1(K)$ is completely determined by the 2-skeleton of K and hence without loss of generality we can assume K to be a 2-complex. Algorithm 8: GENERATOR computes a set $\mathcal{C}$ of 1-cycles from such a complex which includes an optimal basis.

Algorithm 8 GENERATOR(K)

Input:
    A 2-complex K
Output:
    A set of 1-cycles containing an optimal $H_1(K)$-basis

1: Let K^1 be the 1-skeleton of K with vertex set V and edge set E
2: $\mathcal{C} := \{\varnothing\}$
3: **for all** $v \in V$ **do**
4:     Compute a shortest path tree T_v rooted at v in $K^1 = (V, E)$
5:     **for all** $e = (u, w) \in E \setminus T_v$ such that $u, w \in T_v$ **do**
6:         Compute cycle $c_e = \pi_{u,w} \cup \{e\}$ where $\pi_{u,w}$ is the unique path connecting u and w in T_v
7:         $\mathcal{C} := \mathcal{C} \cup \{c_e\}$
8:     **end for**
9: **end for**
10: Output $\mathcal{C}$

Proposition 5.2. GENERATOR(K) *computes an* $H_1(K)$*-generator* $\mathcal{C}$ *with their weights in* $O(n^2 \log n)$ *time for a 2-complex* K *with* n *vertices and edges. Furthermore, the set* $\mathcal{C}$ *contains an optimal basis where* $|\mathcal{C}| = O(n^2)$.

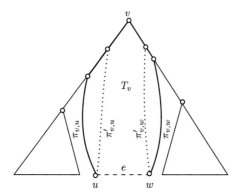

Figure 5.3 Tree T_v and the paths $\pi_{v,u}$, $\pi_{v,w}$, $\pi'_{v,u}$, and $\pi'_{v,w}$.

PROOF. We prove that any cycle c in an optimal H_1-basis $\mathcal{C}^*$ that is not computed by GENERATOR can be replaced by a cycle computed by GENERATOR while keeping $\mathcal{C}^*$ optimal. This proves the claim that the output of GENERATOR contains an optimal basis (and thus $\mathcal{C}$ is necessarily an H_1-generator).

First, assume that $\mathcal{C}^*$ consists of simple cycles because otherwise we can choose such cycles from the cycles of $\mathcal{C}^*$ due to Fact 5.1(b). So, assume that $c \in \mathcal{C}^*$ is simple. Let v be any vertex in c. There exists at least one edge e in c which is not in the shortest path tree T_v. Let $e = \{u, w\}$. Consider the shortest paths $\pi_{v,u}$ and $\pi_{v,w}$ in T_v from the root v to the vertices u and w, respectively. Notice that even though K^1 may be disconnected, vertices u and w are necessarily in T_v. Also, let $\pi'_{v,u}$ and $\pi'_{v,w}$ be the paths from v to u and w, respectively, in the cycle c. If $\pi_{v,u} = \pi'_{v,u}$ and $\pi_{v,w} = \pi'_{v,w}$ we have $c = c_e$ computed by GENERATOR. So, assume that at least one path does not satisfy this condition, say $\pi_{v,u} \neq \pi'_{v,u}$. See Figure 5.3.

Consider the two cycles c_1 and c_2 where c_1 consists of the paths $\pi'_{v,w}$, $\pi_{v,u}$ and e; and c_2 consists of the paths $\pi_{v,u}$ and $\pi'_{v,u}$. Observe that $c = c_1 + c_2$. Also, $w(c_1) \leq w(c)$ and $w(c_2) \leq w(c)$. If both $[c_1]$ and $[c_2]$ are dependent on the classes of the cycles in $\mathcal{C}^* \setminus c$, we will have $[c]$ dependent on them as well. This contradicts that $\mathcal{C}^*$ is an $\mathsf{H}_1(K)$-basis.

If $[c_1]$ is independent of the classes of cycles in $\mathcal{C}^* \setminus c$, obtain a new $\mathsf{H}_1(K)$-basis by replacing c with c_1. Then, apply the same argument on c_1 once more by taking the new vertex v to be the common ancestor of $\pi_{v,u}$ and $\pi'_{v,w}$ and the new edge e to be the old one. We will have a new H_1-basis whose weight is no more than $\mathcal{C}^*$ while replacing one of its cycles that is not computed by GENERATOR with a cycle necessarily computed by GENERATOR.

If $[c_2]$ is independent of the classes of cycles in $\mathcal{C}^* \setminus \{c\}$, obtain a new $\mathsf{H}_1(K)$-basis by replacing c with c_2 and then apply the same argument on c_2 once more by taking the new vertex v to be the common ancestor of $\pi_{v,u}$ and $\pi'_{v,u}$ and the new edge e to be an edge incident to u in c_2. Again, we will have a new H_1-basis whose weight is no more than $\mathcal{C}^*$ while replacing one of its cycles that is not computed by GENERATOR with a cycle necessarily computed by GENERATOR. This completes the claim that the output of GENERATOR contains an optimal basis.

To see that GENERATOR takes time as claimed, observe that each shortest path tree computation takes $O(n \log n)$ time by Dijkstra's algorithm implemented with Fibonacci heap [108]. Summing over $O(n)$ vertices, this gives $O(n^2 \log n)$ time. Each of the $O(n)$ edges in $E \setminus T_v$ for every vertex v gives $O(n)$ cycles in the output accounting for $O(n^2)$ cycles in total giving $|\mathcal{C}| = O(n^2)$. One can save space by representing each such cycle with the edge $E \setminus T_v$ while keeping T_v for all of them without duplicates. Also, observe that the weight of each cycle $w(c_e)$ can be computed as a byproduct of Dijkstra's algorithm because it computes the weights of the shortest paths from the root to any of the vertices. Therefore, in $O(n^2 \log n)$ time, GENERATOR can output an $\mathsf{H}_1(K)$-generator with their weights. $\qquad\square$

5.1.2 Optimal $\mathsf{H}_1(K)$-Basis and Independence Check

To compute an optimal $\mathsf{H}_1(K)$-basis, we first run GENERATOR on K and then feed the output to GREEDYBASIS as presented in Algorithm 9: OPTGEN which outputs an optimal H_1-basis due to Proposition 5.2.

Algorithm 9 OPTGEN(K)

Input:
  A 2-complex K
Output:
  An optimal $\mathsf{H}_1(K)$-basis

1: $\mathcal{C} := $ GENERATOR(K)
2: Output $\mathcal{C}^* := $ GREEDYBASIS($\mathcal{C}$)

However, we need to specify how to check the independence of the cycle classes in step 4 and triviality of cycle c_1 in step 8 of GREEDYBASIS. We do this by using annotations described in Section 4.2.1. Recall that $\mathsf{a}(\cdot)$ denotes the annotation of its argument which is a binary vector. Algorithm 10: ANNOTEDGE is a version of Algorithm 6: ANNOT adapted to edges only.

Algorithm 10 ANNOTEDGE(K)

Input:
  A simplicial 2-complex K
Output:
  Annotations for edges in K

1: Let K^1 be the 1-skeleton of K with edge set E
2: Compute a spanning forest T of K^1; $m = |E| - |T|$
3: For every edge $e \in E \cap T$, assign an m-vector $\mathsf{a}(e)$ where $\mathsf{a}(e) = 0$
4: Index remaining edges in $E \setminus T$ as $e_1, \ldots, e_m$
5: For every edge e_i, assign $\mathsf{a}(e_i)[j] = 1$ if and only if $j = i$
6: **for all** triangles $t \in K$ **do**
7:   **if** $\mathsf{a}(\partial t) \neq 0$ **then**
8:     Pick any nonzero entry b_u in $\mathsf{a}(\partial t)$
9:     Add $\mathsf{a}(\partial t)$ to every edge e so that $\mathsf{a}(e)[u] = 1$
10:    Delete the u-th entry from annotation of every edge
11: **end if**
12: **end for**

Assume that each cycle $c_e \in \mathcal{C}$ output by GENERATOR is represented by e and implicitly by the path $\pi_{u,w}$ in T_v. Assume that an annotation of edges has already been computed. This can be done by the algorithm ANNOTEDGE. A straightforward analysis shows that ANNOTEDGE takes $O(n^3)$ time, where n is the total number of vertices, edges, and triangles in K. However, for achieving better time complexity, we can use the *earliest basis* algorithm described in [60] which runs in time $O(n^\omega)$.

Once the annotations for edges are computed, we need to compute the annotations for the set $\mathcal{C}$ of cycles computed by GENERATOR to check independence among them in GREEDYBASIS. We first describe how we compute the annotations for a cycle in $\mathcal{C}$. We compute an *auxiliary* annotation of the vertices in T_v from the annotations of its edges to facilitate computing $\mathsf{a}(c_e)$ for cycles $c_e \in \mathcal{C}$. We traverse the tree T_v top-down and compute the auxiliary annotation $\mathsf{a}(x)$ of a vertex x in T_v as $\mathsf{a}(x) = \mathsf{a}(y) + \mathsf{a}(e_{xy})$ where y is the parent of x and e_{xy} is the edge connecting x and y. The process is initiated by assigning $\mathsf{a}(v)$ for the root v to be the zero vector. It should be immediately clear that all auxiliary annotations of the vertices can be computed in $O(gn)$ time where g, the length of the annotation vectors, equals $\beta_1(K)$. The annotation of each cycle $c_e \in \mathcal{C}$ is computed as $\mathsf{a}(c_e) = \mathsf{a}(u) + \mathsf{a}(w) + \mathsf{a}(e)$ where $e = (u, w)$. Again, this takes $O(g)$ time per edge e and hence per cycle $c_e \in \mathcal{C}$, giving a time complexity of $O(gn^2)$ in total for the entire set $\mathcal{C}$.

Next, we describe an efficient way of determining the independence of cycles as needed in step 4 of GREEDYBASIS. Independence of the class $[c_e]$ with respect to all classes already chosen by GREEDYBASIS is done in a batch mode. One can do it edge-by edge incurring more cost. We use a divide-and-conquer strategy instead.

Let $c_{e_1}, c_{e_2}, \ldots, c_{e_k}$ be the sorted order of cycles in $\mathcal{C}$ computed by GENERATOR. We construct a matrix A whose i-th column is the vector $\mathsf{a}(c_{e_i})$, and compute the first g columns that are independent called the *earliest basis* of A. Since there are k cycles in $\mathcal{C}$, the matrix A is $g \times k$. We use the following iterative method, based on making blocks, to compute the set J of indices of columns that define the earliest basis. We partition A from left to right into submatrices $A = [A_1 | A_2 | \cdots]$, where each submatrix A_i contains g columns, with the possible exception of the last submatrix, which contains at most g columns. Initially, we set J to be the empty set. We then iterate over the submatrices A_i by increasing index, that is, as they are ordered from left to right. At each iteration we compute the earliest basis for the matrix $[A_J | A_i]$, where A_J is the submatrix whose column indices are in J. We then set J to be the indices from the resulting earliest basis, increase i, and go to the next iteration. At each iteration we need to compute the earliest basis in a matrix with g rows and at most $|J| + g \leq 2g$ columns. Thus, each iteration takes $O(g^\omega)$ time, and there are at most $O(k/g) = O(n^2/g)$ iterations. Summing over all iterations, this gives a time complexity of $O(n^2 g^{\omega-1})$.

Theorem 5.3. *Given a simplicial 2-complex K with n simplices, an optimal $\mathsf{H}_1(K)$-basis can be computed in $O(n^\omega + n^2 g^{\omega-1})$ time.*

PROOF. An H_1-generator containing an optimal (cycle) basis can be computed in $O(n^2 \log n)$ time due to Proposition 5.2. One can compute an optimal H_1-basis from $\mathcal{C}$ by GREEDYBASIS due to Proposition 5.1. However, instead of using GREEDYBASIS, we can apply the divide-and-conquer technique outlined above for computing the cycles output by GREEDYBASIS which takes $O(n^\omega + n^2 g^{\omega-1})$ time. Retaining only the dominant terms, we obtain the claimed complexity for the entire algorithm. □

5.2 Localization

In this section we consider a different optimization problem. Here we are given a p-cycle c in an input complex with nonnegative weights on p-simplices and our goal is to compute a cycle c^* that is of optimal (minimal) weight in the homology class $[c]$; see Figure 5.4. We extend this localization problem from

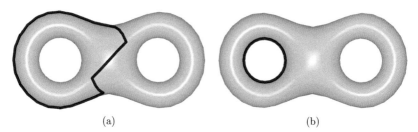

(a)   (b)

Figure 5.4 (a) A nontrivial cycle in a double torus, and (b) an optimal cycle in the class of the cycle in (a).

cycles to chains. For this, first we extend the concept of homologous cycles in Section 2.5 to chains straightforwardly. Two p-chains $c, c' \in \mathsf{C}_p$ are called homologous if and only if they differ by a boundary, that is, $c \in c' + \mathsf{B}_p$. We ask for computing a chain of minimal weight which is homologous to a given chain.

Definition 5.4. Let $w \colon K(p) \to \mathbb{R}_{\geq 0}$ be a nonnegative weight function defined on the set of p-simplices in a simplicial complex K. We extend w to the chain group C_p by defining $w(c) = \sum_i c_i w(\sigma_i)$ where $c = \sum_i c_i \sigma_i$.

Definition 5.5. (OHCP) Given a nonnegative weight function $w \colon K(p) \to \mathbb{R}_{\geq 0}$ defined on the set of p-simplices in a simplicial complex K and a p-chain c in $\mathsf{C}_p(K)$, the *optimal homologous chain problem* (OHCP) is to find a chain c^* which has the minimal weight $w(c^*)$ among all chains homologous to c.

If we use $\mathbb{Z}_2$ as the coefficient ring for defining homology classes, the OHCP becomes NP-hard. We are going to show that it becomes polynomial-time solvable if (i) the coefficient ring is chosen to be integers $\mathbb{Z}$ and (ii) the complex K is such that $\mathsf{H}_p(K)$ does not have a torsion which may be introduced because of using $\mathbb{Z}$ as the coefficient ring.

We will formulate OHCP as an integer program which requires the chains to be represented as an integer vector. Given a p-chain $x = \sum_{i=0}^{m-1} x_i \sigma_i$ with integer coefficients x_i, we use $\mathbf{x} \in \mathbb{Z}^m$ to denote the vector formed by the coefficients x_i. Thus, $\mathbf{x}$ is the representation of the chain x in the elementary p-chain basis, and we will use $\mathbf{x}$ and x interchangeably.

Recall that for a vector $\mathbf{x} \in \mathbb{R}^m$, the 1-*norm* (or ℓ^1-norm) $\|\mathbf{x}\|_1$ is $\sum_i |x_i|$. Let W be any real $m \times m$ diagonal matrix with diagonal entries w_i. Then, the 1-norm of $W\mathbf{x}$, that is, $\|W\mathbf{x}\|_1$, is $\sum_i |w_i||x_i|$. (If W is a general $m \times m$ nonsingular matrix then $\|W\mathbf{x}\|_1$ is called the *weighted 1-norm* of $\mathbf{x}$.)

We now state in words our approach to the optimal homologous chains and later formalize it in Eq. (5.1). The main idea is to cast OHCP as an integer program. Unfortunately, integer programs are in general NP-hard and thus cannot be solved in polynomial time unless P = NP. We solve it by a linear program and identify a class of integer programs called *totally unimodular* for which linear programs give an exact solution. Then, we interpret total unimodularity in terms of topology. Our approach to solve OHCP can be succinctly stated by the following steps:

1. Write OHCP as an integer program involving 1-norm minimization, subject to linear constraints.
2. Convert the integer program into an integer *linear* program by converting the 1-norm cost function to a linear one using the standard technique of introducing some extra variables and constraints.
3. Find the conditions under which the constraint matrix of the integer linear program is totally unimodular.
4. For this class of problems, relax the integer linear program to a linear program by dropping the constraint that the variables be integral. The resulting optimal chain obtained by solving the linear program will be an integer-valued chain homologous to the given chain.

5.2.1 Linear Program

Now we formally pose OHCP as an optimization problem. After showing the existence of solutions, we reformulate the optimization problem as an integer linear program and eventually as a linear program.

Assume that the number of p- and $(p + 1)$-simplices in K is m and n, respectively, and let W be a diagonal $m \times m$ matrix. Using the notation from Section 3.3.1, let D_p represent the boundary matrix for the boundary operator $\partial_p : C_p \to C_{p-1}$ in the elementary chain bases. With this notation, given a p-chain $\mathbf{c}$ represented with an integral vector, the optimal homologous chain problem in dimension p is to solve:

$$\min_{\mathbf{x}, \mathbf{y}} \| W \mathbf{x} \|_1 \quad \text{such that} \quad \mathbf{x} = \mathbf{c} + D_{p+1}\, \mathbf{y}, \text{ and } \mathbf{x} \in \mathbb{Z}^m, \ \mathbf{y} \in \mathbb{Z}^n . \quad (5.1)$$

We assume that W is a diagonal matrix obtained from nonnegative *weights* on simplices. Let w be a nonnegative real-valued weight function on the oriented p-simplices of K and let W be the corresponding diagonal matrix (the i-th diagonal entry of W is $w(\sigma_i) = w_i$).

The resulting objective function $\| W \mathbf{x} \|_1 = \sum_i w_i |x_i|$ in (5.1) is not linear in x_i because it uses the absolute value of x_i. However, it is piecewise-linear in these variables. As a result, Eq. (5.1) can be reformulated as an integer *linear program* by splitting every variable x_i into two parts x_i^+ and x_i^- [27, page 18]:

$$\min \sum_i w_i \, (x_i^+ + x_i^-)$$

$$\text{subject to} \quad \mathbf{x}^+ - \mathbf{x}^- = \mathbf{c} + D_{p+1} \, \mathbf{y}, \qquad (5.2)$$

$$\mathbf{x}^+, \; \mathbf{x}^- \geq \mathbf{0},$$

$$\mathbf{x}^+, \; \mathbf{x}^- \in \mathbb{Z}^m, \; \mathbf{y} \in \mathbb{Z}^n \, .$$

Comparing the above formulation to the standard form integer linear program in Eq. (5.4), we notice that the vector $\mathbf{x}$ in Eq. (5.4) corresponds to $[\mathbf{x}^+, \mathbf{x}^-, \mathbf{y}]^T$ in Eq. (5.2) above. Thus, the minimization is over $\mathbf{x}^+, \mathbf{x}^-$, and $\mathbf{y}$, and the coefficients of x_i^+ and x_i^- in the objective function are w_i, but the coefficients corresponding to y_j are zero. The linear programming relaxation of this formulation just removes the constraints about the variables being integral. The resulting linear program is:

$$\min \sum_i w_i \, (x_i^+ + x_i^-)$$

$$\text{subject to} \quad \mathbf{x}^+ - \mathbf{x}^- = \mathbf{c} + D_{p+1} \, \mathbf{y},$$

$$\mathbf{x}^+, \; \mathbf{x}^- \geq \mathbf{0} \, .$$

To cast the program in standard form [27], we can eliminate the free (unrestricted in sign) variables $\mathbf{y}$ by replacing these by $\mathbf{y}^+ - \mathbf{y}^-$ and imposing the nonnegativity constraints on the new variables. The resulting linear program has the same objective function, and the equality constraints:

$$\min \sum_i w_i \, (x_i^+ + x_i^-)$$

$$\text{subject to} \quad \mathbf{x}^+ - \mathbf{x}^- = \mathbf{c} + D_{p+1} \, (\mathbf{y}^+ - \mathbf{y}^-),$$

$$\mathbf{x}^+, \; \mathbf{x}^-, \; \mathbf{y}^+, \; \mathbf{y}^- \geq \mathbf{0} \, .$$

We can write the above program as

$$\min \mathbf{f}^T \mathbf{z} \quad \text{subject to} \quad A\mathbf{z} = \mathbf{c}, \; \mathbf{z} \geq \mathbf{0}, \qquad (5.3)$$

where $\mathbf{f} = [\mathbf{w}, 0]^T$, $\mathbf{z} = [\mathbf{x}^+, \mathbf{x}^-, \mathbf{y}^+, \mathbf{y}^-]^T$, and the equality constraint matrix is $A = [I \;\; -I \;\; -B \;\; B]$, where $B = D_{p+1}$. This is exactly in the form we want the linear program to be in view of Eq. (5.4). We now prove a result about the total unimodularity of this matrix that allows us to solve the optimization by a linear program.

5.2.2 Total Unimodularity

A matrix is called *totally unimodular* if the determinant of each square submatrix is 0, 1, or -1. The significance of total unimodularity in our setting is due to the following Theorem 5.4, which follows immediately from known results in optimization [201].

Consider an integral vector $\mathbf{b} \in \mathbb{Z}^m$ and a real vector $\mathbf{f} \in \mathbb{R}^n$. Consider the *integer* linear program

$$\min \mathbf{f}^{\mathrm{T}}\mathbf{x} \quad \text{subject to} \quad A\mathbf{x} = \mathbf{b}, \ \mathbf{x} \geq \mathbf{0} \text{ and } \mathbf{x} \in \mathbb{Z}^n . \qquad (5.4)$$

Theorem 5.4. *Let A be an $m \times n$ totally unimodular matrix. Then the integer linear program* (5.4) *can be solved in time polynomial in the dimensions of A.*

Proposition 5.5. *If $B = D_{p+1}$ is totally unimodular, then so is the matrix $[I \quad -I \quad -B \quad B]$.*

PROOF. The proof uses operations that preserve the total unimodularity of a matrix. These are listed in [272, page 280]. If B is totally unimodular then so is the matrix $[-B \quad B]$ since scalar multiples of columns of B are being appended on the left to get this matrix. The full matrix in question can be obtained from this one by appending columns with a single ± 1 on the left, which proves the result. $\square$

As a result of Theorem 5.4 and Proposition 5.5, we have the following *algorithmic* result.

Theorem 5.6. *If the boundary matrix D_{p+1} of a finite simplicial complex of dimension greater than p is totally unimodular, the optimal homologous chain problem* (5.1) *for p-chains can be solved in polynomial time.*

PROOF. We have seen above that a reformulation of OHCP without the integrality constraints leads to the linear program (5.3). By Proposition 5.5, the equality constraint matrix of this linear program is totally unimodular. Then, by Theorem 5.4, the linear program (5.3) can be solved in polynomial time, while achieving an integral solution. $\square$

Manifolds

Our results in the next section (Section 5.2.3) are valid for *any* finite simplicial complex. But first we consider a simpler case – simplicial complexes that are triangulations of manifolds. We show that for finite triangulations of compact p-dimensional *orientable* manifolds, the top nontrivial boundary matrix D_p

is totally unimodular irrespective of the orientations of its simplices. There are examples of non-orientable manifolds where total unimodularity does not hold (Exercise 7). Further examination of why total unimodularity does not hold in these cases leads to the results in Theorem 5.9.

Let K be a finite simplicial complex that triangulates a $(p + 1)$-dimensional compact orientable manifold M.

Theorem 5.7. *For a finite simplicial complex triangulating a $(p + 1)$-dimensional compact orientable manifold, D_{p+1} is totally unimodular irrespective of the orientations of the simplices.*

As a result of the above theorem and Theorem 5.6 we have the following result.

Corollary 5.8. *For a finite simplicial complex triangulating a $(p + 1)$-dimensional compact orientable manifold, the optimal homologous chain problem can be solved for p-dimensional chains in polynomial time.*

5.2.3 Relative Torsion

Now we consider the more general case of simplicial complexes. We characterize the total unimodularity of boundary matrices for arbitrary simplicial complexes. This characterization leads to a torsion-related condition for the complexes; see [242] for the definition of torsion. Since we do not use any conditions about the geometric realization or embedding in $\mathbb{R}^p$ for the complex, the result is also valid for abstract simplicial complexes. As a corollary of the characterization we show that the OHCP can be solved in polynomial time as long as the input complex satisfies a torsion-related condition.

TU and Relative Torsion

Definition 5.6. (Pure simplicial complex) A *pure simplicial complex* of dimension p is a simplicial complex formed by a collection of p-simplices and their faces. Similarly, a *pure subcomplex* is a subcomplex that is a pure simplicial complex.

An example of a pure simplicial complex of dimension p is one that triangulates a p-dimensional manifold. Another example, relevant to our discussion, is a subcomplex formed by a collection of some p-simplices of a simplicial complex and their faces.

Let K be a finite simplicial complex of dimension greater than p. Let $L \subseteq K$ be a pure subcomplex of dimension $p + 1$ and $L_0 \subset L$ be a pure subcomplex of dimension p. Recall the definition of relative boundary operator in Section 2.5.2 used for defining relative homology. Then, the matrix D_{p+1}^{L,L_0} representing the relative boundary operator

$$\partial_{p+1}^{L,L_0} : \mathsf{C}_{p+1}(L, L_0) \to \mathsf{C}_p(L, L_0)$$

is obtained by first *including* the columns of D_{p+1} corresponding to $(p + 1)$-simplices in L and then, from the submatrix so obtained, *excluding* the rows corresponding to the p-simplices in L_0 and any zero rows. The zero rows correspond to p-simplices that are not faces of any of the $(p + 1)$-simplices of L. Then the following holds.

Theorem 5.9. *Matrix D_{p+1} is totally unimodular if and only if $\mathsf{H}_p(L, L_0)$ is torsion-free, for all pure subcomplexes L_0 and L of K of dimensions p and $p + 1$, respectively, where $L_0 \subset L$.*

PROOF. (only if) We show that if $\mathsf{H}_p(L, L_0)$ has torsion for some L and L_0 then D_{p+1} is not totally unimodular. Let D_{p+1}^{L,L_0} be the corresponding relative boundary matrix. Bring D_{p+1}^{L,L_0} to the so-called *Smith normal form* which is a block matrix

$$\begin{bmatrix} \Delta & 0 \\ 0 & 0 \end{bmatrix},$$

where $\Delta = \operatorname{diag}(d_1, \dots, d_l)$ is a diagonal matrix with $d_i \geq 1$ being integers. The row or column of zero matrices in the block shown above may be empty, depending on the dimension of the matrix. This can be done, for example, by using the reduction algorithm [242, pages 55–57]. The construction of the Smith normal form implies that $d_k > 1$ for some $1 \leq k \leq l$ because $\mathsf{H}_p(L, L_0)$ has torsion. Thus, the product $d_1 \cdots d_k$ is greater than 1. By a result of Smith [281] mentioned in [272, page 50], this product is the greatest common divisor of the determinants of all $k \times k$ square submatrices of D_{p+1}^{L,L_0}. It follows that some square submatrix of D_{p+1}^{L,L_0}, and hence of D_{p+1}, has determinant value greater than 1. Then, D_{p+1} is not totally unimodular.

(if) Assume that D_{p+1} is not totally unimodular. We show that, in that case, there exist subcomplexes L_0 and L of dimensions p and $(p + 1)$, respectively, with $L_0 \subset L$, so that $\mathsf{H}_p(L, L_0)$ has torsion. Let S be a square submatrix of D_{p+1} so that $|\det(S)| > 1$. Let L correspond to the columns of D_{p+1} that are *included* in S and let B_L be the submatrix of D_{p+1} formed by these columns. This submatrix B_L may contain zero rows. Those zero rows (if any) correspond to p-simplices that are not a facet of any of the $(p+1)$-simplices in L. To form

S from B_L, we first discard the zero rows to form a submatrix B'_L. This is safe because $\det(S) \neq 0$ and so these zero rows cannot occur in S.

The rows in B'_L correspond to p-simplices that adjoin some $(p+1)$-simplex in L. Let L_0 correspond to rows of B'_L which are *excluded* to form S. Observe that S is the relative boundary matrix D_p^{L,L_0}. Consider the Smith normal form of S. This normal form is a square diagonal matrix obtained by reducing S. Since the elementary row and column operations used for this reduction preserve determinant magnitude, the determinant of the resulting diagonal matrix has magnitude greater than 1. It follows that at least one of the diagonal entries in the normal form is greater than 1. Then, by [242, page 61] $\mathsf{H}_p(L, L_0)$ has torsion. □

Corollary 5.10. *For a simplicial complex K of dimension greater than p, there is a polynomial-time algorithm for answering the following question: Is $\mathsf{H}_p(L, L_0)$ torsion-free for all subcomplexes L_0 and L of dimensions p and $(p+1)$ such that $L_0 \subset L$?*

PROOF. Seymour's decomposition theorem for totally unimodular matrices [272, Theorem 19.6; 273] yields a polynomial-time algorithm for deciding if a matrix is totally unimodular or not [272, Theorem 20.3]. That algorithm applied on the boundary matrix D_{p+1} proves the above assertion. □

A Special Case

In Section 5.2.2, we have seen the special case of compact orientable manifolds. We saw that the top dimensional boundary matrix of a finite triangulation of such a manifold is totally unimodular. Now we show another special case for which the boundary matrix is totally unimodular and hence OHCP is polynomial-time solvable. This case occurs when we ask for optimal p-chains in a simplicial complex K which is embedded in $\mathbb{R}^{p+1}$. In particular, OHCP can be solved by linear programming for 2-chains in 3-complexes embedded in $\mathbb{R}^3$. This follows from the following result.

Theorem 5.11. *Let K be a finite simplicial complex embedded in $\mathbb{R}^{p+1}$. Then, $\mathsf{H}_p(L, L_0)$ is torsion-free for all pure subcomplexes L_0 and L of dimensions p and $p+1$, respectively, such that $L_0 \subset L$.*

Corollary 5.12. *Given a p-chain c in a weighted finite simplicial complex embedded in $\mathbb{R}^{p+1}$, an optimal chain homologous to c can be computed by a linear program.*

5.3 Persistent Cycles

So far, we have considered optimal cycles in a given complex. Now, we consider optimal cycles in the context of a filtration. We know that a filtration of a complex gives rise to persistence of homology classes. An interval module which appears as a bar in the barcode is created by homology classes that get born and die at the endpoints. However, the bar is not associated with the class of a particular cycle because more than one cycle may get born and die at the endpoints. Among all these cycles, we want to identify the cycle that is optimal with respect to a weight assignment as defined earlier. Note that, by Remark 3.4 in Section 3.4, an interval $[b, d-1]$ in the interval decomposition of a persistence module $\mathsf{H}_p(\mathcal{F})$ arising from a simplicial filtration $\mathcal{F}$ corresponds to a closed–open interval $[b, d)$ contributing a point (b, d) in the persistence diagram $\mathrm{Dgm}_p(\mathcal{F})$ as defined in Definition 3.8. We also say that the interval $[b, d)$ belongs to $\mathrm{Dgm}_p(\mathcal{F})$.

Let the cycles be weighted with a weight function $w\colon K(p) \to \mathbb{R}_{\geq 0}$ defined on the set of p-simplices in a simplicial complex K as before.

Definition 5.7. (Persistent cycle) Given a filtration $\mathcal{F}\colon \varnothing = K_0 \hookrightarrow K_1 \hookrightarrow \cdots \hookrightarrow K_n = K$, and a finite interval $[b, d) \in \mathrm{Dgm}_p(\mathcal{F})$, we say a cycle c is a *persistent cycle* for $[b, d)$ if c is born at K_b and becomes a boundary in K_d. For an infinite interval $[b, \infty) \in \mathrm{Dgm}_p(\mathcal{F})$, we say a cycle c is a *persistent cycle* for $[b, \infty)$ if c is born at K_b. In both cases, a persistent cycle is called optimal if it has the least weight among all such cycles for a bar.

Depending on whether the interval is finite or not, we have two cases captured in the following definitions.

Problem 1. (PCYC-FIN$_p$) Given a finite filtration $\mathcal{F}$ and a finite interval $[b, d) \in \mathrm{Dgm}_p(\mathcal{F})$, this problem asks for computing an optimal persistent p-cycle for the bar $[b, d)$.

Problem 2. (PCYC-INF$_p$) Given a finite filtration $\mathcal{F}$ and an infinite interval $[b, \infty) \in \mathrm{Dgm}_p(\mathcal{F})$, this problem asks for computing an optimal persistent p-cycle for the bar $[b, \infty)$.

When $p \geq 2$, computing optimal persistent p-cycles for both finite and infinite intervals is NP-hard in general. We identify a special but important class of simplicial complexes, which we term as *weak $(p+1)$-pseudomanifolds*, whose optimal persistent p-cycles can be computed in polynomial time. A weak

$(p+1)$-pseudomanifold is a generalization of a $(p+1)$-manifold and is defined as follows.

Definition 5.8. (Weak pseudomanifold) A simplicial complex K is a *weak* $(p+1)$-*pseudomanifold* if each p-simplex is a face of no more than two $(p+1)$-simplices in K.

Specifically, it turns out that if the given complex is a weak $(p+1)$-pseudomanifold, the problem of computing optimal persistent p-cycles for finite intervals can be cast into a minimal cut problem (see Section 5.3.1) due to the fact that persistent cycles of such kind are null-homologous in the complex. However, when $p \geq 2$ and intervals are infinite, the computation of the same becomes NP-hard. Nonetheless, for infinite intervals, if we assume that the weak $(p+1)$-pseudomanifold is embedded in $\mathbb{R}^{p+1}$, then the optimal persistent p-cycle problem reduces to a minimal cut problem (see Section 5.3.3) and hence belongs to P. Note that a simplicial complex that can be embedded in $\mathbb{R}^{p+1}$ is necessarily a weak $(p+1)$-pseudomanifold. We also note that while there is an algorithm [93] in the nonpersistence setting which computes an optimal p-cycle by minimal cuts (Exercise 8, the nonpersistence algorithm assumes the $(p+1)$-complex to be embedded in $\mathbb{R}^{p+1}$), the algorithm for finite intervals presented here, to the contrary, does not need the embedding assumption.

Before we present the algorithms for cases where they run in polynomial time, we summarize the complexity results for different cases. In order to make our statements about the hardness results precise, we let WPCYC-FIN$_p$ denote a subproblem[1] of PCYC-FIN$_p$ and let WPCYC-INF$_p$ and WEPCYC-INF$_p$ denote two subproblems of PCYC-INF$_p$, with the subproblems requiring additional constraints on the given simplicial complex. Table 5.1 lists the hardness results for all problems of interest, where the column "Restriction on K" specifies the additional constraints that the subproblems require on the given simplicial complex K. Note that WPCYC-INF$_p$ being NP-hard trivially implies that PCYC-INF$_p$ is NP-hard.

The polynomial-time algorithms for the cases listed in Table 5.1 map the problem of computing optimal persistent cycles into the classic problem of computing minimal cuts in a flow network. The only exception is PCYC-INF$_1$ which can be solved by computing Dijkstra's shortest paths in graphs. We will not consider this special case here whose details can be found in [127].

---

[1] For two problems P_1 and P_2, P_2 is a *subproblem* of P_1 if any instance of P_2 is an instance of P_1 and P_2 asks for computing the same solutions as P_1.

Table 5.1 *Hardness results for optimal persistent cycle problems.*

Problem	Restriction on K	p	Hardness
PCYC-FIN$_p$	—	≥ 1	NP-hard
WPCYC-FIN$_p$	K a weak $(p+1)$-pseudomanifold	≥ 1	Polynomial
PCYC-INF$_p$	—	$= 1$	Polynomial
WPCYC-INF$_p$	K a weak $(p+1)$-pseudomanifold	≥ 2	NP-hard
WEPCYC-INF$_p$	K a weak $(p+1)$-pseudomanifold in $\mathbb{R}^{p+1}$	≥ 2	Polynomial

Undirected Flow Network

An *undirected flow network* (G, s_1, s_2) consists of an undirected graph G with vertex set $V(G)$ and edge set $E(G)$, a capacity function $C \colon E(G) \to [0, +\infty]$, and two non-empty disjoint subsets s_1 and s_2 of $V(G)$. Vertices in s_1 are referred to as *sources* and vertices in s_2 are referred to as *sinks*. A *cut* (S, T) of (G, s_1, s_2) consists of two disjoint subsets S and T of $V(G)$ such that $S \cup T = V(G)$, $s_1 \subseteq S$, and $s_2 \subseteq T$. We define the set of edges *across* the cut (S, T) as

$$E(S, T) = \{e \in E(G) \mid e \text{ connects a vertex in } S \text{ and a vertex in } T\}.$$

The *capacity* of a cut (S, T) is defined as $C(S, T) = \sum_{e \in E(S,T)} C(e)$. A *minimal cut* of (G, s_1, s_2) is a cut with the minimal capacity. How these graphs and cuts appear in our algorithm is explained next with an illustration in Figure 5.5. Note that we allow parallel edges in G (see Figure 5.6) to ease the presentation. These parallel edges can be merged into one edge during computation.

5.3.1 Finite Intervals for Weak $(p+1)$-Pseudomanifolds

In this subsection, we present an algorithm which computes optimal persistent p-cycles for finite intervals given a filtration of a weak $(p+1)$-pseudomanifold when $p \geq 1$. The general approach proceeds as follows: Suppose that the input weak $(p+1)$-pseudomanifold is K which is associated with a simplex-wise filtration $\mathcal{F} \colon \varnothing = K_0 \hookrightarrow K_1 \hookrightarrow \cdots \hookrightarrow K_n$ and the task is to compute an optimal persistent cycle of a finite interval $[b, d) \in \mathrm{Dgm}_p(\mathcal{F})$. Let $\sigma_b^{\mathcal{F}}$ and $\sigma_d^{\mathcal{F}}$ be the creator and destructor pair for the interval $[b, d)$. We first construct an undirected dual graph G for K where vertices of G are dual to $(p+1)$-simplices of K and edges of G are dual to p-simplices of K. One dummy vertex v_∞ termed as the *infinite vertex* which does not correspond to any $(p+1)$-simplices is added to G for graph edges dual to those boundary p-simplices, that is, the p-simplices that are faces of at most one $(p+1)$-simplex. We then build an undirected flow network on top of G where the source is the vertex dual to $\sigma_d^{\mathcal{F}}$ and the sink is the infinite vertex along with the

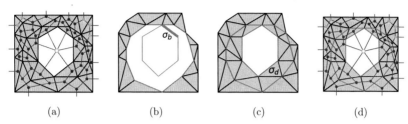

| (a) | (b) | (c) | (d) |

Figure 5.5 An example of the constructions in our algorithm showing the duality between persistent cycles and cuts having finite capacity for $p = 1$. (a) The input weak 2-pseudomanifold K with its dual flow network drawn in blue, where the central hollow vertex denotes the dummy vertex, the red vertex denotes the source, and the orange vertices denote the sinks. All graph edges dual to the outer boundary 1-simplices actually connect to the dummy vertex. (b) The partial complex K_b in the input filtration $\mathcal{F}$, where the bold green 1-simplex denotes $\sigma_b^{\mathcal{F}}$ which creates the green 1-cycle. (c) The partial complex K_d in $\mathcal{F}$, where the 2-simplex $\sigma_d^{\mathcal{F}}$ creates the pink 2-chain killing the green 1-cycle. (d) The green persistent 1-cycle of the interval $[b, d)$ is dual to a cut (S, T) having finite capacity, where S contains all the vertices inside the pink 2-chain and T contains all the other vertices. The red graph edges denote those edges across (S, T) and their dual 1-chain is the green persistent 1-cycle.

set of vertices dual to those $(p + 1)$-simplices which are added to $\mathcal{F}$ after $\sigma_d^{\mathcal{F}}$. If a p-simplex is $\sigma_b^{\mathcal{F}}$ or added to $\mathcal{F}$ before $\sigma_b^{\mathcal{F}}$, we let the capacity of its dual graph edge be its weight; otherwise, we let the capacity of its dual graph edge be $+\infty$. Finally, we compute a minimal cut of this flow network and return the p-chain dual to the edges across the minimal cut as an optimal persistent cycle of the interval.

The intuition of the above algorithm is best explained by an example illustrated in Figure 5.5, where $p = 1$. The key to the algorithm is the duality between persistent cycles of the input interval and cuts of the dual flow network having finite capacity. To see this duality, first consider a persistent p-cycle c of the input interval $[b, d)$. There exists a $(p + 1)$-chain A in K_d created by $\sigma_d^{\mathcal{F}}$ whose boundary equals c, making c killed. We can let S be the set of graph vertices dual to the simplices in A and let T be the set of the remaining graph vertices, then (S, T) is a cut. Furthermore, (S, T) must have finite capacity as the edges across it are exactly dual to the p-simplices in c and the p-simplices in c have indices in $\mathcal{F}$ less than or equal to b. On the other hand, let (S, T) be a cut with finite capacity, then the $(p + 1)$-chain whose simplices are dual to the vertices in S is created by $\sigma_d^{\mathcal{F}}$. Taking the boundary of this $(p + 1)$-chain, we get a p-cycle c. Because p-simplices of c are exactly dual to the edges across (S, T) and each edge across (S, T) has finite capacity, c must reside in K_b. We only need to ensure that c contains $\sigma_b^{\mathcal{F}}$ in order to show that c is a persistent cycle of $[b, d)$. In Section 5.3.2, we argue that c indeed contains $\sigma_b^{\mathcal{F}}$ (proof of Theorem 5.14), so c is a persistent cycle.

In the dual graph, an edge is created for each p-simplex. If a p-simplex has two $(p + 1)$-cofaces, we simply let its dual graph edge connect the two vertices dual to its two $(p + 1)$-cofaces; otherwise, its dual graph edge has to connect to the infinite vertex on one end. A problem about this construction is that some weak $(p + 1)$-pseudomanifolds may have p-simplices being the face of no $(p + 1)$-simplices and these p-simplices may create self-loops around the infinite vertex. To avoid self-loops, we simply ignore these p-simplices. The reason why we can ignore these p-simplices is that they cannot be on the boundary of a $(p + 1)$-chain and hence cannot be on a persistent cycle of minimal weight. Algorithmically, we ignore these p-simplices by constructing the dual graph only from what we call the $(p + 1)$-connected component of K containing $\sigma_d^{\mathcal{F}}$.

Definition 5.9. (q-connected) Let K be a simplicial complex. For $q \geq 1$, two q-simplices σ and σ' of K are *q-connected in K* if there is a sequence of q-simplices of K, $(\sigma_0, \ldots, \sigma_l)$, such that $\sigma_0 = \sigma$, $\sigma_l = \sigma'$, and for all $0 \leq i < l$, σ_i and σ_{i+1} share a $(q - 1)$-face. The property of q-connectedness defines an equivalence relation on q-simplices of K. Each set in the partition induced by the equivalence relation constitutes a *q-connected component* of K. We say K is *q-connected* if any two q-simplices of K are q-connected in K. See Figure 5.6 for an example of 1-connected components and 2-connected components.

We present the pseudocode in Algorithm 11: MINPERSCYCFIN and it works as follows. Lines 1 and 2 set up a complex $\widetilde{K}$ that the algorithm mainly works on, where $\widetilde{K}$ is taken as the closure of the $(p + 1)$-connected component of $\sigma_d^{\mathcal{F}}$. Line 3 constructs the dual graph G from $\widetilde{K}$ and lines 4–15 build the flow network on top of G. Note that we denote the infinite vertex by v_∞. Line 16 computes a minimal cut for the flow network and line 17

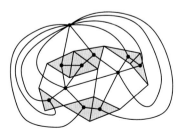

Figure 5.6 A weak 2-pseudomanifold $\widetilde{K}$ embedded in $\mathbb{R}^2$ with three voids. Its dual graph is drawn. The complex has one 1-connected component and four 2-connected components with the 2-simplices in 2-connected components shaded.

returns the p-chain dual to the edges across the minimal cut. In the pseudocodes, to make presentation of algorithms and some proofs easier, we treat a mathematical function as a programming object. For example, the function θ returned by DUALGRAPHFIN in MINPERSCYCFIN denotes the correspondence between the simplices of $\widetilde{K}$ and their dual vertices or edges (see Section 5.3.1 for details). In practice, these constructs can be easily implemented in any programming language.

Algorithm 11 MINPERSCYCFIN($K, p, \mathcal{F}, [b, d)$)

Input:
K: finite p-weighted weak $(p + 1)$-pseudomanifold
p: integer ≥ 1
$\mathcal{F}$: filtration $K_0 \subseteq K_1 \subseteq \cdots \subseteq K_n$ of K
$[b, d)$: finite interval of $\mathrm{Dgm}_p(\mathcal{F})$
Output:
An optimal persistent p-cycle for $[b, d)$

1: $L^{p+1} \leftarrow (p + 1)$-connected component of K containing $\sigma_d^{\mathcal{F}}$ * set up $\widetilde{K}$ *\
2: $\widetilde{K} \leftarrow$ closure of the simplicial set L^{p+1}
3: $(G, \theta) \leftarrow$ DUALGRAPHFIN($\widetilde{K}, p$) * construct dual graph *\
4: **for all** $e \in E(G)$ **do**
5:     **if** index($\theta^{-1}(e)$) $\leq b$ **then**
6:         $C(e) \leftarrow w(\theta^{-1}(e))$ * assign finite capacity *\
7:     **else**
8:         $C(e) \leftarrow +\infty$ * assign infinite capacity *\
9:     **end if**
10: **end for**
11: $s_1 \leftarrow \{\theta(\sigma_d^{\mathcal{F}})\}$ * set the source *\
12: $s_2 \leftarrow \{v \in V(G) \mid v \neq v_\infty, \text{index}(\theta^{-1}(v)) > d\}$ * set the sink *\
13: **if** $v_\infty \in V(G)$ **then**
14:     $s_2 \leftarrow s_2 \cup \{v_\infty\}$
15: **end if**
16: $(S^*, T^*) \leftarrow$ min-cut of (G, s_1, s_2)
17: Output $\theta^{-1}(E(S^*, T^*))$

Complexity

The time complexity of MINPERSCYCFIN depends on the encoding scheme of the input and the data structure used for representing a simplicial complex. For encoding the input, we assume K and $\mathcal{F}$ are represented by a sequence of all the simplices of K ordered by their indices in $\mathcal{F}$, where each simplex is denoted by its set of vertices. We also assume a simple yet reasonable simplicial complex data structure as follows: In each dimension, simplices are mapped to integral identifiers ranging from zero to the number of simplices in that dimension minus one; each q-simplex has an array (or linked list) storing all the id's of its $(q+1)$-cofaces; a hash map for each dimension is maintained for the query of the integral id of each simplex in that dimension based on the spanning vertices of the simplex. We further assume p to be constant. By the above assumptions, let n be the size (number of bits) of the encoded input, then there are no more than n elementary $O(1)$ operations in lines 1 and 2, so the time complexity of lines 1 and 2 is $O(n)$. It is not hard to verify that the flow network construction also takes $O(n)$ time so the time complexity of MINPERSCYCFIN is determined by the minimal cut algorithm. Using the max-flow algorithm by Orlin [249], the time complexity of MINPERSCYCFIN becomes $O(n^2)$.

In the rest of this section, we first describe the subroutine DUALGRAPHFIN, then close the section by proving the correctness of the algorithm.

Dual Graph Construction
We describe the DUALGRAPHFIN subroutine used in Algorithm MINPERSCYCFIN, which returns a dual graph G and a θ denoting two bijections which we use to prove the correctness. Given the input $(\widetilde{K}, p)$, DUALGRAPHFIN constructs an undirected connected graph G as follows:

- Let each vertex v of $V(G)$ correspond to each $(p+1)$-simplex σ^{p+1} of $\widetilde{K}$. If there is any p-simplex of $\widetilde{K}$ which has less than two $(p+1)$-cofaces in $\widetilde{K}$, we add an infinite vertex v_∞ to $V(G)$. Simultaneously, we define a bijection

$$\theta: \{(p+1)\text{-simplices of } \widetilde{K}\} \to V(G) \smallsetminus \{v_\infty\}$$

by letting $\theta(\sigma^{p+1}) = v$. Note that in the above range notation of θ, $\{v_\infty\}$ may not be a subset of $V(G)$.
- Let each edge e of $E(G)$ correspond to each p-simplex σ^p of $\widetilde{K}$. Note that σ^p has at least one $(p+1)$-coface in $\widetilde{K}$. If σ^p has two $(p+1)$-cofaces σ_0^{p+1} and σ_1^{p+1} in $\widetilde{K}$, then let e connect $\theta(\sigma_0^{p+1})$ and $\theta(\sigma_1^{p+1})$; if σ^p has one $(p+1)$-coface σ_0^{p+1} in $\widetilde{K}$, then let e connect $\theta(\sigma_0^{p+1})$ and v_∞. We define another bijection

$$\theta: \{p\text{-simplices of } \widetilde{K}\} \to E(G)$$

using the same notation as the bijection for $V(G)$, by letting $\theta(\sigma^p) = e$.

Note that we can take the image of a subset of the domain under a function. Therefore, if (S, T) is a cut for a flow network built on G, then $\theta^{-1}(E(S, T))$ denotes the set of p-simplices dual to the edges across the cut. Also note that, since simplicial chains with $\mathbb{Z}_2$ coefficients can be interpreted as sets, $\theta^{-1}(E(S, T))$ is also a p-chain.

5.3.2 Algorithm Correctness

In this subsection, we prove the correctness of the algorithm MINPER-SCYCFIN. Some of the symbols we use refer to the pseudocode of the algorithm.

Proposition 5.13. *In the algorithm* MINPERSCYCFIN, s_2 *is not an empty set.*

PROOF. For contradiction, suppose that s_2 is an empty set. Then $v_\infty \notin V(G)$ and $\sigma_d^{\mathcal{F}}$ is the $(p + 1)$-simplex of $\widetilde{K}$ with the greatest index in $\mathcal{F}$. Since $v_\infty \notin V(G)$, any p-simplex of $\widetilde{K}$ must be a face of two $(p + 1)$-simplices of $\widetilde{K}$, so the set of $(p + 1)$-simplices of $\widetilde{K}$ forms a $(p + 1)$-cycle created by $\sigma_d^{\mathcal{F}}$. Then $\sigma_d^{\mathcal{F}}$ must be a positive simplex in $\mathcal{F}$, which is a contradiction. □

The following two propositions specify the duality mentioned at the beginning of Section 5.3.1.

Proposition 5.14. *For any cut (S, T) of (G, s_1, s_2) with finite capacity, the p-chain $c = \theta^{-1}(E(S, T))$ is a persistent p-cycle of $[b, d)$ and $w(c) = C(S, T)$.*

PROOF. Let $A = \theta^{-1}(S)$; it is easy to check that $c = \partial(A)$. The key is to show that c is created by $\sigma_b^{\mathcal{F}}$ which we show now. Suppose that c is created by a p-simplex $\sigma^p \neq \sigma_b^{\mathcal{F}}$. Since $C(S, T)$ is finite, we have that $\text{index}(\sigma^p) < b$. We can let c' be a persistent cycle of $[b, d)$ and $c' = \partial(A')$ where A' is a $(p + 1)$-chain of K_d. Then we have $c + c' = \partial(A + A')$. Since A and A' are both created by $\sigma_d^{\mathcal{F}}$, then $A + A'$ is created by a $(p + 1)$-simplex with an index less than d in $\mathcal{F}$. So $c + c'$ is a p-cycle created by $\sigma_b^{\mathcal{F}}$ which becomes a boundary before $\sigma_d^{\mathcal{F}}$ is added. This means that $\sigma_b^{\mathcal{F}}$ is already paired when $\sigma_d^{\mathcal{F}}$ is added, contradicting the fact that $\sigma_b^{\mathcal{F}}$ is paired with $\sigma_d^{\mathcal{F}}$. Similarly, we can prove that c is not a boundary until $\sigma_d^{\mathcal{F}}$ is added, so c is a persistent cycle of $[b, d)$. Since (S, T) has finite capacity, we must have

$$C(S, T) = \sum_{e \in \theta(c)} C(e) = \sum_{\theta^{-1}(e) \in c} w(\theta^{-1}(e)) = w(c). \qquad \square$$

Proposition 5.15. *For any persistent p-cycle c of $[b, d)$, there exists a cut (S, T) of (G, s_1, s_2) such that $C(S, T) \leq w(c)$.*

PROOF. Let A be a $(p + 1)$-chain in K_d such that $c = \partial(A)$. Note that A is created by $\sigma_d^{\mathcal{F}}$ and c is the set of p-simplices which are the face of exactly one $(p + 1)$-simplex of A. Let $c' = c \cap \widetilde{K}$ and $A' = A \cap \widetilde{K}$; we claim that $c' = \partial(A')$. To prove this, first let σ^p be any p-simplex of c', then σ^p is a face of exactly one $(p + 1)$-simplex σ^{p+1} of A. Since $\sigma^p \in \widetilde{K}$, it is also true that $\sigma^{p+1} \in \widetilde{K}$, and so $\sigma^{p+1} \in A'$. Then σ^p is a face of exactly one $(p + 1)$-simplex of A', so $\sigma^p \in \partial(A')$. On the other hand, let σ^p be any p-simplex of $\partial(A')$, then σ^p is a face of exactly one $(p + 1)$-simplex σ_0^{p+1} of A'. Note that $\sigma_0^{p+1} \in A$, and we want to prove that σ^p is a face of exactly one $(p + 1)$-simplex σ_0^{p+1} of A. Suppose that σ^p is a face of another $(p + 1)$-simplex σ_1^{p+1} of A, then $\sigma_1^{p+1} \in \widetilde{K}$ because $\sigma_0^{p+1} \in \widetilde{K}$. So we have $\sigma_1^{p+1} \in A \cap \widetilde{K} = A'$, contradicting the fact that σ^p is a face of exactly one $(p + 1)$-simplex of A'. Then we have $\sigma^p \in \partial(A)$. Since $\sigma_0^{p+1} \in \widetilde{K}$, we have $\sigma^p \in \widetilde{K}$, which means that $\sigma^p \in c'$.

Let $S = \theta(A')$ and $T = V(G) \smallsetminus S$; then it is true that (S, T) is a cut of (G, s_1, s_2) because A' is created by $\sigma_d^{\mathcal{F}}$. We claim that $\theta^{-1}(E(S, T)) = \partial(A')$. The proof of the equality is similar to the one in the proof of Proposition 5.14. It follows that $E(S, T) = \theta(c')$. We then have that

$$C(S, T) = \sum_{e \in \theta(c')} C(e) = \sum_{\theta^{-1}(e) \in c'} w(\theta^{-1}(e)) = w(c')$$

because each p-simplex of c' has an index less than or equal to b in $\mathcal{F}$.

Finally, since c' is a subchain of c, we must have $C(S, T) = w(c') \leq w(c)$. □

Combining the above results, we conclude:

Theorem 5.16. *Algorithm* MINPERSCYCFIN *computes an optimal persistent p-cycle for the given interval $[b, d)$.*

PROOF. First, the flow network (G, s_1, s_2) constructed by the algorithm MIN-PERSCYCFIN must be valid by Proposition 5.13. Since the interval $[b, d)$ must have a persistent cycle, the flow network (G, s_1, s_2) has a cut with finite capacity by Proposition 5.15. This means that $C(S^*, T^*)$ is finite. By Proposition 5.14, the chain $c^* = \theta^{-1}(E(S^*, T^*))$ is a persistent cycle of $[b, d)$. Suppose that c^* is not an optimal persistent cycle of $[b, d)$ and instead let c' be a minimal persistent cycle of $[b, d)$. Then there exists a cut (S', T') such that $C(S', T') \leq w(c') < w(c^*) = C(S^*, T^*)$ by Propositions 5.14 and 5.15, contradicting the fact that (S^*, T^*) is a minimal cut.   □

5.3.3 Infinite Intervals for Weak $(p + 1)$-Pseudomanifolds Embedded in $\mathbb{R}^{p+1}$

We have already mentioned that computing optimal persistent p-cycles ($p \geq 2$) for infinite intervals is NP-hard even if we restrict to weak $(p + 1)$-pseudomanifolds [128].

However, when the complex is embedded in $\mathbb{R}^{p+1}$, the problem becomes polynomial-time tractable. In this subsection, we present an algorithm for this problem given a weak $(p + 1)$-pseudomanifold embedded in $\mathbb{R}^{p+1}$, when $p \geq 1$. For $p = 1$, the problem is polynomial-time tractable for arbitrary complexes; see Exercise 9. The algorithm uses a similar duality as described in Section 5.3.1. However, a direct use of the approach in Section 5.3.1 does not work. In particular, the dual graph construction is different – previously there is only one dummy vertex corresponding to infinity, now there is one per void. For example, in Figure 5.6, 1-simplices that do not have any 2-cofaces cannot reside in any 2-connected component of the 2-complex. Hence, no cut in the flow network may correspond to a persistent cycle of the infinite interval created by such a 1-simplex. Furthermore, unlike the finite interval case, we do not have a negative simplex whose dual can act as a source in the flow network.

Let $(K, \mathcal{F}, [b, +\infty))$ be an input to the problem where K is a weak $(p + 1)$-pseudomanifold embedded in $\mathbb{R}^{p+1}$, $\mathcal{F} \colon \varnothing = K_0 \hookrightarrow K_1 \hookrightarrow \cdots \hookrightarrow K_n$ is a simplex-wise filtration of K, and $[b, +\infty)$ is an infinite interval of $\mathrm{Dgm}_p(\mathcal{F})$. By the definition of the problem, the task boils down to computing an optimal p-cycle containing $\sigma_b^{\mathcal{F}}$ in K_b. Note that K_b is also a weak $(p + 1)$-pseudomanifold embedded in $\mathbb{R}^{p+1}$.

Generically, assume that $\widetilde{K}$ is an arbitrary weak $(p + 1)$-pseudomanifold embedded in $\mathbb{R}^{p+1}$ and we want to compute an optimal p-cycle containing a p-simplex $\widetilde{\sigma}$ for $\widetilde{K}$. By the embedding assumption, the connected components of $\mathbb{R}^{p+1} \setminus |\widetilde{K}|$ are well defined and we call them the *voids* of $\widetilde{K}$. The complex $\widetilde{K}$ has a natural (undirected) dual graph structure as illustrated by Figure 5.6 for $p = 1$, where the graph vertices are dual to the $(p + 1)$-simplices as well as the voids, and the graph edges are dual to the p-simplices. The duality between cycles and cuts is as follows: Since the ambient space $\mathbb{R}^{p+1}$ is contractible (homotopy equivalent to a point), every p-cycle in $\widetilde{K}$ is the boundary of a $(p + 1)$-dimensional region obtained by pointwise union of certain $(p + 1)$-simplices and/or voids. We can derive a cut[2] of the dual graph

---

[2] The cut mentioned here is defined on a graph without sources and sinks, so a cut is simply a partition of the graph's vertex set into two sets.

by putting all vertices contained in the $(p + 1)$-dimensional region into one vertex set and putting the rest into the other vertex set. On the other hand, for every cut of the graph, we can take the pointwise union of all the $(p + 1)$-simplices and voids dual to the graph vertices in one set of the cut and derive a $(p + 1)$-dimensional region. The boundary of the derived $(p + 1)$-dimensional region is then a p-cycle in $\widetilde{K}$. We observe that by making the source and sink dual to the two $(p + 1)$-simplices or voids that $\widetilde{\sigma}$ adjoins, we can build a flow network where a minimal cut produces an optimal p-cycle in $\widetilde{K}$ containing $\widetilde{\sigma}$.

The efficiency of the above algorithm is in part determined by the efficiency of the dual graph construction. This step requires identifying the voids that the boundary p-simplices are incident on; see Figure 5.6 for an illustration. A straightforward approach would be to first group the boundary p-simplices into p-cycles by local geometry, and then build the nesting structure of these p-cycles to correctly reconstruct the boundaries of the voids. This approach has a quadratic worst-case complexity. To make the void boundary reconstruction faster, we assume that the simplicial complex being worked on is p-connected so that building the nesting structure is not needed. This reconstruction then runs in almost linear time. To satisfy the p-connected assumption, we begin the algorithm by taking $\widetilde{K}$ as a p-connected subcomplex of K_b containing $\sigma_b^{\mathcal{F}}$ and continue only with this $\widetilde{K}$. The computed output is still correct because the minimal cycle in $\widetilde{K}$ is again a minimal cycle in K_b. We skip the details of constructing void boundaries which can be done in $O(n \log n)$ time. Also, we skip the proof of correctness of the following theorem. Interested readers can consult [128] for details.

Theorem 5.17. *Given an infinite interval $[b, \infty) \in \mathrm{Dgm}_p(\mathcal{F})$ for a filtration $\mathcal{F}$ of a weak $(p + 1)$-pseudomanifold K embedded in $\mathbb{R}^{p+1}$, an optimal persistent cycle for $[b, \infty)$ can be computed in $O(n^2)$ time, where n is the number of p and $(p + 1)$-simplices in K.*

5.4 Notes and Exercises

The algorithm to compute an optimal homology basis based on a greedy strategy was first presented by Erickson and Whittlesey [156] who applied it to simplicial 2-manifolds (surfaces). Chen and Freedman [93] showed that the problem is NP-hard for all homology groups of dimensions more than one. It was shown in [135] that an optimal H_1-cycle basis can be computed in $O(n^4)$ time for a simplicial complex with n simplices. The time complexity was improved to $O(n^{\omega+1})$ by Busaryev et al. [60]. Finally, it was settled to $O(n^3)$ in [129]. Borradaile et al. [43] proposed an algorithm for computing an optimal H_1-basis for graphs embedded on surfaces. For a graph with a total of n vertices and edges, the algorithm runs in $O(g^3 n \log n)$ time, where g is the genus plus the number of boundaries in the surface.

The problem of computing a minimal homologous cycle in a given class is NP-hard even in dimension one, as shown by Chambers et al. [73]. They proposed an algorithm for 1-cycles on surfaces utilizing the duality between minimal cuts of a surface-embedded graph and optimal homologous cycles of a dual complex. A better algorithm is proposed in [74]. Both algorithms are fixed-parameter tractable running in time exponential in the genus of the surface. For general dimension, Borradaile et al. [44] showed that the OHCP problem in dimension p can be $O(\sqrt{\log n})$-approximated and is fixed-parameter tractable for weak $(p + 1)$-pseudomanifolds. The only polynomial-time exact algorithm [93] in general dimension for OHCP works for p-cycles in complexes embedded in $\mathbb{R}^{p+1}$, which uses a reduction to minimal (s, t) cuts. Interestingly, when the coefficient is chosen to be $\mathbb{Z}$ instead of $\mathbb{Z}_2$ for the homology groups, the problem becomes polynomial-time solvable if there is no relative torsion as shown in [125]. The material presented in Section 5.2 is taken from that paper.

Persistence added an extra layer of complexity to the problem of computing minimal representative cycles. Escolar and Hiraoka [157] and Obayashi [248] formulated the problem as an integer program by adapting a similar formulation for the nonpersistent case. Wu et al. [302] adapted the algorithm of Busaryev et al. [60] to present an exponential-time algorithm, as well as an A* heuristics in practice. The problem of computing an optimal persistent cycle is NP-hard even for H_1 [127]. The problem becomes polynomial-time solvable for some special cases such as computing optimal persistent 2-cycles in a 3-complex embedded in three dimensions [128]. The materials in Section 5.3 are taken from this last source.

Exercises

1. Show that every cycle in an $H_1(K)$-basis contains a simple cycle which together form an $H_1(K)$-basis themselves.
2. Design an $O(n^2 \log n + n^2 g)$ algorithm to compute the shortest nontrivial 1-cycle in a simplicial 2-complex K with n simplices and $g = \beta_1(K)$. Do the same in $O(n^2 \log n)$ time when K is a 1-complex (a graph).
3. (See [129].) We have given an $O(n^\omega + n^2 g^{\omega-1})$ algorithm for computing an optimal H_1-basis for a complex with n simplices. Taking $g = \Omega(n)$, this runs in $O(n^{\omega+1})$ worst-case time. Give an $O(n^3)$ algorithm for the problem.

4. How can one make the algorithm in [129] more efficient for a weighted graph G with n vertices and edges? For this, show that (i) an annotation for G can be computed in $O(n^2)$ time, (ii) this annotation can be utilized to compute the annotations for $O(n^2)$ candidate cycles in $O(n^3)$ time, and (iii) finally, an optimal basis can be computed in $O(n^3)$ time by the divide-and-conquer greedy algorithm in [129], though more efficiently.

5. Define a minmax basis of $\mathsf{H}_p(K)$ as the set of cycles which generate $\mathsf{H}_p(K)$ and the maximum weight of the cycles is minimized among all such generators. Prove that an optimal H_p-cycle basis defined in Definition 5.3 is also a minmax basis.

6. Prove that a simplicial p-complex embedded in $\mathbb{R}^p$ cannot have torsion in H_{p-1} and hence OHCP for $(p-1)$-cycles can be solved in polynomial time in this case.

7. Take an example of a triangulation of a Möbius strip and show that the integer program formulation of OHCP for it is not totally unimodular.

8. Professor Optimist claims that an optimal H_p-generator for K embedded in $\mathbb{R}^{p+1}$ can be obtained by computing optimal persistent p-cycles for infinite bars in any filtration of K. Show that he is wrong. Give a polynomial-time algorithm for computing a nontrivial p-cycle that has the least weight in K.

9. Consider computing a persistent 1-cycle for a bar $[b, d)$ given a filtration of an edge-weighted complex K. Let c be a cycle created by the edge $e = (u, v)$ at the birth time b where c is formed by the edge e and the shortest path between u and v in the 1-skeleton of the complex K_b. If $[c] = 0$ at K_d, prove that c is an optimal persistent cycle for the bar $[b, d)$.

10. Give an example where the above computed cycle using the shortest path at the birth time is not a persistent cycle.

11. For a finite interval $[b, d) \in \mathrm{Dgm}_p(\mathcal{F})$ of a filtration $\mathcal{F}$ of a weak $(p + 1)$-pseudomanifold, one can take the two vertices of the dual edge of the creator p-simplex σ_b in the algorithm MINPERSCYCFIN (Section 5.3.1) as source and sink, respectively. Give an example to show that this does not work for computing a minimal persistent cycle for $[b, d)$. What about taking the dual vertex of the destroyer simplex σ_d and the infinite vertex as the source and the sink, respectively?

12. (See [92].) For a vertex v in a complex with nonnegative weights on edges, let the discrete geodesic ball B_v^r of radius r be the maximal subcomplex $L \subseteq K$ so that the shortest path from v to every vertex in L is at most r. For a cycle c, let $w(c) = \min\{r \mid c \subseteq B_v^r\}$. Give a polynomial-time algorithm to compute an optimal H_p-cycle basis for any $p \geq 1$ with these weights.

6

Topological Analysis of Point Clouds

In this chapter, we focus on topological analysis of point cloud data (PCD), a common type of input data across a broad range of applications. Often, there is a hidden space of interest, and the PCD we obtain contains only observations/samples from that hidden space. If the sample is sufficiently dense, it should carry information about the hidden space. We are interested in topological information in particular. However, discrete points themselves do not have interesting topology. To impose a connectivity that mimics that of the hidden space, we construct a simplicial complex such as the Rips or Čech complex using the points as vertices. Then, an appropriate filtration is constructed as a proxy for the same on the topological space that the PCD presumably samples. This provides topological summaries such as the persistence diagrams induced by the filtrations. Figure 6.1 [192] shows an example application of this approach. The PCD in this case represents atomic configurations of silica in three different states: liquid, glass, and crystal states. Each atomic configuration can be viewed as a set of weighted points, where each point represents the center of an atom and its weight is the radius of the atom. The persistence diagrams for the three states show distinctive features which can be used for further analysis of the phase transitions. The persistence diagrams can also be viewed as a signature of the input PCD and can be used to compare shapes (e.g., [78]) or provide other analysis.

We mainly focus on PCD consisting of a set of points $P \subseteq (Z, \mathsf{d}_Z)$ embedded in some metric space Z equipped with a metric d_Z. One of the most common choices for (Z, d_Z) in practice is the d-dimensional Euclidean space $\mathbb{R}^d$ equipped with the standard L_p-distance. We review the relevant concepts of constructing Rips and Čech complexes, their filtrations, and describe the properties of the resulting persistence diagrams in Section 6.1. In practice, the size of a filtration can be prohibitively large. In Section 6.2, we discuss

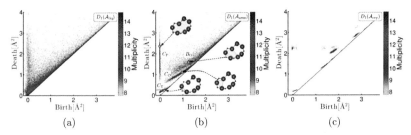

Figure 6.1 Persistence diagrams of silica in liquid (a), glass (b), and crystal (c) states. Image taken from [192], reprinted by permission from Y. Hiraoka et al. (2016, fig. 2).

data sparsification strategies to approximate topological summaries much more efficiently and with theoretical guarantees.

As we have mentioned, a PCD can be viewed as a window through which we can peek at topological properties of the hidden space. In particular, we can infer about the hidden homological information using the PCD at hand if it samples the hidden space sufficiently densely. In Section 6.3, we provide such inference results for the cases when the hidden space is a manifold or is a compact set embedded in the Euclidean space. To obtain theoretical guarantees, we also need to introduce the language of sampling conditions to describe the quality of point samples. Finally, in Section 6.4, we focus on the inference of scalar field topology from a set of point samples P, as well as function values available at these samples. More precisely, we wish to estimate the persistent homology of a real-valued function $f : X \to \mathbb{R}$ from a set of discrete points $P \subset X$ as well as the values of f over P.

6.1 Persistence for Rips and Čech Filtrations

Suppose we are given a finite set of points P in a metric space (Z, d_Z). Consider a closed ball $B_Z(p, r)$ with radius r centered at each point $p \in P$ and consider the space $P^r := \bigcup_{p \in P} B_Z(p, r)$. The Čech complex with respect to P and a parameter $r \geq 0$ is defined as (Definition 2.9)

$$\mathbb{C}_Z^r(P) = \{\sigma = \{p_0, \ldots, p_k\} \mid \bigcap_{i \in [0,k]} B_Z(p, r) \neq \varnothing\}. \quad (6.1)$$

We often omit Z from the subscript when its choice is clear. As mentioned in Section 2.2, the Čech complex $\mathbb{C}^r(P)$ is the *nerve* of the union of balls P^r. If the metric balls centered at points in P in the metric space (Z, d_Z) are convex, then the Nerve Theorem (Theorem 2.1) gives the following corollary.

Corollary 6.1. *For a fixed* $r \geq 0$, *if the metric ball* $B_Z(x, r)$ *is convex for every* $x \in P$, *then* $\mathbb{C}^r(P)$ *is homotopy equivalent to* P^r, *and thus* $\mathsf{H}_k(\mathbb{C}^r(P)) \cong \mathsf{H}_k(P^r)$ *for any dimension* $k \geq 0$.

The above result justifies the utility of Čech complexes. For example, if $P \subseteq \mathbb{R}^d$ and d_Z is the standard L_p-distance for $p > 0$, then the Čech complex $\mathbb{C}^r(P)$ becomes homotopy equivalent to the union of r-radius balls centering at points in P. Later in this chapter, we will also see an example where the points P are taken from a Riemannian manifold X equipped with the Riemannian metric d_X. When the radius r is small enough, the intrinsic metric balls also become convex. In both cases, the resulting Čech complex captures information of the union of r-balls P^r.

In general, it is not clear at which scale (radius r) one should inspect the input PCD. Varying the scale parameter r, we obtain a filtration of spaces $\mathcal{P} := \{P^\alpha \hookrightarrow P^{\alpha'}\}_{\alpha \leq \alpha'}$ as well as a filtered sequence of simplicial complexes $\mathcal{C}(P) := \{\mathbb{C}^\alpha(P) \hookrightarrow \mathbb{C}^{\alpha'}(P)\}_{\alpha \leq \alpha'}$. The homotopy equivalence between P^r and $\mathbb{C}^r$, if it holds, further induces an isomorphism between the persistence modules obtained from these two filtrations.

Proposition 6.2. [90] *If the metric ball* $B(x, r)$ *is convex for every* $x \in P$ *and all* $r \geq 0$, *then the persistence module* $\mathsf{H}_k\mathcal{P}$ *is isomorphic to the persistence module* $\mathsf{H}_k\mathcal{C}(P)$. *This also implies that their corresponding persistence diagrams are identical; that is,* $\mathrm{Dgm}_k\mathcal{P} = \mathrm{Dgm}_k\mathcal{C}(P)$, *for any dimension* $k \geq 0$.

A related persistence-based topological invariant is given by the *Vietoris–Rips filtration* $\mathcal{R}(P) = \{\mathbb{VR}^\alpha(P) \hookrightarrow \mathbb{VR}^{\alpha'}(P)\}_{\alpha \leq \alpha'}$, where the *Vietoris–Rips complex* $\mathbb{VR}^r(P)$ for a finite subset $P \subseteq (Z, \mathsf{d}_Z)$ at scale r is defined as (Definition 2.10)

$$\mathbb{VR}^r(P) = \{\sigma = \{p_0, \ldots, p_k\} \mid \mathsf{d}_Z(p_i, p_j) \leq 2r \text{ for any } i, j \in [0, k]\}. \quad (6.2)$$

Recall from Section 4.1 that the Čech filtration and Vietoris–Rips filtration are multiplicatively 2-interleaved, meaning that their persistence modules are $\log 2$-interleaved at the log scale, and

$$\mathsf{d}_b(\mathrm{Dgm}_{\log}\mathcal{C}(P), \mathrm{Dgm}_{\log}\mathcal{R}(P)) \leq \log 2 \text{ (Corollary 4.4)}. \quad (6.3)$$

Finite Metric Spaces

The above definitions of Čech or Rips complexes assume that P is embedded in an ambient metric space (Z, d_Z). It is possible that $Z = P$ and we simply have a discrete metric space spanned by points in P, which we denote by (P, d_P). Obviously, the construction of Čech and Rips complexes can be extended to this case. In particular, the Čech complex $\mathbb{C}_P^r(P)$ is now defined as

$$\mathbb{C}_P^r(P) = \{\sigma = \{p_0, \dots, p_k\} \mid \bigcap_{i \in [0,k]} B_P(p; r) \neq \varnothing\}, \qquad (6.4)$$

where $B_P(p, r) := \{q \in P \mid \mathsf{d}_P(p, q) \leq r\}$. However, note that when $P \subset Z$ and d_P is the restriction of the metric d_Z to points in P, the Čech complex $\mathbb{C}_P^r(P)$ defined above can be *different* from the Čech complex $\mathbb{C}_Z^r(P)$, as the metric balls (B_P versus B_Z) are different. In particular, in this case, we have the following relation between the two types of Čech complexes:

$$\mathbb{C}_P^r(P) \subseteq \mathbb{C}_Z^r(P) \subseteq \mathbb{C}_P^{2r}(P). \qquad (6.5)$$

On the other hand, in this setting, the two Rips complexes are the same because the definition of Rips complex involves only pairwise distance between input points, not metric balls.

The persistence diagrams induced by the Čech and the Rips filtrations can be used as topological summaries for the input PCD P. We can then, for example, compare input PCDs by comparing these persistence diagram summaries.

Definition 6.1. (Čech, distance; Rips distance) Given two finite point sets P and Q, equipped with appropriate metrics, the *Čech distance* between them is a pseudodistance defined as

$$\mathsf{d}_{Cech}(P, Q) = \max_k \mathsf{d}_B(\mathrm{Dgm}_k \mathcal{C}(P), \mathrm{Dgm}_k \mathcal{C}(Q)).$$

Similarly, the *Rips distance* between P and Q is a pseudodistance defined as

$$\mathsf{d}_{Rips}(P, Q) = \max_k \mathsf{d}_B(\mathrm{Dgm}_k \mathcal{R}(P), \mathrm{Dgm}_k \mathcal{R}(Q)).$$

These distances are stable with respect to the Hausdorff or the Gromov–Hausdorff distance between P and Q depending on whether they are embedded in a common metric space or are viewed as two discrete metric spaces (P, d_P) and (Q, d_Q). We introduce the Hausdorff and Gromov–Hausdorff distances now. Given a point x and a set A from a metric space (X, d), let $\mathsf{d}(x, A) := \inf_{a \in A} \mathsf{d}(x, a)$ denote the closest distance from x to any point in A.

Definition 6.2. (Hausdorff distance) Given two compact sets $A, B \subseteq (Z, \mathsf{d}_Z)$, the *Hausdorff distance* between them is defined as

$$\mathsf{d}_H(A, B) = \max\{ \max_{a \in A} \mathsf{d}_Z(a, B), \ \max_{b \in B} \mathsf{d}_Z(b, A) \}.$$

Note that the Hausdorff distance requires that the input objects are embedded in a common ambient space. In case they are not embedded in any common ambient space, we use the Gromov–Hausdorff distance, which intuitively measures how much two input metric spaces differ from being isometric.

Definition 6.3. (Gromov–Hausdorff distance) Given two metric spaces (X, d_X) and (Y, d_Y), a *correspondence* C is a subset $C \subseteq X \times Y$ so that (i) for every $x \in X$, there exists some $(x, y) \in C$, and (ii) for every $y' \in Y$, there exists some $(x', y') \in C$. The *distortion induced by* C is

$$\text{distort}_C(X, Y) := \frac{1}{2} \sup_{(x,y),(x',y') \in C} |d_X(x, x') - d_Y(y, y')|.$$

The *Gromov–Hausdorff distance between* (X, d_X) *and* (Y, d_Y) is the smallest distortion possible by any correspondence; that is,

$$d_{GH}(X, Y) := \inf_{C \subseteq X \times Y} \text{distort}_C(X, Y).$$

Theorem 6.3. *The Čech and Rips distances satisfy the following stability statements:*

(a) Given two finite sets $P, Q \subseteq (Z, d_Z)$, we have

$$d_{Cech}(P, Q) \leq d_H(P, Q) \quad and \quad d_{Rips}(P, Q) \leq d_H(P, Q).$$

(b) Given two finite metric spaces (P, d_P) and (Q, d_Q), we have

$$d_{Cech}(P, Q) \leq 2d_{GH}((P, d_P), (Q, d_Q)) \quad and$$

$$d_{Rips}(P, Q) \leq d_{GH}((P, d_P), (Q, d_Q)).$$

Note that the bound on $d_{Cech}(P, Q)$ in Theorem 6.3(b) has an extra factor of 2, which comes due to the difference in metric balls – see the discussions after Eq. (6.4). We also remark that Theorem 6.3(b) can be extended to the so-called *totally bounded metric spaces* (which are not necessarily finite) (P, d_P) and (Q, d_Q) defined as follows. First, recall that *an ε-sample* (Definition 2.17) of a metric space (Z, d_Z) is a finite set $S \subseteq Z$ so that for every $z \in Z, d_Z(z, S) \leq \varepsilon$. A metric space (Z, d_Z) is *totally bounded* if there exists a finite ε-sample for every $\varepsilon > 0$. Intuitively, such a metric space can be approximated by a finite metric space for any resolution.

6.2 Approximation via Data Sparsification

One issue with using the Vietoris–Rips or Čech filtrations in practice is that their sizes can become huge, even for a moderate number of points. For example, when the scale r is larger than the diameter of a point set P, the Čech and the Vietoris–Rips complexes of P contain every simplex spanned by points in P, in which case the size of d-skeleton of $\mathbb{C}^r(P)$ or $\mathbb{VR}^r(P)$ is $\Theta(n^{d+1})$ for $n = |P|$.

On the other hand, as shown in Figure 6.2, as the scale r increases, certain points could become "redundant," for example, have no or little contribution to the underlying space of the union of all r-radius balls. Based on this observation, one can approximate these filtrations with *sparsified filtrations* of much smaller size. In particular, as the scale r increases, the point set P with which one constructs a complex is gradually sparsified keeping the total number of simplices in the complex linear in the input size of P where the dimension of the embedding space is assumed to be fixed.

We describe two data sparsification schemes in Sections 6.2.1 and 6.2.2, respectively. We focus on the Vietoris–Rips filtration for points in a Euclidean space $\mathbb{R}^d$ equipped with the standard Euclidean distance d.

6.2.1 Data Sparsification for Rips Filtration via Reweighting

Most of the concepts presented in this section apply to general finite metric spaces though we describe them for finite point sets equipped with a Euclidean metric. The reason for this choice is that the complexity analysis draws upon the specific property of Euclidean space. The reader is encouraged to think about generalizing the definitions and the technique to other metric spaces.

First we restate the definition of δ-sample and δ-sparse sample in Definition 2.17 slightly differently.

Definition 6.4. (Nets; Net-tower) Given a finite set of points $P \subset (\mathbb{R}^d, \mathsf{d})$ and $\gamma, \gamma' \geq 0$, a subset $Q \subseteq P$ is a (γ, γ')-*net of* P if the following two conditions hold.

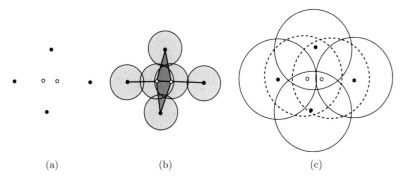

(a)   (b)   (c)

Figure 6.2 Vietoris–Rips complex: (b) at small scale, the Rips complex of points shown in (a) requires the two white points; (c) the two white points become redundant at larger scale.

- Covering condition: Q is a γ-sample for (P, d), that is, for every $p \in P$, $\mathsf{d}(p, Q) \leq \gamma$.
- Packing condition: Q is also γ-sparse, that is, for every $q \neq q' \in Q$, $\mathsf{d}(q, q') \geq \gamma'$.

If $\gamma = \gamma'$, we also refer to Q as a γ-net of P.

A single-parameter family of nets $\{N_\gamma\}_\gamma$ is called a *net-tower* of P if (i) there is a constant $c > 0$ so that for all $\gamma \in \mathbb{R}$, N_γ is a $(\gamma, \gamma/c)$-net for P, and (ii) $N_\gamma \supseteq N_{\gamma'}$ for any $\gamma \leq \gamma'$.

Intuitively, a γ-net approximates a PCD P at resolution γ (covering condition), while also being sparse (packing condition). A net-tower provides a sequence of increasingly sparsified approximation of P.

Net-Tower via Farthest Point Sampling
We now introduce a specific net-tower constructed via the classical strategy of *farthest point sampling*, also called greedy permutation in, for example, [56, 70]. Given a point set $P \subset (\mathbb{R}^d, \mathsf{d})$, choose an arbitrary point p_1 from P and set $P_1 = \{p_1\}$. Pick p_i recursively as $p_i \in \mathrm{argmax}_{p \in P \setminus P_{i-1}} \mathsf{d}(p, P_{i-1})$,[1] and set $P_i = P_{i-1} \cup \{p_i\}$. Now set $\mathsf{t}_{p_i} = \mathsf{d}(p_i, P_{i-1})$, which we refer to as the *exit-time of* p_i. Based on these exit-times, we construct the following two families of sets:

$$\text{open net-tower}\quad \mathcal{N} = \{N_\gamma\}_{\gamma \in \mathbb{R}} \text{ where } N_\gamma := \{p \in P \mid \mathsf{t}_p > \gamma\}, \quad (6.6)$$

$$\text{closed net-tower}\quad \overline{\mathcal{N}} = \{\overline{N}_\gamma\}_{\gamma \in \mathbb{R}} \text{ where } \overline{N}_\gamma := \{p \in P \mid \mathsf{t}_p \geq \gamma\}. \quad (6.7)$$

It is easy to verify that both N_γ and $\overline{N}_\gamma$ are γ-nets, and the families $\mathcal{N}$ and $\overline{\mathcal{N}}$ are indeed two net-towers as γ increases. As γ increases, N_γ and $\overline{N}_\gamma$ can only change when $\gamma = \mathsf{t}_p$ for some $p \in P$. Hence the sequence of subsets $P = P_n \supset P_{n-1} \supseteq \cdots \supseteq P_2 \supseteq P_1$ contain all the distinct sets in the open and closed net-towers $\{N_\gamma\}$ and $\{\overline{N}_\gamma\}$.

In what follows, we discuss a sparsification strategy for the Rips filtration of P using the above net-towers. The approach can be extended to other net-towers, such as the net-tower constructed using the net-tree data structure of [182].

Weights, Weighted Distance, and Sparse Rips Filtration
Given the exit-times t_p for all points $p \in P$, we now associate a *weight* $w_p(\alpha)$ for each point p at a scale α (the graph of this weight function is shown in Figure 6.3): for some constant $0 < \varepsilon < 1$, the weight function is

---

[1] Note that there may be multiple points that maximize $\mathsf{d}(p, P_{i-1})$ making $\mathrm{argmax}_{p \in P \setminus P_{i-1}} \mathsf{d}(p, P_{i-1})$ a set. We can choose p_i to be any point in this set.

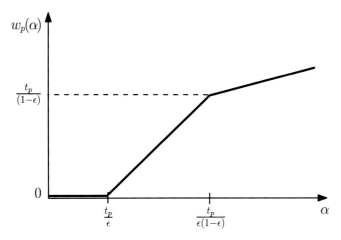

Figure 6.3 Graph of the weight function $w_p(\alpha)$.

$$w_p(\alpha) = \begin{cases} 0 & \dfrac{t_p}{\varepsilon} \geq \alpha, \\[2ex] \alpha - \dfrac{t_p}{\varepsilon} & \dfrac{t_p}{\varepsilon} < \alpha < \dfrac{t_p}{\varepsilon(1-\varepsilon)}, \\[2ex] \varepsilon\alpha & \dfrac{t_p}{\varepsilon(1-\varepsilon)} \leq \alpha. \end{cases}$$

Claim 6.1. *The weight function w_p is a continuous, 1-Lipschitz, and nondecreasing function.*

The parameter ε controls the resolution of the sparsification. The *net-induced distance at scale α* between input points is defined as

$$\widehat{\mathsf{d}}_\alpha(p, q) := \mathsf{d}(p, q) + w_p(\alpha) + w_q(\alpha). \tag{6.8}$$

Definition 6.5. (Sparse (Vietoris–)Rips) Given a set of points $P \subset \mathbb{R}^d$, a constant $0 < \varepsilon < 1$, and the open net-tower $\{N_\gamma\}$ as well as the closed net-tower $\{\overline{N}_\gamma\}$ for P as introduced above, the *open sparse Rips complex at scale α* is defined as

$$\mathsf{Q}^\alpha := \{\sigma \subseteq N_{\varepsilon(1-\varepsilon)\alpha} \mid \forall p, q \in \sigma, \ \widehat{\mathsf{d}}_\alpha(p, q) \leq 2\alpha\}; \tag{6.9}$$

while the *closed sparse Rips at scale α* is defined as

$$\overline{\mathsf{Q}}^\alpha := \{\sigma \subseteq \overline{N}_{\varepsilon(1-\varepsilon)\alpha} \mid \forall p, q \in \sigma, \ \widehat{\mathsf{d}}_\alpha(p, q) \leq 2\alpha\}. \tag{6.10}$$

Set $\mathsf{S}^\alpha := \bigcup_{\beta \le \alpha} \overline{\mathsf{Q}}^\alpha$, which we call the *cumulative complex at scale* α. The *(ε-)sparse Rips filtration* then refers to the $\mathbb{R}$-indexed filtration $\mathsf{S} = \{\mathsf{S}^\alpha \hookrightarrow \mathsf{S}^\beta\}_{\alpha \le \beta}$.

Obviously, $\mathsf{Q}^\alpha \subseteq \overline{\mathsf{Q}}^\alpha$. Note that for $\alpha < \beta$, Q^α is not necessarily included in Q^β (neither is $\overline{\mathsf{Q}}^\alpha$ in $\overline{\mathsf{Q}}^\beta$); while the inclusion $\mathsf{S}^\alpha \subseteq \mathsf{S}^\beta$ always holds.

In what follows, we show that the sparse Rips filtration approximates the standard Vietoris–Rips filtration $\{\mathbb{VR}^r(P)\}$ defined over P, and that the size of the sparse Rips filtration is only linear in n for any fixed dimension d which is assumed to be constant. The main results are summarized in the following theorem.

Theorem 6.4. *Let $P \subset \mathbb{R}^d$ be a set of n points where d is a constant, and $\mathcal{R}(P) = \{\mathbb{VR}^r(P)\}$ be the Vietoris–Rips filtration over P. Given net-towers $\{N_\gamma\}$ and $\{\overline{N}_\gamma\}$ induced by exit-times $\{\mathsf{t}_p\}_{p \in P}$, let $\mathcal{S}(P) = \{\mathsf{S}^\alpha\}$ be its corresponding ε-sparse Rips filtration as defined in Definition 6.5. Then, for a fixed $0 < \varepsilon < 1/3$ the following hold.*

(a) $\mathcal{S}(P)$ and $\mathcal{R}(P)$ are multiplicatively $1/(1 - \varepsilon)$-interleaved at the homology level. Thus, for any $k \ge 0$, the persistence diagram $\mathrm{Dgm}_k \mathcal{S}(P)$ is a $\log(1/(1 - \varepsilon))$-approximation of $\mathrm{Dgm}_k \mathcal{R}(P)$ at the log scale.
(b) For any fixed dimension $k \ge 0$, the total number of k-simplices ever appearing in $\mathcal{S}(P)$ is $\Theta((1/\varepsilon)^{kd} n)$.

In the remainder of this section, we sketch the proof of the above theorem.

PROOF OF THEOREM 6.4(A)
To relate $\mathcal{S}(P)$ to $\mathcal{R}(P)$, we need to go through a sequence of intermediate steps. First, we define the *relaxed Rips complex at scale* α as

$$\widehat{\mathbb{VR}}^\alpha(P) := \{\sigma \subset P \mid \forall p, q \in \sigma, \; \widehat{\mathsf{d}}_\alpha(p, q) \le 2\alpha\}.$$

The following claim ensures that the relaxed Rips complexes form a valid filtration connected by inclusions $\widehat{\mathcal{R}}(P) = \{\widehat{\mathbb{VR}}^\alpha(P) \hookrightarrow \widehat{\mathbb{VR}}^\beta(P)\}_{\alpha \le \beta}$, which we call the *relaxed Rips filtration*.

Claim 6.2. *If $\widehat{\mathsf{d}}_\alpha(p, q) \le 2\alpha \le 2\beta$, then $\widehat{\mathsf{d}}_\beta(p, q) \le 2\beta$.*

PROOF. The weight function w_p is 1-Lipschitz for any $p \in P$ (Claim 6.1). Thus we have that

$$\widehat{\mathsf{d}}_\beta(p,q) = \mathsf{d}(p,q) + w_p(\beta) + w_q(\beta)$$
$$\leq \mathsf{d}(p,q) + w_p(\alpha) + (\beta - \alpha) + w_q(\alpha) + (\beta - \alpha)$$
$$\leq \mathsf{d}(p,q) + w_p(\alpha) + w_q(\alpha) - 2\alpha + 2\beta \leq 2\beta.$$

The last inequality follows from $\mathsf{d}(p,q) + w_p(\alpha) + w_q(\alpha) = \widehat{\mathsf{d}}_\alpha(p,q) \leq 2\alpha$. □

In what follows, we drop the argument P from notation such as in complexes $\mathbb{VR}^\alpha(P)$ or in sparse Rips filtration $\mathcal{S}(P)$ when the point set in question is understood.

Proposition 6.5. *Let* $C = 1/(1 - \varepsilon)$. *Then for any* $\alpha \geq 0$ *we have that* $\mathbb{VR}^{\alpha/C} \subseteq \widehat{\mathbb{VR}}^\alpha \subseteq \mathbb{VR}^\alpha$.

Next, we relate filtrations $\mathcal{S}$ and $\mathcal{R}$ via the relaxed Rips filtration $\widehat{\mathcal{R}}$ by connecting the sparse Rips complexes Q^α and $\overline{\mathsf{Q}}^\alpha$. Consider the following projection of P to points in the net $N_{\varepsilon(1-\varepsilon)\alpha}$ which are also vertices of Q^α:

$$\pi_\alpha(p) = \begin{cases} p & \text{if } p \in N_{\varepsilon(1-\varepsilon)\alpha}, \\ \operatorname{argmin}_{q \in N_{\varepsilon\alpha}} \mathsf{d}(p,q) & \text{otherwise.} \end{cases}$$

Again, if $\operatorname{argmin}_{q \in N_{\varepsilon\alpha}} \mathsf{d}(p,q)$ contains more than one point, we set $\pi_\alpha(p)$ to be an arbitrary one. This projection is well defined as $N_{\varepsilon\alpha} \subseteq N_{\varepsilon(1-\varepsilon)\alpha}$ given that $0 < \varepsilon < 1/3 < 1$. We need several technical results on this projection map, which we rely on later to construct maps between appropriate versions of Rips complexes. First, the following two results are easy to show.

Fact 6.1. *For every* $p \in P$, $\mathsf{d}(p, \pi_\alpha(p)) \leq w_p(\alpha) - w_{\pi_\alpha(p)}(\alpha) \leq \varepsilon\alpha$.

Fact 6.2. *For every pair* $p,q \in P$, *we have that* $\widehat{\mathsf{d}}_\alpha(p, \pi_\alpha(q)) \leq \widehat{\mathsf{d}}_\alpha(p,q)$.

We are now ready to show that inclusion induces an isomorphism between the homology groups of the sparse Rips complex and the relaxed Rips complex.

Proposition 6.6. *For any* $\alpha \geq 0$, *the inclusion* $i : \mathsf{Q}^\alpha \hookrightarrow \widehat{\mathbb{VR}}^\alpha$ *induces an isomorphism at the homology level; that is,* $\mathsf{H}_*(\mathsf{Q}^\alpha) \cong \mathsf{H}_*(\widehat{\mathbb{VR}}^\alpha)$ *under the homomorphism* i_* *induced by* i.

PROOF. First, we consider the projection map π_α and argue that it induces a simplicial map $\pi_\alpha : \widehat{\mathbb{VR}}^\alpha \to \mathsf{Q}^\alpha$ which is in fact a simplicial retraction.[2]

---

[2] A simplicial retraction $f : K \to L$ is a simplicial map from $K \subseteq L$ to L so that $f(\sigma) = \sigma$ for any $\sigma \in K$.

Next, we show that the map $i \circ \pi_\alpha : \widehat{\mathbb{VR}}^\alpha \to \widehat{\mathbb{VR}}^\alpha$ is contiguous to the identity map id: $\widehat{\mathbb{VR}}^\alpha \to \widehat{\mathbb{VR}}^\alpha$. As π_α is a simplicial retraction, it follows that i_* is an isomorphism (Lemma 2 of [275]).

To see that π_α is a simplicial map, apply Fact 6.2 twice to have that

$$\widehat{\mathsf{d}}_\alpha(\pi_\alpha(p), \pi_\alpha(q)) \leq \widehat{\mathsf{d}}_\alpha(p, \pi_\alpha(q)) \leq \widehat{\mathsf{d}}_\alpha(p, q). \tag{6.11}$$

Since both Q^α and $\widehat{\mathbb{VR}}^\alpha$ are clique complexes, this then implies that π_α is a simplicial map. Furthermore, it is easy to see that it is a retraction as $\pi_\alpha(q) = q$ for any q in the vertex set of Q^α (which is $N_{\varepsilon(1-\varepsilon)\alpha}$).

Now to show that $i \circ \pi_\alpha$ is contiguous to id, we observe that for any $p, q \in P$ with $\widehat{\mathsf{d}}_\alpha(p, q) \leq 2\alpha$, all edges among $\{p, q, \pi_\alpha(p), \pi_\alpha(q)\}$ exist and thus all simplices spanned by them exist in $\widehat{\mathbb{VR}}^\alpha$. Indeed, that $\widehat{\mathsf{d}}_\alpha(\pi_\alpha(p), \pi_\alpha(q)) \leq 2\alpha$ is already shown above in Eq. (6.11). Combining Fact 6.1 with the fact that $w_p(\alpha) \leq \varepsilon\alpha$, we have that

$$\widehat{\mathsf{d}}_\alpha(p, \pi_\alpha(p)) = \mathsf{d}(p, \pi_\alpha(p)) + w_p(\alpha) + w_{\pi_\alpha(p)}(\alpha) \leq 2w_p(\alpha) \leq 2\varepsilon\alpha < 2\alpha.$$

Furthermore, by Fact 6.2, $\widehat{\mathsf{d}}_\alpha(p, \pi_\alpha(q)) \leq \widehat{\mathsf{d}}_\alpha(p, q) \leq 2\alpha$. Symmetric arguments can be applied to show that $\widehat{\mathsf{d}}_\alpha(q, \pi_\alpha(q)), \widehat{\mathsf{d}}_\alpha(q, \pi_\alpha(p)) \leq 2\alpha$. This establishes that $i \circ \pi_\alpha$ is contiguous to id. This proves the proposition. $\square$

The closed sparse Rips complex $\overline{\mathsf{Q}}^\alpha$ is the relaxed Rips complex over the vertex set $\overline{N}_{\varepsilon(1-\varepsilon)\alpha}$, which is a superset of the vertex set of Q^α. Hence the above proposition also holds for the inclusion $\mathsf{Q}^\alpha \hookrightarrow \overline{\mathsf{Q}}^\alpha$. It then follows that $\mathsf{H}_*(\mathsf{Q}^\alpha) \cong \mathsf{H}_*(\overline{\mathsf{Q}}^\alpha)$. Finally, we show that the inclusion also induces an isomorphism between $\mathsf{H}_*(\overline{\mathsf{Q}}^\alpha)$ and $\mathsf{H}_*(\mathsf{S}^\alpha)$, which when combined with the above results connects S^α and $\widehat{\mathbb{VR}}^\alpha$.

Proposition 6.7. *For any $\alpha \geq 0$, the inclusion $h : \overline{\mathsf{Q}}^\alpha \hookrightarrow \mathsf{S}^\alpha$ induces an isomorphism at the homology level, that is, $\mathsf{H}_*(\overline{\mathsf{Q}}^\alpha) \cong \mathsf{H}_*(\mathsf{S}^\alpha)$ under h_*.*

PROOF. Consider the sequence $\{\mathsf{S}^\alpha\}_{\alpha \in \mathbb{R}}$. First, we discretize α to have distinct values $\alpha_0 < \alpha_1 < \alpha_2 \cdots \alpha_m$ so that $\mathsf{S}^{\alpha_0} = \varnothing$, and the α_i are exactly the times when the combinatorial structure of S^α changes. As $\mathsf{S}^\alpha = \sum_{\beta \leq \alpha} \overline{\mathsf{Q}}^\beta$, these are also exactly the moments when the combinatorial structure of $\overline{\mathsf{Q}}^\alpha$ changes. Hence we only need to prove the statement for such α_i, and it will then work for all α. Set $\lambda_i := \varepsilon(1 - \varepsilon)\alpha_i$. Note that the vertex set for $\overline{\mathsf{Q}}^{\alpha_i}$ is $\overline{N}_{\lambda_i}$ by the definition of $\overline{\mathsf{Q}}^\alpha$ in Eq. (6.10).

Now fix a $k \geq 0$. We will show that $h : \overline{\mathsf{Q}}^{\alpha_k} \hookrightarrow \mathsf{S}^{\alpha_k}$ induces an isomorphism at the homololgy level. We use some intermediate complexes

$$T_{i,k} := \bigcup_{j=i}^{k} \overline{\mathsf{Q}}^{\alpha_j}, \quad \text{for } i \in [1, k].$$

Obviously, $T_{1,k} = \mathsf{S}^{\alpha_k}$ where $T_{k,k} = \overline{\mathsf{Q}}^{\alpha_k}$. Set $h_i : T_{i+1,k} \hookrightarrow T_{i,k}$. The inclusion $h : \overline{\mathsf{Q}}^{\alpha_k} \hookrightarrow \mathsf{S}^{\alpha_k}$ can then be written as $h = h_1 \circ h_2 \circ \cdots \circ h_{k-1}$. In what follows, we prove that $h_i : T_{i+1,k} \hookrightarrow T_{i,k}$ induces an isomorphism at the homology level for each $i \in [1, k-1]$, which then proves the proposition.

First, note that while $T_{i,k}$ is not necessarily the same as $\overline{\mathsf{Q}}^{\alpha_i}$, they share the same vertex set. Now, because of our choices of the α_i and λ_i, the vertex set of $T_{i+1,k}$, which is the vertex set of $\overline{\mathsf{Q}}^{\alpha_{i+1}}$, namely $\overline{N}_{\lambda_{i+1}}$, equals N_{λ_i}. Hence we can consider the projection $\pi_{\alpha_i} : T_{i,k} \to T_{i+1,k}$ given by the projection of the vertex set $N_{\lambda_{i-1}} = \overline{N}_{\lambda_i}$ of $T_{i,k}$ to the vertex set $N_{\lambda_i} = \overline{N}_{\lambda_{i+1}}$ of $T_{i+1,k}$. To prove that h_i induces an isomorphism at the homology level, by Lemma 2 of [275], it suffices to show that (i) π_{α_i} is a simplicial retraction, and (ii) $h_i \circ \pi_{\alpha_i}$ is contiguous to the identity map $id : T_{i,k} \to T_{i,k}$.

To prove (i), it is easy to verify that π_{α_i} is a retraction. To see that π_{α_i} induces a simplicial map, we need to show that for every $\sigma \in T_{i,k}$, $\pi_{\alpha_i}(\sigma) \in T_{i+1,k}$. As π_{α_i} is a retraction, we only need to prove this for every $\sigma \in T_{i,k} \setminus T_{i+1,k}$. On the other hand, note that by definition, $T_{i,k} \setminus T_{i+1,k} \subseteq \overline{\mathsf{Q}}^{\alpha_i}$. To this end, the argument in Proposition 6.6 also shows that $\pi_{\alpha_i} : \overline{\mathsf{Q}}^{\alpha_i} \to \mathsf{Q}^{\alpha_i}$ is a simplicial map, and furthermore, $h' \circ \pi_{\alpha_i}$ is contiguous to $id' : \overline{\mathsf{Q}}^{\alpha_i} \to \overline{\mathsf{Q}}^{\alpha_i}$, where $h' : \mathsf{Q}^{\alpha_i} \hookrightarrow \overline{\mathsf{Q}}^{\alpha_i}$. Because of our choice of α_i, Q^{α_i} and $\overline{\mathsf{Q}}^{\alpha_{i+1}}$ have the same vertex set, which is N_{λ_i}. Furthermore, for every edge $(p, q) \in \mathsf{Q}^{\alpha_i}$, we have that $\widehat{\mathsf{d}}_{\alpha_i}(p, q) \leq 2\alpha_i$. As $\alpha_i < \alpha_{i+1}$, it follows from Claim 6.2 that $\widehat{\mathsf{d}}_{\alpha_{i+1}}(p, q) \leq 2\alpha_{i+1}$. Hence, the edge (p, q) is in $\overline{\mathsf{Q}}^{\alpha_{i+1}}$. This implies that $\mathsf{Q}^{\alpha_i} \subseteq \overline{\mathsf{Q}}^{\alpha_{i+1}}$. Putting everything together, it follows that, for every $\sigma \in T_{i,k} \setminus T_{i+1,k} \subseteq \overline{\mathsf{Q}}^{\alpha_i}$, we have

$$\pi_{\alpha_i}(\sigma) \in \mathsf{Q}^{\alpha_i} \subseteq \overline{\mathsf{Q}}^{\alpha_{i+1}} \subseteq T_{i+1,k}.$$

Therefore, π_{α_i} is a simplicial map. This finishes the proof of (i).

Now we prove (ii), that is, $h_i \circ \pi_{\alpha_i}$ is contiguous to the identity map $id : T_{i,k} \to T_{i,k}$. This means that we need to show that for every $\sigma \in T_{i,k}$, $\sigma \cup \pi_{\alpha_i}(\sigma) \in T_{i,k}$. Again, as π_{α_i} is a simplicial retraction, we only need to show this for $\sigma \in T_{i,k} \setminus T_{i+1,k} \subseteq \overline{\mathsf{Q}}^{\alpha_i}$. As mentioned above, using the same argument as in Proposition 6.6, we know that $h' \circ \pi_{\alpha_i}$ is contiguous to the identity $id' : \overline{\mathsf{Q}}^{\alpha_i} \to \overline{\mathsf{Q}}^{\alpha_i}$. Hence we have that for every $\sigma \in \overline{\mathsf{Q}}^{\alpha_i}$, $\sigma \cup \pi_{\alpha_i}(\sigma) \in \overline{\mathsf{Q}}^{\alpha_i}$. It follows that $\sigma \cup \pi_{\alpha_i}(\sigma) \in T_{i,k}$ as $\overline{\mathsf{Q}}^{\alpha_i} \subseteq T_{i,k}$. This proves (ii), completing the proof of the proposition.   □

Combining Propositions 6.6 (as well as the discussion after this proposition) and 6.7, we have that $\{\mathsf{S}^\alpha\}$ and $\{\widehat{\mathbb{VR}}^\alpha\}$ induces isomorphic persistence modules. This, together with Proposition 6.5, implies Theorem 6.4(a).

PROOF OF THEOREM 6.4(B)

Let $\mathcal{S}^{(k)}$ denote the set of k-simplices ever appearing in $\mathcal{S}(P)$, which is also the set of k-simplices in the last complex S^∞ of $\mathcal{S}(P)$. To bound the size of $\mathcal{S}^{(k)}$, we charge each simplex in $\mathcal{S}^{(k)}$ to the vertex of it with *smallest* exit-time. Observe that a point $p \in P$ does not contribute to any new edge in the sparse Rips complex $\overline{\mathsf{Q}}^\beta$ for $\beta > \mathsf{t}_p/(\varepsilon(1-\varepsilon))$. This means that to bound the number of simplices charged to p, we only need to bound such simplices in $\overline{\mathsf{Q}}^{\alpha_p}$ with $\alpha_p = \mathsf{t}_p/(\varepsilon(1-\varepsilon))$.

Set $E(p) = \{q \in P \mid (p,q) \in \overline{\mathsf{Q}}^{\alpha_p}$ and $\mathsf{t}_p \leq \mathsf{t}_q\}$. We add p to $E(p)$ too. We claim that $|E(p)| = O((1/\varepsilon)^d)$. In particular, consider the closed net-tower $\{\overline{N}_\gamma\}$; recall that $\overline{N}_\gamma$ is a γ-net. As $E(p) \subseteq \overline{N}_{\mathsf{t}_p}$, the packing condition of the net implies that the closest pair in $E(p)$ has distance at least t_p between them. On the other hand, for each $(p,q) \in \overline{\mathsf{Q}}^{\alpha_p}$, we have $\widehat{\mathsf{d}}_{\alpha_p}(p,q) \leq 2\alpha_p$ implying that $E(p) \subseteq B(p, 2\alpha_p)$. A simple packing argument then implies that the number of points in $E(p)$ is

$$O\left(\left(\frac{2\alpha_p}{\mathsf{t}_p}\right)^d\right) = O\left(\left(\frac{2}{\varepsilon(1-\varepsilon)}\right)^d\right) = O\left(\left(\frac{1}{\varepsilon}\right)^d\right).$$

The last equality follows because $\varepsilon < 1/3$ and thus $1 - \varepsilon \geq 2/3$. The total number of k-simplices charged to p is bounded by $O((1/\varepsilon)^{kd})$, and the total number of k-simplices in $\mathcal{S}(P)$ is $O((1/\varepsilon)^{kd}n)$, proving Theorem 6.4(b).

6.2.2 Approximation via Simplicial Tower

We now describe a different sparsification strategy by directly building a simplicial tower of Rips complexes connected by simplicial maps (Definition 4.1 and the discussion below it) whose persistent homology also approximates that of the standard Rips filtration. This sparsification is conceptually simpler, but its approximation quality is worse than the one introduced in the previous section.

Given a set of points $P \subset \mathbb{R}^d$, $\alpha > 0$, and some $0 < \varepsilon < 1$, we are interested in the following filtration (which is a subsequence of the standard Rips filtration):

$$\mathbb{VR}^\alpha(P) \hookrightarrow \mathbb{VR}^{\alpha(1+\varepsilon)}(P) \hookrightarrow \mathbb{VR}^{\alpha(1+\varepsilon)^2}(P) \hookrightarrow \cdots \hookrightarrow \mathbb{VR}^{\alpha(1+\varepsilon)^m}(P). \quad (6.12)$$

We now construct a sparsified sequence by setting $P_0 := P$, building a sequence of point sets P_k, $k = 0, 1, \ldots, m$, where P_{k+1} is an $(\alpha \varepsilon^2)(1 + \varepsilon)^{k-1}$-net of P_k, and terminating the process when P_m is of constant size.

Consider the vertex map $\pi_k : P_k \to P_{k+1}$, for any $k \in [0, m-1]$, where $\pi_k(v)$ is the nearest neighbor of $v \in P_k$ in P_{k+1}. Define $\widehat{\pi}_k : P_0 \to P_{k+1}$ as $\widehat{\pi}_k := \pi_k \circ \cdots \circ \pi_0$. Based on the fact that P_{k+1} is an $\alpha \varepsilon^2 (1 + \varepsilon)^{k-1}$-net of P_k, it can be verified that π_k induces a simplicial map

$$\pi_k : \mathbb{VR}^{\alpha(1+\varepsilon)^k}(P_k) \to \mathbb{VR}^{\alpha(1+\varepsilon)^{k+1}}(P_{k+1}),$$

which further gives rise to a simplicial map $\widehat{\pi}_k : \mathbb{VR}^{\alpha}(P_0) \to \mathbb{VR}^{\alpha(1+\varepsilon)^{k+1}}(P_{k+1})$. We thus have the following tower of simplicial complexes:

$$\widehat{\mathbb{S}}: \mathbb{VR}^{\alpha}(P_0) \xrightarrow{\pi_0} \mathbb{VR}^{\alpha(1+\varepsilon)}(P_1) \xrightarrow{\pi_1} \cdots \xrightarrow{\pi_{m-1}} \mathbb{VR}^{\alpha(1+\varepsilon)^m}(P_m). \quad (6.13)$$

Claim 6.3. *For any fixed $\alpha \geq 0$, $\varepsilon \geq 0$, and any integer $k \geq 0$, each triangle in the following diagram commutes at the homology level.*

$$
\begin{array}{ccc}
\mathbb{VR}^{\alpha(1+\varepsilon)^k}(P_0) & \xhookrightarrow{\;\; i_k \;\;} & \mathbb{VR}^{\alpha(1+\varepsilon)^{k+1}}(P_0) \\
\Big\uparrow{\scriptstyle j_k} & {\scriptstyle \widehat{\pi}_k} \nearrow & \Big\uparrow{\scriptstyle j_{k+1}} \\
\mathbb{VR}^{\alpha(1+\varepsilon)^k}(P_k) & \xrightarrow{\;\; \pi_k \;\;} & \mathbb{VR}^{\alpha(1+\varepsilon)^{k+1}}(P_{k+1})
\end{array}
$$

Here, the maps i_k and j_k are canonical inclusions.

The above result implies that at the homology level, the sequence in Eq. (6.13) and the sequence in Eq. (6.12) are *weakly $(1 + \varepsilon)$-interleaved* in a multiplicative manner. In particular, different from the interleaving introduced by Definition 4.4 in Section 4.1, here the interleaving relations only hold at *discrete index values* of the filtrations.

Definition 6.6. (Weak interleaving of vector space towers) Let $\mathbb{U} = \left\{ U_a \xrightarrow{u_{a,b}} U_b \right\}_{a_0 \leq a \leq b}$ and $\mathbb{V} = \left\{ V_a \xrightarrow{v_{a,b}} V_b \right\}_{a_0 \leq a \leq b}$ be two vector space towers over an index set $A = \{ a \in \mathbb{R} \mid a \geq a_0 \}$ with resolution $a_0 \geq 0$. For some real number $\varepsilon \geq 0$, we say that they are *weakly ε-interleaved* if there are two families of linear maps $\phi_i : U_{a_0 + i\varepsilon} \to V_{a_0 + (i+1)\varepsilon}$ and $\psi_i : V_{a_0 + i\varepsilon} \to U_{a_0 + (i+1)\varepsilon}$, for any integer $i \geq 0$, such that any subdiagram of the following diagram commutes.

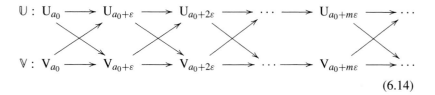

$$(6.14)$$

It turns out that to verify the commutativity of the diagram in Eq. (6.14), it is sufficient to verify it for all subdiagrams of the form as in Eq. (4.3). Furthermore, ε-weakly interleaved persistence modules also have bounded bottleneck distances between their persistence diagrams [77] though the distance bound is relaxed to 3ε; that is, if $\mathbb{U}$ and $\mathbb{V}$ are weakly ε-interleaved, then $\mathsf{d}_b(\mathrm{Dgm}\,\mathbb{U}, \mathrm{Dgm}\,\mathbb{V}) \leq 3\varepsilon$. Analogous results hold for the multiplicative setting. Finally, using a similar packing argument as before, one can also show that the total number of k-simplices that ever appear in the simplicial-map-based sparsification $\widehat{\mathsf{S}}$ is linear in n (assuming that k and the dimension d are both constant). To summarize:

Theorem 6.8. *Given a set of n points $P \subset \mathbb{R}^d$, we can $3\log(1+\varepsilon)$-approximate the persistence diagram of the discrete Rips filtration in Eq. (6.12) by that of the filtration in Eq. (6.13) at the log scale. The number of k-simplices that ever appear in the filtration in Eq. (6.13) is $O((1/\varepsilon)^{O(kd)}n)$.*

6.3 Homology Inference from Point Cloud Data

So far, we have considered the problem of approximating the persistence diagram of a filtration created out of a given PCD. Now we consider the problem of inferring certain homological structure of a (hidden) domain where the input PCD presumably is sampled from. More specifically, the problem we consider is: Given a finite set of points $P \subset \mathbb{R}^d$, residing on or around a hidden domain $X \subseteq \mathbb{R}^d$ of interest, compute or approximate the rank of $\mathsf{H}_*(X)$ using input PCD P. Later in this chapter, X is assumed to be either a smooth Riemannian manifold embedded in $\mathbb{R}^d$, or simply a compact set of $\mathbb{R}^d$.

Main Ingredients

Since points themselves do not have interesting topology, we first construct a certain simplicial complex K, typically a Čech or a Vietoris–Rips complex from P. Next, we compute the homological information of K as a proxy for the same of X. Of course, the approximation becomes faithful only when the given sample P is sufficiently dense and the parameters used for building the complexes are appropriate. The high-level approach works as follows.

Input: A finite point set $P \subset \mathbb{R}^d$ "approximating" a hidden space $X \subset \mathbb{R}^d$.

Step 1: Compute the Čech complex $\mathbb{C}^\alpha(P)$, or a pair of Rips complexes $\mathbb{VR}^\alpha(P)$ and $\mathbb{VR}^{\alpha'}(P)$, for some appropriate $0 < \alpha < \alpha'$.

Step 2: In the case of Čech complex, return $\dim(H_*(\mathbb{C}^\alpha(P)))$ as an approximation of $\dim(H_*(X))$. In the case of Rips complex, return $\mathrm{rank}\,(\mathrm{im}\,(i_*))$, where the homomorphism $i_* : H_*(\mathbb{VR}^\alpha(P)) \to H_*(\mathbb{VR}^{\alpha'}(P))$ is induced by the inclusion $\mathbb{VR}^\alpha(P) \subseteq \mathbb{VR}^{\alpha'}(P)$.

To provide quantitative statements on the approximation quality of the outcome of the above approach, we need to describe first what the quality of the input PCD P is, often referred to as the *sampling conditions*. Intuitively, a better approximation in homology is achieved if the input points P "approximate" or "sample" X better. The quality of input points is often measured by the Hausdorff distance with respect to the Euclidean distances between PCD P and the hidden domain X of interest (Definition 6.2), such as requiring that $d_H(P, X) \leq \varepsilon$ for some $\varepsilon > 0$. Note that points in P do not necessarily lie in X. The approximation guarantee for $\dim(H_*(X))$ relies on relating the distance fields induced by X and by the sample P. We describe the distance field and feature sizes of X in Section 6.3.1. We present how to infer homology for smooth manifolds and compact sets from data in Sections 6.3.2 and 6.3.3, respectively. In Section 6.4, we discuss inferring the persistent homology induced by a scalar function $f : X \to \mathbb{R}$ on X.

6.3.1 Distance Field and Feature Sizes

To describe how well P samples X, we introduce two notions of the so-called "feature size" of X: the local feature size and the weak feature size, both related to the distance field d_X with respect to X.

Definition 6.7. (Distance field) Given a compact set $X \subset \mathbb{R}^d$, the *distance field (w.r.t. X)* is

$$d_X : \mathbb{R}^d \to \mathbb{R}, \; x \mapsto d(x, X),$$

where d is the Euclidean distance associated to $\mathbb{R}^d$. The *α-offset of X* is defined as $X^\alpha := \{x \in \mathbb{R}^d \mid d_X(x) \leq \alpha\}$, which is simply the sublevel set $d_X^{-1}((-\infty, \alpha])$ of d_X.

Given $x \in \mathbb{R}^d$, let $\Pi(x) \in X$ denote the set of closest points of x in X; that is,

$$\Pi(x) = \{y \in X \mid d(x, y) = d_X(x)\}.$$

The *medial axis of* X, denoted by $\mathcal{M}_X$, is the closure of the set of points with more than one closest point in X; that is,

$$\mathcal{M}_X = \text{closure}\{x \in \mathbb{R}^d \mid |\Pi(x)| \geq 2\}.$$

Intuitively, $|\Pi(x)| \geq 2$ implies that the maximal Euclidean ball centered at x whose interior is free of points in X meets X in more than one point on its boundary. Hence, $\mathcal{M}_X$ is the closure of the centers of such maximal empty balls.

Definition 6.8. (Local feature size; Reach) For a point $x \in X$, the *local feature size at* x, denoted by lfs(x), is defined as the minimum distance to the medial axis $\mathcal{M}_X$; that is,

$$\text{lfs}(x) := \mathsf{d}(x, \mathcal{M}_X).$$

The *reach of* X, denoted by $\rho(X)$, is the minimum local feature size of any point in X.

The concept has been primarily developed for the case when X is a smooth manifold embedded in $\mathbb{R}^d$. Indeed, the local feature size can be zero at a nonsmooth point. Consider a planar polygon; its medial axis intersects its vertices, and the local feature size at a vertex is thus zero. The reach of a smoothly embedded manifold could also be zero; see Section 1.2 of [118] for an example. Next, we describe a "weaker" notion of feature size [88, 89], which is more suitable for compact subsets of $\mathbb{R}^d$.

Critical Points of Distance Field

The distance function d_X introduced above is not everywhere differentiable. Its gradient is defined on $\mathbb{R}^d \setminus \{X \cup \mathcal{M}_X\}$. However, one can still define the following vector which extends the notion of the gradient of d_X to include the medial axis $\mathcal{M}_X$: Given any point $x \in \mathbb{R}^d \setminus X$, there exists a unique closed ball with minimal radius that encloses $\Pi(x)$ [226]. Let $c(x)$ denote the center of this minimal enclosing ball, and $r(x)$ its radius. It is easy to see that for any $x \in \mathbb{R}^d \setminus \mathcal{M}_X$, this ball and $c(x)$ degenerate to the unique point in $\Pi(x)$.

Definition 6.9. (Generalized vector field) Define the following vector field $\nabla_d : \mathbb{R}^d \setminus X \to \mathbb{R}^d$ where the *(generalized) gradient vector* at $x \in \mathbb{R}^d \setminus X$ is

$$\nabla_{\mathsf{d}}(x) = \frac{x - c(x)}{\mathsf{d}_X(x)}.$$

The *critical points* of ∇_{d} are points x for which $\nabla_{\mathsf{d}}(x) = 0$. We also call the critical points of ∇_{d} the critical points of the distance function d_X.

This generalized gradient field ∇_{d} coincides with the gradient of the distance function d_X for points in $\mathbb{R}^d \setminus \{X \cup \mathcal{M}_X\}$. The distance field (distance function) and its critical points were previously studied in, for example, [177], and have played an important role in sampling theory and homology inference. In general, a point x is a critical point if and only if $x \in \mathbb{R}^d \setminus X$ is contained in the convex hull of $\Pi(x)$. (The convex hull of a compact set $A \subset \mathbb{R}^d$ is the smallest convex set that contains A.) It is necessary that all critical points of ∇_{d} belong to the medial axis $\mathcal{M}_X$ of X. For the case where X is a finite set of points in $\mathbb{R}^d$, the critical points of d_X are the non-empty intersections of the Delaunay simplices with their dual Voronoi cells (if they exist) [118].

Definition 6.10. (Weak feature size) Let C denote the set of critical points of ∇_{d}. The *weak feature size of X*, denoted by $\mathrm{wfs}(X)$, is the distance between X and C; that is,

$$\mathrm{wfs}(X) = \min_{x \in X} \inf_{c \in C} \mathsf{d}(x, c).$$

Proposition 6.9. *If $0 < \alpha < \alpha'$ are such that there is no critical value of d_X in the closed interval $[\alpha, \alpha']$, then $X^{\alpha'}$ deformation retracts onto X^{α}. In particular, this implies that $\mathsf{H}_*(X^{\alpha}) \cong \mathsf{H}(X^{\alpha'})$.*

In the homology inference frameworks, the reach is usually used for the case when X is a smoothly embedded manifold, while the weak feature size is used for general compact spaces.

6.3.2 Data on a Manifold

We now consider the problem of homology inference from a point sample of a manifold. We first state a standard result from linear algebra (see also the Sandwich Lemma from [90]), which we use several times in homology inference.

Fact 6.3. *Given a sequence $A \to B \to C \to D \to E \to F$ of homomorphisms (linear maps) between finite-dimensional vector spaces over some field, if $\mathrm{rank}\,(A \to F) = \mathrm{rank}\,(C \to D)$, then this quantity also equals $\mathrm{rank}\,(B \to E)$.*

Specifically, if $A \to B \to C \to E \to F$ is a sequence of homomorphisms such that $\mathrm{rank}\,(A \to F) = \dim C$, then $\mathrm{rank}\,(B \to E) = \dim C$.

Let P be a point set sampled from a manifold $X \subset \mathbb{R}^d$. We construct either the Čech complex $\mathbb{C}^{\alpha}(P)$, or a pair of Rips complexes $\mathbb{VR}^{\alpha}(P) \hookrightarrow \mathbb{VR}^{2\alpha}(P)$, for some parameter $\alpha > 0$. The homology groups of these spaces are related as follows:

$$\mathsf{H}(X) \xleftarrow{\quad\text{Prop. 6.10}\quad} \mathsf{H}(P^\alpha) \xleftarrow{\quad\text{Nerve Thm}\quad} \mathsf{H}(\mathbb{C}^\alpha(P)) \xleftarrow{\quad\text{Fact 6.3}\quad} \text{image}\big(\mathsf{H}(\mathbb{VR}^\alpha)\to\mathsf{H}(\mathbb{VR}^{2\alpha})\big).$$

$$(6.15)$$

Specifically, recall that A^r is the r-offset of A which also equals the union of balls $\bigcup_{a\in A} B(a,r)$. The connection between the discrete samples P and the manifold X is made through the union of balls P^α. The following result is a variant of a result by Niyogi, Smale, and Weinberger [246].[3]

Proposition 6.10. *Let $P \subset \mathbb{R}^d$ be a finite point set such that $\mathsf{d}_H(X, P) \le \varepsilon$ where $X \subset \mathbb{R}^d$ is a smooth manifold with reach $\rho(X)$. If $3\varepsilon \le \alpha \le \frac{3}{4}\sqrt{\frac{3}{5}}\,\rho(X)$, then $\mathsf{H}_*(P^\alpha)$ is isomorphic to $\mathsf{H}_*(X)$.*

The Čech complex $\mathbb{C}^\alpha(P)$ is the nerve complex for the set of balls $\{B(p,\alpha), p \in P\}$. As Euclidean balls are convex, the Nerve Theorem implies that $\mathbb{C}^\alpha(P)$ is homotopy equivalent to P^α. It follows that we can use the Čech complex $\mathbb{C}^\alpha(P)$, for an appropriate α, to infer homology of X using the isomorphisms $\mathsf{H}_*(X) \cong \mathsf{H}_*(P^\alpha) \cong \mathsf{H}_*(\mathbb{C}^\alpha(P))$. The first isomorphism follows from Proposition 6.10 and the second one from the homotopy equivalence between the nerve and space.

A stronger statement in fact holds: For any $\alpha \le \beta$, the following diagram commutes.

$$
\begin{array}{ccc}
\mathsf{H}_*(P^\alpha) & \xrightarrow{\;i_*\;} & \mathsf{H}_*(P^\beta) \\
\Big\downarrow{\scriptstyle h_*} & & \Big\downarrow{\scriptstyle h_*} \\
\mathsf{H}_*(\mathbb{C}^\alpha(P)) & \xrightarrow{\;i_*\;} & \mathsf{H}_*(\mathbb{C}^\beta(P))
\end{array}
\qquad(6.16)
$$

Here, i_* stands for the homomorphism induced by inclusions, and h_* is the homomorphism induced by the homotopy equivalence $h : P^\alpha \to \mathbb{C}^\alpha(P)$ given by the Nerve Theorem. This leads to the following theorem on estimating $\mathsf{H}_*(X)$ from a pair of Rips complexes.

Theorem 6.11. *Given a smooth manifold X embedded in $\mathbb{R}^d$, let $\rho(X)$ be its reach. Let $P \subset \mathbb{R}^d$ be a finite sample such that $\mathsf{d}_H(P, X) \le \varepsilon$. For any $3\varepsilon \le \alpha \le \frac{3}{16}\sqrt{\frac{3}{5}}\,\rho(X)$, let $i_* : \mathsf{H}_*(\mathbb{VR}^\alpha) \to \mathsf{H}_*(\mathbb{VR}^{2\alpha})$ be the homomorphism induced by the inclusion $i : \mathbb{VR}^\alpha \to \mathbb{VR}^{2\alpha}$. We have that*

---

[3] The result of [246] assumes that $P \subseteq X$, in which case it shows that P^α deformation retracts to X. In our statement P is not necessarily from X, and the isomorphism follows from results of [246] and Fact 6.3.

$$\text{rank}(\text{im}(i_*)) = \dim(\mathsf{H}_*(\mathbb{C}^\alpha(P))) = \dim(\mathsf{H}_*(X)).$$

PROOF. By Eq. (6.16) and Proposition 6.10, we have that for $3\varepsilon \le \alpha \le \beta \le \frac{3}{4}\sqrt{\frac{3}{5}}\,\rho(X)$,

$$\mathsf{H}_*(X) \cong \mathsf{H}_*(P^\alpha) \cong \mathsf{H}_*(\mathbb{C}^\alpha(P)) \cong \mathsf{H}_*(\mathbb{C}^\beta(P)), \qquad (6.17)$$

where the last isomorphism is induced by inclusion. On the other hand, recall the interleaving relation between the Čech and the Rips complexes:

$$\cdots \subseteq \mathbb{C}^\alpha(P) \subseteq \mathbb{VR}^\alpha(P) \subseteq \mathbb{C}^{2\alpha}(P) \subseteq \mathbb{VR}^{2\alpha}(P) \subseteq \mathbb{C}^{4\alpha}(P) \subseteq \cdots.$$

We thus have the following sequence of homomorphisms induced by inclusion:

$$\mathsf{H}_*(\mathbb{C}^\alpha(P)) \to \mathsf{H}_*(\mathbb{VR}^\alpha(P)) \to \mathsf{H}_*(\mathbb{C}^{2\alpha}(P)) \to \mathsf{H}_*(\mathbb{VR}^{2\alpha}(P)) \to \mathsf{H}_*(\mathbb{C}^{4\alpha}(P)).$$

We have $\mathsf{H}_*(\mathbb{C}^\alpha(P)) \cong \mathsf{H}_*(\mathbb{C}^{2\alpha}(P)) \cong \mathsf{H}_*(\mathbb{C}^{4\alpha}(P))$ by Eq. (6.17). Thus we have

$$\text{rank}(\mathsf{H}_*(\mathbb{C}^\alpha(P)) \to \mathsf{H}_*(\mathbb{C}^{4\alpha}(P))) = \dim(\mathsf{H}_*(\mathbb{C}^\alpha(P))).$$

The theorem then follows from the second part of Fact 6.3. $\qquad\square$

6.3.3 Data on a Compact Set

We now consider the case when we are given a finite set of points P sampling a compact subset $X \subset \mathbb{R}^d$. It is known that an offset X^α for any $\alpha > 0$ may not be homotopy equivalent to X for every compact set X. In fact, there exist compact sets so that $\mathsf{H}_*(X^\lambda)$ is not isomorphic to $\mathsf{H}_*(X)$ no matter how small $\lambda > 0$ is (see Figure 4 of [89]). So, in this case we aim to recover the homology groups of an offset X^λ of X for a sufficiently small $\lambda > 0$.

The high-level framework is in Eq. (6.18). Here we have $0 < \lambda < \text{wfs}(X)$, while $\mathbb{C}^\alpha$ and $\mathbb{VR}^\alpha$ stand for the Čech and Rips complexes $\mathbb{C}^\alpha(P)$ and $\mathbb{VR}^\alpha(P)$ over the point set P. For any $0 < \lambda < \text{wfs}(X)$:

$$\mathsf{H}_*(X^\lambda) \xleftarrow{\text{Prop. 6.12}} \text{image}(\mathsf{H}_*(\mathbb{C}^\alpha) \to \mathsf{H}_*(\mathbb{C}^{2\alpha})) \xleftrightarrow[\text{Fact 6.3}]{\text{Eq. (6.21)}} \text{image}(\mathsf{H}_*(\mathbb{VR}^\alpha) \to \mathsf{H}_*(\mathbb{VR}^{4\alpha})).$$

$$(6.18)$$

This is similar to Eq. (6.15) for the manifold case. However, we no longer have the isomorphism between $\mathsf{H}_*(P^\alpha)$ and $\mathsf{H}_*(X)$. To overcome this difficulty, we leverage Proposition 6.9. This in turn requires us to consider a pair of Čech complexes to infer homology of X^λ, instead of a single Čech complex as in the case of manifolds.

More specifically, suppose that the point set P satisfies that $\mathsf{d}_H(P, X) \le \varepsilon$; then we have the following nested sequence for $\alpha > \varepsilon$ and $\alpha' \ge \alpha + 2\varepsilon$:

$$X^{\alpha-\varepsilon} \subseteq P^{\alpha} \subseteq X^{\alpha+\varepsilon} \subseteq P^{\alpha'} \subseteq X^{\alpha'+\varepsilon}. \tag{6.19}$$

By Proposition 6.9, we know that if it also holds that $\alpha' + \varepsilon < \mathrm{wfs}(X)$, then the inclusions between $X^{\alpha-\varepsilon} \subseteq X^{\alpha+\varepsilon} \subseteq X^{\alpha'+\varepsilon}$ induce isomorphisms between their homology groups, which are also isomorphic to $\mathsf{H}_*(X^{\lambda})$ for $\lambda \in (0, \mathrm{wfs}(X))$. It then follows from the second part of Fact 6.3 that, for $\alpha, \alpha' \in \big(\varepsilon, \mathrm{wfs}(X) - \varepsilon\big)$ and $\alpha' - \alpha \geq 2\varepsilon$, we have

$$\mathsf{H}_*(X^{\lambda}) \cong \mathrm{im}\,(i_*), \text{ where } i_* : \mathsf{H}_*(P^{\alpha}) \to \mathsf{H}_*(P^{\alpha'}) \text{ is induced by inclusion } i : P^{\alpha} \subseteq P^{\alpha'}. \tag{6.20}$$

Combining the above with the commutative diagram in Eq. (6.16), we obtain the following result on inferring homology of X^{λ} using a pair of Čech complexes.

Proposition 6.12. *Let X be a compact set in $\mathbb{R}^d$ and $P \subset \mathbb{R}^d$ a finite set of points with $\mathsf{d}_H(X, P) < \varepsilon$ for some $\varepsilon < \frac{1}{4}\mathrm{wfs}(X)$. Then, for all $\alpha, \alpha' \in \big(\varepsilon, \mathrm{wfs}(X) - \varepsilon\big)$ such that $\alpha' - \alpha \geq 2\varepsilon$, and any $\lambda \in (0, \mathrm{wfs}(X))$, we have $\mathsf{H}_*(X^{\lambda}) \cong \mathrm{im}\,(i_*)$, where $i_* : \mathsf{H}_*(\mathbb{C}^{\alpha}(P)) \to \mathsf{H}_*(\mathbb{C}^{\alpha'}(P))$ is the homomorphism between homology groups induced by the inclusion $i : \mathbb{C}^{\alpha}(P) \hookrightarrow \mathbb{C}^{\alpha'}(P)$.*

Finally, to perform homology inference with the Rips complexes, we again resort to the interleaving relation between Čech and Rips complexes, and apply the first part of Fact 6.3 to the following sequence:

$$\mathsf{H}_*(\mathbb{C}^{\alpha/2}(P)) \to \mathsf{H}_*(\mathbb{VR}^{\alpha/2}(P)) \to \mathsf{H}_*(\mathbb{C}^{\alpha}(P)) \to \mathsf{H}_*(\mathbb{C}^{2\alpha}(P)) \to \mathsf{H}_*(\mathbb{VR}^{2\alpha}(P)) \to \mathsf{H}_*(\mathbb{C}^{4\alpha}(P)). \tag{6.21}$$

If $2\varepsilon \leq \alpha \leq \frac{1}{4}(\mathrm{wfs} - \varepsilon)$, both $\mathsf{H}_*(\mathbb{C}^{\alpha/2}(P)) \to \mathsf{H}_*(\mathbb{C}^{4\alpha}(P))$ and $\mathsf{H}_*(\mathbb{C}^{\alpha}(P)) \to \mathsf{H}_*(\mathbb{C}^{2\alpha}(P))$ have ranks equal to $\dim(\mathsf{H}_*(X^{\lambda}))$ by Proposition 6.12. Applying Fact 6.3, we then obtain the following result.

Theorem 6.13. *Let X be a compact set in $\mathbb{R}^d$ and P a finite point set with $\mathsf{d}_H(X, P) < \varepsilon$ for some $\varepsilon < \frac{1}{9}\mathrm{wfs}(X)$. Then, for all $\alpha \in \big(2\varepsilon, \frac{1}{4}(\mathrm{wfs}(X) - \varepsilon)\big)$ and all $\lambda \in (0, \mathrm{wfs}(X))$, we have $\mathsf{H}_*(X_{\lambda}) \cong \mathrm{im}\,(j_*)$, where j_* is the homomorphism between homology groups induced by the inclusion $j : \mathbb{VR}^{\alpha/2}(P) \hookrightarrow \mathbb{VR}^{2\alpha}(P)$.*

6.4 Homology Inference for Scalar Fields

Suppose we are only given a finite sample $P \subset X$ from a smooth manifold $X \subset \mathbb{R}^d$ together with a potentially noisy version $\hat{f}$ of a smooth function

$f: X \to \mathbb{R}$ presented as a vertex function $\hat{f}: P \to \mathbb{R}$. We are interested in recovering the persistent homology of the sublevel filtration of f from $\hat{f}$. That is, the goal is to approximate the persistent homology induced by f from the discrete sample P and function values $\hat{f}$ on points in P.

6.4.1 Problem Setup

Set $F_\alpha = f^{-1}(-\infty, \alpha] = \{x \in X \mid f(x) \leq \alpha\}$ as the sublevel set of f with respect to α. The sublevel set filtration of X induced by f, denoted by $\mathcal{F}_f = \{F_\alpha; i^{\alpha,\beta}\}_{\alpha \leq \beta}$, is a family of sets F_α totally ordered by inclusion map $i^{\alpha,\beta}: F_\alpha \hookrightarrow F_\beta$ for any $\alpha \leq \beta$ (Section 3.1). This filtration induces the following persistence module:

$$\mathsf{H}_p \mathcal{F}_f = \{\mathsf{H}_p(F_\alpha) \xrightarrow{i_*^{\alpha,\beta}} \mathsf{H}_p(F_\beta)\}_{\alpha \leq \beta}, \quad \text{where } i_*^{\alpha,\beta} \text{ is induced by inclusion map } i^{\alpha,\beta}.$$
(6.22)

For simplicity, we often write the filtration and the corresponding persistence module as $\mathcal{F}_f = \{F_\alpha\}_{\alpha \in \mathbb{R}}$ and $\mathsf{H}_p \mathcal{F}_f = \{\mathsf{H}(F_\alpha)\}_{\alpha \in \mathbb{R}}$, when the choices of maps connecting their elements are clear.

Our goal is to approximate the persistence diagram $\mathrm{Dgm}_p(\mathcal{F}_f)$ from point samples P and $\hat{f}: P \to \mathbb{R}$. Intuitively, we construct a specific Čech (or Rips) complex $\mathbb{C}^r(P)$, use $\hat{f}$ to induce a filtration of $\mathbb{C}^r(P)$, and then use its persistent homology to approximate $\mathrm{Dgm}_p(\mathcal{F}_f)$. More specifically, we need to consider the *nested pair filtration* for either $\mathbb{C}^r(P)$ or $\mathbb{VR}^r(P)$.

Nested Pair Filtration
Let $P_\alpha = \{p \in P \mid \hat{f}(p) \leq \alpha\}$ be the set of sample points with the function value for $\hat{f}$ at most α, which presumably samples the sublevel set F_α of X with respect to f. To estimate the topology of F_α from these discrete samples P_α, we consider either the Čech complex $\mathbb{C}^r(P_\alpha)$ or the Rips complex $\mathbb{VR}^r(P_\alpha)$. For the time being, consider $\mathbb{VR}^r(P_\alpha)$. As we already saw in previous sections, the topological information of F_α can be inferred from a pair of nested complexes $\mathbb{VR}^r(P_\alpha) \xrightarrow{j_\alpha} \mathbb{VR}^{r'}(P_\alpha)$ for some appropriate $r < r'$. To study $\mathcal{F}_f$, we need to inspect $F_\alpha \to F_\beta$ for $\alpha \leq \beta$. To this end, fixing r and r', for any $\alpha \leq \beta$, consider the following commutative diagram induced by inclusions.

$$
\begin{array}{ccc}
\mathsf{H}_*(\mathbb{VR}^r(P_\alpha)) & \longrightarrow & \mathsf{H}_*(\mathbb{VR}^r(P_\beta)) \\
\downarrow{\scriptstyle i_{\alpha*}} & & \downarrow{\scriptstyle i_{\beta*}} \\
\mathsf{H}_*(\mathbb{VR}^{r'}(P_\alpha)) & \xrightarrow{\ j_{\alpha*}^\beta\ } & \mathsf{H}_*(\mathbb{VR}^{r'}(P_\beta))
\end{array}
$$
(6.23)

Set $\phi_\alpha^\beta : \text{im}\,(i_{\alpha*}) \to \text{im}\,(i_{\beta*})$ to be $\phi_\alpha^\beta = j_{\alpha}^\beta{}_* |_{\text{im}\,(i_{\alpha*})}$, that is, the restriction of $j_{\alpha}^\beta{}_*$ to $\text{im}\,(i_{\alpha*})$. This map is well defined as the diagram above commutes. This gives rise to a persistence module $\{\text{im}\,(i_{\alpha*}); \phi_\alpha^\beta\}_{\alpha \le \beta}$, that is, a family of totally ordered vector spaces $\{\text{im}\,(i_\alpha)\}$ with commutative homomorphisms ϕ_α^β between any two elements. We formalize and generalize the above construction below.

Definition 6.11. (Nested pair filtration) A *nested pair filtration* is a sequence of pairs of complexes $\{AB_\alpha = (A_\alpha, B_\alpha)\}_{\alpha \in \mathbb{R}}$ where (i) $A_\alpha \overset{i_\alpha}{\hookrightarrow} B_\alpha$ is inclusion for every α and (ii) $AB_\alpha \hookrightarrow AB_\beta$ for $\alpha \le \beta$ is given by $A_\alpha \hookrightarrow A_\beta$ and $B_\alpha \overset{j_\alpha^\beta}{\hookrightarrow} B_\beta$. The p-th persistence module of the filtration $\{AB_\alpha\}_{\alpha \in \mathbb{R}}$ is given by the homology module $\{\text{im}\,(\mathsf{H}_p(A_\alpha) \to \mathsf{H}_p(B_\alpha)); \phi_\alpha^\beta\}_{\alpha \le \beta}$ where ϕ_α^β is the restriction of $j_\alpha^\beta{}_*$ on the $\text{im}\,i_{\alpha*}$. For simplicity, we say the module is induced by the nested pair filtration $\{A_\alpha \hookrightarrow B_\alpha\}$.

The high-level approach of inferring persistent homology of a scalar field $f : X \to \mathbb{R}$ from a set of points P equipped with $\hat{f} : P \to \mathbb{R}$ involves the following steps:

Step 1. Sort all points of α_i in nondecreasing $\hat{f}$-values, $P = \{p_1, \dots, p_n\}$. Set $\alpha_i = \hat{f}(p_i)$ for $i \in [1, n]$.

Step 2. Compute the persistence diagram induced by the filtration of nested pairs $\{\mathbb{VR}^r(P_{\alpha_i}) \hookrightarrow \mathbb{VR}^{r'}(P_{\alpha_i})\}_{i \in [1,n]}$ (or $\{\mathbb{C}^r(P_{\alpha_i}) \hookrightarrow \mathbb{C}^{r'}(P_{\alpha_i})\}_{i \in [1,n]}$) for appropriate parameters $0 < r < r'$.

The persistent homology (as well as persistence diagram) induced by the filtration of nested pairs is computed via the algorithm in [104]. To obtain an approximation guarantee for the above approach, we consider an intermediate object defined by the intrinsic Riemannian metric on the manifold X. Indeed, note that the filtration of X with respect to f is intrinsic in the sense that it is independent of how X is embedded in $\mathbb{R}^d$. Hence it is more natural to approximate its persistent homology with an object defined intrinsically for X.

Given a compact Riemannian manifold $X \subset \mathbb{R}^d$ embedded in $\mathbb{R}^d$, let d_X be the Riemannian metric of X inherited from the Euclidean metric d_E of $\mathbb{R}^d$. Let $B_X(x, r) := \{y \in X \mid \mathsf{d}_X(x, y) \le r\}$ be the *geodesic ball on* X centered at x and with radius r, and $B_X^o(x, r)$ be the open geodesic ball. In contrast, $B_E(x, r)$ (or simply $B(x, r)$) denotes the Euclidean ball in $\mathbb{R}^d$. A ball $B_X^o(x, r)$ is *strongly convex* if for every pair $y, y' \in B_X(x, r)$, there exists a unique minimizing geodesic between y and y' whose interior is contained within $B^o_X(x, r)$. For details on these concepts, see [76, 164].

Definition 6.12. (Strong convexity) For $x \in X$, let $\rho_c(x; X)$ denote the supremum of radius r such that the geodesic ball $B^o_X(x, r)$ is strongly convex. The *strong convexity radius of* (X, d_X) is defined as $\rho_c(X) := \inf_{x \in X} \rho_c(x; X)$.

Let $\mathsf{d}_X(x, P) := \inf_{p \in P} \mathsf{d}_X(x, p)$ denote the closest geodesic distance between x and the set $P \subseteq X$.

Definition 6.13. (ε-geodesic sample) A point set $P \subset X$ is an *ε-geodesic sample of* (X, d_X) if for all $x \in X$, $\mathsf{d}_X(x, P) \leq \varepsilon$.

Recall that P_α is the set of points in P with $\hat{f}$-value at most α. The union of geodesic balls $P^{\delta;X}_\alpha = \bigcup_{p \in P_\alpha} B_X(p, \delta)$ is intuitively the "δ-thickening" of P_α within the manifold X. We use two kinds of Čech and Rips complexes. One is defined with the metric d_E of the ambient Euclidean space which we call *(extrinsic) Čech complex* $\mathbb{C}^\delta(P_\alpha)$ and *(extrinsic) Rips complex* $\mathbb{VR}^\delta(P_\alpha)$. The other is *intrinsic Čech complex* $\mathbb{C}^\delta_X(P_\alpha)$ and *intrinsic Rips complex* $\mathbb{VR}^\delta_X(P_\alpha)$ that are defined with the intrinsic metric d_X. Note that $\mathbb{C}^\delta_X(P_\alpha)$ is the nerve complex of the union of geodesic balls forming $P^{\delta;X}_\alpha$. Also the interleaving relation between the Čech and Rips complexes remains the same as for general geodesic spaces; that is, $\mathbb{C}^\delta_X(P_\alpha) \subseteq \mathbb{VR}^\delta_X(P_\alpha) \subseteq \mathbb{C}^{2\delta}_X(P_\alpha)$ for any α and δ.

6.4.2 Inference Guarantees

Recall from Section 4.1 that two ε-interleaved filtrations lead to ε-interleaved persistence modules, which further means that the bottleneck distance between their persistence diagrams is bounded by ε. Here we first relate the space filtration with the intrinsic Čech filtrations and then relate these intrinsic ones with the extrinsic Čech or Rips filtrations of nested pairs as illustrated below:

$$\{F_\alpha\} \longleftrightarrow \{P^{r;X}_\alpha\} \longleftrightarrow \{\mathbb{C}^r_X(P_\alpha)\} \longleftrightarrow \{\mathbb{C}^r(P_\alpha) \hookrightarrow \mathbb{C}^{r'}(P_\alpha)\} \text{ or } \{\mathbb{VR}^r(P_\alpha) \hookrightarrow \mathbb{VR}^{r'}(P_\alpha)\}.$$
(6.24)

Proposition 6.14. *Let $X \subset \mathbb{R}^d$ be a compact Riemannian manifold with intrinsic metric d_X, and let $f : X \to \mathbb{R}$ be a C-Lipschitz function. Suppose $P \subset X$ is an ε-geodesic sample of X, equipped with $\hat{f} : P \to \mathbb{R}$ so that $\hat{f} = f|_P$. Then, for any fixed $\delta \geq \varepsilon$, the filtration $\{F_\alpha\}_\alpha$ and the filtration $\{P^{\delta;X}_\alpha\}_\alpha$ are $(C\delta)$-interleaved with respect to inclusions.*

The intrinsic Čech complex $\mathbb{C}^\delta_X(P_\alpha)$ is the nerve complex for $\{B_X(p, \delta)\}_{p \in P_\alpha}$. Furthermore, for $\delta < \rho_c(X)$, the family of geodesic balls in $\{B_X(p, \delta)\}_{p \in P_\alpha}$ form a cover of the union $P_\alpha^{\delta;X}$ that satisfies the condition of the Nerve Theorem (Theorem 2.1). Hence, there is a homotopy equivalence between the nerve complex $\mathbb{C}_X^\delta(P_\alpha)$ and $P_\alpha^{\delta;X}$. Furthermore, using the same argument for showing that the diagram in Eq. (6.16) commutes (Lemma 3.4 of [90]), one can show that the following diagram commutes for any $\alpha \leq \beta \in \mathbb{R}$ and $\delta \leq \xi < \rho_c(X)$.

$$
\begin{array}{ccc}
\mathsf{H}_*(P_\alpha^{\delta;X}) & \xrightarrow{\;\;i_*\;\;} & \mathsf{H}_*(P_\beta^{\xi;X}) \\
\downarrow{\scriptstyle h_*} & & \downarrow{\scriptstyle h_*} \\
\mathsf{H}_*(\mathbb{C}_X^\delta(P_\alpha)) & \xrightarrow{\;\;i_*\;\;} & \mathsf{H}_*(\mathbb{C}_X^\xi(P_\beta))
\end{array}
\qquad (6.25)
$$

Here the horizontal homomorphisms are induced by inclusions, and the vertical ones are isomorphisms induced by the homotopy equivalence between a union of geodesic balls and its nerve complex. The above diagram leads to the following result (see Lemma 2 of [86] for details).

Corollary 6.15. *Let X, f, and P be as in Proposition 6.14 (although f does not need to be C-Lipschitz). For any $\delta < \rho_c(X)$, $\{P_\alpha^{\delta;X}\}_{\alpha \in \mathbb{R}}$ and $\{\mathbb{C}_X^\delta(P_\alpha)\}_{\alpha \in \mathbb{R}}$ are 0-interleaved. Hence they induce isomorphic persistence modules which have identical persistence diagrams.*

Combining with Proposition 6.14, this implies that the filtration $\{\mathbb{C}_X^\delta(P_\alpha)\}_\alpha$ and the filtration $\{F_\alpha\}_\alpha$ are Cδ-interleaved for $\varepsilon \leq \delta < \rho_c(X)$.

However, we cannot access the intrinsic metric d_X of the manifold X and thus cannot directly construct intrinsic Čech complexes. It turns out that for points that are sufficiently close, their Euclidean distance forms a constant factor approximation of the geodesic distance between them on X.

Proposition 6.16. *Let $X \subset \mathbb{R}^d$ be an embedded Riemannian manifold with reach ρ_X. For any two points $x, y \in X$ with $\mathsf{d}_E(x, y) \leq \rho_X/2$, we have that*

$$
\mathsf{d}_E(x, y) \leq \mathsf{d}_X(x, y) \leq \left(1 + \frac{4\mathsf{d}_E^2(x, y)}{3\rho_X^2}\right) \mathsf{d}_E(x, y) \leq \frac{4}{3}\mathsf{d}_E(x, y).
$$

This implies the following nested relation between the extrinsic and intrinsic Čech complexes:

$$
\mathbb{C}_X^\delta(P_\alpha) \subseteq \mathbb{C}^\delta(P_\alpha) \subseteq \mathbb{C}_X^{\frac{4}{3}\delta}(P_\alpha) \subseteq \mathbb{C}^{\frac{4}{3}\delta}(P_\alpha) \subseteq \mathbb{C}_X^{\frac{16}{9}\delta}(P_\alpha), \quad \text{for any } \delta < \frac{3}{8}\rho_X. \quad (6.26)
$$

Note that a similar relation also holds between the intrinsic Čech filtration and the extrinsic Rips complexes due to the nested relation between extrinsic Čech and Rips complexes. To infer persistent homology from nested pair filtrations for complexes constructed under the Euclidean metric, we use the following key lemma from [86], which can be thought of as a persistent version as well as a generalization of Fact 6.3.

Proposition 6.17. *Let X, f, and P be as in Proposition 6.14. Suppose that there exist $\varepsilon' \leq \varepsilon'' \in [\varepsilon, \rho_c(X))$ and two filtrations $\{G_\alpha\}_\alpha$ and $\{G'_\alpha\}_\alpha$, so that*

$$\text{for all } \alpha \in \mathbb{R}, \quad \mathbb{C}^\varepsilon_X(P_\alpha) \subseteq G_\alpha \subseteq \mathbb{C}^{\varepsilon'}_X(P_\alpha) \subseteq G'_\alpha \subseteq \mathbb{C}^{\varepsilon''}_X(P_\alpha).$$

Then the persistence module induced by the filtration $\{F_\alpha\}_\alpha$ for f and that induced by the nested pairs of filtrations $\{G_\alpha \hookrightarrow G'_\alpha\}_\alpha$ are $C\varepsilon''$-interleaved, where f is C-Lipschitz.

Combining this proposition with the sequences in Eq. (6.26), we obtain the following results on inferring the persistent homology induced by a function $f : X \to \mathbb{R}$.

Theorem 6.18. *Let $X \subset \mathbb{R}^d$ be a compact Riemannian manifold with intrinsic metric d_X, and $f : X \to \mathbb{R}$ a C-Lipschitz function on X. Let ρ_X and $\rho_c(X)$ be the reach and the strong convexity radius of (X, d_X), respectively. Suppose $P \subset X$ is an ε-geodesic sample of X, equipped with $\hat{f} : P \to \mathbb{R}$ such that $\hat{f} = f|_P$. Then:*

(a) *for any fixed r such that $\varepsilon \leq r \leq \min\{\frac{9}{16}\rho_c(X), \frac{9}{32}\rho_X\}$, the persistent homology module induced by the sublevel set filtration of $f : X \to \mathbb{R}$ and that induced by the filtration of nested pairs $\{\mathbb{C}^r(P_\alpha) \hookrightarrow \mathbb{C}^{\frac{4}{3}r}(P_\alpha)\}_\alpha$ are $\frac{16}{9}Cr$-interleaved; and*

(b) *for any fixed r such that $2\varepsilon \leq r \leq \min\{\frac{9}{32}\rho_c(X), \frac{9}{64}\rho_X\}$, the persistent homology module induced by the sublevel set filtration of f and that induced by the filtration of nested pairs $\{\mathbb{VR}^r(P_\alpha) \hookrightarrow \mathbb{VR}^{\frac{8}{3}r}(P_\alpha)\}_\alpha$ are $\frac{32}{9}Cr$-interleaved.*

In particular, in each case above, the bottleneck distance between their respective persistence diagrams is bounded by the stated interleaving distance between persistence modules.

6.5 Notes and Exercises

Part of Theorem 6.3 is proved in [77, 78]. A complete proof as well as a thorough treatment for geometric complexes such as Rips and Čech complexes can be found in [81]. The first approach on data sparsification for Rips filtrations is proposed by Sheehy [274]. The presentation of Section 6.2.1 is based on a combination of the treatments of sparsification in [56] and [275] (in [275], a net-tower created via net-tree data structure (e.g., [182]) is used for constructing sparse Rips filtration). Extension of such sparsification to Čech complexes and a geometric interpretation are provided in [70]. The Rips sparsification is extended to handle weighted Rips complexes derived from distance to measures in [56]. Sparsification via simplicial towers is introduced in [124]. This is an application of the algorithm we presented in Section 4.2 for computing persistent homology for a simplicial tower. Simplicial maps allow batch collapse of vertices and leads to more aggressive sparsification. However, in practice, it is observed that it also has the overconnection issues as one collapses the vertices. This issue is addressed in [134]. In particular, the SimBa algorithm of [134] exploits the simplicial maps for sparsification, but connects vertices at sparser levels based on a certain distance between two sets (each of which intuitively is the set of original points mapped to a vertex at the present sparsified level). While SimBa has similar approximation guarantees in sparsification, in practice, the sparsified sequence of complexes has much smaller size compared to prior approaches.

Much of the materials in Section 6.3 are taken from [81, 86, 90, 246]. We remark that there have been different variations of the medial axis in the literature. We follow the notation from [118]. We also note that there exists a robust version of the medial axis, called the λ-medial axis, proposed in [88]. The concept of the local feature size was originally proposed in [270] in the context of mesh generation, and a different version that we describe in this chapter was introduced in [8] in the context of curve/surface reconstruction. The local feature size has been widely used in the field of surface reconstruction and mesh generation; see the books [97, 118]. Critical points of the distance field were originally studied in [177]. See [88, 89, 226] for further studies as well as the development on weak feature sizes.

In homology inference for manifolds, we note that Niyogi, Smale, and Weinberger in [246] provide two deformation retract results from the union of balls over P to a manifold X; Proposition 3.1 holds for the case when $P \subset X$, while Proposition 7.1 holds when P is within a tubular neighborhood of X. The latter has much stronger requirement on the radius α. In our presentation, Proposition 6.10 uses a corollary of Proposition 3.1 of [246] to obtain an isomorphism between the homology groups of union of balls and of X. This allows a better range of the parameter α – however, we lose the deformation retraction here; see the footnote above Proposition 6.10. Results in Section 6.4 are mostly based on the work in [86].

This chapter focuses on presenting the main framework behind homology (or persistent homology) inference from point cloud data. The current theoretical guarantees hold when input points sample the hidden domain well within Hausdorff distance. For more general noise models that include outliers and statistical noise, we need a more robust notion of distance field than what we used in Section 6.3.1. To this end, an elegant concept called *distance to measures* (DTM) has been proposed in [79], which has many nice properties and can lead to more robust homological inferences; see, for example, [82]. An alternative approach using kernel distance is proposed in [257]. See also [56, 79, 248] for data sparsification or homology inference for points corrupted with more general noise, and [55] for persistent homology inference under more general noise for input scalar fields.

Exercises

1. Prove Theorem 6.3(a).
2. Prove the bound on the Rips pseudodistance $\mathsf{d}_{Rips}(P, Q)$ in Theorem 6.3(b).
3. Given two finite sets of points $P, Q \subset \mathbb{R}^d$, let d_P and d_Q denote the restriction of the Euclidean metric over P and Q, respectively. Consider the Hausdorff distance $\delta_H = \mathsf{d}_H(P, Q)$ between P and Q, as well as the Gromov–Hausdorff distance $\delta_{GH} = \mathsf{d}_{GH}((P, \mathsf{d}_P), (Q, \mathsf{d}_Q))$.

   (a) Prove that $\delta_{GH} \leq \delta_H$.

   (b) Assume $P, Q \subset \mathbb{R}^2$. Let $\mathcal{T}$ stand for the set of rigid transformations over $\mathbb{R}^2$ (rotation, reflection, translations, and their combinations). Let $\delta_H^* := \inf_{t \in \mathcal{T}} \delta_H(P, t(Q))$ denote the smallest Hausdorff distance possible between P and a copy of Q under rigid transformation. Give an example of $P, Q \subset \mathbb{R}^2$ such that δ_H^* is much larger than δ_{GH}, say $\delta_H^* \geq 10\delta_{GH}$ (in fact, this can hold for any fixed constant).

4. Prove Proposition 6.5.
5. Consider the greedy permutation approach introduced in Section 6.2, and the assignment of exit-times for points $p \in P$. Construct the open tower $\{N_\gamma\}$ and closed tower $\{\overline{N}_\gamma\}$ as described in the chapter. Prove that both N_γ and $\overline{N}_\gamma$ are γ-nets for P.

6. Suppose we are given $P_0 \supset P_1$ sampled from a metric space (Z, d) where P_1 is a γ-net of P_0. Define $\pi : P_0 \to P_1$ as $\pi(p) \mapsto \operatorname{argmin}_{q \in P_1} \mathsf{d}(p, q)$ (if $\operatorname{argmin}_{q \in P_1} \mathsf{d}(p, q)$ contains more than one point, then set $\pi(p)$ to be any point q that minimizes $\mathsf{d}(p, q)$).

(a) Prove that the vertex map π induces a simplicial map $\pi : \mathbb{VR}^\alpha(P_0) \to \mathbb{VR}^{\alpha + \gamma}(P_1)$.

(b) Consider the following diagram. Prove that the map $j \circ \pi$ is contiguous to the inclusion map i.

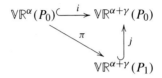

7. Let P be a set of points in $\mathbb{R}^d$. Let d_2 and d_1 denote the distance metric under L_2-norm and under L_1-norm, respectively. Let $\mathbb{C}_2(P)$ and $\mathbb{C}_1(P)$ be the Čech filtration over P induced by d_2 and d_1, respectively. Show the relation between the log-scaled version of persistence diagrams $\mathrm{Dgm}_{\log}\mathbb{C}_2(P)$ and $\mathrm{Dgm}_{\log}\mathbb{C}_1(P)$, that is, bound $\mathsf{d}_b(\mathrm{Dgm}_{\log}\mathbb{C}_2(P), \mathrm{Dgm}_{\log}\mathbb{C}_1(P))$ (see the discussion above Corollary 4.4 in Chapter 4).

8. Prove Proposition 6.14. Using the fact that the diagram in Eq. (6.25) commutes, prove Corollary 6.15.

7

Reeb Graphs

Topological persistence provides an avenue to study a function $f : X \to \mathbb{R}$ on a space X. *Reeb graphs* provide another avenue to do the same; although the summarizations produced by the two differ in a fundamental way. Topological persistence produces barcodes as a simplified signature of the function. Reeb graphs instead provide a one-dimensional (skeleton) structure which represents a simplification of the input domain X while taking the function into account for this simplification. Of course, one loses higher-dimensional homological information in the Reeb graphs, but at the same time, it offers a much lighter and computationally inexpensive transformation of the original space which can be used as a signature for tasks such as shape matching and functional similarity. An example from [190] is given in Figure 7.1, where a multiresolutional representation of the Reeb graph is used to match surface models.

We define the Reeb graph and introduce some properties of it in Section 7.1. We also describe efficient algorithms to compute it for the piecewise-linear setting in Section 7.2. For comparing Reeb graphs, we need to define distances among them. In Section 7.3, we present two equivalent distance measures for the Reeb graphs and give a stability result of these distances with respect to changes in the input function that define the Reeb graph. In particular, we note that a Reeb graph can also be viewed as a graph equipped with a "height" function on it which is induced by the original function $f : X \to \mathbb{R}$ on the input domain. This height function provides a natural metric on the Reeb graph, rendering a view of the Reeb graph as a specific metric graph. This further leads to a distance measure for Reeb graphs based on the Gromov–Hausdorff distance idea, which we present in Section 7.3. An alternative way to define a distance for Reeb graphs is based on the interleaving idea, which we also introduce in Section 7.3. It turns out that these two versions of distances for Reeb graphs are strongly equivalent, meaning that they are within a constant factor of each other.

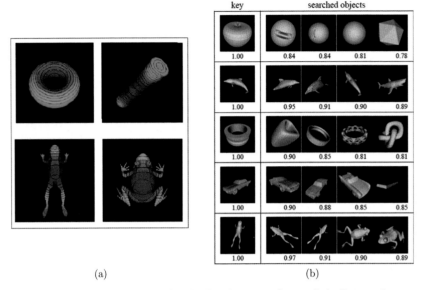

(a)   (b)

Figure 7.1 (a) A description function based on averaging geodesic distances is shown on different models, together with some isocontours of this function. This function is robust with respect to near-isometric deformation of shapes. (b) The Reeb graph of the descriptor function (from the left) is used to compare different shapes. Here, given a query shape (called "key"), the most similar shapes retrieved from a database are shown on the right. Images taken from [190], reprinted by permission from ACM: M. Hilaga et al. (2001).

7.1 Reeb Graph: Definitions and Properties

Before we give a formal definition of the Reeb graph, let us recall some relevant definitions from Section 1.1. A topological space X is *disconnected* if there are two *disjoint* open sets U and V so that $X = U \cup V$. It is called *connected* otherwise. A *connected component* of X is a maximal subset (subspace) that is connected. Given a continuous function $f : X \rightarrow \mathbb{R}$ on a finitely triangulable topological space X, for each $a \in \mathbb{R}$, consider the levelset $f^{-1}(a) = \{x \in X : f(x) = a\}$ of f. It is a subspace of X and we can talk about its connected components in this subspace topology (see Figure 7.2).

Definition 7.1. (Reeb graph) Define an equivalence relation $\sim$ on X by asserting $x \sim y$ if and only if (i) $f(x) = f(y) = \alpha$ and (ii) x and y belong to the same connected component of the levelset $f^{-1}(\alpha)$. Let $[x]$ denote the equivalent class to which a point $x \in X$ belongs. The *Reeb graph* R_f of $f : X \rightarrow \mathbb{R}$ is the quotient space $X/\sim$, that is, the set of equivalent classes equipped with

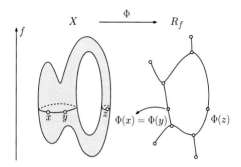

Figure 7.2 Reeb graph R_f of the function $f : X \to \mathbb{R}$.

the quotient topology (see Figure 7.2). Let $\Phi \colon X \to \mathsf{R}_f$, $x \mapsto [x]$ denote the quotient map.

If the input is "nice," for example, if f is a Morse function on a compact manifold, or a PL-function on a compact polyhedron, then R_f indeed has the structure of a finite one-dimensional regular CW complex which is a graph, and this is why it is commonly called a Reeb graph. In particular, from now on, we tacitly assume that the input function $f \colon X \to \mathbb{R}$ is *levelset tame*, meaning that (i) each levelset $f^{-1}(a)$ has a finite number of components, and each component is path connected,[1] and (ii) f is of Morse type (Definition 4.14). It is known that Morse functions on a compact smooth manifold and PL-functions on finite simplicial complexes are both levelset tame.

A levelset may consist of several connected components, each of which is called a *contour*. Intuitively, the Reeb graph R_f is obtained by collapsing contours (connected components) in each levelset $f^{-1}(a)$ continuously. In particular, as we vary a, R_f tracks the changes (e.g., creation, deletion, splitting, and merging) of connected components in the levelsets $f^{-1}(a)$, and thus is a meaningful topological summary of $f \colon X \to \mathbb{R}$.

As the function f is constant on each contour in a levelset, $f \colon X \to \mathbb{R}$ also induces a continuous function $\tilde{f} \colon \mathsf{R}_f \to \mathbb{R}$ defined as $\tilde{f}(z) = f(x)$ for any pre-image $x \in \Phi^{-1}(z)$ of z. To simplify notation, we often write $f(z)$ instead of $\tilde{f}(z)$ for $z \in \mathsf{R}_f$ when there is no ambiguity, and use $\tilde{f}$ mostly to emphasize the different domains of the functions. In all illustrations of this chapter, we plot the Reeb graph with the vertical coordinate of a point z to be the function value $f(z)$.

---

[1] As introduced in Exercise 3 of Chapter 1, a topological space $\mathbb{T}$ is path connected if any two points $x, y \in \mathbb{T}$ can be joined by a path, that is, there exists a continuous map $f : [0, 1] \to \mathbb{T}$ of the segment $[0, 1] \subset \mathbb{R}$ onto $\mathbb{T}$ so that $f(0) = x$ and $f(1) = y$.

Critical Points

As we describe above, the Reeb graph can be viewed as the underlying space
of a one-dimensional cell complex, where there is also a function $\tilde{f}: \mathsf{R}_f \to \mathbb{R}$
defined on R_f. We can further assume that the function $\tilde{f}$ is monotone along
each 1-cell of R_f – if not, we simply insert a new node where this condition
fails, and the tameness of $f: X \to \mathbb{R}$ guarantees that we only need to add a
finite number of nodes. Hence we can view the Reeb graph as the underlying
space of a one-dimensional simplicial complex (graph) (V, E) associated with
a function $\tilde{f}$ that is monotone along each edge $e \in E$. Note that we can further insert more nodes into an edge in E, breaking it into multiple edges; see,
for example, the augmented Reeb graph in Figure 7.4(c) later. We now continue with this general view of the Reeb graph, whose underlying space is a
graph equipped with a function $\tilde{f}$ that is monotone along each edge. We can
then talk about the induced critical points as in Definition 3.23. An alternative
(and simpler) way to describe such critical points is as follows: Given a node
$x \in V$ in the vertex set $V := V(\mathsf{R}_f)$ of the Reeb graph R_f, let *up-degree*
(resp. *down-degree*) of x denote the number of edges incident to x that have
higher (resp. lower) values of $\tilde{f}$ than x. A node is *regular* if both its up-degree
and down-degree are equal to 1, and *critical* otherwise. A critical point is a
minimum (maximum) if it has down-degree 0 (up-degree 0), and a down-fork
(up-fork) if it has down-degree (up-degree) larger than 1. A critical point can
be degenerate, having more than one type of criticality: for example, a point
with down-degree 0 and up-degree 2 is both a minimum and an up-fork.

Note that because of the monotonicity of $\tilde{f}$ at regular points, the Reeb
graph together with its associated function is completely described, up to
homeomorphisms preserving the function, by the function values at the critical
points.

Now imagine that one sweeps the domain X in increasing order of f-values,
and tracks the changes in the connected components during this process. New
components appear (at down-degree 0 nodes), existing components vanish (at
up-degree 0 nodes), or components merge or split (at down/up-forks). The
Reeb graph R_f encodes such changes thereby making it a simple but meaningful topological summary of the function $f: X \to \mathbb{R}$. However, it only tracks
the connected components in the levelset, thus cannot capture complete information about f. Nevertheless, it reflects certain aspects about both the domain
X itself and the function f defined on it, which we describe in Section 7.2.3.

Variants of Reeb Graphs

Treating a Reeb graph as a simplicial 1-complex, we can talk about 1-cycles
(loops) in it. A loop-free Reeb graph is also called a *contour tree*, which itself
has found many applications in computer graphics and visualization. Instead of

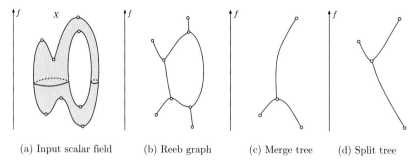

(a) Input scalar field   (b) Reeb graph   (c) Merge tree   (d) Split tree

Figure 7.3 Examples of the Reeb graph, the merge tree, and the split tree of an input scalar field.

tracking the connected components within a *levelset*, one can also track them within the *sublevel set* while sweeping X along increasing f-values, or track them within the *superlevel set* while sweeping X along decreasing f-values. The resulting topological summaries are called the *merge tree* and the *split tree*, respectively. See the precise definition below and examples in Figure 7.3.

Definition 7.2. (Merge tree; Split tree) Define $x \sim_M y$ if and only if $f(x) = f(y) = a$ and x is connected to y within the sublevel set $f^{-1}((-\infty, a])$. Then the quotient space $T_M = X/\sim_M$ is the *merge tree* with respect to f.

Alternatively, if we define $x \sim_S y$ if and only if $f(x) = f(y) = a$ and x is connected to y within the superlevel set $f^{-1}([a, +\infty))$, then the quotient space $T_S = X/\sim_S$ is the *split tree* with respect to f.

Indeed, for the levelset tame functions we consider, T_M and T_S are both finite trees. If R_f is loop-free (thus a tree), then this contour tree is uniquely decided by, and can be computed from, the merge and split trees of f.

Finally, instead of real-valued functions, one can define a similar quotient space $X/\sim$ for a continuous map $f: X \to Z$ to a general metric space (e.g., $Z = \mathbb{R}^d$), where $\sim$ is the equivalence relation $x \sim y$ if and only if $f(x) = f(y) = a$ and x is connected to y within the levelset $f^{-1}(a)$. The resulting structure is called the *Reeb space*. See Section 9.3 where we consider this generalization in the context of another structure called *mapper*.

7.2 Algorithms in the PL-Setting

Piecewise-Linear Setting

Consider a simplicial complex K and a PL-function $f: |K| \to \mathbb{R}$ on it. Since R_f depends only on the connectivity of each levelset, for a generic function

f (where no two vertices have the same function value), the Reeb graph of f depends only on the 2-skeleton of K. From now on, we assume that f is generic and $K = (V, E, T)$ is a simplicial 2-complex with vertex set V, edge set E, and triangle set T. Let n_v, n_e, and n_t denote the size of V, E, and T, respectively, and set $m = n_v + n_e + n_t$. We sketch algorithms to compute the Reeb graph for the PL-function f. Sometimes, they output the so-called *augmented Reeb graph*, which is essentially a refinement of the Reeb graph R_f with certain additional degree-two vertices inserted in arcs of R_f.

Definition 7.3. (Augmented Reeb graph) Given a PL-function $f: |K| \to \mathbb{R}$ defined on a simplicial complex $K = (V, E, T)$, let R_f be its Reeb graph and $\Phi_f: |K| \to \mathsf{R}_f(K)$ be the associated quotient map. The *augmented Reeb graph* of $f: |K| \to \mathbb{R}$, denoted by $\widehat{\mathsf{R}}_f$, is obtained by inserting each point in $\Phi_f(V) := \{\Phi_f(v) \mid v \in V\}$ as graph nodes to R_f (if it is not already in).

For a PL-function, each critical point of the Reeb graph R_f (w.r.t. $\tilde{f}: \mathsf{R}_f \to \mathbb{R}$ induced by f) is necessarily the image of some vertex in K, and thus the critical points form a subset of points in $\Phi_f(V)$. The augmented Reeb graph $\widehat{\mathsf{R}}_f$ then includes all remaining points in $\Phi_f(V)$ as (degree-two) graph nodes. See Figure 7.4 for an example, where as a convention, we plot a node $\Phi_f(v)$ at the same height (function value) as v.

We now sketch the main ideas behind two algorithms that compute the Reeb graph for a PL-function with the best time complexity, one deterministic and the other randomized.

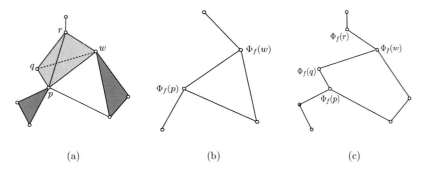

(a)   (b)   (c)

Figure 7.4 (a) A simplicial complex K. The set of 2-simplices of K include $\triangle rpq$, $\triangle rpw$, and $\triangle rqw$, as well as the two dark-colored triangles incident to p and to w, respectively. (b) Reeb graph of the height function on $|K|$. (c) Its augmented Reeb graph.

7.2.1 An $O(m \log m)$-Time Algorithm via Dynamic Graph Connectivity

Here we describe an $O(m \log m)$-time algorithm [253] for computing the Reeb graph of a PL-function $f : |K| \to \mathbb{R}$, whose time complexity is the best among all existing algorithms for Reeb graph computation. We assume for simplicity that no two vertices in V share the same f-value.

As $K = (V, E, T)$ is a simplicial 2-complex, the levelset $f^{-1}(a)$ for any function value a consists of nodes (intersection of the levelset $f^{-1}(a)$ with edges in E) and edges (intersection of the levelset $f^{-1}(a)$ with some triangles in T). This can be viewed as yet another graph, which we denote by $G_a = (W_a, F_a)$ and refer to as the *pre-image graph*: Each vertex in W_a corresponds to some edge in E. Each edge in F_a connects two vertices in W_a and thus can be associated to a pair of edges in E adjoining a certain triangle in T. See Figure 7.5 for an example. Obviously, connected components in G_a correspond to connected components in $f^{-1}(a)$, and under the quotient map Φ, each component is mapped to a single point in the Reeb graph R_f.

A natural idea to construct the Reeb graph R_f of $f : |K| \to \mathbb{R}$ is to sweep the domain K with increasing value of a, track the connected components in G_a during the course, and record the changes (merging or splitting of components, or creation and removal of components) in the resulting Reeb graph.

Furthermore, as f is a PL-function, the combinatorial structure of G_a can only change when we sweep past a vertex $v \in V$. When that happens, only edges/triangles from K incident to v can incur changes in G_a; see Figure 7.5. Let s_v denote the total number of simplices incident on v. It is easy to see that as one sweeps through the vertex v, only $O(s_v)$ number of insertions and deletions are needed to update the pre-image graph G_a. To be able to build

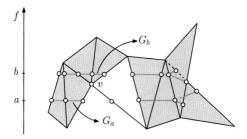

Figure 7.5 As one sweeps past v, the combinatorial structure of the pre-image graph changes. G_a has three connected components (one of which contains a single point only), while G_b has only two components.

the Reeb graph R_f, we simply need to maintain the connectivity of G_a as we sweep. Assuming we have a data structure to achieve this, the high-level framework of the sweep algorithm is then summarized in Algorithm 12: REEB-SWEEPALG.

Algorithm 12 REEB-SWEEPALG(K, f)

Input:
A simplicial 2-complex K and a vertex function $f : V(K) \to \mathbb{R}$
Output:
The Reeb graph of the PL-function induced by f

1: Sort vertices in $V = \{v_1, \ldots, v_{n_v}\}$ in increasing order of f-values
2: Initialize the Reeb graph R and the pre-image graph G_a to be empty
3: **for** $i = 1$ to n_v **do**
4:     $LC = \text{LOWERCOMPS}(v_i)$
5:     UPDATEPREIMAGE(v_i)   \∗ update the pre-image graph G_a ∗\
6:     $UC = \text{UPPERCOMPS}(v_i))$
7:     UPDATEREEBGRAPH(R, LC, UC, v_i)
8: **end for**
9: Output R as the Reeb graph

In particular, suppose we have a data structure, denoted by DYNSF, that maintains a spanning forest of the pre-image graph at any moment. Each connected component in the pre-image graph is associated with a certain vertex v from V, called the *representative vertex* of this component, which indicates that this component is created when passing through v. We assume that the data structure DYNSF allows the following operations: First, assume that a graph node $e_a \in W_a$ in the pre-image graph G_a is generated by edge $e \in K$, that is, e_a is the intersection of e with the levelset $f^{-1}(a)$.

- FIND(e): given an edge $e \in E$, returns the representative vertex of the component in the current pre-image graph G_a containing the node $e_a \in W_a$ generated by e.
- INSERT(e, e'), DELETE(e, e'): inserts an edge (e_a, e'_a) into G_a and deletes (e_a, e'_a) from G_a respectively while still maintaining a spanning forest for G_a under these operations.

Using these operations, the pseudocodes for the subroutines called in the algorithm REEB-SWEEPALG are given in Algorithms 13: LOWERCOMPS, 14: UPDATEPREIMAGE, and 15: UPDATEREEBGRAPH. (The routine UPPER-COMPS is symmetric to LOWERCOMPS and thus omitted.) These codes assume that edges of K not intersecting the levelsets are still in the pre-image graphs as isolated nodes; hence there is no need to add or remove isolated nodes.

Algorithm 13 LOWERCOMPS(v)

Input:
　A vertex $v \in K$
Output:
　A list L_c of connected components in the pre-image graph generated by the lower star of v

1: LC = empty list
2: **for** all edges e in the lower star of v **do**
3: 　c = DYNSF.FIND(e)
4: 　**if** c is not marked "listed" **then**
5: 　　LC.add(c);　and mark c as "listed"
6: 　**end if**
7: **end for**

Time Complexity Analysis

Suppose the input simplicial 2-complex $K = (V, E, T)$ has n vertices and m simplices in total. Sorting the vertices takes $O(n \log n)$ time. Then steps 4 to 7 of the algorithm REEB-SWEEPALG performs $O(m)$ numbers of FIND, INSERT, and DELETE operations using the data structure DYNSF.

One could use a state-of-the-art data structure for dynamic graph connectivity as DYNSF – indeed, this is the approach taken in [146]. However, note that this is an offline version of the dynamic graph connectivity problem, as all insertions/deletions are known in advance and thus can be precomputed. To this end, we assign each edge in the pre-image graph a weight, which is the time (f-value) it will be deleted from the pre-image graph G_a. We then maintain a maximum spanning forest of G_a during the sweeping to maintain connectivity. In general, a deletion of a maximum spanning tree edge (u, v) can incur an expensive search in the pre-image graph for a replacement edge (as u and v may still be connected). However, because of the specific assignment of edge weights, this expensive search is avoided in this case. If a maximum spanning tree edge is to be deleted, it will simply break the tree in the maximum spanning forest containing this edge, and no replacement edge needs to be identified. One can use a standard dynamic tree data structure, such as the Link-Cut trees [280], to maintain the maximum spanning forest efficiently in $O(\log m)$ amortized time for each find/insertion/deletion operation. Putting everything together, it takes $O(m \log m)$ time to compute the Reeb graph by the sweep.

Algorithm 14 UPDATEPREIMAGE(v)

Input:
   A vertex $v \in K$
Output:
   Update the pre-image graph after sweeping past v

1: **for** all triangles uvw incident on v **do**
2:       * without loss of generality assume $f(u) < f(w)$ *\\
3:       **if** $f(v) < f(u)$ **then**
4:           DYNSF.INSERT(vu, vw)
5:       **else**
6:           **if** $f(v) > f(w)$ **then**
7:               DYNSF.DELETE(vu, vw)
8:           **else**
9:               DYNSF.DELETE(uv, uw)
10:              DYNSF.INSERT(vw, uw)
11:          **end if**
12:      **end if**
13: **end for**

Algorithm 15 UPDATEREEBGRAPH(R, LC, UC, v)

Input:
   Current Reeb graph R for $f^{-1}(-\infty, f(v))$, a vertex v, the list LC (resp. UC) of components in the lower star (resp. upper star) of v
Output:
   Update Reeb graph R to be that for sublevel set $f^{-1}(-\infty, f(v) + \varepsilon]$ for an infinitesimally small $\varepsilon > 0$

1: Create a new node $\hat{v}$ in R corresponding to v
2: Assign node $\hat{v}$ to each component in UC
3: Create an arc in R between $\hat{v}$ and the Reeb graph node corresponding to the representative vertex of each c in LC
4: Return updated Reeb graph R

Theorem 7.1. *Given a PL-function* $f : |K| \to \mathbb{R}$, *let* m *denote the total number of simplices in the 2-skeleton of* K. *One can compute the (augmented) Reeb graph* R_f *of* f *in* $O(m \log m)$ *time.*

7.2.2 A Randomized Algorithm with $O(m \log m)$ Expected Time

In this section we describe a randomized algorithm [185] whose expected time complexity matches the previous algorithm. However, it uses a strategy different from sweeping: Intuitively, it directly models the effect of the quotient map Φ, but does so in a randomized manner so as to obtain a good (expected) running time.

In general, given $f : X \to \mathbb{R}$ and associated quotient map $\Phi : X \to \mathsf{R}_f$, each connected component (contour) C within a levelset $f^{-1}(a)$ is mapped (collapsed) to a single point $\Phi(C)$ in R_f. For the case where $X = |K|$ and f is piecewise-linear over simplices in K, the image of the collection of contours passing through every vertex in V decides the nodes in the augmented Reeb graph $\widehat{\mathsf{R}}$, and intuitively contains sufficient information for constructing $\widehat{\mathsf{R}}$. The high-level algorithm to compute the augmented Reeb graph $\widehat{\mathsf{R}}$ is given in Algorithm 16: REEB-RANDOMALG. See Figure 7.6 for an illustration of the algorithm.

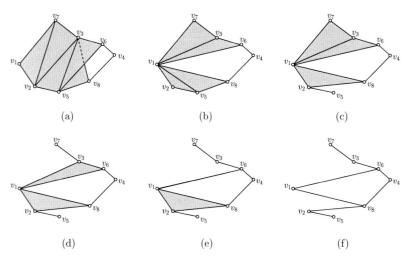

Figure 7.6 The vertices are randomly ordered. Starting from the initial simplicial complex in (a), the algorithm performs vertex collapse for vertices in this random order, as shown in (b)–(f).

Algorithm 16 REEB-RANDOMALG(K, f)

Input:

A simplicial 2-complex K and a vertex function $f : V(K) \to \mathbb{R}$

Output:

The augmented Reeb graph of the PL-function induced by f

1: Let $V = \{v_1, \ldots, v_{n_v}\}$ be a random permutation of vertices in V

2: Set $K_0 = K$ and $f_0 = f$

3: **for** $i = 1$ to n_v **do**

4:     Collapse the contour of $f_{i-1} : |K_{i-1}| \to \mathbb{R}$ passing through
        (incident to) v_i and obtain complex K_i

5:     $f_i : |K_i| \to \mathbb{R}$ is the PL-function on K_i induced from f_{i-1}

6: **end for**

7: Output the final complex K_{n_v} as the augmented Reeb graph

---

In particular, algorithm REEB-RANDOMALG starts with function $f_0 = f$ defined on the original simplicial complex $K_0 = K$. Take a random permutation of all vertices in $V = V(K)$. At the beginning of the i-th iteration, it maintains a PL-function $f_{i-1} : |K_{i-1}| \to \mathbb{R}$ over a partially collapsed simplicial complex K_{i-1} whose augmented Reeb graph is the same as that of f. It then "collapses" the contour of f_{i-1} passing through the vertex v_i and obtains a new PL-function $f_i : |K_i| \to \mathbb{R}$ over a further collapsed simplicial complex K_i that maintains the augmented Reeb graph.

The key is to implement this "collapse" step (lines 4 and 5). To see the effect of collapsing the contour incident to a vertex, see Figure 7.7(a) and (b). To see how the collapse is implemented, first consider the triangle qp_1p_2 incident to vertex q as in Figure 7.7(c), and assume that q is the *mid-vertex* of this triangle, that is, its height value ranks second among the three vertices of the triangle. Intuitively, we need to map each horizontal segment (part of a contour at different height) to the corresponding point along the edges qp_1 and qp_2. If this triangle incident to q that we are collapsing has one or more triangles sharing the edge p_1p_2 as shown in Figure 7.7(d), then for each such incident triangle, we need to process it appropriately. In particular, see one such triangle (p_1, p_2, r) in Figure 7.7(d); then, as q' is sent to q, the dotted edge rq' becomes edge rq as shown. Thus, the triangle rp_1p_2 is now split into two new triangles qrp_1 and qrp_2. In this case, it is easy to see that at most one of the new triangles will have q as the mid-vertex. We collapse this triangle and continue the process until no triangle with q as the mid-vertex is left (Figure 7.7b). Triangle(s)

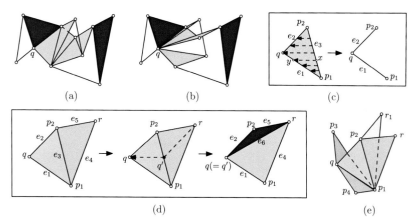

Figure 7.7 The function f is the height function. The contour incident to point q for the complex in (a) is collapsed, resulting in a new complex in (b); and (c) the collapse of the contour within a single triangle incident to q. (d) An example where this triangle is bordering another triangle. (e) There are two triangles incident to q that has q being the mid-vertex; and they both need to be processed. The triangle qp_1p_4 does not have q as mid-vertex, and it is not touched while processing q.

incident to q but not having q as the mid-vertex are not processed, for example, triangle qp_1p_4 in Figure 7.7(e). At this point, the entire contour passing through q is collapsed into a single point, and lines 4 and 5 of the algorithm are executed.

After processing each vertex as described above, the algorithm REEB-RANDOMALG in the end computes the final complex K_{n_v} in line 7. It is necessarily a simplicial 1-complex because no vertex can be the mid-vertex of any triangle, implying that there is no triangle left. It is easy to see that, by construction, K_{n_v} is the augmented Reeb graph with respect to $f : |K| \to \mathbb{R}$.

Time Complexity

For each vertex v, the time complexity of the collapse is proportional to the number of triangles T_v intersected by the contour C_v passing through v. In the worst case, $T_v = |n_t|$, giving rise to $O(n_v n_t)$ worst-case running time for algorithm REEB-RANDOMALG. This worst-case time complexity turns out to be tight. However, if one processes the vertices in a random order, then the worst-case behavior is unlikely to happen, and the expected running time can be proven to be $O(m \log n_v) = O(m \log m)$. Essentially, one argues that an original triangle from the input simplicial complex is split only $O(\log n_v) = O(\log m)$ expected number of times, thus creating $O(\log m)$ expected number of intermediate triangles which takes $O(\log m)$ expected time to collapse. The argument is in spirit similar to the analysis of the path length in a randomly built binary search tree [108].

Theorem 7.2. *Given a PL-function* $f : |K| \to \mathbb{R}$ *defined on a simplicial 2-complex K with m simplices, one can compute the (augmented) Reeb graph in* $O(m \log m)$ *expected time.*

7.2.3 Homology Groups of Reeb Graphs

Homology groups for a graph can have nontrivial ranks only in dimensions zero and one. Therefore, for a Reeb graph R_f, we only need to consider $\mathsf{H}_0(\mathsf{R}_f)$ and $\mathsf{H}_1(\mathsf{R}_f)$. In particular, their ranks $\beta_0(\mathsf{R}_f)$ and $\beta_1(\mathsf{R}_f)$ are simply the number of connected components and the number of independent loops in R_f, respectively.

Fact 7.1. *For a tame function* $f : X \to \mathbb{R}$, $\beta_0(X) = \beta_0(\mathsf{R}_f)$ *and* $\beta_1(X) \geq \beta_1(\mathsf{R}_f)$.

The equality $\beta_0(X) = \beta_0(\mathsf{R}_f)$ in the above statement follows from the fact that R_f is the quotient space $X/\sim$ and each equivalent class itself is connected (it is a connected component in some levelset). The relation on β_1 can be proven directly, and it is also a byproduct of Theorem 7.4 below (combined with Fact 7.2). The above statement also implies that if X is simply connected, then R_f is loop-free.

For the case where X is a 2-manifold, more information about X can be recovered from the Reeb graph of a Morse function defined on it.

Theorem 7.3. [106] *Let* $f : X \to \mathbb{R}$ *be a Morse function defined on a connected and compact 2-manifold. Then,*

(a) if X is orientable, $\beta_1(\mathsf{R}_f) = \beta_1(X)/2$; *and*
(b) if X is non-orientable, $\beta_1(\mathsf{R}_f) \leq \beta_1(X)/2$.

We now present a result that characterizes $\mathsf{H}_1(\mathsf{R}_f)$ with respect to $\mathsf{H}_1(X)$ in a more precise manner, which also generalizes Theorem 7.3.

Horizontal and Vertical Homology
Given a continuous function $f : X \to \mathbb{R}$, let $X_{=a} := f^{-1}(a)$ and $X_I := f^{-1}(I)$ denote its levelset and interval set as before for $a \in \mathbb{R}$ and for an open or closed interval $I \subseteq \mathbb{R}$, respectively. We first define the so-called horizontal and vertical homology groups with respect to f.

A p-th homology class $h \in H_p(X)$ is *horizontal* if there exists a finite set of values $\{a_i \in \mathbb{R}\}_{i \in A}$, where A is a finite index set, such that h has a pre-image under the map $H_p(\bigcup_{i \in A} X_{=a_i}) \rightarrow H_p(X)$ induced by inclusion. The set of horizontal homology classes form a subgroup $\overline{H}_p(X)$ of $H_p(X)$ since the trivial homology class is horizontal, and the addition of any two horizontal homology classes is still horizontal. We call this subgroup $\overline{H}_p(X)$ the *horizontal homology group of X with respect to* f. The *vertical homology group of X with respect to* f is then defined as

$$\check{H}_p(X) := H_p(X)/\overline{H}_p(X), \text{ the quotient of } H_p(X) \text{ with } \overline{H}_p(X).$$

The coset $\omega + \overline{H}_p(X)$ for every class $\omega \in H_p(X)$ provides an equivalence class in $\check{H}_p(X)$. We call h a *vertical homology class* if $h + \overline{H}_p(X)$ is not zero in $\check{H}_p(X)$. In other words, $h \notin \overline{H}_p(X)$. Two homology classes h_1 and h_2 are *vertically homologous* if $h_1 \in h_2 + \overline{H}_p(X)$.

Fact 7.2. *By definition,* rank $(H_p(X))$ = rank $(\overline{H}_p(X))$ + rank $(\check{H}_p(X))$.

Let I be a *closed interval* of $\mathbb{R}$. We define the *height of* $I = [a, b]$ to be $height(I) = |b - a|$; note that the height could be zero. Given a homology class $h \in H_p(X)$ and an interval I, we say that h is *supported by* I if $h \in$ im (i_*) where $i_* \colon H_p(X_I) \rightarrow H_p(X)$ is the homomorphism induced by the canonical inclusion $X_I \hookrightarrow X$. In other words, X_I contains a p-cycle γ from the homology class h. We define the *height of a homology class* $h \in H_p(X)$ to be

$$height(h) = \inf_{I \text{ supports } h} height(I).$$

Isomorphism Between $\check{H}_1(X)$ and $H_1(R_f)$
The surjection $\Phi \colon X \rightarrow R_f(X)$ induces a chain map $\Phi_\#$ from the one-dimensional singular chain group of X to the one-dimensional singular chain group of $R_f(X)$ which eventually induces a homomorphism $\Phi_* \colon H_1(X) \rightarrow H_1(R_f(X))$. For the horizontal subgroup $\overline{H}_1(X)$, we have that $\Phi_*(\overline{H}_1(X)) = 0 \in \overline{H}_1(R_f(X))$. Hence Φ_* induces a well-defined homomorphism between the quotient groups,

$$\check{\Phi} \colon \check{H}_1(X) = \frac{H_1(X)}{\overline{H}_1(X)} \rightarrow \frac{H_1(R_f(X))}{\overline{H}_1(R_f(X))} = H_1(R_f(X)).$$

The right equality above follows from the fact that $\overline{H}_1(R_f(X)) = 0$, which holds because every levelset of $R_f(X)$ consists only of a finite set of disjoint points due to the levelset tameness of the function $f \colon X \rightarrow \mathbb{R}$. It turns out that $\check{\Phi}$ is an isomorphism. Intuitively, this is not surprising as Φ maps each contour in the levelset to a single point, which in turn collapses every horizontal cycle.

Theorem 7.4. *Given a levelset tame function* $f : X \to \mathbb{R}$, *let* $\check{\Phi} : \check{\mathsf{H}}_1(X) \to \mathsf{H}_1(\mathsf{R}_f(X))$ *be the homomorphism induced by the surjection* $\Phi : X \to \mathsf{R}_f(X)$ *as defined above. Then the map* $\check{\Phi}$ *is an isomorphism. Furthermore, for any vertical homology class* $h \in \check{\mathsf{H}}_1(X)$, *we have that* $height(h) = height(\check{\Phi}(h))$.

Persistent Homology for $f : \mathsf{R}_f \to \mathbb{R}$

We have discussed earlier that the Reeb graph of a levelset tame function $f : X \to \mathbb{R}$ can be represented by a graph whose edges have monotone function values. Then, the function $f : \mathsf{R}_f \to \mathbb{R}$ can be treated as a PL-function on the simplicial 1-complex R_f. This gives rise to the standard setting where a PL-function f is defined on a simplicial 1-complex R_f whose persistence is to be computed. We can apply algorithm ZEROPERDG from Section 3.5.3 to compute the zeroth persistence diagram $\mathrm{Dgm}_0(f)$. For computing the one-dimensional persistence diagram $\mathrm{Dgm}_1(f)$, one can modify this algorithm slightly by registering the function values of the edges that create cycles. These are edges that connect vertices in the same component. The function values of these edges are the birth points of the 1-cycles that never die. This algorithm takes $O(n \log n + m\alpha(n))$ time, where m and n are the number of vertices and edges, respectively, in R_f.

We can also compute the levelset zigzag persistence of f (Section 4.5) using the zigzag persistence algorithm in Section 4.3.2. However, taking advantage of the graph structures, one can compute the levelset zigzag persistence for a Reeb graph with n vertices and edges in $O(n \log n)$ time using an algorithm of [5] that takes advantage of the mergeable tree data structure [169]. Only the zeroth persistence diagram $\mathrm{Dgm}_0(f)$ is nontrivial in this case. We can read the zeroth persistence diagram for the standard persistence using Theorem 4.15 from this levelset persistence diagram. Furthermore, for every infinite bar $[a_i, \infty)$ in the standard one-dimensional persistence diagram, we get a pairing (a_j, a_i) (open–open bar) in the zeroth levelset diagram $\mathrm{Dgm}_0(f)$.

Reeb graphs can be a useful tool to compute the zeroth levelset zigzag persistence diagram of a function on a topological space. Let $f : X \to \mathbb{R}$ be a continuous function whose zeroth persistence diagram we want to compute. We have already observed that the function f induces a continuous function on the Reeb graph R_f. To distinguish the two domains more explicitly, we denote the former function as f^X and the latter as f^R. The following observation helps computing the zeroth levelset zigzag persistence diagram $\mathrm{Dgm}_0(f^X)$ because computationally it is much harder to process a space, say the underlying space of a simplicial complex, than only a graph (simplicial 1-complex).

Proposition 7.5. *One has* $\mathrm{Dgm}_0(f^X) = \mathrm{Dgm}_0(f^R)$ *where the diagrams are for the zeroth levelset zigzag persistence.*

The result follows from the following observation. Consider the levelset zigzag filtrations $\mathcal{F}^X$ and $\mathcal{F}^R$ for the two functions as in sequence (4.15).

$$\mathcal{F}^X : X_{(a_0,a_2)} \leftarrow \cdots \hookrightarrow X_{(a_{i-1},a_{i+1})} \hookleftarrow X_{(a_i,a_{i+1})} \hookrightarrow X_{(a_i,a_{i+2})} \hookleftarrow \cdots \hookrightarrow X_{(a_{n-1},a_{n+1})}$$

$$\mathcal{F}^R : \mathsf{R}_{f_{(a_0,a_2)}} \leftarrow \cdots \hookrightarrow \mathsf{R}_{f_{(a_{i-1},a_{i+1})}} \hookleftarrow \mathsf{R}_{f_{(a_i,a_{i+1})}} \hookrightarrow \mathsf{R}_{f_{(a_i,a_{i+2})}} \hookleftarrow \cdots \hookrightarrow \mathsf{R}_{f_{(a_{n-1},a_{n+1})}}$$

Using the notation for interval sets $X_i^j = X_{(a_i,a_j)}$ and $\mathsf{R}_i^j = \mathsf{R}_{f_{(a_i,a_j)}}$, we have the following commutative diagram between the zeroth levelset zigzag persistence modules.

$$\mathsf{H}_0\mathcal{F}^X : \qquad \mathsf{H}_0(X_0^0) \longrightarrow \mathsf{H}_0(X_0^1) \longleftarrow \mathsf{H}_0(X_1^1) \cdots \longrightarrow \mathsf{H}_0(X_{n-1}^n) \longleftarrow \mathsf{H}_0(X_n^n)$$
$$\Big\Vert \qquad\qquad \Big\Vert \qquad\qquad \Big\Vert \qquad\qquad \Big\Vert \qquad\qquad \Big\Vert$$
$$\mathsf{H}_0\mathcal{F}^R : \qquad \mathsf{H}_0(\mathsf{R}_0^0) \longrightarrow \mathsf{H}_0(\mathsf{R}_0^1) \longleftarrow \mathsf{H}_0(\mathsf{R}_1^1) \cdots \longrightarrow \mathsf{H}_0(\mathsf{R}_{n-1}^n) \longleftarrow \mathsf{H}_0(\mathsf{R}_n^n)$$

All vertical maps are isomorphisms because the number of components in X_j^i is exactly equal to the number of components in the quotient space $\mathsf{R}_j^i = X_j^i / \sim$ which is used to define the Reeb graph. All horizontal maps are induced by inclusions. It follows that every square in the above diagram commutes. Therefore, the above two modules are isomorphic.

7.3 Distances for Reeb Graphs

Several distance measures have been proposed for Reeb graphs. In this section, we introduce two distances, one based on a natural interleaving idea, and the other based on the Gromov–Hausdorff distance idea. It has been shown that these two distance measures are strongly equivalent, that is, they are within a constant factor of each other for general Reeb graphs. For the special case of *merge trees*, the two distance measures are exactly the same.

So far, we have used R_f to denote the Reeb graph of a function f. For notational convenience, in the following we use a different notation $\mathbb{F}$ for R_f. Suppose we are given two Reeb graphs $\mathbb{F}$ and $\mathbb{G}$ with the functions $f : \mathbb{F} \to \mathbb{R}$

and $g: \mathbb{G} \to \mathbb{R}$ associated to them. To emphasize the associated functions we write $(\mathbb{F}, f)$ and $(\mathbb{G}, g)$ in place of $\mathbb{F}$ and $\mathbb{G}$ when convenient. Again, we assume that each Reeb graph is a finite simplicial 1-complex and the function is strictly monotone on each edge. Our goal is to develop a concept of distance $\mathsf{d}(\mathbb{F}, \mathbb{G})$ between them. Intuitively, if two Reeb graphs are "the same," then they are isomorphic and the function value of each point is also preserved under the isomorphism. If two Reeb graphs are not the same, we aim to measure how far they deviate from being "isomorphic." The two distances we introduce below both follow this intuition, but measure the "deviation" differently.

7.3.1 Interleaving Distance

We borrow the idea of interleaving between persistence modules (Section 3.4) to define a distance between Reeb graphs. Roughly speaking, instead of requiring that there is an isomorphism between the two Reeb graphs, which would give rise to a pair of maps between them, $\phi: \mathbb{F} \to \mathbb{G}$ and $\phi^{-1}: \mathbb{G} \to \mathbb{F}$ that are function-preserving, we look for the existence of a pair of "compatible" maps between appropriately "thickened" versions of $\mathbb{F}$ and $\mathbb{G}$ and the distance is measured by the minimum amount of the "thickening" needed. We make this more precise below. First, given any space X, set $X_\varepsilon := X \times [-\varepsilon, \varepsilon]$.

Definition 7.4. (ε-smoothing) Given a Reeb graph $(\mathbb{F}, f)$, its ε-smoothing, denoted by $\mathsf{S}_\varepsilon(\mathbb{F}, f)$, is the Reeb graph of the function $f_\varepsilon: \mathbb{F}_\varepsilon \to \mathbb{R}$ where $f_\varepsilon(x, t) = f(x) + t$ for $x \in \mathbb{F}$ and $t \in [-\varepsilon, \varepsilon]$. In other words, $\mathsf{S}_\varepsilon(\mathbb{F}, f) = \mathbb{F}_\varepsilon / \sim_{f_\varepsilon}$, where $\sim_{f_\varepsilon}$ denotes the equivalence relation where $x \sim_{f_\varepsilon} y$ if and only if $x, y \in \mathbb{F}_\varepsilon$ are from the same contour of f_ε.

See Figure 7.8 for an example. As $\mathsf{S}_\varepsilon(\mathbb{F}, f)$ is the quotient space $\mathbb{F}_\varepsilon / \sim_{f_\varepsilon}$, we use $[x, t]$, $x \in \mathbb{F}, t \in [-\varepsilon, \varepsilon]$, to denote a point in $\mathsf{S}_\varepsilon(\mathbb{F}, f)$, which is

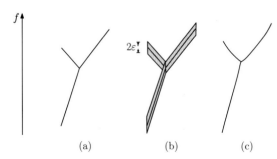

Figure 7.8 Here, we have (a) the Reeb graph $(\mathbb{F}, f)$, (b) its ε-thickening $(\mathbb{F}_\varepsilon, f_\varepsilon)$, and (c) the Reeb graph $\mathsf{S}_\varepsilon(\mathbb{F}, f)$ of $f_\varepsilon: \mathbb{F}_\varepsilon \to \mathbb{R}$.

the equivalent class of $(x, t) \in \mathbb{F}_\varepsilon$ under the equivalence relation $\sim_{f_\varepsilon}$. Also, note that there is a natural *"quotiented-inclusion" map* $\iota\colon (\mathbb{F}, f) \to S_\varepsilon(\mathbb{F}, f)$ defined as $\iota(x) = [x, 0]$, for any $x \in \mathbb{F}$.

Suppose we have two Reeb graphs (A, f_a) and (B, f_b). A map μ : $(A, f_a) \to (B, f_b)$ between them is *function-preserving* if $f_a(x) = f_b(\mu(x))$ for each $x \in A$. A function-preserving map μ between (A, f_a) and $S_\varepsilon(B, f_b)$ induces a function-preserving map μ_ε between $S_\varepsilon(A, f_a)$ and $S_{2\varepsilon}(B, f_b)$ as follows:

$$\mu_\varepsilon\colon S_\varepsilon(A, f_a) \to S_{2\varepsilon}(B, f_b) \qquad \text{such that} \quad [x, t] \mapsto [\mu(x), t].$$

Now consider the "quotiented-inclusion" map ι introduced earlier, and suppose we also have a pair of function-preserving maps $\phi\colon (\mathbb{F}, f) \to S_\varepsilon(\mathbb{G}, g)$ and $\psi\colon (\mathbb{G}, g) \to S_\varepsilon(\mathbb{F}, f)$. Using the above construction, we then obtain the following maps:

$$\iota_\varepsilon\colon S_\varepsilon(\mathbb{F}, f) \to S_{2\varepsilon}(\mathbb{F}, f), \quad [x, t] \mapsto [x, t],$$

$$\phi_\varepsilon\colon S_\varepsilon(\mathbb{F}, f) \to S_{2\varepsilon}(\mathbb{G}, g), \quad [x, t] \mapsto [\phi(x), t],$$

$$\psi_\varepsilon\colon S_\varepsilon(\mathbb{G}, g) \to S_{2\varepsilon}(\mathbb{F}, f), \quad [y, t] \mapsto [\psi(y), t].$$

Definition 7.5. (Reeb graph interleaving) A pair of continuous maps $\phi\colon (\mathbb{F}, f) \to S_\varepsilon(\mathbb{G}, g)$ and $\psi\colon (\mathbb{G}, g) \to S_\varepsilon(\mathbb{F}, f)$ are *ε-interleaved* if (i) both of them are function-preserving, and (ii) the following diagram commutes.

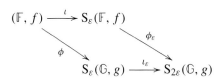

One can recognize that the above requirements of commutativity mirror the rectangular and triangular commutativity in the case of persistence modules (Definition 3.16). It is easy to verify the rectangular commutativity, that is, to verify that the following diagram (and its symmetric version involving maps ψ and ψ_ε) commutes.

$$
\begin{array}{ccc}
(\mathbb{F}, f) & \xrightarrow{\ \iota\ } & S_\varepsilon(\mathbb{F}, f) \\
& \searrow{\scriptstyle \phi} & \downarrow{\scriptstyle \phi_\varepsilon} \\
& & S_\varepsilon(\mathbb{G}, g) \xrightarrow{\ \iota_\varepsilon\ } S_{2\varepsilon}(\mathbb{G}, g)
\end{array}
$$

Rectangular commutativity, however does not embody the interaction between maps ϕ and ψ. The key technicality lies in verifying the triangular commutativity, that is, ϕ and ψ make the diagram below (and its symmetric version) commute.

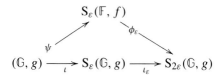

For sufficiently large ε, $S_\varepsilon(A, f_a)$ for any Reeb graph becomes a single segment with monotone function values on it. Hence one can always find maps ϕ and ψ that are ε-interleaved for sufficiently large ε. On the other hand, if $\varepsilon = 0$, then this implies $\psi = \phi^{-1}$. Hence the smallest ε accommodating ε-interleaved maps indicates how far the input Reeb graphs are from being identical. This forms the intuition behind defining the following distance between Reeb graphs.

Definition 7.6. (Interleaving distance) Given two Reeb graphs $(\mathbb{F}, f)$ and $(\mathbb{G}, g)$, the *interleaving distance* between them is defined as follows:

$$\mathsf{d}_I(\mathbb{F}, \mathbb{G}) = \inf\{\varepsilon \mid \text{there exists a pair of } \varepsilon\text{-interleaved maps between } (\mathbb{F}, f)$$
$$\text{and } (\mathbb{G}, g) \}. \tag{7.1}$$

7.3.2 Functional Distortion Distance

We now define another distance between Reeb graphs called the *functional distortion distance* which takes a metric space perspective. It views a Reeb graph as an appropriate metric space, and measures the distance between two Reeb graphs via a construction similar to what is used for defining Gromov–Hausdorff distances.

Definition 7.7. (Function-induced metric) Given a path π from u to v in a Reeb graph (A, f_a), the *height of* π is defined as

$$height(\pi) = \max_{x \in \pi} f_a(x) - \min_{x \in \pi} f_a(x).$$

Let $\Pi(u, v)$ denote the set of all paths between two points $u, v \in A$. The *function-induced metric* $\mathsf{d}_{f_a} : A \times A \to \mathbb{R}$ *on* A induced by f_a is defined as

$$\mathsf{d}_{f_a}(u, v) = \min_{\pi \in \Pi(u,v)} height(\pi).$$

In other words, $\mathsf{d}_{f_a}(u, v)$ is the minimum length of any closed interval $I \subset \mathbb{R}$ such that u and v are in the same path component of $f_a^{-1}(I)$. It is easy to verify for a finite Reeb graph that the function-induced distance d_{f_a} is indeed a proper metric on it, and hence we can view the Reeb graph (A, f_a) as a metric space (A, d_{f_a}). Refer to Chapter 9, Definition 9.6, for a generalized version of this metric.

Definition 7.8. (Functional distortion distance) Given two Reeb graphs $(\mathbb{F}, f)$ and $(\mathbb{G}, g)$, and a pair of continuous maps $\Phi : \mathbb{F} \to \mathbb{G}$ and $\Psi : \mathbb{G} \to \mathbb{F}$, set

$$C(\Phi, \Psi) = \{(x, y) \in \mathbb{F} \times \mathbb{G} \mid \Phi(x) = y, \text{ or } x = \Psi(y)\}$$

and

$$D(\Phi, \Psi) = \sup_{(x,y),(x',y') \in C(\Phi, \Psi)} \frac{1}{2} \left| \mathsf{d}_f(x, x') - \mathsf{d}_g(y, y') \right|.$$

The *functional distortion distance* between $(\mathbb{F}, f)$ and $(\mathbb{G}, g)$ is defined as

$$\mathsf{d}_{FD}(\mathbb{F}, \mathbb{G}) = \inf_{\Phi, \Psi} \max\{ D(\Phi, \Psi), \|f - g \circ \Phi\|_\infty, \|g - f \circ \Psi\|_\infty \}.$$

$$(7.2)$$

Note that the maps Φ and Ψ are not required to preserve function values; however, the terms $\|f - g \circ \Phi\|_\infty$ and $\|g - f \circ \Psi\|_\infty$ bound the difference in function values under the maps Φ and Ψ. If we ignore these two terms $\|f - g \circ \Phi\|_\infty$ and $\|g - f \circ \Psi\|_\infty$, and if we do not assume that Φ and Ψ have to be continuous, then d_{FD} is the simply the Gromov–Hausdorff distance between the metric spaces $(\mathbb{F}, \mathsf{d}_f)$ and $(\mathbb{G}, \mathsf{d}_g)$ [175]. The above definition is thus a function-adapted version of the continuous Gromov–Hausdorff distance.[2]

Properties of the Distances
The two distances we introduced turn out to be strongly equivalent.

Theorem 7.6. (Bi-Lipschitz equivalence) *One has* $\mathsf{d}_{FD} \leq 3\mathsf{d}_I \leq 3\mathsf{d}_{FD}$.

Furthermore, it is known that for Reeb graphs $\mathbb{F}$, $\mathbb{G}$ derived from two "nice" functions $f, g \colon X \to \mathbb{R}$ defined on the same domain X, both distances are *stable* [20, 115].

Definition 7.9. (Stable distance) Given $f, g \colon X \to \mathbb{R}$, let $(\mathbb{F}, \tilde{f})$ and $(\mathbb{G}, \tilde{g})$ be the Reeb graphs of f and g, respectively. We say that a Reeb graph distance d_R is *stable* if

---

[2] It turns out that if one removes the requirement of continuity on Φ and Ψ, the resulting functional distortion distance takes values within a constant factor of d_{FD} that we defined for the case of Reeb graphs.

$$\mathsf{d}_R\big((\mathbb{F}, \tilde{f}), (\mathbb{G}, \tilde{g})\big) \leq \|f - g\|_\infty.$$

Finally, it is also known that these distances are bounded from below (up to a constant factor) by the bottleneck distance between the persistence diagrams associated to the two input Reeb graphs. In particular, given $(\mathbb{F}, f)$ (and similarly for $(\mathbb{G}, g)$), consider the zeroth persistence diagram $\mathrm{Dgm}_0(f)$ induced by the levelset zigzag filtration of f as in the previous section. We consider only the zeroth persistence homology as each levelset $f^{-1}(a)$ consisting of only a finite set of points. We have the following result (see Theorem 3.2 of [32]).

Theorem 7.7. *One has* $\mathsf{d}_b(\mathrm{Dgm}_0(f), \mathrm{Dgm}_0(g)) \leq 2\mathsf{d}_I(\mathbb{F}, \mathbb{G}) \leq 2\mathsf{d}_{FD}(\mathbb{F}, \mathbb{G})$.

Universal Reeb Graph Distance
We introduced two Reeb graph distances above. There are other possible distances for Reeb graphs, such as the *edit distance* originally developed for Reeb graphs induced by functions on curves and surfaces. All these distances are stable, which is an important property to have. The following concept allows one to identify the most "discriminative" Reeb graph distance among all stable distances.

Definition 7.10. (Universal Reeb graph distance) A Reeb graph distance d_U is *universal* if and only if (i) d_U is stable, and (ii) for any other stable Reeb graph distance d_S, we have $\mathsf{d}_S \leq \mathsf{d}_U$.

It has been shown that neither the interleaving distance nor the functional distortion distance is universal. On the other hand, for Reeb graphs of piecewise-linear functions defined on compact triangulable spaces, such a universal Reeb graph distance indeed exists. In particular, one can construct a universal Reeb graph distance via a pullback idea to a common space; see [21]. The authors of [21] propose two further edit-like distances for Reeb graphs, both of which are universal.

Computation
Unfortunately, except for the bottleneck distance d_b, the computation of any of the distances mentioned above is at least as hard as graph isomorphism. In fact, even for merge trees (which are a simpler variant of the Reeb graph, described in Definition 7.2 at the end of Section 7.1), it is NP-hard to compute the interleaving distance between them [6]. But for this special case, a fixed-parameter tractable algorithm exists [289].

7.4 Notes and Exercises

The Reeb graph was originally introduced for Morse functions [261]. It was naturally extended to more general spaces as it does not rely on smooth/differential structures. This graph, as a summary of a scalar field, has found many applications in graphics, visualization, and more recently data analysis; see, for example, [30, 31, 87, 122, 167, 189, 190, 276, 285, 290, 301]. Its loop-free version, the contour tree, has many applications of its own. Properties of the Reeb graph have been studied in [106, 139]. The concept of Reeb space was introduced in [150]. The relations of merge, split and contour trees are studied in [66, 296].

An $O(m \log m)$ algorithm to compute the Reeb graph of a function on a triangulation of a 2-manifold is given in [106], where m is the size of the triangulation: In particular, it follows a similar high-level framework as in Algorithm 12: REEB-SWEEPALG. For the case where K represents the triangulation of a 2-manifold, the pre-image graph G_a has a simpler structure (a collection of disjoint loops for a generic value a). Hence the connectivity of the G_a can be maintained efficiently in $O(\log n_v)$ time, rendering an $O(m \log n_v) = O(m \log m)$ time algorithm to compute the Reeb graph [106]. Several subsequent algorithms are proposed to handle more general cases; see, for example, [145, 146, 254, 287]. The best existing algorithm for computing the Reeb graph of a PL-function defined on a simplicial complex, as described in Section 7.2.1, was proposed by Parsa in [253]. The randomized algorithm with the same time complex (in expectation) described in Section 7.2.2 was given in [185]. The loop-free version of the Reeb graph, namely, the contour tree, can be computed much more efficiently in $O(n \log n)$ time, where n is the total number of vertices and edges in the input simplicial complex domain [66]. As a byproduct, this algorithm also computes both the merge tree and split tree of the input PL-function within the same time complexity.

The concepts of horizontal and vertical homology groups were originally introduced in [102] for any dimensions. The specific connection of the one-dimensional case to the Reeb graphs (e.g., Theorem 7.4) was described in [139]. The zeroth levelset zigzag persistence (or equivalently, the zeroth and first extended persistence) for the Reeb graph can be computed in $O(n \log n)$ time using an algorithm of Agarwal et al. [5] originally proposed for computing persistence of functions on surfaces based on mergeable tree data structures [169]. For the correctness proof of this algorithm, see [126].

The interleaving distance of merge trees was originally introduced by Morozov et al. in [238]. The interleaving distance for Reeb graphs is more complicated, and was introduced by de Silva et al. [115]. There is also an equivalent cosheave-theoretical way of defining the interleaving distance. Its description involves the sheaf theory [111]. The functional distortion distance for Reeb graphs was originally introduced in [20], and its relation to interleaving distance was studied in [24]. The lower bound in Theorem 7.7 was proven in [32]; while some weaker bounds were earlier given in [24, 47]. An interesting distance between Reeb graphs can be defined by mapping its levelset zigzag persistence module to a two-parameter persistence module. See the notes in Chapter 12 for more details. The edit distance for Reeb graphs induced by functions on curves or surfaces has been proposed in [158, 159]. Finally, the universality of Reeb graph distance and universal (edit-like) distance for Reeb graphs was proposed and studied in [21]. It remains an interesting open question whether the interleaving distance (and thus functional distortion distance) is within a constant factor of the universal Reeb graph distance.

Exercises

1. Suppose we are given a triangulation K of a two-dimensional square. Let $f: |K| \to \mathbb{R}$ be a PL-function on K induced by a vertex function $f: V(K) \to \mathbb{R}$. Assume that all vertices have distinct function values.
   (a) Given a value $a \in \mathbb{R}$, describe the topology of the contour $f^{-1}(a)$.
   (b) As we vary a continuously from $-\infty$ to $+\infty$, show that the connectivity of $f^{-1}(a)$ can only change when a equals $f(v)$ for some $v \in V(K)$.
   (c) Enumerate all cases of topological changes of contours when a passes through $f(v)$ for some $v \in V$.
2. Given a finite simplicial complex K and a PL-function f induced by $f: V(K) \to \mathbb{R}$, let $\mathsf{R}_f(K)$ be the Reeb graph with respect to f. Suppose we add a new simplex σ of dimension one or two to K, and let K' be the new simplicial complex. Describe how to obtain the new Reeb graph $\mathsf{R}_f(K')$ from $\mathsf{R}_f(K)$.
3. Recall the vertical homology group introduced in Section 7.2.3. Suppose we are given compact spaces $X \subset Y$ and a function $f: Y \to \mathbb{R}$; without loss of generality, denote the restriction of f over X also by $f: X \to \mathbb{R}$. Prove that the inclusion induces a well-defined homomorphism $\iota_p^*: \check{\mathsf{H}}_p(X) \to \check{\mathsf{H}}_p(Y)$ between the vertical homology groups $\check{\mathsf{H}}_p(X)$ and $\check{\mathsf{H}}_p(Y)$ with respect to f.
4. Recall the concept of merge tree introduced in Definition 7.2 and Figure 7.3(c). An alternative way to define interleaving distance for merge trees is as follows [238]:
   First, a merge tree (T, h) can be treated as a rooted tree where the function h serves as the height function, and the function value from the root to

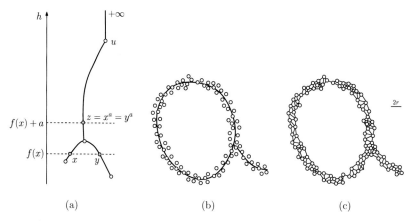

Figure 7.9 (a) The point z is the a-shift of both x and y. (b) An example of input points sampling a hidden graph (Q-shaped curve). (c) The r-Rips complex spanned by these points "approximates" a thickened version of the hidden graph $G \subset \mathbb{R}^2$. The Reeb graph for distance to a basepoint will then aim to recover this hidden graph.

any leaf is monotonically decreasing. We also extend the root upward to $+\infty$. See Figure 7.9(a). Given any point $x \in |T|$, we can then refer to any point along the path from x to $+\infty$ as its ancestor; in particular, we define x^a, called the *a-shift of x*, as the ancestor of x with function value $h(x) + a$.

Consider two merge trees $\mathcal{T}_f = (T_1, f)$ and $\mathcal{T}_g = (T_2, g)$. A pair of continuous maps $\alpha \colon |T_1| \to |T_2|$ and $\beta \colon |T_2| \to |T_1|$ are ε-compatible if the following conditions are satisfied:

(i) $g(\alpha(x)) = f(x) + \varepsilon$ for any $x \in |T_1|$;   (ii) $f(\beta(y)) = g(y) + \varepsilon$ for any $y \in |T_2|$;
(iii) $\beta \circ \alpha(x) = x^{2\varepsilon}$ for any $x \in |T_1|$.   (iv) $\alpha \circ \beta(y) = y^{2\varepsilon}$ for any $y \in |T_2|$.

The interleaving distance between merge trees can then also be defined as:

$$\mathsf{d}_I(\mathcal{T}_f, \mathcal{T}_g) := \inf_{\varepsilon} \{\text{there exists a pair of } \varepsilon\text{-compatible maps between } \mathcal{T}_f \text{ and } \mathcal{T}_g\}.$$

(a) Show that for merge trees, $\mathsf{d}_I(\mathcal{T}_f, \mathcal{T}_g) = \mathsf{d}_{FD}(\mathcal{T}_f, \mathcal{T}_g)$.

(b) Suppose T_1 and T_2 have complexity n and m, respectively. Given a threshold δ, design an algorithm to check whether there exists a δ-compatible map between $\mathcal{T}_1$ and $\mathcal{T}_2$. (Note that the time complexity of your algorithm may depend exponentially on n and m.) (Hint: Due to conditions (iii) and (iv) above for δ-compatible maps, knowing the image $\alpha(x)$ for a point $x \in |T_1|$ will determine the image of all ancestors of x under α.)

5. Given a finite simplicial complex K, let n_d denote the number of d-dimensional simplices in K. Let f be a PL-function on K induced by $f : V(K) \to \mathbb{R}$, and assume that all n_0 vertices in $V(K)$ are already sorted in nondecreasing order of f. Describe an algorithm to compute the merge tree for K with respect to f, and give the time complexity of your algorithm. (Make your algorithm as efficient as possible.)

6. [Programming exercise] Let P be a set of points in $\mathbb{R}^d$. Imagine that points in P are sampled around a hidden graph $G \subset \mathbb{R}^d$; in particular, P is an ε-sample of G. See Figure 7.9(b) and (c). Implement the following algorithm to compute a graph from P as an approximation of the hidden graph G.

   Step 1. Compute the Rips complex $K := \mathbb{VR}^r(P)$ for a parameter r. Assume K is connected. (If not, perform the following for each connected component of K.) Assign the weight of each edge in the 1-skeleton K^1 of K to be its length.

   Step 2. Choose a point $\mathbf{q} \in P$ as the base point. Let $f : P \to \mathbb{R}$ be the shortest path distance function from any point $p \in P$ to the base point $\mathbf{q}$ in the weighted graph K^1.

   Step 3. Compute the Reeb graph $\widehat{G}$ of the PL-function induced by f, and return $\widehat{G}$.

   The returned Reeb graph $\widehat{G}$ can serve as an approximation of the hidden graph G. See [87, 167] for analysis of variants of the above procedure.

8

Topological Analysis of Graphs

In this chapter, we present some examples of topological tools that help analyze or summarize graphs. In the previous chapter, we discussed one specific type of graph, the Reeb graph, obtained by quotienting a space with the connected components of levelsets of a given function. Abstractly, a Reeb graph can also be considered as a graph equipped with a height function. In this chapter, we focus on general graphs. Structures such as cliques in a graph correspond to simplices as we have seen in Vietoris–Rips complexes. They can help summarizing or characterizing graph data. See Figure 8.1 for an example [262], where a directed graph is used to model the synaptic network of neurons built by taking neurons as the vertices and the synaptic connections directed from pre- to postsynaptic neurons as the directed edges. It is observed that there are unusually high numbers of directed cliques (viewed as a simplex as we show in Section 8.3.1) in such networks, compared to other biological networks or random graphs. Topological analysis such as the one described in Section 8.3 can facilitate such applications.

Before considering directed graphs, we focus on topological analysis of undirected graphs in Sections 8.1 and 8.2. We present topological approaches to summarize and compare undirected graphs. In Section 8.3, we discuss how to obtain topological invariants for directed graphs. In particular, we describe two ways of defining homology for directed graphs. The first approach constructs an appropriate simplicial complex over an input directed graph and then takes the corresponding simplicial homology of this simplicial complex (Section 8.3.1). The second approach considers the so-called path homology for directed graphs, which differs from the simplicial homology. It is based on constructing a specific chain complex directly from directed paths in the input graph, and defining a homology group using the boundary operators associated with the resulting chain complex (Section 8.3.2). It turns out that both

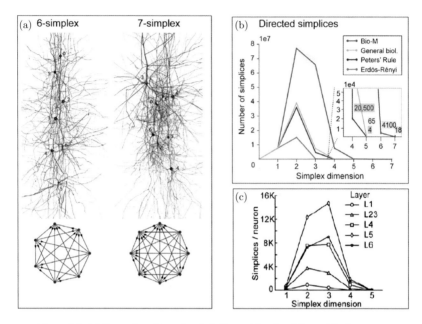

Figure 8.1 (a) Examples of two directed cliques (simplices) formed in the synaptic network. (b) The number of p-simplices for different types of graphs, where "Bio-M" is the synaptic network from reconstructed neurons. Note that this neuronal network has far more directed cliques than other biological or random graphs. (c) The count of directed cliques further differs depending on which layers neurons reside in. Image taken from [262], licensed by M. W. Reimann et al. (2017) under CC BY 4.0 (https://creativecommons.org/licenses/by/4.0/).

path homology and the persistent version of it can be computed via a matrix reduction algorithm similar to the one used in the standard persistence algorithm for simplicial filtrations, though with some key differences. We describe this algorithm in Section 8.3.3, and mention an improved algorithm for the first homology.

8.1 Topological Summaries for Graphs

We have seen graphs in various contexts so far. Here we consolidate some of the persistence results that specifically involve graphs. Sometimes graphs appear as abstract objects devoid of any geometry where they are described only combinatorially. At other times, graphs are equipped with a function or a metric. Reeb graphs studied in the previous chapter fall into this latter category. Combinatorial graphs (weighted or unweighted) can also be viewed as metric graphs, by associating them with an appropriate shortest path metric (Section 8.1.2).

8.1.1 Combinatorial Graphs

A graph G is combinatorially presented as a pair $G = (V, E)$, where V is a set called nodes/vertices of G and $E \subseteq V \times V$ is a set of pairs of vertices called edges of G. We now introduce two common ways to obtain a persistence-based topological summary for G.

Graphs Viewed as a Simplicial 1-Complex
We can view G as a simplicial 1-complex with V and E being the set of 0-simplices and 1-simplices, respectively. Using tools such as the persistence algorithm for graph filtration mentioned in Chapter 3, we can summarize G with respect to a given PL-function $f \colon |G| \to \mathbb{R}$ by the persistence diagram $\mathrm{Dgm} f$. This is what was done in Section 7.2.3 in the previous chapter while describing persistent homology for Reeb graphs. In practice, the chosen PL-functions are sometimes called *descriptor functions*. For example, we can choose $f \colon |G| \to \mathbb{R}$ to be given by a vertex function called the degree function, where $f(v)$ equals the degree of the graph node v in G. Some other choices for the descriptor function include the heat-kernel signature function [284] used in [67] and the Ollivier–Ricci curvature of graphs [227] used in [305]. Note that, under this view, given that the domain is a simplicial 1-complex, there is only zeroth and first persistent homology to consider.

Clique-Complex View
Given a graph $G = (V, E)$, its induced *clique complex*, also called the *flag complex*, is defined as follows.

Definition 8.1. (Clique complex) Given a graph $G = (V, E)$, a *clique simplex* σ of dimension k is

$$\sigma = \{v_{i_0}, \ldots, v_{i_k}\} \text{ where either } k = 0 \text{ or for any } j \neq j' \in [0, k], \ (v_{i_j}, v_{i_{j'}}) \in E.$$

By definition, every face of a clique simplex is also a clique simplex. Therefore, the collection of all clique simplices forms a simplicial complex C_G called the *clique complex* of G. In other words, the vertices of any $(k + 1)$-clique in G spans a k-simplex in C_G.

Given a weighted graph $G = (V, E, \omega)$ with $\omega \colon E \to \mathbb{R}$, let G^a denote the subgraph of G spanned by all edges with weight at most a; that is, $G^a = (V^a, E^a)$ where $E^a = \{(u, v) \mid \omega(u, v) \leq a\}$ and V^a is the vertex set adjoining E^a. Let C_{G^a} be the clique complex induced by G^a. It is easy to see that $\mathsf{C}_{G^a} \subseteq \mathsf{C}_{G^b}$ for any $a \leq b$. Assuming all edges $E = \{e_1, \ldots, e_m\}$ are sorted in nondecreasing order of their weights and setting $a_i = \omega(e_i)$, we thus obtain the following clique-complex filtration:

$$\mathsf{C}_{G^{a_1}} \hookrightarrow \mathsf{C}_{G^{a_2}} \hookrightarrow \cdots \hookrightarrow \mathsf{C}_{G^{a_m}}.$$

The persistent homology induced by the clique-complex filtration can be used to summarize the weighted graph $G = (V, E, \omega)$. Here one can consider the k-th homology groups for k up to $|V| - 1$.

8.1.2 Graphs Viewed as Metric Spaces

A finite *metric graph* is a metric space $(|G|, \mathsf{d}_G)$ where the space is the underlying space of a finite graph G, equipped with a length metric d_G [58]. We have already seen metric graphs in the previous chapter where Reeb graphs are equipped with a metric induced by a function (Definition 7.7). We can also obtain a metric graph from a (positively) weighted combinatorial graph.

Given a graph $G = (V, E, \omega)$ where the weight of each edge is positive,[1] we can view it as a metric graph $(|G|, \mathsf{d}_G)$ obtained by gluing a set of length segments (edges), where intuitively d_G is the shortest path metric on $|G|$ induced by edge lengths $\omega(e)$.

Fact 8.1. *A positively weighted graph* $G = (V, E, \omega)$ *induces a metric graph* $(|G|, \mathsf{d}_G)$.

Indeed, viewing G as a simplicial 1-complex, let $|G|$ be the underlying space of G. For every edge $e \in E$, consider the arclength parameterization $e \colon [0, \omega(e)] \to |e|$, and define $\mathsf{d}_G(x, y) = |e^{-1}(y) - e^{-1}(x)|$ for every pair $x, y \in |e|$. The length of any path $\pi(u, v)$ between two points $u, v \in |G|$ is the sum of the lengths of the restrictions of π to edges in G. The distance $\mathsf{d}_G(u, v)$ between any two points $u, v \in |G|$ is the minimum length of any path connecting u to v in $|G|$ which is a metric. The metric space $(|G|, \mathsf{d}_G)$ is the metric graph of G.

Intrinsic Čech and Vietoris–Rips Filtrations

Given a metric graph $(|G|, \mathsf{d}_G)$, let $B^o_{|G|}(x; r) := \{y \in |G| \mid \mathsf{d}_G(x, y) < r\}$ denote the radius-r *open* metric ball centered at $x \in |G|$. Following Definitions 2.9 and 2.10,[2] the intrinsic Čech complex $\mathbb{C}^r(|G|)$ and intrinsic Vietoris–Rips complex $\mathbb{VR}^r(|G|)$ are defined as:

$$\mathbb{C}^r(|G|) := \left\{ \{x_0, \ldots, x_p\} \,\middle|\, \bigcap_{i \in [0, p]} B^o_{|G|}(x_i; r) \neq \varnothing \right\};$$

$$\mathbb{VR}^r(|G|) := \left\{ \{x_0, \ldots, x_p\} \mid \mathsf{d}_G(x_i, x_j) < 2r \text{ for any } i \neq j \in [0, p] \right\}.$$

---

[1] If G is unweighted, then $\omega \colon E \to \mathbb{R}$ is the constant function $\omega(e) = 1$ for any $e \in E$.
[2] Note that here we use open metric balls instead of closed metric balls to define the Čech and Rips complexes, so that the theoretical result in Theorem 8.1 is cleaner to state.

Remark 8.1. *Observe that intrinsic Čech and Vietoris–Rips complexes as defined above are infinite complexes because we consider all points in the underlying space. Alternatively, $G = (V, E, \omega)$ can also be viewed as a discrete metric space $(V, \hat{d})$ where $\hat{d} : V \times V \to \mathbb{R}^+ \cup \{0\}$ is the restriction of d_G to graph nodes V of G. We can thus build* discrete intrinsic Čech or Vietoris–Rips complexes *spanned by only vertices in G. If G is a complete graph, then the discrete Vietoris–Rips complex at scale r is equivalent to the clique complex for G^r as introduced in Section 8.1.1. Most of our discussions below apply to analogous results for the discrete case.*

We now consider the intrinsic Čech filtration $\mathcal{C} := \{\mathbb{C}^r\}_{r \in \mathbb{R}}$ and intrinsic Vietoris–Rips filtration $\mathcal{R} := \{\mathbb{VR}^r\}_{r \in \mathbb{R}}$, and their induced persistence modules $\mathsf{H}_p\mathcal{C} := \{\mathsf{H}_p(\mathbb{C}^r)\}_{r \in \mathbb{R}}$ and $\mathsf{H}_p\mathcal{R} := \{\mathsf{H}_p(\mathbb{VR}^r)\}_{r \in \mathbb{R}}$. We have (see [81]) the following.

Fact 8.2. *Given a finite metric graph $(|G|, d_G)$ induced by $G = (V, E, \omega)$, the persistence modules $\mathsf{H}_p\mathcal{C}$ and $\mathsf{H}_p\mathcal{R}$ are both q-tame (recall the definition of q-tame in Section 3.4).*

Hence both the intrinsic Čech and intrinsic Vietoris–Rips filtrations induce well-defined persistence diagrams, which can be used as summaries (signatures) for the input graph $G = (V, E, \omega)$.

In what follows, we present some results on the homotopy types of these simplicial complexes, as well as their induced persistent homology.

Topology of Čech and Vietoris–Rips Complexes
The intrinsic Čech and Vietoris–Rips complexes induced by a metric graph may have nontrivial high-dimensional homology groups. The following results from [2] provide a precise characterization of the homotopy groups of these complexes for a metric graph whose underlying space is a circle. Specifically, let $\mathbb{S}^1$ denote the circle of unit circumference which is assumed for simplicity; the results below can be extended to a circle of any length by appropriate scaling. Let $\mathbb{S}^d$ denote the d-dimensional sphere.

Theorem 8.1. *Let $0 < r < \frac{1}{2}$. There are homotopy equivalences: for $\ell = 0, 1, \ldots$,*

$$\mathbb{C}^r(\mathbb{S}^1) \simeq \mathbb{S}^{2\ell+1} \quad if \quad \frac{\ell}{2(\ell+1)} < r \le \frac{\ell+1}{2(\ell+2)}$$

and

$$\mathbb{VR}^{r/2}(\mathbb{S}^1) \simeq \mathbb{S}^{2\ell+1} \quad if \quad \frac{\ell}{2\ell+1} < \frac{r}{2} \le \frac{\ell+1}{2\ell+3}.$$

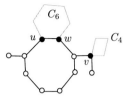

Figure 8.2 A 4-cycle C_4 is attached to the base graph along vertex v; while a 6-cycle C_6 is attached to the base graph along edge (u, w).

We remark that if one uses the closed ball to define these complexes, then the statements are similar and but involve some additional technicalities; see [2].

Much less is known for more general metric graphs. Below we present two sets of results: Theorem 8.2 characterizes the intrinsic Vietoris–Rips complexes for a certain family of metric graphs [3]; while Theorem 8.3 characterizes only the first persistent homology induced by the intrinsic Čech complexes, but for any finite metric graph [166]. Recall that $\tilde{H}_p$ denotes the p-th reduced homology group.

Theorem 8.2. *Let G be a finite metric graph, with each edge of length one, that can be obtained from a vertex by iteratively attaching (i) an edge along a vertex or (ii) a k-cycle graph along a vertex or a single edge for $k > 2$ (see, e.g., Figure 8.2). Then we have that $\tilde{H}_p(\mathbb{VR}(G; r)) \approx \bigoplus_{i=1}^{n} \tilde{H}_p(\mathbb{VR}(C_{k_i}; r))$ where $\bigoplus$ stands for the direct sum, n is the number of times operation (ii) is performed, and C_{k_i} is a loop of k_i edges (and thus C_{k_i} is of length k_i) which was attached in the i-th time that operation (ii) is performed.*

The above theorem can be relaxed to allow for different edge lengths, though one needs to define the "gluing" more carefully in that case. See [3] for details. The graphs described in Theorem 8.2 are intuitively generated by iteratively gluing a simple loop along a "short" simple path in the existing graph. Note that the above theorem implies that the Vietoris–Rips complex for a connected metric tree has isomorphic reduced homology groups as a point.

Persistent Homology Induced by Čech Complexes

Instead of a fixed scale, Theorem 8.3 below provides a complete characterization for the first persistent homology of intrinsic Čech complex filtration of a general finite metric graph. To present the result, we recall the concept of the shortest cycle basis (optimal basis) for $H_1(G)$ while treating $G = (V, E, \omega)$ as a simplicial 1-complex (Definition 5.3). Specifically, in our setting, given any 1-cycle $\gamma = e_{i_1} + e_{i_2} + \cdots + e_{i_s}$, define *the length of γ* to be $length(\gamma) = \sum_{j=1}^{s} \omega(e_{i_j})$. A *cycle basis* of G refers to a set of g 1-cycles $\Gamma = \{\gamma_1, \ldots, \gamma_g\}$ that form a basis for the one-dimensional cycle group $Z_1(G)$. Notice that we can replace $H_1(G)$ with the cycle group $Z_1(G)$ because the two are isomorphic in the case of graphs. Given a cycle basis Γ, its *length sequence* is the sequence of lengths of elements in the basis in nondecreasing order. A cycle basis of G is a *shortest cycle basis* if its length sequence is lexicographically minimal among all cycle bases of G.

Theorem 8.3. *Let* $G = (V, E, \omega)$ *be a finite graph with positive weight function* $\omega\colon E \to \mathbb{R}$. *Let* $\{\gamma_1, \ldots, \gamma_g\}$ *be a shortest cycle basis of* G *where* $g = \operatorname{rank}(Z_1(G))$, *and for each* $i = 1, \ldots, g$, *let* $\ell_i = length(\gamma_i)$. *Then, the first persistence diagram* $\operatorname{Dgm}_1 \mathcal{C}$ *induced by the intrinsic Čech filtration* $\mathcal{C} := \{\mathbb{C}^r(|G|)\}_{r \in \mathbb{R}}$ *on the metric graph* $(|G|, \mathsf{d}_G)$ *consists of the following set of points on the y-axis:*

$$\operatorname{Dgm}_1 \mathcal{C} = \{(0, \ell_i/4) \mid 1 \le i \le g\}.$$

Unfortunately, no such characterization is available for high-dimensional cases. Some partial results on the higher-dimensional persistent homology induced by intrinsic Čech filtration are given in [132].

8.2 Graph Comparison

The topological invariants described in the previous section can be used as signatures to compare graphs. For example, given two graphs $G_1 = (V_1, E_1, \omega_1)$ and $G_2 = (V_2, E_2, \omega_2)$ with positive weight functions, let $\mathcal{C}(G_1)$ and $\mathcal{C}(G_2)$ denote the intrinsic Čech filtrations for $(|G_1|, \mathsf{d}_{G_1})$ and $(|G_2|, \mathsf{d}_{G_2})$, respectively. We can then define $\mathsf{d}_{IC}(G_1, G_2) = \mathsf{d}_b(\operatorname{Dgm}_1 \mathcal{C}(G_1), \operatorname{Dgm}_1 \mathcal{C}(G_2))$ and d_{IC} gives rise to a pseudodistance (a metric for which the first axiom may hold only with the "if" condition) for the family of finite graphs with positive weights. Furthermore, this pseudodistance is stable with respect to the Gromov–Hausdorff distance by a generalization of Theorem 6.3(b) to totally bounded metric spaces (see the discussion after Theorem 6.3).

Persistence-Distortion Distance
In what follows, we introduce another pseudodistance for metric graphs, called the *persistence distortion* distance, which, instead of mapping the entire graph into a single persistence diagram, maps each point in the graph to such a summary. This distance can thus compare (metric) graphs at a more refined level.

First, given a finite metric graph $(|G|, \mathsf{d}_G)$, for any point $s \in |G|$, consider the shortest path distance function $f_s\colon |G| \to \mathbb{R}$ defined as: $x \mapsto \mathsf{d}_G(x, s)$. Let

$$P_s := \operatorname{Dgm}_0 f_s,$$ the zeroth persistence diagram induced by the function f_s.

$$(8.1)$$

Let $\mathbb{G}$ and $\mathbb{D}$ denote the space of finite metric graphs and the space of finite persistence diagrams, respectively; and let $2^{\mathbb{D}}$ denote the space of all subsets of $\mathbb{D}$. We define:

$$\phi\colon \mathbb{G} \to 2^{\mathbb{D}} \quad \text{where for any } |G| \in \mathbb{G}, \ \phi(|G|) \mapsto \{ P_s \mid s \in |G| \}. \quad (8.2)$$

In other words, ϕ maps a metric graph $|G|$ to a set of (infinitely many) points $\phi(|G|)$ in the space of persistence diagrams $\mathbb{D}$. The image $\phi(|G|)$ is another graph in the space of persistence diagrams, though this map ϕ is not necessarily injective.

Now let $(|G_1|, d_{G_1})$ and $(|G_2|, d_{G_2})$ denote the metric graphs induced by finite graphs $G_1 = (V_1, E_1, \omega_1)$ and $G_2 = (V_2, E_2, \omega_2)$ with positive edge weights.

Definition 8.2. (Persistence distortion distance) Given finite metric graphs $(|G_1|, d_{G_1})$ and $(|G_2|, d_{G_2})$, the *persistence distortion distance* between them, denoted by $d_{PD}(G_1, G_2)$, is the Hausdorff distance $d_H(\phi(|G_1|), \phi(|G_2|))$ between the two image sets $\phi(|G_1|)$ and $\phi(|G_2|)$ in the space of persistence diagrams $(\mathbb{D}, d_b)$ equipped with the bottleneck distance d_b. In other words, setting $\mathcal{A} := \phi(|G_1|)$ and $\mathcal{B} := \phi(|G_2|)$, we have

$$d_{PD}(G_1, G_2) := d_H\big(\phi(|G_1|), \phi(|G_2|)\big)$$

$$= \max \left\{ \max_{P \in \mathcal{A}} \min_{Q \in \mathcal{B}} d_b(P, Q); \ \max_{Q \in \mathcal{B}} \min_{P \in \mathcal{A}} d_b(P, Q) \right\}.$$

The persistence distortion d_{PD} is a pseudometric. It can be computed in polynomial time for finite input graphs. It is stable with respect to the Gromov–Hausdorff distance between the two input metric graphs.

Theorem 8.4. *One has $d_{PD}(G_1, G_2) \le 6 d_{GH}(|G_1|, |G_2|)$.*

One can also define a *discrete persistence distortion* distance $\widehat{d}_{PD} = d_H(\widehat{\phi}(G_1), \widehat{\phi}(G_2))$, where $\widehat{\phi}(G) := \{P_s \mid s \in V\}$ for a graph $G = (V, E, \omega)$. Both the persistence distortion distance and its discrete variant can be computed in time polynomial in the size (number of vertices and edges) of the combinatorial graphs G_1 and G_2 generating the metric graphs $|G_1|$ and $|G_2|$, respectively.

8.3 Topological Invariants for Directed Graphs

In this section, we assume that we are given a directed graph $G = (V, \vec{E}, \omega)$ where $\vec{E} \subseteq V \times V$ is the directed edge set, and $\omega\colon \vec{E} \to \mathbb{R}$ is the edge weight function (if the input graph is unweighted, we assume that all weights equal to 1). Each directed edge (u, v) is an ordered pair, and thus edge $(u, v) \ne (v, u)$.

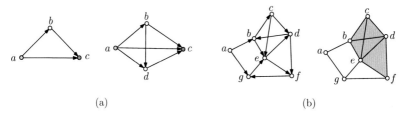

(a)   (b)

Figure 8.3 (a) A 3-clique and a 4-clique with source a and sink c. (b) A directed
graph (left) and its directed clique complex (right). The set of triangles in this
complex are: $\{bce, ced, edf\}$. There are no higher-dimensional simplices. Note
that if the edge (b, d) is also in the directed graph in (b), then the tetrahedron
$bcde$ will be in its corresponding directed clique complex.

For simplicity, we assume that there is no self-loop (v, v) in $\vec{E}$, and also there
is at most one directed edge between an ordered pair of nodes. Given a node
$v \in V$, its in-degree is $indeg(v) = |\{u \mid (u, v) \in \vec{E}\}|$, and its out-degree is
$outdeg(v) = |\{u \mid (v, u) \in \vec{E}\}|$.

8.3.1 Simplicial Complexes for Directed Graphs

Treating a directed graph as an asymmetric network (as it may be that
$\omega(u, v) \neq \omega(v, u)$), one can extend ideas in the previous section to this asymmetric setting. We give two examples below: both cases lead to simplicial
complexes from an input directed graph (weighted or unweighted), and one
can then compute (persistent) homological information of (filtrations of) this
simplicial complex as summaries of input directed graphs.

Directed Clique Complex

A node in a directed graph is a *source node* if it has in-degree 0; and it is a *sink
node* if it has out-degree 0. A directed cycle is a sequence of directed edges
$(v_0, v_1), (v_1, v_2), \ldots, (v_k, v_0)$. A graph is a *directed acyclic graph (DAG)* if it
does not contain any directed cycle. A graph $(\{v_1, \ldots, v_k\}, E')$ is a *directed k-
clique* if (i) there is exactly one edge between any pair of (unordered) vertices
(thus there are $\binom{k}{2}$ edges in E'), and (ii) it is a DAG. See Figure 8.3(a) for
examples. A set of vertices $\{v_{i_1}, \ldots, v_{i_k}\}$ spans a directed clique in $G = (V, \vec{E})$
if there is a subset of edges of $E' \subseteq \vec{E}$ such that $(\{v_{i_1}, \ldots, v_{i_k}\}, E')$ is a directed
k-clique. It is easy to see that given a directed clique, any subset of its vertices
also forms a directed clique (Exercise 5).

Definition 8.3. (Directed clique complex) Given a directed graph $G = (V, \vec{E})$,
the *directed clique complex* induced by G is a simplicial complex K defined as

$$\widehat{C}(G) := \{\sigma = \{v_{i_1}, \ldots, i_{i_k}\} \mid \{v_{i_1}, \ldots, i_{i_k}\} \text{ spans a directed } k\text{-clique in } G\}.$$

Hence a k-clique spans a $(k\text{-}1)$-simplex in the directed clique complex. See Figure 8.3(b) for a simple example. Now given a weighted directed graph $G = (V, \vec{E}, \omega)$, for any $a \geq 0$, let G^a be the subgraph of G spanned by all directed edges whose weight is at most a. Assuming all edges $e_1, \ldots, e_m$, $m = |\vec{E}|$, are sorted by their weights in a nondecreasing order, set $a_i = \omega(e_i)$. Similar to the clique complex filtration for undirected graphs introduced in Section 8.1.1, this gives rise to the following filtration of simplicial complexes induced by the directed clique complexes:

$$\widehat{\mathsf{C}}(G^{a_1}) \hookrightarrow \widehat{\mathsf{C}}(G^{a_2}) \hookrightarrow \cdots \hookrightarrow \widehat{\mathsf{C}}(G^{a_m}).$$

One can then use the persistence diagram induced by the above filtration as a topological invariant for the input directed graph G.

Definition 8.4. (Dowker complex) Given a weighted directed graph $G = (V, \vec{E}, \omega)$ and a threshold δ, the *Dowker δ-sink complex* is the following simplicial complex:

$$D_\delta^{si}(G) := \{\sigma = \{v_{i_0}, \ldots, v_{i_d}\} \mid \text{there exists } v \in V$$
$$\text{so that } \omega(v_{i_j}, v) \leq \delta \text{ for any } j \in [0, d]\}. \qquad (8.3)$$

In the above definition, v is called a δ-sink for the simplex σ. In the example on the right of Figure 8.3(a), assume all edges have weight 1. If we now remove edge (b, d), then abd is not a 3-clique any more in $G^{\delta=1}$. However, abd still forms a 2-simplex in the Dowker sink complex D_1^{si} with sink c.

In general, as δ increases, we obtain a sequence of Dowker complexes connected by inclusions, called the *Dowker sink filtration* $\mathcal{D}^{si}(G) = \{D_\delta^{si} \hookrightarrow D_{\delta'}^{si}\}_{\delta \leq \delta'}$.

Alternatively, one can define the *Dowker δ-source complex* in a symmetric manner:

$$D_\delta^{so}(G) := \{\sigma = \{v_{i_0}, \ldots, v_{i_d}\} \mid \text{there exists } v \in V$$
$$\text{so that } \omega(v, v_{i_j}) \leq \delta \text{ for any } j \in [0, d]\}, \qquad (8.4)$$

resulting in a *Dowker source filtration* $\mathcal{D}^{so}(G) = \{D_\delta^{so} \hookrightarrow D_{\delta'}^{so}\}_{\delta \leq \delta'}$. It turns out that by the duality theorem of Dowker [147], the two Dowker complexes have isomorphic homology groups. It can be further shown that the choice of Dowker complexes does not matter when persistent homology is considered [98].

Theorem 8.5. (Dowker duality) *Given a directed graph $G = (V, \vec{E}, \omega)$, for any threshold $\delta \in \mathbb{R}$ and dimension $p \geq 0$, we have $\mathsf{H}_p(D_\delta^{si}) \cong \mathsf{H}_p(D_\delta^{so})$. Furthermore, the persistence modules induced by the Dowker sink and the Dowker source filtrations are isomorphic as well, that is,*

$$\text{Dgm}_p \, \mathcal{D}^{si} = \text{Dgm}_p \, \mathcal{D}^{so}, \; \text{for any } p \geq 0.$$

8.3.2 Path Homology for Directed Graphs

In this subsection, we introduce the so-called *path homology*, which is different from the simplicial homology that we defined for clique complex and Dowker complex. Instead of constructing a simplicial complex from an input directed graph and considering its simplicial homology group, here, we use the directed graph to define a chain complex directly. The resulting path homology group has interesting mathematical structures behind; for example, there is a concept of homotopy in directed graphs under which the path homology is preserved, and it accommodates the Künneth formula [186].

Note that in this chapter, we have assumed that a given directed graph $G = (V, \vec{E})$ does not contain self-loops (where a *self-loop* is an edge (u, u) from u to itself). For notational simplicity, below we sometimes use index i to refer to vertex $v_i \in V = \{v_1, \ldots, v_n\}$.

Let $\mathbf{k}$ be a field with 0 and 1 being the additive and multiplicative identities, respectively. We use $-a$ to denote the additive inverse of a in $\mathbf{k}$. An *elementary p-path* on V is an ordered sequence $v_{i_0}, v_{i_1}, \ldots, v_{i_p}$ of $p + 1$ of the vertices of V, which we denote by $e_{v_{i_0}, v_{i_1}, \ldots, v_{i_d}}$, or just $e_{i_0, i_1, \cdots, i_p}$ for simplicity. Let $\Lambda_p = \Lambda_p(G, \mathbf{k})$ denote the $\mathbf{k}$-linear space of all linear combinations of elementary p-paths with coefficients from $\mathbf{k}$. The set $\{e_{i_0, \ldots, i_p} \mid i_0, \ldots, i_p \in V\}$ forms a basis for Λ_p. Each element c of Λ_d is called a *p-path* or *p-chain*, and it can be written as

$$c = \sum\nolimits_{i_0, \cdots, i_p \in V} a_{i_0 \cdots i_p} e_{i_0 \cdots i_p}, \; \text{where } a_{i_0 \cdots i_p} \in \mathbf{k}.$$

Similar to the case of simplicial complexes, we can define the *boundary map* $\partial_p : \Lambda_p \to \Lambda_{p-1}$ as

$$\partial_p \, e_{i_0 \cdots i_p} = \sum_{j \in [0, p]} (-1)^j e_{i_0 \cdots \hat{i}_j \cdots i_p}, \; \text{for any elementary } p\text{-path } e_{i_0 \cdots i_p},$$

where $\hat{i}_k$ means the removal of index i_k. The boundary of a p-path $c = \sum a_{i_0 \cdots i_p} \cdot e_{i_0 \cdots i_p}$ is thus $\partial_p c = \sum a_{i_0 \cdots i_p} \cdot \partial_p e_{i_0 \cdots i_p}$. For convenience, we set $\Lambda_{-1} = 0$ and note that Λ_0 is the set of $\mathbf{k}$-linear combinations of vertices in V. It is easy to show that $\partial_{p-1} \cdot \partial_p = 0$, for any $p > 0$. In what follows, we often omit the dimension p from ∂_p when it is clear from the context.

Next, we restrict the consideration to real paths in directed graphs formed by consecutive directed edges. Specifically, given a directed graph $G = (V, \vec{E})$, we call an elementary p-path $e_{i_0, \ldots, i_p}$ *allowed* if there is an edge from i_k to i_{k+1} for all $k \in [0, p-1]$. Define $\mathcal{A}_p$ as the space spanned by all allowed elementary p-paths, that is, $\mathcal{A}_p := \text{span}\{e_{i_0 \cdots i_p} : e_{i_0 \cdots i_p} \text{ is allowed}\}$. An elementary p-path

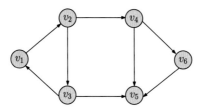

Figure 8.4 A directed graph G.

$i_0 \cdots i_p$ is called *regular* if $i_k \neq i_{k+1}$ for all k, and is *irregular* otherwise. Clearly, every allowed path is regular since there is no self-loop. However, applying the boundary map ∂ to Λ_p may create irregular paths. For example, $\partial e_{uvu} = e_{vu} - e_{uu} + e_{uv}$ is irregular because of the term e_{uu}. To deal with this case, the term containing consecutive repeated vertices is taken as 0. Thus, for the previous example, we have $\partial e_{uvu} = e_{vu} - 0 + e_{uv} = e_{vu} + e_{uv}$. The boundary map ∂ on $\mathcal{A}_p$ is now taken to be the boundary map for Λ_p restricted on $\mathcal{A}_p$ with this modification, where all terms with consecutive repeated vertices created by the boundary map ∂ are replaced with 0s. For simplicity, we still use the same symbol ∂ to represent this modified boundary map on the space of allowed paths.

After restricting the boundary operator to the space of allowed paths $\mathcal{A}_p$, the inclusion that $\partial \mathcal{A}_p \subset \mathcal{A}_{p-1}$ may not hold; that is, the boundary of an allowed p-path is not necessarily an allowed $(p - 1)$-path. To this end, we adopt a stronger notion of allowed paths: a path c is ∂-*invariant* if both c and ∂c are allowed. Let $\Omega_p := \{c \in \mathcal{A}_p \mid \partial c \in \mathcal{A}_{p-1}\}$ be the space generated by all ∂-invariant p-paths. Note that $\partial \Omega_p \subset \Omega_{p-1}$ (as $\partial^2 = 0$). This gives rise to the following *chain complex of ∂-invariant allowed paths*:

$$\cdots \xrightarrow{\partial} \Omega_p \xrightarrow{\partial} \Omega_{p-1} \xrightarrow{\partial} \cdots \xrightarrow{\partial} \Omega_1 \xrightarrow{\partial} \Omega_0 \xrightarrow{\partial} 0.$$

We can now define the homology groups of this chain complex.

Definition 8.5. (Path homology) The p-th cycle group is defined as $\mathsf{Z}_p = \ker \partial|_{\Omega_p}$, and elements in Z_p are called p-*cycles*. The p-th boundary group is defined as $\mathsf{B}_p = \operatorname{Im} \partial|_{\Omega_{p+1}}$, with elements of B_p called p-*boundary cycles* (or simply p-*boundaries*). The *p-th path homology group* is defined as $\mathsf{H}_p(G, \mathbf{k}) = \mathsf{Z}_p / \mathsf{B}_p$.

Examples

Consider the directed graph in Figure 8.4, and assume that the coefficient field $\mathbf{k} \neq \mathbb{Z}_2$. Examples of elementary 1-path include: $e_{12}, e_{24}, e_{13}, e_{14}$, and so on. However, e_{13} and e_{14} are not an allowed 1-path. More examples of allowed 1-paths include: $e_{12} + e_{46}, e_{12} + e_{31}, e_{46} + e_{65} + e_{45}$ and $e_{46} + e_{65} - e_{45}$. Note that any allowed 1-path is also ∂-invariant; that is, $\Omega_1 = \mathcal{A}_1$, as all 0-paths are allowed. Observe that $\partial(e_{46} + e_{65} + e_{45}) = e_6 - e_4 + e_5 - e_6 + e_5 - e_4 = 2e_5 - 2e_4$, which is not 0 (unless the coefficient field $\mathbf{k} = \mathbb{Z}_2$). However, $\partial(e_{46} + e_{65} - e_{45}) = 0$, meaning that $e_{46} + e_{65} - e_{45} \in \mathsf{Z}_1$. Other 1-cycle examples include

$$e_{12} + e_{23} + e_{31}, e_{24} + e_{45} - e_{23} - e_{35}, \text{ and } e_{12} + e_{24} + e_{45} - e_{53} + e_{31} \in \mathsf{Z}_1.$$

Examples of elementary 2-paths include: $e_{123}, e_{245}, e_{256}$ and e_{465}. However, e_{256} is not allowed. Consider the allowed 2-path e_{245}: its boundary $\partial e_{245} = e_{45} - e_{25} + e_{24}$ is not allowed as e_{25} is not allowed. Hence the allowed 2-path e_{245} is not ∂-invariant; similarly, we can see that neither e_{235} nor e_{123} is in Ω_2. It is easy to check that $e_{465} \in \Omega_2$ as $\partial e_{465} = e_{65} - e_{45} + e_{46}$. Also note that while neither e_{235} nor e_{245} is in Ω_2, the allowed 2-path $e_{245} - e_{235}$ is ∂-invariant as

$$\partial(e_{245} - e_{235}) = e_{45} - e_{25} + e_{24} - e_{35} + e_{25} - e_{23} = e_{45} + e_{24} - e_{35} - e_{23} \in \mathcal{A}_1.$$

This example suggests that elementary ∂-invariant p-paths *do not* necessarily form a basis for Ω_p – this is rather different from the case of simplicial complex, where the set of p-simplices form a basis for the p-th chain group.

The above discussion also suggests that $e_{46} + e_{65} - e_{45}, e_{24} + e_{45} - e_{23} - e_{35} \in \mathsf{B}_1$.

For the example in Figure 8.4,

$$\{e_{12} + e_{23} + e_{31}, e_{46} + e_{65} - e_{45}, e_{24} + e_{45} - e_{23} - e_{35}\}$$
is a basis for the 1-cycle group Z_1;
$$\{e_{46} + e_{65} - e_{45}, e_{24} + e_{45} - e_{23} - e_{35}\}$$
is a basis for the 1-boundary group B_1; while
$$\{e_{245} - e_{235}, e_{465}\} \text{ is a basis for the space of } \partial\text{-invariant 2-paths } \Omega_2.$$

Persistent Path Homology for Directed Graphs
Given a weighted directed graph $G = (V, \vec{E}, \omega)$, let G^a denote the subgraph of G containing all directed edges with weight at most a. This gives rise to a filtration of graphs $\mathcal{G} \colon \{G^a \hookrightarrow G^b\}_{a \le b}$. Let $\mathsf{H}_p(G^a)$ denote the p-th path homology induced by graph G^a. It can be shown [99] that the inclusion $G^a \hookrightarrow G^b$ induces a well-defined homomorphism $\xi_p^{a,b} \colon \mathsf{H}_p(G^a) \to \mathsf{H}_p(G^b)$, and the sequence $\mathcal{G} \colon \{G^a \hookrightarrow G^b\}_{a \le b}$ leads to a persistence module $\mathsf{H}_p\mathcal{G} \colon \{\mathsf{H}_p(G^a) \to \mathsf{H}_p(G^b)\}_{a \le b}$.

8.3.3 Computation of (Persistent) Path Homology

The example in the previous section illustrates the challenge for computing path homology induced by a directed graph G in comparison to simplicial homology. In particular, the set of elementary allowed d-paths may no longer form a basis for the space of the ∂-invariant d-paths Ω_d: Indeed, recall that for the graph in Figure 8.4, $\{e_{465}, e_{245} - e_{235}\}$ form a basis for Ω_2, yet, neither e_{245} nor e_{235} belongs to Ω_2.

We now present an algorithm to compute the persistent path homology of a given weighted directed graph $G = (V, \vec{E}, \omega)$. Note that as a byproduct, this algorithm can also compute the path homology of a directed graph.

Algorithm Setup

Given a p-path τ, its *allowed-time* is set to be the smallest value (weight) a when it belongs to $\mathcal{A}_p(G^a)$; and we denote it by $\text{at}(\tau) = a$. Let $A_p = \{\tau_1, \ldots, \tau_t\}$ denote the set of elementary allowed p-paths, sorted by their allowed-times in a nondecreasing order. Similarly, set $A_{p-1} = \{\sigma_1, \ldots, \sigma_s\}$ to be the sequence of elementary allowed $(p - 1)$-paths sorted by their allowed-times in a nondecreasing order. Let $a_1 < a_2 < \cdots < a_{\hat{t}}$ be the sequence of *distinct* allowed-times of elementary p-paths in A_p in increasing order. Obviously, $\hat{t} \leq t = |A_p|$. Similarly, let $b_1 < b_2 < \cdots < b_{\hat{s}}$ be the sequence of *distinct* allowed-times for $(p - 1)$-paths in A_{p-1} sorted in increasing order.

Note that A_p (resp. A_{p-1}) forms a basis for $\mathcal{A}_p(G)$ (resp. $\mathcal{A}_{p-1}(G)$). In fact, for any i, set $A_p^{a_i} := \{\tau_j \mid \text{at}(\tau_j) \leq a_i\}$. It is easy to see that $A_p^{a_i}$ equals $\{\tau_1, \ldots, \tau_{\rho_i}\}$, where

$$\rho_i \in [1, t] \text{ is the largest index of any elementary } p\text{-path whose}$$
$$\text{allowed-time is at most } a_i, \quad (8.5)$$

and $A_p^{a_i}$ forms a basis for $\mathcal{A}_p(G^{a_i})$. Note that the cardinality of $A_p^{a_i} \setminus A_p^{a_{i-1}}$ could be larger than 1 and that is why ρ_i is not necessarily equal to i. A symmetric statement holds for $A_{p-1}^{b_j}$ and $\mathcal{A}_{p-1}(G^{b_j})$.

From now on, we fix a dimension p. At high level, the algorithm for computing the p-th persistent path homology has the following three steps, which looks similar to the algorithm that computes standard persistent homology for simplicial complexes. However, there are key differences in the implementation of these steps.

Step 1. Set up a "boundary matrix" $M = M_p$.
Step 2. Perform left-to-right matrix reduction to transform M to a reduced form $\widehat{M}$.
Step 3. Construct the persistence diagram from the reduced matrix $\widehat{M}$.

The details of these steps are given as follows.

Description of Step 1

The columns of M correspond to A_p, ordered by their allowed-times. We would like $\text{col}_M[i] = \partial \tau_i$. However, the boundary of an allowed path may not be allowed. Hence the rows of the matrix need to correspond to not only the

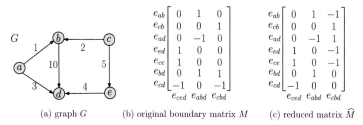

(a) graph G (b) original boundary matrix M (c) reduced matrix $\widehat{M}$

Figure 8.5 The input is the weighted directed graph in (a). Its one-dimensional boundary matrix M as constructed in step 1 is shown in (b). Note that $\mathtt{at}(e_{cd}) = +\infty$ (so $e_{cd} \notin \mathcal{A}_1(G)$). For each edge (i.e., elementary allowed 1-path) in G, its allowed-time is simply its weight. There are only three elementary allowed 2-paths, and their allowed-times are: $\mathtt{at}(e_{ced}) = 5$, $\mathtt{at}(e_{abd}) = 10$, and $\mathtt{at}(e_{cbd}) = 10$. (c) The reduced matrix. From this matrix, we can deduce that the first persistence diagram (for path homology) includes two points: $(10, 10)$ and $(5, 10)$ (generated by the second and third columns). Note that for the first column (corresponding to e_{ced}), as $\mathtt{at}(\mathtt{col}_{\widehat{M}}[1]) = \infty$, hence the corresponding γ_1 is not ∂-invariant.

elementary allowed $(p-1)$-paths in A_{p-1} (ordered by their allowed-times), but also any elementary (non-allowed) $(p-1)$-path that appears in the boundary of any $\tau_j \in A_p$: we assign the allowed-times for such paths to be $+\infty$. The rows of M are ordered in nondecreasing allowed-times from top to bottom. Let $\widehat{A}_{p-1} = \{\sigma_1, \ldots, \sigma_s, \sigma_{s+1}, \ldots, \sigma_\ell\}$ be the final set of elementary $(p-1)$-paths corresponding to rows of M. Note that the first s elements are from A_{p-1}, while those in $\{\sigma_{s+1}, \ldots, \sigma_\ell\}$ are not allowed, and have allowed-time $+\infty$. See the example in Figure 8.5(a) and (b).

The matrix M represents the boundary operator ∂_p restricted to A_p. In other words, the i-th column of M, denoted by $\mathtt{col}_M[i]$, contains the boundary of τ_i, represented using the basis elements in $\widehat{A}_{p-1}$; that is, $\partial_p \tau_i = \sum_{j=1}^{\ell} \mathtt{col}_M[i][j]\sigma_j$. From a vector representation point of view, we will also simply say that $\partial_p \tau_i = \mathtt{col}_M[i]$. The allowed-time for the $(p-1)$-path represented by a column vector C is simply the allowed-time $\mathtt{at}(\sigma_j)$ associated to the low-row index $j = lowId(C)$ in this vector. It is important to note that the rows of M are ordered in increasing indices from top down. Hence $lowId$ of a column means the largest index in $\widehat{A}_{p-1}$ for which this column contains a nonzero entry. We further associate a p-path γ_i with the i-th column of M for each $i \in [1, t]$, with the property $\partial_p \gamma_i = \mathtt{col}_M[i]$. At the beginning of the algorithm, γ_i is initialized to be τ_i and later will be updated through the reduction process in step 2 below.

Description of Step 2
We now perform the standard left-to-right matrix reduction to M, where the only allowed operation is to add a column to some column on its right. We

convert M to its *reduced form* $\widehat{M}$ (Definition 3.13); and through this process, we also update γ_i accordingly so that at any moment, $\partial_p \gamma_i = \mathrm{col}_{M'}[i]$ where M' is the updated boundary matrix at that point. In particular, if we add column j to column $i > j$, then we will update $\gamma_i = \gamma_i + \gamma_j$. We note that other than the additional maintenance of the γ, this reduction step of M is the same as the reduction in Algorithm 3: MATPERSISTENCE given in Section 3.3. The following claim follows easily from the fact that there are only left-to-right column additions, and that the allowed-times of the γ_i are initially sorted in nondecreasing order.

Claim 8.1. *For any $i \in [1, t]$, the allowed-time of γ_i remains the same through any sequence of left-to-right column additions.*

Let Ω_p^i denote the space of ∂-invariant p-paths with respect to G^{a_i}; that is, $\Omega_p^i = \Omega_p(G^{a_i})$. Given a p-path τ, let $\mathrm{ent}(\tau)$ be its *entry-time*, which is the smallest value a such that $\tau \in \Omega_p(G^a)$. It is easy to see that for any p-path τ, we have that

$$\mathrm{ent}(\tau) = \max\{\mathrm{at}(\tau), \mathrm{at}(\partial_p \tau)\}. \tag{8.6}$$

Recall that each column vector $\mathrm{col}_{\widehat{M}}[i]$ is in fact the vector representation of a $(p-1)$-path (with respect to basis elements in $\widehat{A} = \{\sigma_1, \ldots, \sigma_\ell\}$). Also, the allowed-time for a column $\mathrm{col}_{\widehat{M}}[i]$ is given by $\mathrm{at}(\mathrm{col}_{\widehat{M}}[i]) = \mathrm{at}(\sigma_h)$ where $h = lowId(\mathrm{col}_{\widehat{M}}[i])$.

Claim 8.2. *Given a reduced matrix $\widehat{M}$, let $C = \sum_{i=1}^t c_i \mathrm{col}_{\widehat{M}}[i]$ be a $(p-1)$-path. Let $\mathrm{col}_{\widehat{M}}[j]$ be the column with lowest (i.e., largest) lowId among all columns $\mathrm{col}_{\widehat{M}}[i]$s such that $c_i \neq 0$, and set $h = lowId(\mathrm{col}_{\widehat{M}}[j])$. It then follows that $\mathrm{at}(C) = \mathrm{at}(\sigma_h)$.*

Now for the reduced matrix $\widehat{M}$, given any $i \in [1, \hat{t}]$, we set ρ_i to be the largest index $j \in [1, t]$ such that $\mathrm{at}(\gamma_j) \leq a_i$. By Claim 8.1, for each j there is a fixed allowed-time associated to the p-path γ_j associated to it, which stays invariant through the reduction process. So this quantity ρ_i is well defined, consistent with what we defined earlier in Eq. (8.5), and remains invariant through the reduction process. Now set

$$\Gamma^i := \{\gamma_1, \ldots, \gamma_{\rho_i}\},$$
$$I_i := \{j \leq \rho_i \mid \mathrm{at}(\mathrm{col}_{\widehat{M}}[j]) \leq a_i\},$$

and

$$\Sigma^i := \{\gamma_j \mid \mathrm{ent}(\gamma_j) \leq a_i\} = \{\gamma_j \mid j \in I_i\}.$$

Theorem 8.6. *For any* $k \in [1, \hat{t}]$, Γ^k *forms a basis for* $\mathcal{A}_p^k := \mathcal{A}(G^{a_k})$; *while* Σ^k *forms a basis for* $\Omega_p^k = \Omega_p(G^{a_k})$.

PROOF. That Γ^k forms a basis for $\mathcal{A}_p^k$ follows easily from the facts that originally $\{\tau_1, \ldots, \tau_{\rho_k}\}$ form a basis for $\mathcal{A}_p^k$, and the left-to-right column additions maintain this. In what follows, we prove that Σ^k forms a basis for Ω_p^k. First, note that all elements in Σ^k represent paths in Ω_p^k and they are linearly independent by construction (as their low-row indices are distinct). So we only need to show that any element in Ω_p^k can be represented by a linear combination of vectors in Σ^k.

Let ξ_k denote the largest index $j \in [1, s]$ such that $\mathrm{at}(\sigma_j) \leq a_k$. In other words, an equivalent formulation for I_k is that $I_k = \{j \leq \rho_k \mid lowId(\mathrm{col}_{\widehat{M}}[j]) \leq \xi_k\}$.

Now consider any $\gamma \in \Omega_p^k \subseteq \mathcal{A}_p^k$. As Γ^k forms a basis for $\mathcal{A}_p^k$, we have that

$$\gamma = \sum_{i=1}^{\rho_k} c_i \gamma_i \quad \text{and} \quad \partial\gamma = \sum_{i=1}^{\rho_k} c_i \partial\gamma_i = \sum_{i=1}^{\rho_k} c_i \mathrm{col}_{\widehat{M}}[i].$$

As $\gamma \in \Omega_p^k$ and $\mathrm{ent}(\gamma) = \max\{\mathrm{at}(\gamma), \mathrm{at}(\partial\gamma)\}$ (see Eq. (8.6)), we have $\mathrm{at}(\gamma) \leq a_k$ and $\mathrm{at}(\partial\gamma) \leq a_k$. By Claim 8.2, it follows that for any $j \in [1, \rho_k]$ with $c_j \neq 0$, its *lowId* satisfies $lowId(\mathrm{col}_{\widehat{M}}[j]) \leq \xi_k$. Hence each such index j with $c_j \neq 0$ must belong to I_k, and as a result, γ can be written as a linear combination of p-paths in Σ^k. Combined with the fact that all vectors in Σ^k are in Ω_p^k and are linearly independent, it follows that Σ^k forms a basis for Ω_p^k. □

Corollary 8.7. *Set* $J_k := \{j \in I_k \mid \mathrm{col}_{\widehat{M}}[j] \text{ is all zeros}\}$. *Further we set* $Z^k := \{\gamma_j \mid j \in J_k\}$; *and* $B^k := \{\mathrm{col}_{\widehat{M}}[j] \mid j \in I_k \setminus J_k\}$. *Then (i)* Z^k *forms a basis for the p-dimensional cycle group* $\mathsf{Z}_p(G^{a_k})$; *and (ii)* B^k *forms a basis for the (p − 1)-dimensional boundary group* $\mathsf{B}_{p-1}(G^{a_k})$.

PROOF. Let $\hat{\partial}_p$ denote the restriction of ∂_p over Ω_p. Recall that $\mathsf{Z}_p = \mathrm{Ker}\,\hat{\partial}_p$, while $\mathsf{B}_{p-1} = \mathrm{Im}\hat{\partial}_p$. It is easy to see that by construction of Z^k, we have $Z^k \subseteq \mathrm{Span}(Z^k) \subseteq \mathsf{Z}_p(G^{a_k})$. Since all the Γ_i are linearly independent, we thus have that vectors in Z^k are linearly independent. It then follows that $|Z^k| \leq \mathrm{rank}\,(\mathsf{Z}_p(G^{a_k}))$ where $|Z^k|$ stands for the cardinality of Z^k.

Similarly, as the matrix $\widehat{M}$ is reduced, all nonzero columns of $\widehat{M}$ are linearly independent, and thus vectors in B^k are linearly independent. Furthermore, by Theorem 8.6, each vector in B^k is in $\mathsf{B}_{p-1}(G^{a_k})$ (as it is the boundary of a p-path from Ω_p^k). Hence we have that $\mathrm{Span}(B^k) \subseteq \mathsf{B}_{p-1}(G^{a_k})$, and $|B^k| \leq \mathrm{rank}\,(\mathsf{B}_{p-1}(G^{a_k}))$.

On the other hand, let $\hat{\partial}_p|_{\Omega_p^k}$ denote the restriction of $\hat{\partial}_p$ to only $\Omega_p^k \subseteq \Omega_p$. Note that by the Rank Nullity Theorem,

$$|\Sigma_p^k| = \text{rank}\,(\Omega_p^k) = \text{rank}\,(\ker(\hat{\partial}_p|_{\Omega_p^k})) + \text{rank}\,(\text{im}\,(\hat{\partial}_p|_{\Omega_p^k})) = \text{rank}\,(Z_p(G^{a_k})) + \text{rank}\,(B_{p-1}(G^{a_k})).$$

As $\text{rank}\,(\Sigma_p^k) = |Z^k| + |B^k|$, and combining the above equation with the inequalities obtained in the previous paragraphs, it follows that it must be that $|Z^k| = \text{rank}\,(Z_p(G^{a_k}))$ and $|B^k| = \text{rank}\,(B_{p-1}(G^{a_k}))$. The claim then follows. □

Description of Step 3: Constructing Persistence Diagram from the Reduced Matrix $\widehat{M}$

Given a weighted directed graph $G = (V, \vec{E}, \omega)$, for each dimension $p \geq 0$, construct the boundary matrix M_{p+1} as described above in step 1. Perform the left-to-right column reduction to M_{p+1} to obtain a reduced form $\widehat{M} = \widehat{M}_{p+1}$ as in step 2. The p-th persistence diagram $\text{Dgm}_p \mathcal{G}$ where $\mathcal{G} : \{G^a \hookrightarrow G^b\}_{a \leq b}$ can be computed as follows.

Let $\mu_p^{a,b}$ denote the persistence pairing function: that is, the persistence point (a, b) is in $\text{Dgm}_p \mathcal{G}$ with multiplicity $\mu_p^{a,b}$ if and only if $\mu_p^{a,b} > 0$. At the beginning, $\mu_p^{a,b}$ is initialized to be zero for all $a, b \in \mathbb{R}$. We then inspect every nonzero column $\text{col}_{\widehat{M}}[i]$, and take the following actions.

- If $\text{at}(\text{col}_{\widehat{M}}[i]) \neq \infty$, then we increase the pairing function $\mu^{\text{at}(\text{col}_{\widehat{M}}[i]),\text{ent}(\gamma_i)}$ by 1, where γ_i is the allowed elementary $(p+1)$-path corresponding to this column. Observe that $\text{at}(\text{col}_{\widehat{M}}[i]) \leq \text{ent}(\gamma_i)$ because

  $$\text{ent}(\gamma_i) = \max\{\text{at}(\gamma_i), \text{at}(\partial\gamma_i)\} = \max\{\text{at}(\tau_i), \text{at}(\text{col}_{\widehat{M}}[i])\}.$$

- Otherwise, the path γ_i that corresponds to this column is not ∂-invariant (i.e., not in Ω_p), and we do nothing.
- Finally, consider the reduced matrix $\widehat{M}_p$ for the p-th boundary matrix M_p as constructed in step 1. Recall the construction of J_k as in Corollary 8.7. For any $j \in J_k$ such that j does not appear as the low-row index of any column in $\widehat{M}_{p+1}$, we increase the pairing function $\mu^{\text{at}(\tau),\infty}$ by 1, where τ is the elementary p-path corresponding to this column.

See Figure 8.5 for an example. Let N_p denote the number of allowed elementary p-paths in G: obviously, $N_p = O(n^{p+1})$. However, as we saw earlier, the number of rows of M_{p+1} is not necessarily bounded by N_p; and we can only bound it by the number of elementary p-paths in G, which we denote by $\widehat{N}_p$. If we use the standard Gaussian elimination for the column reduction as in Algorithm 3: MATPERSISTENCE, then the time complexity to compute the

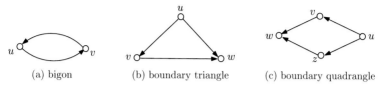

Figure 8.6 (a) Boundary bigon, (b) triangle, and (c) quadrangle. Such boundary cycles generate all one-dimensional boundary cycles.

reduced matrix $\widehat{M}_{p+1}$ is $O(\widehat{N}_p^2 N_{p+1})$. One can further improve it using the fast matrix multiplication time.

We note that due to Theorem 8.6 and Corollary 8.7, the above algorithm is rather similar to the matrix reduction for the standard persistent homology induced by simplicial complexes. However, the example in Figure 8.5 shows the difference.

Improved Computation for First Persistent Path Homology
The time complexity can be improved for computing the zeroth and first persistent path homology. In particular, the zeroth persistence path homology coincides with the zeroth persistent homology induced by the persistence of clique complexes, and thus can be computed in $O(m\alpha(n) + n \log n)$ time using the union–find data structure, where $n = |V|$ and $m = |\vec{E}|$.

For the one-dimensional case, it turns out that the boundary group has further structures. In particular, the one-dimensional boundary group is generated by only the specific forms of bigons, triangles, and quadrangles as shown in Figure 8.6. The first persistent path homology can thus be computed more efficiently by a different algorithm (from the above matrix reduction algorithm) by enumerating a certain family of boundary cycles of small cardinality which generates the boundary group. In particular, the cardinality of this family depends on the so-called *arboricity* a(G) *of* G. Ignoring the direction of edges in graph G (i.e., viewing it as an undirected graph), its arboricity a(G) is the minimum number of edge-disjoint spanning forests into which G can be decomposed [183]. An alternative definition of the arboricity is that

$$a(G) = \max_{H \text{ is a subgraph of } G} \frac{|E(H)|}{|V(H)| - 1}. \tag{8.7}$$

Without describing the algorithm developed in [130], we present its computational complexity for the first persistent path homology in the following theorem.

Theorem 8.8. *Given a directed weighted graph* $G = (V, \vec{E}, w)$ *with* $n = |V|$, $m = |\vec{E}|$, *and* $N_p = O(n^{p+1})$ *the number of allowed elementary p-paths,* *assume that the time to compute the rank of an $r \times r$ matrix is r^{ω}. Let $d_{in}(v)$ and $d_{out}(u)$ denote the in-degree and out-degree of a node $v \in V$, and $\mathrm{a}(G)$ be the arboricity of G. Set $K = \min\{\mathrm{a}(G)m, \sum_{(u,v) \in \vec{E}} (d_{in}(u) + d_{out}(u))\}$. Then we can compute the p-th persistent path homology:*

(a) in $O(m\alpha(n) + n \log n)$ time when $p = 0$, and
(b) in $O(Km^{\omega-1} + \mathrm{a}(G)m)$ time when $p = 1$.

In particular, the arboricity $\mathrm{a}(G) = O(1)$ for planar graphs; thus it takes $O(n^{\omega})$ time to compute the first persistent path homology for a planar directed graph G.

8.4 Notes and Exercises

The clique complex (also called the flag complex) is one of the most common ways to construct a simplicial complex from a graph. Recent years have seen much work on using the topological profiles associated with the clique complex for network analysis; for example, one of the early applications in [256]. Most materials covered in Section 8.1.2 come from [2, 3, 81]; note that [81] provides a detailed exposition for the intrinsic Čech and Vietoris–Rips filtrations of general metric spaces (beyond merely metric graphs). Theorem 8.2 comes as a corollary of Proposition 4 in [3], which is a stronger result than this theorem: in particular, Proposition 4 of [3] characterizes the homotopy type of the family of graphs described in Theorem 8.2.

The comparison of graphs via persistence distortion was proposed in [133].

Topological analysis for directed graphs and asymmetric networks is more recent. Nevertheless, the clique complex for directed graphs has already found applications in practical domains; see, for example, [144, 230, 262]. The path homology was originally proposed in [171], and further studied and developed in [172,173,174]. Its persistent version is proposed and studied in [99]. Note that as mentioned earlier, the path homology is not a simplicial homology. Nevertheless, we have shown in this chapter that there is still a matrix reduction algorithm to compute it for any dimension, with the same time complexity required for computing the homology groups for simplicial complexes. The path homology also has a rich mathematical structure. There is a concept of homotopy theory for digraphs under which the path homology is preserved [172], and it is also dual to the cohomology theory of diagrams introduced in [173]. Note that in this chapter, we have assumed that the input directed graph does not have self-loops. Additional care is needed to handle such loops.

The matrix reduction algorithm for computing the persistent path homology that we described in Section 8.3.3 is based on the work in [99]. The algorithm of [99] assumes that the input graph is a complete and weighted directed graph; or equivalently, is a finite set V with a weight function $w : V \times V \to \mathbb{R}$ that may be asymmetric. We modify it so that the algorithm works with an arbitrary weighted directed graph. Finally, a hypergraph $G = (V, E)$ consists of a finite set of nodes V, and a collection $E \subseteq 2^V$ of subsets of V, each such subset called a hyperedge. (In other words, a graph is a hypergraph where every hyperedge has cardinality 2.) We remark that the idea behind path homology has also been extended to defining the so-called *embedded homology* for hypergraphs [48].

Exercises

1. Consider a metric tree $(|T|, \mathsf{d}_T)$ induced by a positively weighted finite tree $T = (V, E, w)$. Suppose the largest edge weight is w_0. Consider the discrete intrinsic Čech complex $\mathbb{C}^r(V)$ spanned by vertices in V. That is, let $B_T(x; r) := \{y \in |T| \mid \mathsf{d}_T(x, y) < r\}$ denote the open radius-r ball around a point x. Then, we have

$$\mathbb{C}^r(V) := \left\{ \langle v_0, \ldots, v_p \rangle \, \middle| \, v_i \in V \text{ for } i \in [0, p], \text{ and } \bigcap_{i \in [0, p]} B_T(v_i; r) \neq \varnothing \right\}.$$

Prove that for any $r > w_0$, $\mathbb{C}^r(V)$ is homotopy equivalent to a point.

2. Consider a finite graph $G = (V, E)$ with unit edge length, and its induced metric d_G on it. For a base point $v \in V$, let $f_v : |G| \to R$ be the shortest path distance function to v; that is, for any $x \in |G|$, $f_v(x) = \mathsf{d}_G(x, v)$.
   (a) Characterize the maxima of this function f_v.
   (b) Show that the total number of critical values of f_v is bounded from above by $O(n + m)$.
   (c) Show that this shortest path distance function can be described by $O(n + m)$ functions whose total descriptive complexity is $O(n + m)$.

3. Consider a finite metric graph $G = (V, E, \omega)$ induced by positive edge weight $\omega : E \to \mathbb{R}^+$. Recall that for each basepoint $s \in |G|$, it is mapped to the persistence diagram P_s as in Eq. (8.1) (which is a point in the space of persistence diagrams). Show that this map is 1-Lipschitz with respect to the bottleneck metric on the space of persistence diagrams; that is, $\mathsf{d}_b(\mathsf{P}_s, \mathsf{P}_t) \leq \mathsf{d}_G(s, t)$ for any two $s, t \in |G|$.

4. Given two finite metric graphs $\mathcal{G}_1 = (|G_1|, \mathsf{d}_{G_1})$ and $\mathcal{G}_2 = (|G_2|, \mathsf{d}_{G_2})$, pick an arbitrary point $v \in |G_1|$ and consider its associated shortest path

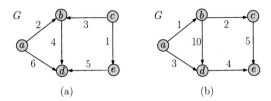

(a)   (b)

Figure 8.7 (a) Graph for Exercise 6. (b) Graph for Exercise 7. Edge weights are marked.

distance function $f_v \colon |G_1| \to \mathbb{R}$ to this point; that is, $f_v(x) = \mathsf{d}_{G_1}(x, v)$ for any $x \in |G_1|$. For any point $w \in |G_2|$, let $g_w \colon |G_2| \to \mathbb{R}$ denote the shortest path distance function to w in $|G_2|$ via d_{G_2}. Let $\mathrm{Dgm}_0 f_v$ (resp. $\mathrm{Dgm}_0 g_w$) denote the zeroth persistence diagram induced by the *super-level set filtration* of f_v (resp. of g_w). Argue that there exists some point $w^* \in |G_2|$ such that $\mathsf{d}_b(\mathrm{Dgm}_0 f_v, \mathrm{Dgm}_0 g_{w^*}) \leq C \cdot \mathsf{d}_{GH}(\mathcal{G}_1, \mathcal{G}_2)$ for some constant $C > 0$, where d_{GH} is the Gromov–Hausdorff distance.

5. Show that given a directed clique, any subset of its vertices spans a directed subgraph with a unique source and a unique sink.

6. Consider the graph in Figure 8.7(a). Compute the zeroth and first persistence diagrams for the filtrations induced by:
   (a) the directed clique complexes;
   (b) the Dowker sink complexes; and
   (c) the Dowker source complexes.

7. Consider the graph in Figure 8.7(b). Compute the first persistence diagram for the filtrations:
   (a) induced by directed clique complexes; and
   (b) induced by path homology.

8. Consider a pair of directed graphs $G = (V, E)$ and $G' = (V, E')$ spanned by the same set of vertices V, and $E' = E \cup \{(u, v)\}$; that is, G' equals G with an additional directed edge $e = (u, v)$. Consider path homology. Consider the first cycle and boundary groups for G and for G'.
   (a) Show that rank $(\mathsf{Z}_1(G')) \leq \mathrm{rank}\,(\mathsf{Z}_1(G)) + 1$.
   (b) Give an example of G and G' where rank $(\mathsf{B}_1(G')) - \mathrm{rank}\,(\mathsf{B}_1(G)) \in \Theta(n)$, where $n = |V|$.

9

Cover, Nerve, and Mapper

Data can be complex both in terms of the domain where they come from and in terms of properties/observations associated with them which are often modeled as functions/maps. For example, we can have a set of patients, where each patient is associated with multiple biological markers, giving rise to a multivariate function from the space of patients to an image domain that may or may not be the Euclidean space. To this end, we need to analyze not only real-valued scalar fields as we did so far in the book, but also more complex maps defined on a given domain, such as multivariate, circle-valued, sphere-valued maps, etc.

One way to analyze complex maps is to use the *mapper* methodology introduced by Singh et al. in [277]. In particular, given a map $f : X \rightarrow Z$, the mapper $M(f, \mathcal{U})$ creates a topological metaphor for the structure behind f by pulling back a cover $\mathcal{U}$ of the space Z to a cover on X through f. This mapper methodology can work with any (reasonably tame) continuous maps between two topological spaces. It converts complex maps and covers of the target space into simplicial complexes, which are much easier to process computationally. One can view the map f and a finite cover of the space Z as the lens through which the input data X is examined. It is in some sense related to Reeb graphs which also summarizes f but without any particular attention to a cover of the codomain. Figure 9.1 shows a mapper construction where the reader can see its similarity to the Reeb graph. The choice of different maps and covers allows the user to capture different aspects of the input data. The mapper methodology has been successfully applied to analyzing various types of data; we have shown an example in Figure 2(e) in the Prelude, for others see, for example, [228, 245].

To understand the mapper and its multiscale version *multiscale mapper* better, we study first some properties of nerves, as they are at the core of these constructions. We already know the Nerve Theorem (Theorem 2.1) which

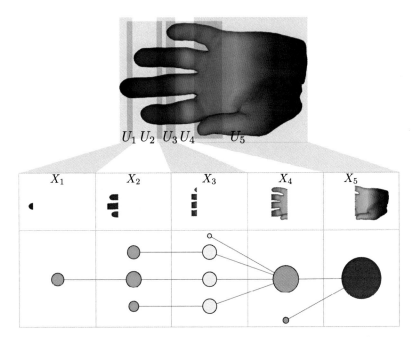

Figure 9.1 The function values on a hand model are binned into intervals as indicated by different colors. The mapper [277] corresponding to these intervals (cover) is shown with the graph below. Image courtesy of Facundo Mémoli and Gurjeet Singh.

states that if every intersection of cover elements in a cover $\mathcal{U}$ is contractible, then the nerve $N(\mathcal{U})$ is homotopy equivalent to the space $X = \bigcup \mathcal{U}$. However, we cannot hope for such a good cover all the time and need to investigate what happens if the cover is not good. Sections 9.1 and 9.2 are devoted to this study. Specifically, we show that if every cover element satisfies a weaker property that it is only path connected, then the nerve may not preserve homotopy, but satisfies a surjectivity property in one-dimensional homology.

One limitation of the mapper is that it is defined with respect to a fixed cover of the target space. Naturally, the behavior of the mapper under a change of cover is of interest because it has the potential to reveal the property of the map at different scales. Keeping this in mind, we study a multiscale version of mapper, which we refer to as *multiscale mapper*. It is capable of producing a multiscale summary in the form of a persistence diagram using a cover of the codomain at different scales. In Section 9.4, we discuss the stability of the multiscale mapper under changes in the input map and/or in the tower $\mathcal{U}$ of covers. An efficient algorithm for computing mapper and multiscale mapper for a real-valued PL-function is presented in Section 9.5. In Section 9.6, we consider the more general case of a map $f : X \to Z$ where X is a simplicial complex but Z is not necessarily Euclidean. We show that we can use an even simpler combinatorial version of the multiscale mapper, which only acts on *vertex sets* of X with connectivity given by the 1-skeleton graph of X. The cost we pay here is that the resulting persistence diagram *approximates* (instead of computing exactly) the persistence diagram of the standard multiscale mapper if the tower of covers of Z is "good" in a certain sense.

9.1 Covers and Nerves

In this section we present several facts about covers of a topological space and their nerves. Specifically, we focus on maps between covers and the maps they induce between nerves and their homology groups.

Let X denote a path connected topological space. Recall that by this we mean that there exists a continuous function called path $\gamma : [0, 1] \to X$ connecting every pair of points $\{x, x'\} \in X \times X$, where $\gamma(0) = x$ and $\gamma(1) = x'$. Also recall that for a topological space X, a collection $\mathcal{U} = \{U_\alpha\}_{\alpha \in A}$ of open sets such that $\bigcup_{\alpha \in A} U_\alpha = X$ is called an open cover of X (Definition 1.6). Although it is not required in general, we will always assume that each open cover U_α is path connected.

Maps Between Covers

If we have two covers $\mathcal{U} = \{U_\alpha\}_{\alpha \in A}$ and $\mathcal{V} = \{V_\beta\}_{\beta \in B}$ of a space X, a *map of covers* from $\mathcal{U}$ to $\mathcal{V}$ is a set map $\xi : A \to B$ so that $U_\alpha \subseteq V_{\xi(\alpha)}$ for every $\alpha \in A$. We abuse the notation ξ to also indicate the map $\mathcal{U} \to \mathcal{V}$. The following proposition connects a map between covers to a simplicial map between their nerves.

Proposition 9.1. *Given a map of covers $\xi : \mathcal{U} \to \mathcal{V}$, there is an induced simplicial map $N(\xi) : N(\mathcal{U}) \to N(\mathcal{V})$ given on vertices by the map ξ.*

PROOF. Write $\mathcal{U} = \{U_\alpha\}_{\alpha \in A}$ and $\mathcal{V} = \{V_\beta\}_{\beta \in B}$. Then, for all $\alpha \in A$ we have $U_\alpha \subseteq V_{\xi(\alpha)}$. Now take any $\sigma \in N(\mathcal{U})$. We need to prove that $\xi(\sigma) \in N(\mathcal{V})$. For this observe that

$$\bigcap_{\beta \in \xi(\sigma)} V_\beta = \bigcap_{\alpha \in \sigma} V_{\xi(\alpha)} \supseteq \bigcap_{\alpha \in \sigma} U_\alpha \neq \varnothing,$$

where the last step follows because $\sigma \in N(\mathcal{U})$. □

An example is given in Figure 9.2, where both maps $N(\xi)$ and $N(\zeta)$ are simplicial. Furthermore, if $\mathcal{U} \xrightarrow{\xi} \mathcal{V} \xrightarrow{\zeta} \mathcal{W}$ are three different covers of a

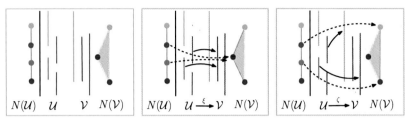

Figure 9.2 Cover maps ξ and ζ indicated by solid arrows induce simplicial maps $N(\xi)$ and $N(\zeta)$ whose corresponding vertex maps are indicated by dashed arrows.

topological space with the intervening maps of covers between them, then $N(\zeta \circ \xi) = N(\zeta) \circ N(\xi)$ as well.

The following fact will be very useful later for defining multiscale mappers.

Proposition 9.2. (Induced maps are contiguous) *Let $\zeta, \xi : \mathcal{U} \to \mathcal{V}$ be any two maps of covers. Then, the simplicial maps $N(\zeta)$ and $N(\xi)$ are contiguous.*

PROOF. Write $\mathcal{U} = \{U_\alpha\}_{\alpha \in A}$ and $\mathcal{V} = \{V_\beta\}_{\beta \in B}$. Then, for all $\alpha \in A$ we have both

$$U_\alpha \subseteq V_{\zeta(\alpha)} \quad \text{and} \quad U_\alpha \subseteq V_{\xi(\alpha)} \quad \Rightarrow \quad U_\alpha \subseteq V_{\zeta(\alpha)} \cap V_{\xi(\alpha)}.$$

Now take any $\sigma \in N(\mathcal{U})$. We need to prove that $\zeta(\sigma) \cup \xi(\sigma) \in N(\mathcal{V})$. For this write

$$\bigcap_{\beta \in \zeta(\sigma) \cup \xi(\sigma)} V_\beta = \left(\bigcap_{\alpha \in \sigma} V_{\zeta(\alpha)} \right) \cap \left(\bigcap_{\alpha \in \sigma} V_{\xi(\alpha)} \right)$$

$$= \bigcap_{\alpha \in \sigma} \left(V_{\zeta(\alpha)} \cap V_{\xi(\alpha)} \right) \supseteq \bigcap_{\alpha \in \sigma} U_\alpha \neq \varnothing,$$

where the last step follows from assuming that $\sigma \in N(\mathcal{U})$. It implies that the vertices in $\zeta(\sigma) \cup \xi(\sigma)$ span a simplex in $N(\mathcal{V})$. $\quad\square$

In Figure 9.2, the two maps $N(\xi)$ and $N(\zeta)$ can be verified to be contiguous (Definition 2.7). Furthermore, contiguous maps induce identical maps at the homology level (Fact 2.11). Proposition 9.2 implies that the map $\mathsf{H}_*(N(\mathcal{U})) \to \mathsf{H}_*(N(\mathcal{V}))$ thus induced can be deemed canonical.

Maps at Homology Level

Now we focus on establishing various maps at the homology levels for covers and their nerves. We first establish a map $\phi_\mathcal{U}$ between X and the geometric realization $|N(\mathcal{U})|$ of a nerve complex $N(\mathcal{U})$. This helps us to define a map $\phi_{\mathcal{U}*}$ from the singular homology groups of X to the simplicial homology

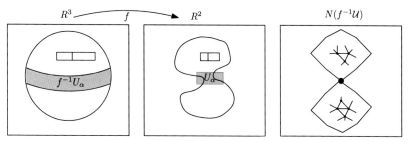

Figure 9.3 The map $f : \mathbb{S}^2 \subset \mathbb{R}^3 \to \mathbb{R}^2$ takes the sphere to $\mathbb{R}^2$. The pullback of the cover element U_α makes a band surrounding the equator which causes the nerve $N(f^{-1}\mathcal{U})$ to pinch in the middle creating two 2-cycles. This shows that the map $\phi_\mathcal{U} : X \to N(\mathcal{U})$ may not induce a surjection in H_2.

groups of $N(\mathcal{U})$ (through the singular homology of $|N(\mathcal{U})|$). The Nerve Theorem (Theorem 2.1) says that if the elements of $\mathcal{U}$ intersect only in contractible spaces, then $\phi_\mathcal{U}$ is a homotopy equivalence and hence $\phi_{\mathcal{U}*}$ is an isomorphism between $\mathsf{H}_*(X)$ and $\mathsf{H}_*(N(\mathcal{U}))$. The contractibility condition can be weakened to a *homology ball* condition to retain the isomorphism between the two homology groups [220]. In the absence of such conditions of the cover, simple examples exist to show that $\phi_{\mathcal{U}*}$ could be neither a monomorphism (injection) nor an epimorphism (surjection). Figure 9.3 gives an example where $\phi_{\mathcal{U}*}$ is not surjective in H_2. However, for one-dimensional homology groups, the map $\phi_{\mathcal{U}*}$ is necessarily a surjection when each element in the cover $\mathcal{U}$ is path connected. We call such a cover $\mathcal{U}$ *path connected*. The simplicial maps arising out of cover maps between path connected covers induce a surjection between the first homology groups of two nerve complexes.

Blow-up Space
The proof of the Nerve Theorem given by Hatcher in [186] uses a construction that connects the two spaces X and $|N(\mathcal{U})|$ via a blow-up space $X_\mathcal{U}$ that is a product space of $\mathcal{U}$ and the geometric realization $|N(\mathcal{U})|$. In our case $\mathcal{U}$ may not satisfy the contractibility condition as in that proof. Nevertheless, we use a similar construction to define three maps, $\zeta : X \to X_\mathcal{U}$, $\pi : X_\mathcal{U} \to |N(\mathcal{U})|$, and $\phi_\mathcal{U} : X \to |N(\mathcal{U})|$, where $\phi_\mathcal{U} = \pi \circ \zeta$ is referred to as the *nerve map*; see Figure 9.4(a). Details about the construction of these maps follow.

Denote the elements of the cover $\mathcal{U}$ as U_α for α taken from some indexing set A. The vertices of $N(\mathcal{U})$ are denoted by $\{u_\alpha, \alpha \in A\}$, where each u_α corresponds to the cover element U_α. For each finite non-empty intersection $U_{\alpha_0,\dots,\alpha_n} := \bigcap_{i=0}^n U_{\alpha_i}$ consider the product $U_{\alpha_0,\dots,\alpha_n} \times \Delta^n_{\alpha_0,\dots,\alpha_n}$, where $\Delta^n_{\alpha_0,\dots,\alpha_n}$ denotes the n-dimensional simplex with vertices $u_{\alpha_0}, \dots, u_{\alpha_n}$. Consider now the disjoint union

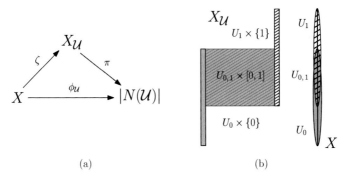

Figure 9.4 (a) Various maps used for blow-up space; (b) example of a blow-up space.

$$M := \bigsqcup_{\alpha_0,\dots,\alpha_n \in A:\, U_{\alpha_0,\dots,\alpha_n} \neq \varnothing} U_{\alpha_0,\dots,\alpha_n} \times \Delta^n_{\alpha_0,\dots,\alpha_n},$$

together with the following identification: each point $(x, y) \in M$, with $x \in U_{\alpha_0,\dots,\alpha_n}$ and $y \in [\alpha_0, \dots, \widehat{\alpha_i}, \dots, \alpha_n] \subset \Delta^n_{\alpha_0,\dots,\alpha_n}$ is identified with the corresponding point in the product $U_{\alpha_0,\dots,\widehat{\alpha_i},\dots,\alpha_n} \times \Delta_{\alpha_0,\dots,\widehat{\alpha_i},\dots,\alpha_n}$ via the inclusion $U_{\alpha_0,\dots,\alpha_n} \subset U_{\alpha_0,\dots,\widehat{\alpha_i},\dots,\alpha_n}$. Here $[\alpha_0, \dots, \widehat{\alpha_i}, \dots, \alpha_n]$ denotes the i-th face of the simplex $\Delta^n_{\alpha_0,\dots,\alpha_n}$. Denote by $\sim$ this identification and now define the space $X_{\mathcal{U}} := M/\!\!\sim$. An example for the case when X is a line segment and $\mathcal{U}$ consists of only two open sets is shown in Figure 9.4(b).

In what follows we assume that the space X is compact. The main motivation behind restricting X to such spaces is that they admit a condition called partition of unity which we use to establish further results.

Definition 9.1. (Locally finite) An open cover $\{\mathcal{U}_\alpha, \alpha \in A\}$ of X is called a *refinement* of another open cover $\{\mathcal{V}_\beta, \beta \in B\}$ of X if every element $U_\alpha \in \mathcal{U}$ is contained in an element $V_\beta \in \mathcal{V}$. Furthermore, $\mathcal{U}$ is called *locally finite* if every point $x \in X$ has a neighborhood contained in finitely many elements of $\mathcal{U}$.

Definition 9.2. (Partition of unity) A collection of real-valued continuous functions $\{\varphi_\alpha \colon X \to [0,1], \alpha \in A\}$ is called a *partition of unity* if (i) $\sum_{\alpha \in A} \varphi_\alpha(x) = 1$ for all $x \in X$, and (ii) for every $x \in X$, there are only finitely many $\alpha \in A$ such that $\varphi_\alpha(x) > 0$.

If $\mathcal{U} = \{U_\alpha, \alpha \in A\}$ is any open cover of X, then a partition of unity $\{\varphi_\alpha, \alpha \in A\}$ is *subordinate* to $\mathcal{U}$ if the support[1] $\operatorname{supp}(\varphi_\alpha)$ of φ_α is contained in U_α for each $\alpha \in A$.

---

[1] The support of a real-valued function is the subset of the domain whose image is nonzero.

Fact 9.1. [259] *For any open cover* $\mathcal{U} = \{U_\alpha,\ \alpha \in A\}$ *of a compact space X, there exists a partition of unity* $\{\varphi_\alpha, \alpha \in A\}$ *subordinate to* $\mathcal{U}$.

We assume that X is compact and hence for an open cover $\mathcal{U} = \{U_\alpha\}_\alpha$ of X, we can choose any partition of unity $\{\varphi_\alpha, \alpha \in A\}$ subordinate to $\mathcal{U}$ according to Fact 9.1. For each $x \in X$ such that $x \in U_\alpha$, denote by x_α the corresponding copy of x residing in $X_{\mathcal{U}}$. For our choice of $\{\varphi_\alpha, \alpha \in A\}$, define the map $\zeta : X \to X_{\mathcal{U}}$ as

$$\text{for any } x \in X, \quad \zeta(x) := \sum_{\alpha \in A} \varphi_\alpha(x)\, x_\alpha.$$

The map $\pi : X_{\mathcal{U}} \to |N(\mathcal{U})|$ is induced by the individual projection maps

$$U_{\alpha_0,\dots,\alpha_n} \times \Delta^n_{\alpha_0,\dots,\alpha_n} \to \Delta^n_{\alpha_0,\dots,\alpha_n}.$$

Then, it follows that $\phi_{\mathcal{U}} = \pi \circ \zeta : X \to |N(\mathcal{U})|$ satisfies, for $x \in X$,

$$\phi_{\mathcal{U}}(x) = \sum_{\alpha \in A} \varphi_\alpha(x)\, u_\alpha. \tag{9.1}$$

We have the following fact [259, p. 108]:

Fact 9.2. *The map ζ is a homotopy equivalence.*

9.1.1 Special Case of H_1

Now, we show that the nerve maps at the homology level are surjective for one-dimensional homology groups, namely all homology classes in $N(\mathcal{U})$ arise from those in $X = \bigcup \mathcal{U}$. Furthermore, if we assume that X is equipped with a pseudometric, we can define a size for cycles with this pseudometric and show that all homology classes with representative cycles having a large enough size survive in the nerve $N(\mathcal{U})$. Note that the result is not true beyond one-dimensional homology (recall Figure 9.3).

To prove this result for H_1, first, we make a simple observation that connects the classes in singular homology of $|N(\mathcal{U})|$ to those in the simplicial homology of $N(\mathcal{U})$. The result follows immediately from the isomorphism between singular and simplicial homology induced by the geometric realization; see [242]. Recall that $[c]$ denotes the class of a cycle c. If c is simplicial, $|c|$ denotes its underlying space.

Proposition 9.3. *Every 1-cycle γ in $|N(\mathcal{U})|$ has a 1-cycle γ' in $N(\mathcal{U})$ so that $[\gamma] = [|\gamma'|]$.*

Proposition 9.4. *If $\mathcal{U}$ is path connected, $\phi_{\mathcal{U}*} \colon \mathsf{H}_1(X) \to \mathsf{H}_1(|N(\mathcal{U})|)$ is a surjection, where $\phi_{\mathcal{U}*}$ is the homomorphism induced by the nerve map defined in Eq.* (9.1).

PROOF. Let $[\gamma]$ be any class in $\mathsf{H}_1(|N(\mathcal{U})|)$. Because of Proposition 9.3, we can assume that $\gamma = |\gamma'|$, where γ' is a 1-cycle in the 1-skeleton of $N(\mathcal{U})$. We will construct a 1-cycle $\gamma_{\mathcal{U}}$ in $X_{\mathcal{U}}$ so that $\pi(\gamma_{\mathcal{U}}) = \gamma$. Assume first that such a $\gamma_{\mathcal{U}}$ can be constructed. Then, consider the map $\zeta \colon X \to X_{\mathcal{U}}$ in the construction of the nerve map $\phi_{\mathcal{U}}$ where $\phi_{\mathcal{U}} = \pi \circ \zeta$. There exists a class $[\gamma_X]$ in $\mathsf{H}_1(X)$ so that $\zeta_*([\gamma_X]) = [\gamma_{\mathcal{U}}]$ because ζ_* is an isomorphism by Fact 9.2. Then, $\phi_{\mathcal{U}*}([\gamma_X]) = \pi_*(\zeta_*([\gamma_X]))$ because $\phi_{\mathcal{U}*} = \pi_* \circ \zeta_*$. It follows that $\phi_{\mathcal{U}*}([\gamma_X]) = \pi_*([\gamma_{\mathcal{U}}]) = [\gamma]$, showing that $\phi_{\mathcal{U}*}$ is surjective.

Therefore, it remains only to show that a 1-cycle $\gamma_{\mathcal{U}}$ can be constructed given γ' in $N(\mathcal{U})$ so that $\pi(\gamma_{\mathcal{U}}) = \gamma = |\gamma'|$. Let $e_0, e_1, \ldots, e_{r-1}, e_r = e_0$ be an ordered sequence of edges on γ'. Recall the construction of the space $X_{\mathcal{U}}$. In that terminology, let $e_i = \Delta^n_{\alpha_i \alpha_{(i+1) \bmod r}}$. Let $v_i = e_{(i-1) \bmod r} \cap e_i$ for $i \in [0, r-1]$. The vertex $v_i = v_{\alpha_i}$ corresponds to the cover element U_{α_i} where $U_{\alpha_i} \cap U_{\alpha_{(i+1) \bmod r}} \neq \varnothing$ for every $i \in [0, r-1]$. Choose a point x_i in the common intersection $U_{\alpha_i} \cap U_{\alpha_{(i+1) \bmod r}}$ for every $i \in [0, r-1]$. Then, the *edge* path $\tilde{e}_i = e_i \times x_i$ is in $X_{\mathcal{U}}$ by construction. Also, letting x_{α_i} be the lift of x_i in the lifted U_{α_i}, we can choose a *vertex* path $x_{\alpha_i} \rightsquigarrow x_{\alpha_{(i+1) \bmod r}}$ residing in the lifted U_{α_i} and hence in $X_{\mathcal{U}}$ because U_{α_i} is path connected. Consider the following cycle obtained by concatenating the edge and vertex paths:

$$\gamma_{\mathcal{U}} = \tilde{e}_0 x_{\alpha_0} \rightsquigarrow x_{\alpha_1} \tilde{e}_1 \cdots \tilde{e}_{r-1} x_{\alpha_{r-1}} \rightsquigarrow x_{\alpha_0}.$$

By projection, we have $\pi(\tilde{e}_i) = e_i$ for every $i \in [0, r-1]$ and $\pi(x_{\alpha_i} \rightsquigarrow x_{\alpha_{(i+1) \bmod r}}) = v_{\alpha_i}$ and thus $\pi(\gamma_{\mathcal{U}}) = \gamma$ as required. $\qquad\square$

Since we are eventually interested in the simplicial homology groups of the nerves rather than the singular homology groups of their geometric realizations, we make one more transition using the known isomorphism between the two homology groups (Theorem 2.10). Specifically, if $\iota_{\mathcal{U}} \colon \mathsf{H}_p(|N(\mathcal{U})|) \to \mathsf{H}_p(N(\mathcal{U}))$ denotes this isomorphism, we let

$$\bar{\phi}_{\mathcal{U}*} \colon \mathsf{H}_1(X) \to \mathsf{H}_1(N(\mathcal{U})) \quad \text{denote the composition } \iota_{\mathcal{U}} \circ \phi_{\mathcal{U}*}. \tag{9.2}$$

As a corollary to Proposition 9.4, we obtain the following.

Theorem 9.5. *If $\mathcal{U}$ is path connected, $\bar{\phi}_{\mathcal{U}*} \colon \mathsf{H}_1(X) \to \mathsf{H}_1(N(\mathcal{U}))$ is a surjection.*

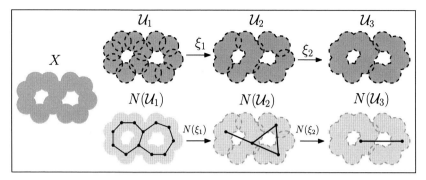

Figure 9.5 Sequence of cover maps induce a simplicial tower and hence a persistence module: classes in H_1 can only die.

From Nerves to Nerves

We now extend the result in Theorem 9.5 to simplicial maps between two nerves induced by cover maps. Figure 9.5 illustrates this fact. The following proposition is key to establishing the result.

Proposition 9.6. (Coherent partitions of unity) *Suppose* $\{U_\alpha\}_{\alpha \in A} = \mathcal{U} \xrightarrow{\theta} \mathcal{V} = \{V_\beta\}_{\beta \in B}$ *are open covers of a compact topological space* X *and* $\theta : A \to B$ *is a map of covers. Then there exists a partition of unity* $\{\varphi_\alpha\}_{\alpha \in A}$ *subordinate to the cover* $\mathcal{U}$ *such that, if for each* $\beta \in B$ *we define*

$$\psi_\beta := \begin{cases} \sum_{\alpha \in \theta^{-1}(\beta)} \varphi_\alpha & \text{if } \beta \in \text{im}(\theta), \\ 0 & \text{otherwise,} \end{cases}$$

then the set of functions $\{\psi_\beta\}_{\beta \in B}$ *is a partition of unity subordinate to the cover* $\mathcal{V}$.

PROOF. The proof closely follows that of [259, Corollary, p. 97]. Since X is compact, there exists a partition of unity $\{\varphi_\alpha\}_{\alpha \in A}$ subordinate to $\mathcal{U}$. The fact that the sum in the expression of ψ_β is well defined and continuous follows from the fact that the family $\{\text{supp}(\varphi_\alpha)\}_\alpha$ is locally finite. Let $C_\beta := \bigcup_{\alpha \in \theta^{-1}(\beta)} \text{supp}(\varphi_\alpha)$. The set C_β is closed, $C_\beta \subset U_\beta$, and $\psi_\beta(x) = 0$ for $x \notin C_\beta$ so that $\text{supp}(\psi_\beta) \subset C_\beta \subset V_\beta$. Now, to check that the family $\{C_\beta\}_{\beta \in B}$ is locally finite pick any point $x \in X$. Since $\{\text{supp}(\varphi_\alpha)\}_\alpha$ is locally finite there is an open set O containing x such that O intersects only finitely many elements in $\mathcal{U}$. Denote these cover elements by $U_{\alpha_1}, \ldots, U_{\alpha_\ell}$. Now, notice that if $\beta \in B$ and $\beta \notin \{\theta(\alpha_i), i = 1, \ldots, \ell\}$, then O does not intersect C_β. Then, the family $\{\text{supp}(\psi_\beta)\}_{\beta \in B}$ is locally finite. It then follows that for $x \in X$ one has

$$\sum_{\beta \in B} \psi_\beta(x) = \sum_{\beta \in B} \sum_{\alpha \in \theta^{-1}(\beta)} \varphi_\alpha(x) = \sum_{\alpha \in A} \varphi_\alpha(x) = 1.$$

We have obtained that $\{\psi_\beta\}_{\beta \in B}$ is a partition of unity subordinate to $\mathcal{V}$ as needed by the proposition. $\qquad\square$

Let $\{U_\alpha\}_{\alpha \in A} = \mathcal{U} \xrightarrow{\theta} \mathcal{V} = \{V_\beta\}_{\beta \in B}$ be two open covers of X connected by a map of covers $\theta \colon A \to B$. Apply Proposition 9.6 to obtain coherent partitions of unity $\{\varphi_\alpha\}_{\alpha \in A}$ and $\{\psi_\beta\}_{\beta \in B}$ subordinate to $\mathcal{U}$ and $\mathcal{V}$, respectively. Let the nerve maps $\phi_{\mathcal{U}} \colon X \to |N(\mathcal{U})|$ and $\phi_{\mathcal{V}} \colon X \to |N(\mathcal{V})|$ be defined as in Eq. (9.1) using these coherent partitions of unity. Let $N(\mathcal{U}) \xrightarrow{\tau} N(\mathcal{V})$ be the simplicial map induced by the cover map θ. The map τ can be extended to a (linear) continuous map $\hat{\tau} \colon |N(\mathcal{U})| \to |N(\mathcal{V})|$ by assigning $y \in |N(\mathcal{U})|$ to $\hat{\tau}(y) \in |N(\mathcal{V})|$ where

$$y = \sum t_\alpha u_\alpha \implies \hat{\tau}(y) = \sum t_\alpha \hat{\tau}(u_\alpha), \text{ with } \sum t_\alpha = 1.$$

Claim 9.1. *The map $\hat{\tau}$ satisfies the property that, for $x \in X$, $\hat{\tau}(\phi_{\mathcal{U}}(x)) = \phi_{\mathcal{V}}(x)$.*

PROOF. For any point $x \in X$, one has $\phi_{\mathcal{U}}(x) = \sum_{\alpha \in A} \varphi_\alpha(x) u_\alpha$ where u_α is the vertex corresponding to $U_\alpha \in \mathcal{U}$ in $|N(\mathcal{U})|$. Then,

$$\hat{\tau} \circ \phi_{\mathcal{U}}(x) = \hat{\tau}\left(\sum_{\alpha \in A} \varphi_\alpha(x) u_\alpha\right) = \sum_{\alpha \in A} \varphi_\alpha(x) \tau(u_\alpha) = \sum_{\alpha \in A} \varphi_\alpha(x) v_{\theta(\alpha)}$$

$$= \sum_{\beta \in B} \sum_{\alpha \in \theta^{-1}(\beta)} \varphi_\alpha(x) v_{\theta(\alpha)} = \sum_{\beta \in B} \psi_\beta(x) v_\beta = \phi_{\mathcal{V}}(x) \qquad\square$$

An immediate corollary of the above claim follows.

Corollary 9.7. *The induced maps of $\phi_{\mathcal{U}*} \colon \mathsf{H}_p(X) \to \mathsf{H}_p(|N(\mathcal{U})|)$, $\phi_{\mathcal{V}*} \colon \mathsf{H}_p(X) \to \mathsf{H}_p(|N(\mathcal{V})|)$, and $\hat{\tau}_* \colon \mathsf{H}_p(|N(\mathcal{U})|) \to \mathsf{H}_p(|N(\mathcal{V})|)$ commute, that is, $\phi_{\mathcal{V}*} = \hat{\tau}_* \circ \phi_{\mathcal{U}*}$.*

With the fact that isomorphism between singular and simplicial homology commutes with simplicial maps and their linear continuous extensions, Corollary 9.7 implies the next proposition.

Proposition 9.8. *(One has that) $\bar{\phi}_{\mathcal{V}*} = \tau_* \circ \bar{\phi}_{\mathcal{U}*}$ where $\bar{\phi}_{\mathcal{V}*} \colon \mathsf{H}_p(X) \to \mathsf{H}_p(N(\mathcal{V}))$, $\bar{\phi}_{\mathcal{U}*} \colon \mathsf{H}_p(X) \to \mathsf{H}_p(N(\mathcal{U}))$, and $\tau \colon N(\mathcal{U}) \to N(\mathcal{V})$ is the simplicial map induced by a cover map $\mathcal{U} \to \mathcal{V}$.*

PROOF. Consider the diagram in Figure 9.6. The upper triangle commutes by Corollary 9.7. The bottom square commutes by the property of simplicial

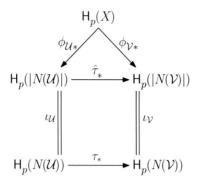

Figure 9.6 Maps relevant for Proposition 9.8; $\bar{\phi}_{\mathcal{V}*} = \iota_{\mathcal{V}} \circ \phi_{\mathcal{V}*}$ and $\bar{\phi}_{\mathcal{U}*} = \iota_{\mathcal{U}} \circ \phi_{\mathcal{U}*}$. The triangular "roof" and the square "room" commute, so does the entire "house."

maps; see Theorem 34.4 in [242]. The claim in the proposition follows by combining these two commuting subdiagrams.   □

Proposition 9.8 extends Theorem 9.5 to the simplicial maps between two nerves.

Theorem 9.9. *Let* $\tau : N(\mathcal{U}) \to N(\mathcal{V})$ *be a simplicial map induced by a cover map* $\mathcal{U} \to \mathcal{V}$ *where both* $\mathcal{U}$ *and* $\mathcal{V}$ *are path connected. Then,* $\tau_* : \mathsf{H}_1(N(\mathcal{U})) \to \mathsf{H}_1(N(\mathcal{V}))$ *is a surjection.*

PROOF. Consider the maps

$$\mathsf{H}_1(X) \overset{\bar{\phi}_{\mathcal{U}*}}{\to} \mathsf{H}_1(N(\mathcal{U})) \overset{\tau_*}{\to} \mathsf{H}_1(N(\mathcal{V})) \quad \text{and} \quad \mathsf{H}_1(X) \overset{\bar{\phi}_{\mathcal{V}*}}{\to} \mathsf{H}_1(N(\mathcal{V})).$$

By Proposition 9.8, $\tau_* \circ \bar{\phi}_{\mathcal{U}*} = \bar{\phi}_{\mathcal{V}*}$. By Theorem 9.5, the map $\bar{\phi}_{\mathcal{V}*}$ is a surjection. It follows that τ_* is a surjection.   □

9.2 Analysis of Persistent H_1-Classes

Using the language of persistent homology, the results in the previous section imply that one-dimensional homology classes can die in the nerves, but they cannot be born. In this section, we further characterize the classes that survive. The distinction among the classes is made via a notion of "size." Intuitively, we show that the classes with "size" much larger than the "size" of the cover survive. The "size" is defined using a pseudometric that the space X is assumed to be equipped with. Precise statements are made in the subsections. Let (X, d) be a pseudometric space, meaning that d satisfies the axioms of a metric

(Definition 1.8) except the first axiom, that is, $d(x, x') = 0$ may not necessarily imply $x = x'$. Assume X is compact. We define a "size" for a homology class that reflects how big the smallest cycle in the class is with respect to the metric d.

Definition 9.3. (Size; Cycle basis) The *size* $s(X')$ of a subset X' of the pseudometric space (X, d) is defined to be its diameter, that is, $s(X') = \sup_{x,x' \in X' \times X'} d(x, x')$. The size of a class $c \in H_p(X)$ is defined as $s(c) = \inf_{z \in c} s(z)$. According to Definition 5.3, a set of p-cycles $z_1, z_2, \ldots, z_n$ of $H_p(X)$ is called a *cycle basis* if the classes $[z_1], [z_2], \ldots, [z_n]$ together form a basis of $H_p(X)$. It is called an optimal cycle basis if $\sum_{i=1}^{n} s(z_i)$ is minimal among all cycle bases.

Lebesgue Number of a Cover
Our goal is to characterize the classes in the nerve of $\mathcal{U}$ with respect to the sizes of their pre-images in X via the map $\phi_\mathcal{U}$. The *Lebesgue number* of a cover $\mathcal{U}$ becomes useful in this characterization. It is the largest real number $\lambda(\mathcal{U})$ so that any subset of X with size at most $\lambda(\mathcal{U})$ is contained in at least one element of $\mathcal{U}$. Formally, the Lebesgue number $\lambda(\mathcal{U})$ of $\mathcal{U}$ is defined as:

$$\lambda(\mathcal{U}) = \sup\{\delta \mid \forall\, X' \subseteq X \text{ with } s(X') \le \delta, \exists\, U_\alpha \in \mathcal{U} \text{ where } X' \subseteq U_\alpha\}.$$

As we will see below, a homology class of size no more than $\lambda(\mathcal{U})$ cannot survive in the nerve (Proposition 9.12). Further, the homology classes whose sizes are significantly larger than the maximum size of a cover do necessarily survive, where we define the *maximum size of a cover* as

$$s_{max}(\mathcal{U}) := \max_{U \in \mathcal{U}}\{s(U)\}.$$

Theorem 9.10 summarizes these observations. Its proof follows after the proof of Proposition 9.12.

Let $z_1, z_2, \ldots, z_g$ be a nondecreasing sequence of the cycles with respect to their sizes in an optimal cycle basis of $H_1(X)$. Consider the map $\phi_\mathcal{U} : X \to |N(\mathcal{U})|$ as introduced in Eq. (9.1), and the map $\bar{\phi}_{\mathcal{U}*}$ as defined by Eq. (9.2). We have the following result.

Theorem 9.10. *Let $\mathcal{U}$ be a path connected cover of X and $z_1, z_2, \ldots, z_g$ be a sequence of an optimal cycle basis of $H_1(X)$ as stated above.*

(a) *Let $\ell = g + 1$ if $\lambda(\mathcal{U}) > s(z_g)$. Otherwise, let $\ell \in [1, g]$ be the smallest integer so that $s(z_\ell) > \lambda(\mathcal{U})$. If $\ell \neq 1$, then we have that the class $\bar{\phi}_{\mathcal{U}*}[z_j] = 0$ for $j = 1, \ldots, \ell - 1$. Moreover, if $\ell \neq g + 1$, then the classes $\{\bar{\phi}_{\mathcal{U}*}[z_j]\}_{j=\ell,\ldots,g}$ generate $H_1(N(\mathcal{U}))$.*

(b) *The classes* $\{\bar{\phi}_{\mathcal{U}*}[z_j]\}_{j=\ell',...,g}$ *are linearly independent where* $s(z_{\ell'}) > 4s_{max}(\mathcal{U})$.

The result above says that only the classes of $H_1(X)$ generated by cycles of large enough size survive in the nerve. To prove this result, we use a map ρ that sends each 1-cycle in $N(\mathcal{U})$ to a 1-cycle in X. We define a chain map $\rho : C_1(N(\mathcal{U})) \to C_1(X)$ among one-dimensional chain groups as follows. It is sufficient to exhibit the map for an elementary chain of an edge, say $e = \{u_\alpha, u_{\alpha'}\} \in C_1(N(\mathcal{U}))$. Since e is an edge in $N(\mathcal{U})$, the two cover elements U_α and $U_{\alpha'}$ in X have a common intersection. Let $a \in U_\alpha$ and $b \in U_{\alpha'}$ be two points that are arbitrary but fixed for U_α and $U_{\alpha'}$, respectively. Pick a path $\xi(a, b)$ (viewed as a singular chain) in the union of U_α and $U_{\alpha'}$ which is path connected as both U_α and $U_{\alpha'}$ are. Then, define $\rho(e) = \xi(a, b)$. A cycle γ when pushed back by ρ and then pushed forward by $\phi_{\mathcal{U}}$ remains in the same class. The following proposition states this fact whose proof appears in [132].

Proposition 9.11. *Let* γ *be any* 1-*cycle in* $N(\mathcal{U})$. *Then,* $[\phi_{\mathcal{U}}(\rho(\gamma))] = [|\gamma|]$.

The following proposition provides a sufficient characterization of the cycles whose classes become trivial after the pushforward.

Proposition 9.12. *Let* z *be a* 1-*cycle in* $C_1(X)$. *Then,* $[\phi_{\mathcal{U}}(z)] = 0$ *if* $\lambda(\mathcal{U}) > s(z)$.

PROOF. It follows from the definition of the Lebesgue number that there exists a cover element $U_\alpha \in \mathcal{U}$ such that $z \subseteq U_\alpha$ because $s(z) < \lambda(\mathcal{U})$. We claim that there is a homotopy equivalence that sends $\phi_{\mathcal{U}}(z)$ to a vertex in $N(\mathcal{U})$ and hence $[\phi_{\mathcal{U}}(z)]$ is trivial.

Let x be any point in z. Recall that $\phi_{\mathcal{U}}(x) = \Sigma_i \varphi_i(x) u_{\alpha_i}$. Since U_α has a common intersection with each U_{α_i} so that $\varphi_{\alpha_i}(x) \neq 0$, we can conclude that $\phi_{\mathcal{U}}(x)$ is contained in a simplex with the vertex u_α. Continuing this argument with all points of z, we observe that $\phi_{\mathcal{U}}(z)$ is contained in simplices that share the vertex u_α. It follows that there is a homotopy that sends $\phi_{\mathcal{U}}(z)$ to u_α, a vertex of $N(\mathcal{U})$. □

PROOF OF THEOREM 9.10. (a) By Proposition 9.12, we have $\phi_{\mathcal{U}*}[z] = [\phi_{\mathcal{U}}(z)] = 0$ if $\lambda(\mathcal{U}) > s(z)$. This establishes the first part of the assertion because $\bar{\phi}_{\mathcal{U}*} = \iota \circ \phi_{\mathcal{U}*}$ where ι is an isomorphism between the singular homology of $|N(\mathcal{U})|$ and the simplicial homology of $N(\mathcal{U})$. To see the second part, notice that $\bar{\phi}_{\mathcal{U}*}$ is a surjection by Theorem 9.5. Therefore, the classes $\bar{\phi}_{\mathcal{U}*}(z)$ where $s(z) \geq \lambda(\mathcal{U})$ contain a basis for $H_1(N(\mathcal{U}))$. Hence they generate it.

(b): For a contradiction, assume that there is a subsequence $\{\ell_1, \ldots, \ell_t\} \subset \{\ell', \ldots, g\}$ so that $\sum_{j=1}^{t} [\phi_{\mathcal{U}}(z_{\ell_j})] = 0$. Let $z = \sum_{j=1}^{t} \phi_{\mathcal{U}}(z_{\ell_j})$. Let γ be a 1-cycle in $N(\mathcal{U})$ so that $[z] = [|\gamma|]$ whose existence is guaranteed by Proposition 9.3. As $\sum_{j=1}^{t} [\phi_{\mathcal{U}}(z_{\ell_j})] = 0$, it must be that there is a 2-chain D in $N(\mathcal{U})$ so that $\partial D = \gamma$. Consider a triangle $t = \{u_{\alpha_1}, u_{\alpha_2}, u_{\alpha_3}\}$ contributing to D. Let $a_i' = \phi_{\mathcal{U}}^{-1}(u_{\alpha_i})$. Since t appears in $N(\mathcal{U})$, the covers $U_{\alpha_1}, U_{\alpha_2}$, and U_{α_3} containing a_1', a_2', and a_3', respectively, have a common intersection in X. This also means that each of the paths $a_1' \rightsquigarrow a_2'$, $a_2' \rightsquigarrow a_3'$, and $a_3' \rightsquigarrow a_1'$ has size at most $2s_{max}(\mathcal{U})$. Then, $\rho(\partial t)$ is mapped to a 1-cycle in X of size at most $4s_{max}(\mathcal{U})$. It follows that $\rho(\partial D)$ can be written as a linear combination of cycles of size at most $4s_{max}(\mathcal{U})$. Since $z_1, \ldots, z_g$ form an optimal cycle basis of $H_1(X)$, each of the 1-cycles of size at most $4s_{max}(\mathcal{U})$ is generated by basis elements $z_1, \ldots, z_k$ where $s(z_k) \leq 4s_{max}(\mathcal{U})$. Therefore, the class of $z' = \phi_{\mathcal{U}}(\rho(\gamma))$ is generated by a linear combination of the basis elements whose pre-images have size at most $4s_{max}(\mathcal{U})$. The class $[z']$ is the same as the class $[|\gamma|]$ by Proposition 9.11. But, by assumption, $[|\gamma|] = [z]$ is generated by a linear combination of the basis elements whose sizes are larger than $4s_{max}(\mathcal{U})$, reaching a contradiction. Hence the assumption cannot hold and (ii) is true. $\square$

9.3 Mapper and Multiscale Mapper

In this section we extend the previous results to the structures called mapper and multiscale mapper. Recall that X is assumed to be compact. Consider a cover of X obtained indirectly as a pullback of a cover of another space Z. This gives rise to the so-called *mapper*. More precisely, let $f : X \rightarrow Z$ be a continuous map where Z is equipped with an open cover $\mathcal{U} = \{U_\alpha\}_{\alpha \in A}$ for some index set A. Since f is continuous, the sets $\{f^{-1}(U_\alpha), \alpha \in A\}$ form an open cover of X. For each α, we can now consider the decomposition of $f^{-1}(U_\alpha)$ into its path connected components, and we write $f^{-1}(U_\alpha) = \bigcup_{i=1}^{j_\alpha} V_{\alpha,i}$, where j_α is the number of path connected components $V_{\alpha,i}$ in $f^{-1}(U_\alpha)$. We write $f^*\mathcal{U}$ for the cover of X obtained this way from the cover $\mathcal{U}$ of Z and refer to it as the *pullback cover* of X induced by $\mathcal{U}$ via f. By construction, every element in this pullback cover $f^*\mathcal{U}$ is path connected.

Notice that there are pathological examples of f where $f^{-1}(U_\alpha)$ may shatter into infinitely many path components. This motivates us to consider *well-behaved* functions f: we require that for every path connected open set $U \subseteq Z$, the pre-image $f^{-1}(U)$ has *finitely* many open path connected components. Consequently, all nerves of pullbacks of finite covers become finite.

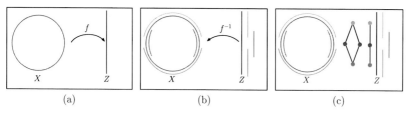

Figure 9.7 Mapper construction: (a) a map $f: X \to Z$ from a circle to a subset $Z \subset \mathbb{R}$; (b) the inverse map f^{-1} induces a cover of the circle from a cover $\mathfrak{U}$ of Z; and (c) the nerves of the two covers of X and Z, where the nerve on the left (quadrangle shaped) is the mapper induced by f and $\mathfrak{U}$.

Definition 9.4. (Mapper) Let X and Z be topological spaces and let $f: X \to Z$ be a well-behaved and continuous map. Let $\mathfrak{U} = \{U_\alpha\}_{\alpha \in A}$ be a finite open cover of Z. The *mapper* arising from these data is defined to be the nerve of the pullback cover $f^*(\mathfrak{U})$ of X; that is, $M(\mathfrak{U}, f) := N(f^*(\mathfrak{U}))$. See an illustration in Figure 9.7.

Notice that we define the mapper using finite covers which allows us to extend the definitions of persistence modules and persistence diagrams from previous chapters to the case of mappers. However, in the next remark and later we allow infinite covers for simplicity. The definition of mapper remains valid with infinite covers.

Remark 9.1. *The construction of mapper is quite general if we allow the cover $\mathfrak{U}$ to be infinite. For example, it can encompass both the Reeb graph and merge trees: consider X a topological space and $f: X \to \mathbb{R}$. Then, consider the following two options for $\mathfrak{U} = \{U_\alpha\}_{\alpha \in A}$, the other ingredient of the construction:*

- $U_\alpha = (-\infty, \alpha)$ *for $\alpha \in A = \mathbb{R}$. This corresponds to sublevel sets which in turn lead to merge trees. See, for example, the construction in Figure 9.8(b).*
- $U_\alpha = (\alpha - \varepsilon, \alpha + \varepsilon)$ *for $\alpha \in A = \mathbb{R}$, for some fixed $\varepsilon > 0$. This corresponds to (ε-thick) levelsets, which induce a relaxed notion of Reeb graphs. See the description in "Mapper for PCD" below and Figure 9.8(a).*

In these two examples, for simplicity of presentation, the set A is allowed to have infinite cardinality. Also, note that one can take any open cover of $\mathbb{R}$ in this definition. This may give rise to other constructions beyond merge trees or Reeb graphs. For instance, using the infinite setting for simplicity again, one may choose any point $r \in \mathbb{R}$ and let $U_\alpha = (r - \alpha, r + \alpha)$ for each $\alpha \in A = \mathbb{R}$ or other constructions.

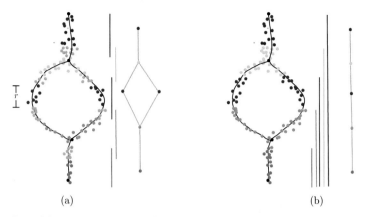

(a)   (b)

Figure 9.8 Mapper construction for a point cloud, a map $f : P \to Z$ from PCD P to a subset $Z \subset \mathbb{R}$; the graph G^r is not shown. (a) The covers are intervals, points are colored with the interval colors, gray points have values in two overlapping intervals, and the mapper is a discretized Reeb graph. (b) The covers are sublevel sets, points are colored with the smallest levelset they belong to, and the discretized Reeb graph does not have the central loop any more.

Mapper for PCD

Consider a finite metric space (P, d_P), that is, a point set P with distances between every pair of points. For a real $r \geq 0$, one can construct a graph $G^r(P)$ with every point in P as a vertex where an edge (p, p') is in $G^r(P)$ if and only if $\mathsf{d}_P(p, p') \leq r$. Let $f : P \to \mathbb{R}$ be a real-valued function on the point set P. For a set of intervals $\mathcal{U}$ covering $\mathbb{R}$, we can construct the mapper as follows. For every interval $(a, b) \in \mathcal{U}$, let $P_{(a,b)} = f^{-1}((a, b))$ be the set of points with function values in the range (a, b). Each such set consists of a partition $P_{(a,b)} = \bigsqcup P_{(a,b)}^i$ determined by the graph connectivity of $G^r(P)$. Each set $P_{(a,b)}^i$ consists of the vertices of a connected component of the subgraph of $G^r(P)$ spanned by the vertices in $P_{(a,b)}$. The vertex sets $\bigcup_{(a,b) \in \mathcal{U}} \{P_{(a,b)}^i\}$ thus obtained over all intervals constitute a cover $f^{-1}(\mathcal{U})$ of P. The nerve of this cover is the mapper $\mathrm{M}(P, f)$. Here the intersection between cover elements is determined by the intersection of discrete sets.

Observe that, in the above construction, if one takes the intervals of $\mathcal{U} = \{U_i\}_{i \in \mathbb{Z}}$ where $U_i = (i - \varepsilon, i + \varepsilon)$ for some $\varepsilon \in (0, 1)$ causing only two consecutive intervals to overlap partially, then we get a discretized approximation of the Reeb graphs of the function that f approximates on the discretized sample P. Figure 9.8 illustrates this observation. In the limit that each interval degenerates to a point, the discretized Reeb graph converges to the original Reeb graph as shown in [132, 241].

9.3.1 Multiscale Mapper

A mapper $M(\mathcal{U}, f)$ is a simplicial complex encoding the structure of f through the lens of Z. However, the simplicial complex $M(\mathcal{U}, f)$ provides only *one* snapshot of X at a *fixed* scale determined by the scale of the cover $\mathcal{U}$. Using the idea of persistent homology, we study the evolution of the mapper $M(f, \mathcal{U}_a)$ for a *tower of covers* $\mathcal{U} = \{\mathcal{U}_a\}_{a \in A}$. The tower by definition coarsens the cover with increasing indices and hence provides mappers at multiple scales.

As an intuitive example, consider a real-valued function $f : X \to \mathbb{R}$, and a cover $\mathcal{U}_\varepsilon$ of $\mathbb{R}$ consisting of all possible intervals of length ε. Intuitively, as ε tends to 0, the corresponding mapper $M(f, \mathcal{U}_\varepsilon)$ approaches the Reeb graph of f. As ε increases, we look at the Reeb graph at coarser and coarser resolution. The multiscale mapper in this case roughly encodes this simplification process.

The idea of multiscale mapper requires a sequence of covers of the target space connected by cover maps. Through pullbacks, it generates a sequence of covers on the domain. In particular, first we have the next result.

Proposition 9.13. *Let* $f : X \to Z$, *and* $\mathcal{U}$ *and* $\mathcal{V}$ *be two covers of* Z *with a map of covers* $\xi : \mathcal{U} \to \mathcal{V}$. *Then, there is a corresponding map of covers between the respective pullback covers of* X, *that is,* $f^*(\xi) : f^*(\mathcal{U}) \longrightarrow f^*(\mathcal{V})$.

PROOF. Indeed, we only need to note that if $U \subseteq V$, then $f^{-1}(U) \subseteq f^{-1}(V)$, and therefore it is clear that each path connected component of $f^{-1}(U)$ is included in exactly one path connected component of $f^{-1}(V)$. More precisely, let $\mathcal{U} = \{U_\alpha\}_{\alpha \in A}$ and $\mathcal{V} = \{V_\beta\}_{\beta \in B}$, with $U_\alpha \subseteq V_{\xi(\alpha)}$ for $\alpha \in A$. Let $\widehat{U}_{\alpha, i}$, $i \in \{1, \ldots, n_\alpha\}$, denote the connected components of $f^{-1}(U_\alpha)$ and $\widehat{V}_{\beta, j}$, $j \in \{1, \ldots, m_\beta\}$, denote the connected components of $f^{-1}(V_\beta)$. Then, the map of covers $f^*(\xi)$ from $f^*(\mathcal{U})$ to $f^*(\mathcal{V})$ is given by requiring that each set $\widehat{U}_{\alpha, i}$ is sent to the *unique* set of the form $\widehat{V}_{\xi(\alpha), j}$ so that $\widehat{U}_{\alpha, i} \subseteq \widehat{V}_{\xi(\alpha), j}$. □

Furthermore, observe that if $\mathcal{U} \xrightarrow{\xi} \mathcal{V} \xrightarrow{\zeta} \mathcal{W}$ are three different covers of a topological space with the intervening maps of covers between them, then $f^*(\zeta \circ \xi) = f^*(\zeta) \circ f^*(\xi)$.

The above result for three covers easily extends to multiple covers and their pullbacks. The sequence of pullbacks connected by cover maps and the corresponding sequence of nerves connected by simplicial maps define multiscale mappers. Recall the definition of towers (Definition 4.1) to designate a sequence of *objects* connected with maps. Let $\mathfrak{U} = \{\mathcal{U}_a \xrightarrow{u_{a,a'}} \mathcal{U}_{a'}\}_{r \leq a \leq a'}$ denote a tower, where $r = \mathrm{res}(\mathfrak{U})$ refers to its resolution. The objects here can be covers, simplicial complexes, or vector spaces. The notion of resolution and

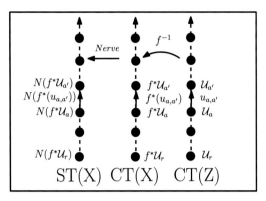

Figure 9.9 Illustrating construction of multiscale mapper from a cover tower; CT
and ST denote cover and simplicial towers, respectively, that is, $CT(Z) = \mathfrak{U}$,
$CT(X) = f^*(\mathfrak{U})$, and $ST(X) = N(f^*(\mathfrak{U}))$.

the variable a intuitively specify the granularity of the covers and the simplicial
complexes induced by them.

The pullback property given by Proposition 9.13 makes it possible to take
the pullback of a given tower of covers of a space via a given continuous
function into another space as stated in the proposition below.

Proposition 9.14. *Let $\mathfrak{U}$ be a cover tower of Z and $f \colon X \to Z$ be a
continuous function. Then, $f^*(\mathfrak{U})$ is a cover tower of X.*

In general, given a cover tower $\mathfrak{W}$ of a space X, the nerve of each cover in $\mathfrak{W}$
together with simplicial maps induced by each map of $\mathfrak{W}$ provides a simplicial
tower which we denote by $N(\mathfrak{W})$.

Definition 9.5. (Multiscale mapper) Let X and Z be topological spaces and
$f \colon X \to Z$ be a continuous map. Let $\mathfrak{U}$ be a cover tower of Z. Then, the
multiscale mapper is defined to be the simplicial tower obtained by the nerve
of the pullback:

$$\mathrm{MM}(\mathfrak{U}, f) := N(f^*(\mathfrak{U})),$$

where the simplicial maps are induced by the respective cover maps. See
Figure 9.9 for an illustration.

Consider for example a sequence $\mathrm{res}(\mathfrak{U}) \le a_1 < a_2 < \cdots < a_n$ of n distinct
real numbers. Then, the definition of multiscale mapper $\mathrm{MM}(\mathfrak{U}, f)$ gives rise
to the following simplicial tower:

$$N(f^*(\mathfrak{U}_{a_1})) \to N(f^*(\mathfrak{U}_{a_2})) \to \cdots \to N(f^*(\mathfrak{U}_{a_n})), \qquad (9.3)$$

which is a sequence of simplicial complexes connected by simplicial maps.

Applying to them the homology functor $H_p(\cdot)$, $p = 0, 1, 2, \ldots$, with coefficients in a field, one obtains a persistence module: a tower of vector spaces connected by linear maps,

$$H_p\big(N(f^*(\mathcal{U}_{a_1}))\big) \to \cdots \to H_p\big(N(f^*(\mathcal{U}_{a_n}))\big). \tag{9.4}$$

Given our assumptions that the covers are finite and that the function f is well behaved, we obtain that the homology groups of all nerves have finite dimensions. Thus, we get a persistence module which is pointwise finite-dimensional (p.f.d.; see Section 3.4). Now one can summarize the persistence module induced by $MM(\mathcal{U}, f)$ with its persistent diagram $\mathrm{Dgm}_p MM(\mathcal{U}, f)$ for each dimension $p \in \mathbb{N}$. The diagram $\mathrm{Dgm}_p MM(\mathcal{U}, f)$ can be viewed as a topological summary of f through the lens of $\mathcal{U}$.

9.3.2 Persistence of H_1-Classes in Mapper and Multiscale Mapper

To apply the results for nerves in Section 9.2 to mappers and multiscale mappers, we need a "size" measure on X. For this, we assume that Z is a metric space and we pull back the metric to X via $f \colon X \to Z$. Assuming that X is path connected, let $\Gamma_X(x, x')$ denote the set of all continuous paths $\gamma \colon [0, 1] \to X$ between any two given points $x, x' \in X$ so that $\gamma(0) = x$ and $\gamma(1) = x'$.

Definition 9.6. (Pullback metric) Given a metric space (Z, d_Z), we define its *pullback metric* as the following pseudometric d_f on X: for $x, x' \in X$,

$$\mathsf{d}_f(x, x') := \inf_{\gamma \in \Gamma_X(x,x')} \mathrm{diam}_Z(f \circ \gamma).$$

Consider the Lebesgue number of the pullback covers of X. The following observation in this respect is useful.

Proposition 9.15. *Let $\mathcal{U}$ be a cover for the codomain Z and $\mathcal{U}'$ be its restriction to $f(X)$. Then, the pullback cover $f^*\mathcal{U}$ has the same Lebesgue number as that of $\mathcal{U}'$; that is, $\lambda(f^*\mathcal{U}) = \lambda(\mathcal{U}')$.*

PROOF. First, observe that, for any path connected cover of X, a subset of X that realizes the Lebesgue number can be taken as path connected because, if not, this subset can be connected by a path entirely lying within the cover element containing it. Let $X' \subseteq X$ be any subset where $s(X') \leq \lambda(\mathcal{U}')$. Then, $f(X') \subseteq Z$ has a diameter at most $\lambda(\mathcal{U}')$ by the definitions of size (Definition 9.3) and pullback metric. Therefore, by the definition of Lebesgue number,

$f(X')$ is contained in a cover element $U' \in \mathcal{U}'$. Since X' is path connected, a path connected component of $f^{-1}(U')$ contains X'. It follows that there is a cover element in $f^*\mathcal{U}$ that contains X'. Since X' was chosen as an arbitrary path connected subset of size at most $\lambda(\mathcal{U}')$, we have $\lambda(f^*\mathcal{U}) \geq \lambda(\mathcal{U}')$. At the same time, it is straightforward from the definition of size that each cover element in $f^{-1}(U')$ has at most the size of U' for any $U' \in \mathcal{U}'$. Combining with the fact that $\mathcal{U}'$ is the restriction of $\mathcal{U}$ to $f(X)$, we have $\lambda(f^*\mathcal{U}) \leq \lambda(\mathcal{U}')$, establishing the equality as claimed. $\square$

Given a cover $\mathcal{U}$ of Z, consider the mapper $N(f^*\mathcal{U})$. Let $z_1, \ldots, z_g$ be an optimal cycle basis for $H_1(X)$ where the metric used to define optimality is the pullback metric d_f. Then, as a consequence of Theorem 9.10 we have the following.

Theorem 9.16. *Let $f : X \to Z$ be a map from a path connected space X to a metric space Z equipped with a cover $\mathcal{U}$ (parts (a) and (b) below) or a tower of covers $\{\mathcal{U}_a\}$ (part (c) below). Let $\mathcal{U}'$ be the restriction of $\mathcal{U}$ to $f(X)$.*

(a) *Let $\ell = g + 1$ if $\lambda(\mathcal{U}') > s(z_g)$. Otherwise, let $\ell \in [1, g]$ be the smallest integer so that $s(z_\ell) > \lambda(\mathcal{U}')$. If $\ell \neq 1$, the class $\phi_{\mathcal{U}*}[z_j] = 0$ for $j = 1, \ldots, \ell - 1$. Moreover, if $\ell \neq g+1$, the classes $\{\phi_{\mathcal{U}*}[z_j]\}_{j=\ell,\ldots,g}$ generate $H_1(N(f^*\mathcal{U}))$.*
(b) *The classes $\{\phi_{\mathcal{U}*}[z_j]\}_{j=\ell',\ldots,g}$ are linearly independent where $s(z_{\ell'}) > 4s_{max}(\mathcal{U})$.*
(c) *Consider an H_1-persistence module of a multiscale mapper induced by a tower of path connected covers:*

$$H_1\big(N(f^*\mathcal{U}_{a_0})\big) \xrightarrow{s_{1*}} H_1\big(N(f^*\mathcal{U}_{a_1})\big) \xrightarrow{s_{2*}} \cdots \xrightarrow{s_{n*}} H_1\big(N(f^*\mathcal{U}_{a_n})\big). \quad (9.5)$$

Let $\hat{s}_{i*} = s_{i*} \circ s_{(i-1)*} \circ \cdots \circ \phi_{\mathcal{U}_{a_0}*}$. Then, the assertions in (a) and (b) hold for $H_1(N(f^*\mathcal{U}_{a_i}))$ with the map $\hat{s}_{i*} : X \to N(f^*\mathcal{U}_{a_i})$.

9.4 Stability

To be useful in practice, the multiscale mapper should be stable against perturbations in the maps and the covers. We show that such stability is enjoyed by the multiscale mapper under some natural condition on the tower of covers. Recall that previous stability results for towers as described in Section 4.1 were drawn on the notion of interleaving. We identify compatible notions of interleaving for cover towers as a way to measure the "closeness" between two cover towers.

9.4.1 Interleaving of Cover Towers and Multiscale Mappers

In this section we consider cover and simplicial towers indexed over $\mathbb{R}$. In practice, we often have a cover tower $\mathfrak{U} = \left\{ \mathfrak{U}_a \xrightarrow{u_{a,a'}} \mathfrak{U}_{a'} \right\}_{a \le a'}$ indexed by a discrete set in $A \subset \mathbb{R}$. Any such tower can be extended to a cover tower indexed over $\mathbb{R}$ by taking $\mathfrak{U}_\varepsilon = \mathfrak{U}_a$ for each index $\varepsilon \in (a, a')$ where a and a' are any two consecutive indices in the ordered set A.

Definition 9.7. (Interleaving of cover towers) Let $\mathfrak{U} = \{\mathfrak{U}_a\}$ and $\mathfrak{V} = \{\mathcal{V}_a\}$ be two cover towers of a topological space X so that $\mathrm{res}(\mathfrak{U}) = \mathrm{res}(\mathfrak{V}) = r$. Given $\eta \ge 0$, we say that $\mathfrak{U}$ and $\mathfrak{V}$ are η-*interleaved* if one can find cover maps $\zeta_a : \mathfrak{U}_a \to \mathcal{V}_{a+\eta}$ and $\xi_{a'} : \mathcal{V}_{a'} \to \mathfrak{U}_{a'+\eta}$ for all $a, a' \ge r$. See the diagram below.

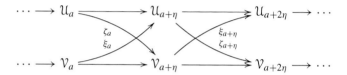

Analogously, if we replace the operator "+" by the multiplication "·" in the above definition, then we say that $\mathfrak{U}$ and $\mathfrak{V}$ are *multiplicatively η-interleaved*.

Proposition 9.17. The following hold:

(a) If $\mathfrak{U}$ and $\mathfrak{V}$ are (multiplicatively) η_1-interleaved and $\mathfrak{V}$ and $\mathfrak{W}$ are (multiplicatively) η_2-interleaved, then $\mathfrak{U}$ and $\mathfrak{W}$ are (multiplicatively $(\eta_1\eta_2)$-) $(\eta_1 + \eta_2)$-interleaved.

(b) Let $f : X \to Z$ be a continuous function and $\mathfrak{U}$ and $\mathfrak{V}$ be two (multiplicatively) η-interleaved towers of covers of Z. Then, $f^*(\mathfrak{U})$ and $f^*(\mathfrak{V})$ are also (multiplicatively) η-interleaved.

Note that in the definition of interleaving cover towers, we do not have an explicit requirement that maps need to make subdiagrams commute, unlike the interleaving between simplicial towers (Definition 4.2). However, it follows from Proposition 9.2 that interleaving cover towers lead to interleaving between simplicial towers for $N(\mathfrak{U})$ and $N(\mathfrak{V})$ as shown in the proposition below.

Proposition 9.18. *Let $\mathfrak{U}$ and $\mathfrak{V}$ be two (multiplicatively) η-interleaved cover towers of X with $\mathrm{res}(\mathfrak{U}) = \mathrm{res}(\mathfrak{V})$. Then, $N(\mathfrak{U})$ and $N(\mathfrak{V})$ are also (multiplicatively) η-interleaved.*

PROOF. We prove the proposition for additive interleaving. Replacing the "+" operator with "·" gives the proof for multiplicative interleaving. Let r denote the common resolution of $\mathfrak{U}$ and $\mathfrak{V}$. Write $\mathfrak{U} = \left\{\mathcal{U}_a \xrightarrow{u_{a,a'}} \mathcal{U}_{a'}\right\}_{r \leq a \leq a'}$ and $\mathfrak{V} = \left\{\mathcal{V}_a \xrightarrow{v_{a,a'}} \mathcal{V}_{a'}\right\}_{r \leq a \leq a'}$, and for each $a \geq r$ let $\zeta_a : \mathcal{U}_a \to \mathcal{V}_{a+\eta}$ and $\xi_a : \mathcal{V}_a \to \mathcal{U}_{a+\eta}$ be given as in Definition 9.7. To define interleaving between the towers of nerves arising out of covers, we consider similar diagrams to (4.3) at the level of covers involving covers of the form $\mathcal{U}_a$ and $\mathcal{V}_a$, and apply the nerve construction. This operation yields diagrams identical to those in (4.3) where for every a, a' with $a' \geq a \geq r$ one has:

- $K_a := N(\mathcal{U}_a)$ and $L_a := N(\mathcal{V}_a)$;
- $x_{a,a'} := N(u_{a,a'})$, for $r \leq a \leq a'$; $y_{a,a'} := N(v_{a,a'})$, for $r \leq a \leq a'$; $\varphi_a := N(\zeta_a)$; and $\psi_a := N(\xi_a)$.

To satisfy Definition 4.2, it remains to verify conditions (i) to (iv) therein. We only verify (i), since the proofs of the others follow the same arguments. For this, notice that both the composite map $\xi_{a+\eta} \circ \zeta_a$ and $u_{a,a+2\eta}$ are maps of covers from $\mathcal{U}_a$ to $\mathcal{U}_{a+2\eta}$. By Proposition 9.2 we then have that $N(\xi_{a+\eta} \circ \zeta_a)$ and $N(u_{a,a+2\eta}) = f_{a,a+2\eta}$ are contiguous. But, by the properties of the nerve construction, $N(\xi_{a+\eta} \circ \zeta_a) = N(\xi_{a+\eta}) \circ N(\zeta_a) = \psi_{a+\eta} \circ \varphi_a$, which completes the claim. $\qquad\square$

Combining Propositions 9.17 and 9.18, we get that the two multiscale mappers under cover perturbations remain stable, which is the first part of Corollary 9.19. Recall from Chapter 4 that, for a finite simplicial tower $\mathcal{S}$ and $p \in \mathbb{N}$, we denote by $\mathrm{Dgm}_p(\mathcal{S})$ the p-th persistence diagram of the tower $\mathcal{S}$ with coefficients in a fixed field. Using Proposition 9.18 and Theorem 4.3, we have a stability result for $\mathrm{Dgm}_p\mathrm{MM}(\mathfrak{U}, f)$ when f is kept fixed but the cover tower $\mathfrak{U}$ is perturbed, which is the second part of the corollary below.

Corollary 9.19. *For $\eta \geq 0$, let $\mathfrak{U}$ and $\mathfrak{V}$ be two finite cover towers of Z with $\mathrm{res}(\mathfrak{U}) = \mathrm{res}(\mathfrak{V}) > 0$. Let $f : X \to Z$ be well behaved and $\mathfrak{U}$ and $\mathfrak{V}$ be η-interleaved. Then, $\mathrm{MM}(\mathfrak{U}, f)$ and $\mathrm{MM}(\mathfrak{V}, f)$ are η-interleaved. In particular, the bottleneck distance between the persistence diagrams $\mathrm{Dgm}_p\mathrm{MM}(\mathfrak{U}, f)$ and $\mathrm{Dgm}_p\mathrm{MM}(\mathfrak{V}, f)$ is at most η for all $p \in \mathbb{N}$.*

9.4.2 (c, s)-Good Covers

Although $\mathrm{Dgm}_p\mathrm{MM}(\mathfrak{U}, f)$ is stable under perturbations of the covers $\mathfrak{U}$ as we showed, it is not necessarily stable under perturbations of the map f. To address this issue, we introduce a special family of covers called (c, s)-good covers. To define these covers, we use the index value of the covers to denote their scales. The notation ε for indexing is chosen to emphasize this meaning.

Definition 9.8. ((c, s)-Good cover tower) Given a cover tower $\mathfrak{U} = \{\mathfrak{U}_\varepsilon\}_{\varepsilon \geq s > 0}$, we say that it is *(c,s)-good* if for any $\varepsilon \geq s > 0$, we have that (i) $s_{max}(\mathfrak{U}_\varepsilon) \leq \varepsilon$ and (ii) $\lambda(\mathfrak{U}_{c\varepsilon}) \geq \varepsilon$.

As an example, consider the cover tower $\mathfrak{U} = \{\mathfrak{U}_\varepsilon\}_{\varepsilon \geq s}$ with $\mathfrak{U}_\varepsilon := \{B_{\varepsilon/2}(z) \mid z \in Z\}$. It is a $(2, s)$-good cover tower of the metric space (Z, d_Z).

We now characterize the persistent homology of multiscale mappers induced by (c, s)-good cover towers. Theorem 9.20 states that the multiscale mappers induced by any two (c, s)-good cover towers interleave with each other, implying that their respective persistence diagrams are also close under the bottleneck distance. From this point of view, the persistence diagrams induced by any two (c, s)-good cover towers contain roughly the same information.

Theorem 9.20. *Given a map* $f \colon X \to Z$, *let* $\mathfrak{U} = \left\{\mathfrak{U}_\varepsilon \xrightarrow{u_{\varepsilon,\varepsilon'}} \mathfrak{U}_{\varepsilon'}\right\}_{\varepsilon \leq \varepsilon'}$ *and* $\mathfrak{V} = \left\{\mathcal{V}_\varepsilon \xrightarrow{v_{\varepsilon,\varepsilon'}} \mathcal{V}_{\varepsilon'}\right\}_{\varepsilon \leq \varepsilon'}$ *be two* (c, s)-*good cover towers of* Z. *Then the corresponding multiscale mappers* $\mathrm{MM}(\mathfrak{U}, f)$ *and* $\mathrm{MM}(\mathfrak{V}, f)$ *are multiplicatively* c-*interleaved.*

PROOF. First, before we can prove the theorem, we make the following observation.

Claim 9.2. *Any two* (c, s)-*good cover towers* $\mathfrak{U}$ *and* $\mathfrak{V}$ *are multiplicatively* c-*interleaved.*

PROOF. This follows easily from the definitions of (c, s)-good cover tower. Specifically, first we construct $\zeta_\varepsilon \colon \mathfrak{U}_\varepsilon \to \mathcal{V}_{c\varepsilon}$. For any $U \in \mathfrak{U}_\varepsilon$, we have that $\mathrm{diam}(U) \leq \varepsilon$. Furthermore, since $\mathfrak{V}$ is (c, s)-good, there exists $V \in \mathcal{V}_{c\varepsilon}$ such that $U \subseteq V$. Set $\zeta_\varepsilon(U) = V$; if there are multiple choices of V, we can choose an arbitrary one. We can construct $\xi_{\varepsilon'} \colon \mathcal{V}_{\varepsilon'} \to \mathfrak{U}_{c\varepsilon'}$ in a symmetric manner, and the claim then follows. ☐

This claim, combined with Propositions 9.17 and 9.18, prove Theorem 9.20. ☐

We also need the following definition in order to state the stability results precisely.

Definition 9.9. Given a tower of covers $\mathfrak{U} = \{\mathfrak{U}_\varepsilon\}$ and $\varepsilon_0 \geq \mathrm{res}(\mathfrak{U})$, we define the ε_0-*truncation* of $\mathfrak{U}$ as the tower $\mathrm{Tr}_{\varepsilon_0}(\mathfrak{U}) := \left\{\mathfrak{U}_\varepsilon\right\}_{\varepsilon_0 \leq \varepsilon}$. Observe that, by definition, $\mathrm{res}(\mathrm{Tr}_{\varepsilon_0}(\mathfrak{U})) = \varepsilon_0$.

Proposition 9.21. *Let* X *be a compact topological space,* (Z, d_Z) *be a compact path connected metric space, and* $f, g \colon X \to Z$ *be two continuous functions such that for some* $\delta \geq 0$ *one has that* $\delta = \max_{x \in X} \mathsf{d}_Z(f(x), g(x))$.

Let $\mathfrak{W}$ be any (c, s)-good cover tower of Z. Let $\varepsilon_0 = \max(1, s)$. Then, the ε_0-truncations of $f^*(\mathfrak{W})$ and $g^*(\mathfrak{W})$ are multiplicatively $\left(2c \max(\delta, s) + c\right)$-interleaved.

PROOF. For notational convenience write $\eta := 2c \max(\delta, s) + c$, $\{\mathfrak{U}_t\} = \mathfrak{U} := f^*(\mathfrak{W})$, and $\{\mathcal{V}_t\} = \mathfrak{V} := g^*(\mathfrak{W})$. With regards to satisfying Definition 4.2 for $\mathfrak{U}$ and $\mathfrak{V}$, for each $\varepsilon \geq \varepsilon_0$ we need only exhibit maps of covers $\zeta_\varepsilon : \mathfrak{U}_\varepsilon \to \mathcal{V}_{\eta\varepsilon}$ and $\xi_\varepsilon : \mathcal{V}_\varepsilon \to \mathfrak{U}_{\eta\varepsilon}$. We first establish the following claim, where we recall that the offset O^r is defined as $O^r := \{z \in Z \mid \mathsf{d}_Z(z, O) \leq r\}$.

Claim 9.3. *For all $O \subset Z$, and all $\delta' \geq \delta$, $f^{-1}(O) \subseteq g^{-1}(O^{\delta'})$.*

PROOF. Let $x \in f^{-1}(O)$, then $\mathsf{d}_Z(f(x), O) = 0$. Thus,

$$\mathsf{d}_Z(g(x), O) \leq \mathsf{d}_Z(f(x), O) + \mathsf{d}_Z(g(x), f(x)) \leq \delta,$$

which implies the claim.   □

Now, pick any $\varepsilon \geq \varepsilon_0$, any $U \in \mathfrak{U}_\varepsilon$, and fix $\delta' := \max(\delta, s)$. Then, there exists $W \in \mathcal{W}_\varepsilon$ such that $U \in \mathrm{cc}(f^{-1}(W))$, where $\mathrm{cc}(Y)$ stands for the set of path connected components of Y. Claim 9.3 implies that $f^{-1}(W) \subseteq g^{-1}(W^{\delta'})$. Since $\mathfrak{W}$ is a (c, s)-good cover of the connected space Z and $s \leq \max(\delta, s) \leq 2\delta' + \varepsilon$, there exists at least one set $W' \in \mathcal{W}_{c(2\delta'+\varepsilon)}$ such that $W^{\delta'} \subseteq W'$. This means that U is contained in some element of $\mathrm{cc}(g^{-1}(W'))$ where $W' \in \mathcal{W}_{c(2\delta'+\varepsilon)}$. But, also, since $c(2\delta'+\varepsilon) \leq c(2\delta'+1)\varepsilon$ for $\varepsilon \geq \varepsilon_0 \geq 1$, there exists $W'' \in \mathcal{W}_{c(2\delta'+1)\varepsilon}$ such that $W' \subseteq W''$. This implies that U is contained in some element of $\mathrm{cc}(g^{-1}(W''))$ where $W'' \in \mathcal{W}_{c(2\delta'+1)\varepsilon}$. This process, when applied to all $U \in \mathfrak{U}_\varepsilon$, for all $\varepsilon \geq \varepsilon_0$, defines a map of covers $\zeta_t : \mathfrak{U}_t \to \mathcal{V}_{(2c\delta'+c)\varepsilon}$. A similar observation produces for each $\varepsilon \geq \varepsilon_0$ a map of covers ξ_ε from $\mathcal{V}_\varepsilon$ to $\mathcal{V}_{(2c\delta'+c)\varepsilon}$.

So we have in fact proved that ε_0-truncations of $\mathfrak{U}$ and $\mathfrak{V}$ are multiplicatively η-interleaved, completing the proof of Proposition 9.21.   □

Applying Proposition 9.21, Proposition 9.18, and Corollary 4.4, we get the following result, where $\mathrm{Dgm}_{\log}$ stands for the persistence diagram at the log scale (of coordinates).

Corollary 9.22. *Let $\mathfrak{W}$ be a (c, s)-good cover tower of the compact connected metric space Z and let $f, g : X \to Z$ be any two well-behaved continuous functions such that $\max_{x \in X} \mathsf{d}_Z(f(x), g(x)) = \delta$. Then, the bottleneck distance between the persistence diagrams satisfies*

$$\mathsf{d}_b(\mathrm{Dgm}_{\log}\mathrm{MM}(\mathfrak{W}, f), \mathrm{Dgm}_{\log}\mathrm{MM}(\mathfrak{W}, g)) \leq \log(2c \max(s, \delta) + c)$$
$$+ \max(0, \log(1/s)).$$

PROOF. We use the notation of Proposition 9.21. Let $\mathfrak{U} = f^*(\mathfrak{W})$ and $\mathfrak{V} = g^*(\mathfrak{W})$. If $\max(1, s) = s$, then $\mathfrak{U}$ and $\mathfrak{V}$ are multiplicatively $(2c \max(s, \delta) + c)$-interleaved by Proposition 9.21, which gives a bound on the bottleneck distance of $\log(2c \max(s, \delta) + c)$ between the corresponding persistence diagrams at the log scale by Corollary 4.4. In the case when $s < 1$, the bottleneck distance remains the same only for the 1-truncations of $\mathfrak{U}$ and $\mathfrak{V}$. Shifting the starting point of the two families to the left by at most s can introduce barcodes of lengths at most $\log(1/s)$ or can stretch the existing barcodes to the left by at most $\log(1/s)$ for the respective persistence modules at the log scale. To see this, consider the persistence module below where $\varepsilon_1 = s$:

$$\mathsf{H}_k\big(N(f^*(\mathfrak{U}_{\varepsilon_1}))\big) \to \mathsf{H}_k\big(N(f^*(\mathfrak{U}_{\varepsilon_2}))\big) \to \cdots \to \mathsf{H}_k\big(N(f^*(\mathfrak{U}_1))\big) \to \cdots \to \mathsf{H}_k\big(N(f^*(\mathfrak{U}_{\varepsilon_n}))\big).$$

A homology class born at any index in the range $[s, 1)$ either dies at or before the index 1 or is mapped to a homology class of $\mathsf{H}_k\big(N(f^*(\mathfrak{U}_1))\big)$. In the first case, we have a barcode of length at most $|\log s| = \log(1/s)$ at the log scale. In the second case, a barcode of the persistence module

$$\mathsf{H}_k\big(N(f^*(\mathfrak{U}_{\varepsilon_1}))\big) \to \cdots \to \mathsf{H}_k\big(N(f^*(\mathfrak{U}_{\varepsilon_n}))\big)$$

starting at index 1 gets stretched to the left by at most $|\log s| = \log(1/s)$. The same conclusion can be drawn for the persistence module induced by $\mathfrak{V}$. Therefore the bottleneck distance between the respective persistence diagrams at log scale changes by at most $\log(1/s)$. $\square$

9.4.3 Relation to Intrinsic Čech Filtration

In Section 9.3.2, we have seen that given a tower of covers $\mathfrak{U}$ and a map $f \colon X \to Z$ there exists a natural pullback pseudometric d_f defined on the input domain X (Definition 9.6). With such a pseudometric on X, we can now construct the standard (intrinsic) Čech filtration $\mathcal{C}(X) = \{\mathbb{C}_\varepsilon(X)\}_\varepsilon$ (or Rips filtration) in X directly, instead of computing the nerve complex of the pullback covers as required by mapper. The resulting filtration $\mathcal{C}(X)$ is *connected by inclusion maps* instead of *simplicial maps*. This is easier for computational purposes even though one has a method to compute the persistence diagram of a tower involving arbitrary simplicial maps (Sections 4.2 and 4.4). Furthermore, it turns out that the resulting sequence of Čech complexes $\mathcal{C}$ interleaves with the sequence of complexes $\mathrm{MM}(\mathfrak{U}, f)$, implying that their corresponding persistence diagrams approximate each other. Specifically, in Theorem 9.23, we show that when the codomain of the function $f \colon X \to Z$ is a *metric space*

(Z, d_Z), the multiscale mapper induced by any (c, s)-good cover tower inter-leaves (at the homology level) with an intrinsic Čech filtration of X defined below. We have already considered Čech filtrations before (Section 6.1). How-ever, we considered only a finite subset of a metric space to define the Čech complex (Definition 2.9). Here we redefine it again to account for the fact that each point of the (pseudo)metric space is considered and call it the intrinsic Čech complex (see an earlier example of intrinsic Čech complex when we analyze graphs in Section 8.1.2).

Definition 9.10. (Intrinsic Čech complex) Given a (pseudo)metric space (Y, d_Y), its *intrinsic Čech complex* $\mathbb{C}^r(Y)$ at scale r is defined as the nerve complex of the set of intrinsic r-balls $\{B(y; r)\}_{y \in Y}$ defined using (pseudo) metric d_Y.

The above definition gives way to defining a Čech filtration.

Definition 9.11. (Intrinsic Čech filtration) The *intrinsic Čech filtration* of the (pseudo)metric space (Y, d_Y) is

$$\mathcal{C}(Y) = \{\mathbb{C}^r(Y) \hookrightarrow \mathbb{C}^{r'}(Y)\}_{0 < r < r'}.$$

The intrinsic Čech filtration at resolution s is defined as $\mathcal{C}_s(Y) = \{\mathbb{C}^r(Y) \hookrightarrow \mathbb{C}^{r'}(Y)\}_{s \le r < r'}$.

Recall the definition of the pseudometric d_f on X (Definition 9.6) induced from a metric on Z. Applying Definition 9.10 on the pseudometric space (X, d_f), we obtain its intrinsic Čech complex $\mathbb{C}^r(X)$ at scale r and then its Čech filtration $\mathcal{C}_s(X)$.

Theorem 9.23. *Let $\mathcal{C}_s(X)$ be the intrinsic Čech filtration of (X, d_f) starting with resolution s. Let $\mathfrak{U} = \{\mathfrak{U}_\varepsilon \xrightarrow{u_{\varepsilon, \varepsilon'}} \mathfrak{U}_{\varepsilon'}\}_{s \le \varepsilon \le \varepsilon'}$ be a (c, s)-good cover tower of the compact connected metric space Z. Then the multiscale mapper $\mathrm{MM}(\mathfrak{U}, f)$ and $\mathcal{C}_s(X)$ are multiplicatively $2c$-interleaved.*

By Corollary 4.4 on multiplicative interleaving, the following result is deduced immediately from Theorem 9.23.

Corollary 9.24. *Given a continuous map $f : X \to Z$ and a (c, s)-good cover tower $\mathfrak{U}$ of Z, let $\mathrm{Dgm}_{\log} \mathrm{MM}(\mathfrak{U}, f)$ and $\mathrm{Dgm}_{\log} \mathcal{C}_s$ denote the log-scaled per-sistence diagrams of the persistence modules induced by $\mathrm{MM}(\mathfrak{U}, f)$ and by the intrinsic Čech filtration $\mathcal{C}_s$ of (X, d_f), respectively. We have that*

$$\mathsf{d}_b(\mathrm{Dgm}_{\log} \mathrm{MM}(\mathfrak{U}, f), \mathrm{Dgm}_{\log} \mathcal{C}_s) \le 2c.$$

9.5 Exact Computation for PL-Functions on Simplicial Domains

The stability result in Theorem 9.23 further motivates us to design efficient algorithms for constructing multiscale mapper or its approximation in practice. *A priori*, the construction of the mapper and multiscale mapper may seem clumsy. Even for PL-functions defined on a simplicial complex, the standard algorithm needs to determine for each simplex the subset (partial simplex) on which the function value falls within a certain range. We observe that for such an input, it is sufficient to consider the restriction of the function to the 1-skeleton of the complex for computing the mapper and the multiscale mapper. Since the 1-skeleton (a graph) is typically much smaller in size than the full complex, this helps improving the time efficiency of computing the mapper and multiscale mapper.

Consider one of the most common types of input in practice, a real-valued PL-function $f : |K| \to \mathbb{R}$ defined on the underlying space $|K|$ of a simplicial complex K given as a vertex function. In what follows, we consider this PL-setting, and show that interestingly, if the input function satisfies a mild "minimum diameter" condition, then we can compute both mapper and multiscale mapper from simply the 1-skeleton (graph structure) of K. This makes the computation of the multiscale mapper from a PL-function significantly faster and simpler, as its time complexity depends on the size of the 1-skeleton of K, which is typically orders of magnitude smaller than the total number of simplices (such as triangles, tetrahedra, etc.) in K.

Recall that K^1 denotes the 1-skeleton of a simplicial complex K: that is, K^1 contains the set of vertices and edges of K. Define $\tilde{f} : |K^1| \to \mathbb{R}$ to be the restriction of f to $|K^1|$; that is, $\tilde{f}$ is the PL-function on $|K^1|$ induced by function values at vertices.

Condition 9.1. (Minimum diameter condition) For a cover tower $\mathfrak{W}$ of a compact connected metric space (Z, d_Z), let

$$\kappa(\mathfrak{W}) := \inf\{\mathrm{diam}(W); \ W \in \mathcal{W} \in \mathfrak{W}\}$$

denote the minimum diameter of any element of any cover of the tower $\mathfrak{W}$. Given a simplicial complex K with a function $f : |K| \to Z$ and a tower of covers $\mathfrak{W}$ of the metric space Z, we say that $(K, f, \mathfrak{W})$ satisfies the *minimum diameter* condition if $\mathrm{diam}(f(\sigma)) \leq \kappa(\mathfrak{W})$ for every simplex $\sigma \in K$.

In our case, f is a PL-function, and thus satisfying the minimum diameter condition means that for every edge $e = (u, v) \in K^1$, $|f(u) - f(v)| \leq \kappa(\mathfrak{W})$.

In what follows we assume that K is connected. We do not lose any generality by this assumption because the arguments below can be applied to each connected component of K.

Definition 9.12. (Isomorphic simplicial towers) Two simplicial towers $\mathcal{S} = \left\{ S_\varepsilon \xrightarrow{s_{\varepsilon,\varepsilon'}} S_{\varepsilon'} \right\}$ and $\mathcal{T} = \left\{ T_\varepsilon \xrightarrow{t_{\varepsilon,\varepsilon'}} T_{\varepsilon'} \right\}$ are *isomorphic*, denoted $\mathcal{S} \cong \mathcal{T}$, if $\mathrm{res}(\mathcal{S}) = \mathrm{res}(\mathcal{T})$, and there exist simplicial isomorphisms η_ε and $\eta_{\varepsilon'}$ such that the diagram below commutes for all $\mathrm{res}(\mathcal{S}) \leq \varepsilon \leq \varepsilon'$.

$$
\begin{array}{ccc}
S_\varepsilon & \xrightarrow{\;s_{\varepsilon,\varepsilon'}\;} & S_{\varepsilon'} \\
\eta_\varepsilon \, \big\| & & \big\| \, \eta_{\varepsilon'} \\
T_\varepsilon & \xrightarrow{\;t_{\varepsilon,\varepsilon'}\;} & T_{\varepsilon'}
\end{array}
$$

Our main result in this section is the following theorem which enables us to compute the mapper, multiscale mapper, as well as the persistence diagram for the multiscale mapper of a PL-function f from its restriction $\tilde{f}$ to the 1-skeleton of the respective simplicial complex.

Theorem 9.25. *Given a PL-function $f : |K| \to \mathbb{R}$ and a tower of covers $\mathfrak{W}$ of the image of f with $(K, f, \mathfrak{W})$ satisfying the minimum diameter condition, we have that* $\mathrm{MM}(\mathfrak{W}, f) \cong \mathrm{MM}(\mathfrak{W}, \tilde{f})$*, where $\tilde{f}$ is the restriction of f to $|K^1|$.*

We show in Proposition 9.26 that the two mapper outputs $\mathrm{M}(\mathcal{W}, f)$ and $\mathrm{M}(\mathcal{W}, \tilde{f})$ are identical up to a relabeling of their vertices (hence simplicially isomorphic) for every $\mathcal{W} \in \mathfrak{W}$. Also, since the simplicial maps in the filtrations $\mathrm{MM}(\mathfrak{W}, f)$ and $\mathrm{MM}(\mathfrak{W}, \tilde{f})$ are induced by the pullback of the same tower of covers $\mathfrak{W}$, they are identical again up to the same relabeling of the vertices. This then establishes the theorem.

In what follows, for clarity of exposition, we use X and X^1 to denote the underlying space $|K|$ and $|K^1|$ of K and K^1, respectively. Also, we do not distinguish between a simplex $\sigma \in K$ and its image $|\sigma| \subseteq X$ and thus freely say $\sigma \subseteq X$ when it actually means that $|\sigma| \subseteq X$ for a simplex $\sigma \in K$.

Proposition 9.26. *If $(K, f, \mathfrak{W})$ satisfies the minimum diameter condition, then for every $\mathcal{W} \in \mathfrak{W}$, $\mathrm{M}(\mathcal{W}, f)$ is identical to $\mathrm{M}(\mathcal{W}, \tilde{f})$ up to relabeling of the vertices.*

PROOF. Let $\mathcal{U} = f^*\mathcal{W}$ and $\tilde{\mathcal{U}} = \tilde{f}^*\mathcal{W}$. By definition of $\tilde{f}$, each $\tilde{U} \in \tilde{\mathcal{U}}$ is a connected component of some $U \cap X^1$ for some $U \in \mathcal{U}$. In Proposition 9.27, we show that $U \cap X^1$ is connected for every $U \in \mathcal{U}$. Therefore, for every element $U \in \mathcal{U}$, there is a unique element $\tilde{U} = U \cap X^1$ in $\tilde{\mathcal{U}}$ and vice versa. It is not hard to show that $\bigcap_{i=1}^{k} U_i \neq \varnothing$ if and only if $\bigcap_{i=1}^{k} \tilde{U}_i \neq \varnothing$. This finishes the proof.   □

Proposition 9.27. *If $(X, f, \mathfrak{W})$ satisfies the minimum diameter condition, then for every $W \in \mathfrak{W}$ and every $U \in f^*(W)$, the set $U \cap X^1$ is connected.*

PROOF. Fix $U \in f^*(W)$. If $U \cap X^1$ is not connected, let $C_1, \ldots, C_k$ denote its $k \geq 2$ connected components. First, we show that each C_i contains at least one vertex of X^1. Let $e = (u, v)$ be any edge of X^1 that intersects U. If both ends u and v lie outside U, then $|f(u) - f(v)| > |\max_U f - \min_U f| \geq \kappa(\mathfrak{W})$. But this violates the minimum diameter condition. Thus, at least one vertex of e is contained in U. It immediately follows that C_i contains at least one vertex of X^1.

Let Δ be the set of all simplices $\sigma \subseteq X$ so that $\sigma \cap U \neq \varnothing$. Fix $\sigma \in \Delta$ and let x be any point in $\sigma \cap U$. We defer the proof of the following claim as an exercise.

Claim 9.4. *There exists a point y in an edge of σ so that $f(x) = f(y)$.*

Since σ contains an edge e that is intersected by U, it contains a vertex of e that is contained in U. This means every simplex $\sigma \in \Delta$ has a vertex contained in U. For each $i = 1, \ldots, k$ let $\Delta_i := \{\sigma \subseteq X \mid \mathrm{V}(\sigma) \cap C_i \neq \varnothing\}$. Since every simplex $\sigma \in \Delta$ has a vertex contained in U, we have $\Delta = \bigcup_i \Delta_i$. We argue that the sets $\Delta_1, \ldots, \Delta_k$ are disjoint from each other. Otherwise, there exist $i \neq j$ and a simplex σ with a vertex u in Δ_i and another vertex v in Δ_j. Then, the edge (u, v) must be in U because f is PL. But this contradicts that C_i and C_j are disjoint. This establishes that each Δ_i is disjoint from each other and hence Δ is not connected, contradicting that U is connected. Therefore, our initial assumption that $U \cap X^1$ is disconnected is wrong.   □

9.6 Approximating Multiscale Mapper for General Maps

While the results in the previous section concern real-valued PL-functions, we now provide a significant generalization for the case where f maps the underlying space of K into an arbitrary compact metric space Z. We present a "combinatorial" version of the (multiscale) mapper where each connected component of a pullback $f^{-1}(W)$ for any cover W in the cover of Z consists of only vertices of K. Hence, the construction of the nerve complex for this modified (multiscale) mapper is purely combinatorial, simpler, and more efficient to implement. But we lose the "exactness," that is, in contrast with the guarantees provided by Theorem 9.25, the combinatorial mapper only *approximates* the actual multiscale mapper at the homology level. Also, it requires a (c, s)-good tower of covers of Z. One more caveat is that the towers of simplicial complexes arising in this case interleave not in the (strong) sense of Definition 4.4 but in a weaker sense (Definition 6.6). This limitation worsens the approximation result by a factor of 3.

In what follows, as before, $cc(O)$ for a set O denotes the set of all path connected components of O.

Given a map $f: |K| \to Z$ defined on the underlying space $|K|$ of a simplicial complex K, to construct the mapper and multiscale mapper, one needs to compute the pullback cover $f^*(\mathcal{W})$ for a cover $\mathcal{W}$ of the compact metric space Z. Specifically, for any $W \in \mathcal{W}$ one needs to compute the pre-image $f^{-1}(W) \subset |K|$ and shatter it into connected components. Even in the setting adopted in Section 9.5, where we have a PL-function $\tilde{f}: |K^1| \to \mathbb{R}$ defined on the 1-skeleton K^1 of K, the connected components in $cc(\tilde{f}^{-1}(W))$ may contain vertices, edges, and also *partial edges*: say, for an edge $e \in K^1$, its intersection $e_W = e \cap f^{-1}(W) \subseteq e$, that is, $f(e_W) = f(e) \cap W$, is a partial edge. See Figure 9.10 for an example. In general for more complex maps, $\sigma \cap f^{-1}(W)$ for any k-simplex σ may be partial triangles, tetrahedra, etc., which can be a nuisance for computations. The combinatorial version of mapper and multiscale mapper sidesteps this problem by ensuring that each connected component in the pullback $f^{-1}(W)$ consists of *only vertices of K*. It is thus simpler and faster to compute.

9.6.1 Combinatorial Mapper and Multiscale Mapper

Let G be a graph with vertex set $V(G)$ and edge set $E(G)$. Suppose we are given a map $f: V(G) \to Z$ and a finite open cover $\mathcal{W} = \{W_\alpha\}_{\alpha \in A}$ of the metric space (Z, d_Z). For any $W_\alpha \in \mathcal{W}$, the pre-image $f^{-1}(W_\alpha)$ consists of a set of vertices which is shattered into subsets by the connectivity of the graph G. These subsets are taken as connected components. We now formalize this:

Definition 9.13. (*G*-induced connected component) Given a set of vertices $O \subseteq V(G)$, *the set of connected components of O induced by G*, denoted by $cc_G(O)$, is the partition of O into a maximal subset of vertices connected in $G_O \subseteq G$, the subgraph spanned by vertices in O. We refer to each such maximal subset of vertices as a *G-induced connected component of O*. We

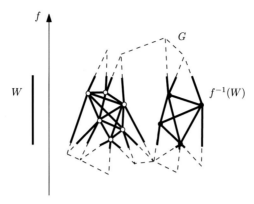

Figure 9.10 Partial thickened edges belong to the two connected components in $f^{-1}(W)$. Note that each set in $\mathrm{cc}_G(f^{-1}(W))$ contains only the set of vertices of a component in $\mathrm{cc}(f^{-1}(W))$.

define $f^{*G}(\mathcal{W})$, the *G-induced pullback via the function* f, as the collection of all G-induced connected components $\mathrm{cc}_G(f^{-1}(W_\alpha))$ for all $\alpha \in A$.

Definition 9.14. (*G*-induced multiscale mapper) Similar to the mapper construction, we define the *G-induced mapper* $\mathrm{M}_G(\mathcal{W}, f)$ as the nerve complex $N(f_G^*(\mathcal{W}))$.

Given a tower of covers $\mathfrak{W} = \{\mathcal{W}_\varepsilon\}$ of Z, we define the *G-induced multiscale mapper* $\mathrm{MM}_G(\mathfrak{W}, f)$ as the tower of G-induced nerve complexes $\{N(f_G^*(\mathcal{W}_\varepsilon)) \mid \mathcal{W}_\varepsilon \in \mathfrak{W}\}$.

Given a map $f : |K| \to Z$ defined on the underlying space $|K|$ of a simplicial complex K, let $f_V : \mathrm{V}(K) \to \mathbb{R}$ denote the restriction of f to the vertices of K. Consider the 1-skeleton graph K^1 that provides the connectivity information for vertices in $\mathrm{V}(K)$. Given any cover tower $\mathfrak{W}$ of the metric space Z, the K^1-induced multiscale mapper $\mathrm{MM}_{K^1}(\mathfrak{W}, f_V)$ is called the *combinatorial multiscale mapper of f with respect to* $\mathfrak{W}$.

9.6.2 Advantage of Combinatorial Multiscale Mapper

A simple description of the computation of the combinatorial mapper is given in Algorithm 17. For the simple PL example in Figure 9.10, $f^{-1}(W)$ contains two connected components: one consists of the set of white dots, while the other consists of the set of black dots. More generally, the construction of the pullback cover needs to inspect only the 1-skeleton K^1 of K, which is typically of significantly smaller size. Furthermore, the construction of the

---

Algorithm 17 MMAPPER(f, K, $\mathfrak{W}$)

Input:

$f : |K| \to Z$ given by $f_V : V(K) \to Z$, a cover tower $\mathfrak{W} = \{W_1, \ldots, W_t\}$

Output:

Persistence diagram $\mathrm{Dgm}_*(\mathrm{MM}_{K^1}(\mathfrak{W}, f_V))$ induced by the combinatorial MM of f with respect to $\mathfrak{W}$

1: **for** $i = 1, \ldots, t$ **do**
2:    Compute $V_W \subseteq V(K)$ where $f(V_W) = f(V(K)) \cap W$ and $\{V_W^j\}_j = \mathrm{cc}_{K^1}(V_W)$, for all $W \in W_i$
3:    Compute nerve complex $N_i = N(\{V_W^j\}_{j,W})$
4: **end for**
5: Compute the filtration $\mathcal{F} : \{N_i \to N_{i+1}, i \in [1, t-1]\}$
6: Compute $\mathrm{Dgm}_*(\mathcal{F})$

---

nerve complex N_i as in Algorithm 17 is also much simpler: We simply remember, for each vertex $v \in V(K)$, the set I_v of id's of connected components $\{V_W^j\}_{j, W \in W_i}$ which contain it. Any subset of I_v gives rise to a simplex in the nerve complex N_i.

Let $\mathrm{MM}(\mathfrak{W}, f)$ denote the standard multiscale mapper as introduced in Section 9.3.1. Our main result in this section is that if $\mathfrak{W}$ is a (c, s)-good cover tower of Z, then the resulting two simplicial towers, $\mathrm{MM}(\mathfrak{W}, f)$ and $\mathrm{MM}_{K^1}(\mathfrak{W}, f_V)$, weakly interleave (Definition 6.6), and admit a bounded distance between their respective persistence diagrams as a consequence of the weak interleaving result of [77]. This weaker setting of interleaving worsens the approximation by a factor of 3.

Theorem 9.28. *Assume that (Z, d_Z) is a compact and connected metric space. Given a map $f : |K| \to Z$, let $f_V : V(K) \to Z$ be the restriction of f to the vertex set $V(K)$ of K.*

Given a (c, s)-good cover tower $\mathfrak{W}$ of Z such that $(K, f, \mathfrak{W})$ satisfies the minimum diameter condition (cf. Condition 9.1), the bottleneck distance between the persistence diagrams $\mathrm{Dgm}_{\log} \mathrm{MM}(\mathfrak{W}, f)$ and $\mathrm{Dgm}_{\log} \mathrm{MM}_{K^1}(\mathfrak{W}, f_V)$ is at most $3 \log(3c) + 3 \max(0, \log(1/s))$ for all $k \in \mathbb{N}$.

9.7 Notes and Exercises

A corollary of the nerve theorem is that the space and the nerve have isomorphic homology groups if all intersections of cover elements are homotopically trivial. This chapter studies a case when covers do not necessarily satisfy this property. The result that for path connected covers, no new one-dimensional homology class is created in the nerve is proved in [132]. The materials in Sections 9.1 and 9.2 are taken from there. This result can be generalized for other dimensions; see Exercise 5.

The concept of mapper was introduced by Singh, Mémoli, and Carlsson [277], and has since been used in diverse applications; see, for example, [225, 228, 245, 288]. The authors of [277] showed for the first time that a cover for the codomain in addition to domains can be useful for data analysis. The mapper in some sense is connected to Reeb graphs (spaces) where the cover elements degenerate to points in the codomain; see [241], for example. The structure and stability of one-dimensional mapper is studied in great detail by Carrière and Oudot in [69]. They showed that given a real-valued function $f : X \to \mathbb{R}$ and an appropriate cover $\mathcal{U}$, the extended persistence diagram of a mapper $M(\mathcal{U}, f)$ is a subset of the same of the Reeb graph R_f. Furthermore, they characterized the features of the Reeb graph that may disappear from the mapper. The mapper (for a real-valued function f) can also be viewed as a Reeb graph $\mathsf{R}_{f'}$ of a perturbed function $f' : X' \to \mathbb{R}$. It is shown in [69] how one can track the changes between R_f and the mapper by computing the functional distortion distance (Definition 7.8) between R_f and $\mathsf{R}_{f'}$. In [15], the author established a convergence result between mapper for a real-valued f and the Reeb graph R_f. Specifically, the mapper is characterized with a zigzag persistence module that is a coarsening of the zigzag persistence module for R_f. It is shown that the mapper converges to R_f in the bottleneck distance of the corresponding zigzag persistence diagrams as the lengths of the intervals in the cover approach zero. Munch and Wang [241] showed a similar convergence in interleaving distance (Definition 7.6) using sheaf theory [111].

The multiscale mapper which works on the notion of a filtration of covers was developed in [131]. Most of the materials in this chapter are taken from this paper. The results on the class of 1-cycles that persist through multiscale mapper are taken from [132].

Exercises

1. For a simplicial complex K, simplices with no cofacet are called maximal simplices. Consider a closed cover of $|K|$ with the closures of the maximal simplices as the cover elements. Let $N(K)$ denote the nerve of this cover. Prove that $N(N(K))$ is isomorphic to a subcomplex of K.

2. (See [17].) A vertex v in K is said to be dominated by a vertex v' if every maximal simplex containing v also contains v' [17]. We say K collapses strongly to a complex L if L is obtained by a series of deletions of

dominated vertices with all their incident simplices. Show that K strongly collapses to $N(N(K))$.

3. We say a cover $\mathcal{U}$ of a metric space (Y, d) is an (α, β)-cover if $\alpha \leq \lambda(\mathcal{U})$ and $\beta \geq s_{max}(\mathcal{U})$.
   (a) Consider a δ-sample P of Y, that is, every metric ball $B(y; \delta)$, $y \in Y$, contains a point in P. Prove that the cover $\mathcal{U} = \{B(p; 2\delta)\}_{p \in P}$ is a $(\delta, 4\delta)$-cover of Y.
   (b) Prove that the infinite cover $\mathcal{U} = \{B(y; \delta)\}_{y \in Y}$ is a $(\delta, 2\delta)$-cover of Y.

4. Theorem 9.5 requires that the cover be path connected. Show that this condition is necessary by presenting a counterexample otherwise.

5. One may generalize Theorem 9.5 as follows: If for any $k \geq 0$, t-wise intersections of cover elements for all $t > 0$ have trivial reduced homology for H_{k-t}, then the nerve map induces a surjection in H_k. Prove or disprove it.

6. Consider a function $f : X \to Z$ from a path connected space X to a metric space Z. Define the equivalence relation $\sim_f$ such that $x \sim_f x'$ holds if and only if $f(x) = f(x')$ and there exists a continuous path $\gamma \in \Gamma_X(x, x')$ such that $f \circ \gamma$ is constant. The Reeb space R_f is the quotient of X under this equivalence relation.
   (a) Prove that the quotient map $q : X \to \mathsf{R}_f$ is surjective and also induces a surjection $q_* : \mathsf{H}_1(X) \to \mathsf{H}_1(\mathsf{R}_f)$.
   (b) Call a class $[c] \in \mathsf{H}_1(X)$ vertical if and only if there is no $c' \in \mathsf{C}_1(X)$ so that $[c] = [c']$ and $f \circ \sigma$ is constant for every $\sigma \in c'$. Show that $q_*([c]) \neq 0$ if and only if c is vertical.
   (c) Let $z_1, \ldots, z_g$ be an optimal cycle basis (Definition 9.3) of $\mathsf{H}_1(X)$ defined with respect to the pseudometric d_f (Definition 9.6). Let $\ell \in [1, g]$ be the smallest integer so that $s(z_\ell) \neq 0$. Prove that if no such ℓ exists, $\mathsf{H}_1(\mathsf{R}_f)$ is trivial, otherwise $\{[q(z_i)]\}_{i=\ell,\ldots,g}$ is a basis for $\mathsf{H}_1(\mathsf{R}_f)$.

7. Let us endow R_f with a distance $\tilde{\mathsf{d}}_f$ that descends via the map q: for any equivalence classes $r, r' \in \mathsf{R}_f$, pick $x, x' \in X$ with $r = q(x)$ and $r' = q(x')$, then define

$$\tilde{\mathsf{d}}_f(r, r') := \mathsf{d}_f(x, x').$$

   Prove that $\tilde{\mathsf{d}}_f$ is a pseudometric.

8. Prove Proposition 9.17.

9. Prove Theorem 9.23.

10. Prove Claim 9.4.

10

Discrete Morse Theory and Applications

Discrete Morse theory is a combinatorial version of the classical Morse theory. Invented by Forman [161], the theory combines topology with the combinatorial structure of a cell complex. Specifically, much like the fact that critical points of a smooth Morse function on a manifold determine its topological entities such as homology groups and Euler characteristics, an analogous concept called critical simplices of a discrete Morse function also determine similar structures for the complex it is defined on. Gradient vectors associated with smooth Morse functions give rise to integral lines and eventually the notion of stable and unstable manifolds [233]. Similarly, a discrete Morse function defines discrete gradient vectors leading to V-paths analogous to the integral lines. Using these V-paths, one can define the analogues of stable and unstable manifolds of the critical simplices.

It turns out that an acyclic pairing between simplices and their faces so that every simplex participates in at most one pair provides a discrete Morse function and conversely a discrete Morse function defines such a pairing. This pairing, termed a Morse matching, is a main building block of the discrete Morse theory. In this chapter, we connect this matching with the pairing obtained through the persistence algorithm. Specifically, we present an algorithm for computing a Morse matching and hence a discrete Morse vector field by connecting persistence pairs through V-paths. This requires an operation called critical pair cancellation which may not succeed all the time. However, for 1-complexes and simplicial 2-manifolds (pseudo-manifolds), it always succeeds. Sections 10.1 and 10.2 are devoted to these results.

In Section 10.4, we apply our persistence-based discrete Morse vector field to reconstruct geometric graphs from their noisy samples. Here we show that unstable manifolds of critical edges can recover a graph with guarantees from density data that captures the hidden graph reasonably well. We provide two

289

applications of using this graph reconstruction algorithm, one for road network reconstructions from Global Positioning System (GPS) trajectories and satellite images, and another for neuron reconstructions from their images. Section 10.5 describes these applications.

10.1 Discrete Morse Function

Following Forman [161] we define a discrete Morse function (henceforth called Morse function in this chapter).

Definition 10.1. (Discrete Morse function) A function $f : K \to \mathbb{R}$ on a simplicial complex K is called a *discrete Morse function* (*Morse function*) if for every p-simplex $\sigma^p \in K$ the following two conditions hold.[1] Recall that every $(p-1)$-face of σ^p is called its facet and every $(p+1)$-simplex adjoining σ^p is called its cofacet. The conditions are:

- $\#\{\sigma^{p-1} \mid \sigma^{p-1} \text{ is a facet of } \sigma^p \text{ and } f(\sigma^{p-1}) \geq f(\sigma^p)\} \leq 1$;
- $\#\{\sigma^{p+1} \mid \sigma^{p+1} \text{ is a coface of } \sigma^p \text{ and } f(\sigma^{p+1}) \leq f(\sigma^p)\} \leq 1$.

The first condition says that at most one facet of a simplex σ has function value higher than or equal to $f(\sigma)$; and the second condition says that at most one cofacet of a simplex σ can have function value lower than or equal to $f(\sigma)$. By a result of Chari [75], the two conditions imply that the two sets above are disjoint, that is, if a pair (σ^{p-1}, σ^p) satisfies the first condition, there is no pair (σ^p, σ^{p+1}) satisfying the second condition, and vice versa. This means that a Morse function f induces a matching.

Definition 10.2. (Matching) A set of ordered pairs $M = \{(\sigma, \tau)\}$ is a *matching* in K if the following conditions hold:

- For any $(\sigma, \tau) \in M$, σ is a facet of τ.
- Any simplex in K can appear in at most one pair in M.

Such a matching M defines two disjoint subsets $L \subseteq K$ and $U \subseteq K$ where there is a bijection $\mu : L \to U$ such that $M = \{(\sigma, \mu(\sigma)) \mid \sigma \in L\}$.

In Figure 10.1, we indicate a matching by putting an arrow from the lower-dimensional simplex to the higher-dimensional simplex. Observe that the source of each arrow is a facet of the target of the arrow.

---

[1] Forman formulated the discrete Morse function for more general cell complexes.

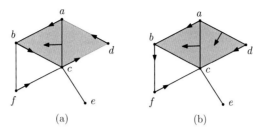

Figure 10.1 Two DMVFs: (a) the matching is not Morse because the sequence $a \prec ab \prec b \prec bc \prec c \prec cd \prec d \prec da$ is cyclic; (b) the matching is Morse, and there is no cyclic sequence.

Note, however, that the matching in K defined by a Morse function has an additional property of acyclicity which we show next. First, let us define a relation $\sigma_i \prec \sigma_{i+1}$ if $\sigma_{i+1} = \mu(\sigma_i)$ or σ_{i+1} is a facet of σ_i but $\sigma_i \neq \mu(\sigma_{i+1})$.

Definition 10.3. (*V*-path and Morse matching) Given a matching M in K, for $k > 0$, a *V-path* π is a sequence

$$\pi : \sigma_0 \prec \sigma_1 \prec \cdots \prec \sigma_{i-1} \prec \sigma_i \prec \sigma_{i+1} \prec \cdots \prec \sigma_k, \qquad (10.1)$$

where for $0 < i < k$, $\sigma_i \neq \mu(\sigma_{i-1})$ implies $\sigma_{i+1} = \mu(\sigma_i)$. In other words, a *V*-path is an alternating sequence of facets and cofacets thus alternating in dimensions where every consecutive pair also alternates between matched and unmatched pairs. A *V*-path is *cyclic* if the first simplex σ_0 is a facet of the last simplex σ_k or $\sigma_0 = \mu(\sigma_k)$ and the matching M is called *cyclic* if there is such a path in it. Otherwise, M is called *acyclic*. An acyclic matching in K is called a *Morse matching*.

In Figure 10.1(a), the matching indicated by the arrows is not a Morse matching whereas the matching in Figure 10.1(b) is a Morse matching. Observe that in a sequence like (10.1), the function values on facets of the matched pairs strictly decrease. This observation leads to the following fact.

Fact 10.1. *The matching induced by a Morse function on K is acyclic, thus is a Morse matching.*

We also have the following relation in the opposite direction.

Fact 10.2. *A Morse matching M in K defines a Morse function on K.*

PROOF. First, order those simplices which are in some pair of M. A simplex σ^{p-1} is ordered before σ^p if $(\sigma^{p-1}, \sigma^p) \in M$ and it is ordered after σ^p if it is a facet of σ^p but $(\sigma^{p-1}, \sigma^p) \notin M$. Such an ordering is possible because M is acyclic. Then, simply order the rest of the simplices not in any pair of M according to their increasing dimensions. Assign the order numbers as the function values of the simplices, which can easily be verified to satisfy the two conditions of a (discrete) Morse function on K in Definition 10.1.   □

Since a given Morse matching M in K can be associated with a Morse function f on K, we call the simplices not covered by M the *critical simplices* of f. Let $c_i = c_i(M)$ denote the number of i-dimensional critical simplices. Recall that $\beta_i = \beta_i(K)$ denotes the i-th Betti number, the dimension of the homology group $\mathsf{H}_i(K)$. Assume that $c_i, \beta_i = 0$ for $i > p$ where K is p-dimensional. The following result is due to Forman [161]. It is analogous to Theorem 1.5 for a smooth Morse function in the smooth setting.

Proposition 10.1. *Given a Morse function f on K with its induced Morse matching M, let the c_i and β_i be defined as above. We have:*

(a) weak Morse inequality
  (i) $c_i \geq \beta_i$ for all $i \geq 0$,
  (ii) $c_p - c_{p-1} + \cdots \pm c_0 = \beta_p - \beta_{p-1} + \cdots \pm \beta_0$ where K is p-dimensional;
(b) strong Morse inequality
$$c_i - c_{i-1} + c_{i-2} - \cdots \pm c_0 \geq \beta_i - \beta_{i-1} + \beta_{i-2} - \cdots \pm \beta_0 \text{ for all } i \geq 0.$$

The weak Morse inequality can be derived from the strong Morse inequality (Exercise 7).

10.1.1 Discrete Morse Vector Field

Morse matchings can be interpreted naturally as a discrete counterpart of a vector field.

Definition 10.4. (Discrete Morse vector field) A *discrete Morse vector field* (DMVF) V in a simplicial complex K is a partition $V = C \sqcup L \sqcup U$ of K where L is the set of facets paired with a unique cofacet in U in a Morse matching M giving $\mu(L) = U$ and C is the set of unpaired simplices called *critical simplices*. We also say that V is induced by matching M in this case.

We interpret each pair $(\sigma, \tau = \mu(\sigma))$ as a vector originating at σ and terminating at τ and draw the vector by an arrow with tail in σ and head in τ; see Figures 10.1 and 10.2. The critical simplices are treated as critical points of the vector field, justifying their names. The vertex e and edge ce in both

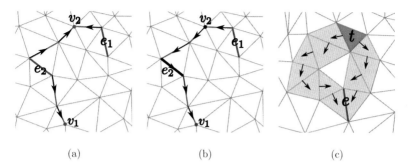

(a)   (b)   (c)

Figure 10.2 Critical vertices and edges are marked red. (a) Before cancellation of edge–vertex pair (v_2, e_2). (b) After cancellation, the path from e_2 to v_2 is inverted, giving rise to a critical V-path from e_1 to v_1, making (v_1, e_1) now potentially cancellable. (c) The edge–triangle pair (e, t), if cancelled, creates a cycle, as there are two V-paths between them.

Figure 10.1(a) and (b) are critical whereas the vertex c is critical only in Figure 10.1(b) and the edge bf is only critical in Figure 10.1(a).

In analogy to the integral lines for smooth vector fields, we define the so-called critical V-paths for discrete Morse vector fields.

Definition 10.5. (Critical V-path) Given a DMVF $V = C \sqcup L \sqcup U$ induced by a matching M, a V-path $\pi : \sigma_0 \prec \sigma_1 \prec \cdots \prec \sigma_{i-1} \prec \sigma_i \prec \sigma_{i+1} \prec \cdots \prec \sigma_k$ is critical in M if both σ_0 and σ_k are critical.

Observe that σ_0 and σ_k in the above definition are necessarily a p- and $(p-1)$-simplex, respectively, if the V-path alternates between p- and $(p-1)$-simplices. The V-path corresponding to a critical V-path cannot be cyclic due to this observation. The critical triangle cda with any of its edges in Figure 10.1(a) forms a noncritical V-path whereas the pair $ce \prec e$ forms a critical V-path in Figure 10.1(b).

In a critical V-path π, the pairs $(\sigma_1, \sigma_2), \ldots, (\sigma_{2i-1}, \sigma_{2i}), \ldots, (\sigma_{k-2}, \sigma_{k-1})$ are matched. We can cancel the pairs of critical simplices (σ_0, σ_k) by reversing the matched pairs.

Definition 10.6. (Cancellation) Let (σ_0, σ_k) be a pair of critical simplices with a critical V-path $\pi : \sigma_0 \prec \sigma_1 \prec \cdots \prec \sigma_{i-1} \prec \sigma_i \prec \sigma_{i+1} \prec \cdots \prec \sigma_k$. The pair (σ_0, σ_k) is *cancelled* if one modifies the matching by shifting the matched pairs by one position, that is, by asserting that the pairs $(\sigma_k, \sigma_{k-1}), \ldots, (\sigma_{i+1}, \sigma_i), \ldots, (\sigma_1, \sigma_0)$ are matched instead – we refer to this as the *(Morse) cancellation* on (σ_0, σ_k). Observe that a cancellation essentially reverses the vectors in the V-path π and additionally converts critical simplices

σ_0 and σ_k to be noncritical; see Figure 10.2. We say that the pair (σ_0, σ_k) is *(Morse) cancellable* if there exists a *unique V*-path between them.

Observe that a cancellation preserves the property of matching, that is, the new pairs together with the undisturbed pairs indeed form a matching. *Uniqueness* of the critical *V*-path connecting a pair of critical simplices ensures that the resulting new matching remains Morse. If there is more than one such critical *V*-path, the new matching may become cyclic – for example, in Figure 10.2(c), the cancellation of one critical *V*-path between the triangle–edge pair creates a cyclic *V*-path. The uniqueness of critical *V*-path is sufficient to ensure that such cyclic matching cannot be produced. In particular, we have the following result.

Proposition 10.2. *Given a Morse matching M, suppose we cancel a pair of critical simplices σ and σ' in a DMVF V via a critical V-path to obtain a new matching M'. Then M' remains a Morse matching if and only if this V-path is the only critical V-path connecting σ and σ' in V (i.e., the pair (σ, σ') is cancellable as per Definition 10.6).*

PROOF.   First, assume that there are two *V*-paths π and π' originating at σ and ending at σ'. Since π and π' are distinct and have common simplices σ at the beginning and σ' at the end, there are simplices τ and τ' where the two paths differ for the first time after τ and join again for the first time at τ'. Reversing one *V*-path, say π, creates a *V*-path from τ' to τ. This subpath along with the *V*-path from τ to τ' on π' creates a cyclic *V*-path, thus proving the "only if" part.

Next, suppose that there is only a single *V*-path from σ to σ'. After reversing this path, we claim that no cyclic *V*-path is created. For contradiction, assume that a cyclic *V*-path is created as the result of reversal of π. Let the maximal subpath of the reversed π on this cyclic path start at τ and end at τ'. We have $\tau \neq \tau'$ because otherwise the original matching needs to be cyclic in the first place. But then the cyclic *V*-path has a subpath from τ' to τ that is not in π. Since the reversed *V*-path π has a subpath from τ to τ', the original path has a subpath from τ' to τ. This means that the DMVF *V* originally had two *V*-paths from σ to σ', with one of them being π and the other one containing a subpath not in π. This forms a contradiction that there is a single *V*-path from σ to σ'. Hence the assumption that a cyclic *V*-path is created is wrong, which completes the proof of the "if" part.   □

10.2 Persistence-Based Discrete Morse Vector Fields

Given a simplicial complex K, one can set up a trivial DMVF where every simplex is critical, that is, $V = K \sqcup \varnothing \sqcup \varnothing$. Then, one may use cancellations to build the vector field further by constructing more matchings. The key to the success of this approach is to identify pairs of critical simplices that can be cancelled without creating cyclic paths. One way to do this is by taking advantage of persistence pairs among simplices.

10.2.1 Persistence-Guided Cancellation

First, we consider the case of simplicial 1-complexes which consist of only vertices and edges. Such a complex admits a DMVF obtained by cancelling the persistence pairs successively. Here we consider pairs with finite persistence only. Recall that some of the creator simplices are never paired with a destructor because the class created by them never dies. They are paired with ∞. Such *essential pairs* are not considered in the following proposition.

Proposition 10.3. *Let $(v_1, e_1), (v_2, e_2), \ldots, (v_n, e_n)$ be the sequence of all non-essential persistence pairs of vertices and edges sorted in increasing order of the appearance of the edges e_i in a filtration of a 1-complex K. Let V_0 be the DMVF in K with all simplices being critical. Suppose DMVF V_{i-1} can be obtained by cancelling successively $(v_1, e_1), (v_2, e_2), \ldots, (v_{i-1}, e_{i-1})$. Then, (v_i, e_i) can be cancelled in V_{i-1} providing a DMVF V_i for all $i \geq 1$.*

PROOF. Inductively assume that (i) V_{i-1} is a DMVF obtained as claimed in the proposition and (ii) any matched edge in V_{i-1} is a paired edge in a persistence pair. We argue that these two hypotheses hold for V_i, proving the claim due to the hypothesis (i).

The base case for $i = 1$ is true trivially because V_0 is a DMVF and there is no matched edge. Inductively assume that V_{i-1} satisfies the inductive hypothesis for $i > 1$. Consider the persistence pair (v_i, e_i). First, we observe that a V-path $e_i = e_{i_1} \prec v_{i_1} \prec \cdots \prec e_{i_n} \prec v_{i_n} = v_i$ exists in V_{i-1}. If not, starting from the two endpoints of e_i, we attempt to follow the two V-paths and let $v, v' \neq v_i$ be the first two critical vertices encountered during this construction. Without loss of generality, assume that v' appears before v in the filtration. Then, the zero-dimensional class $[v + v']$ is born when v is introduced. It is destroyed by e_i. It follows that (v, e_i) is a persistence pair (Fact 3.3) contradicting that actually (v_i, e_i) is a persistence pair. For induction, consider the V-path $e_i = e_{i_1} \prec v_{i_1} \prec \cdots \prec e_{i_n} \prec v_{i_n} = v_i$ in V_{i-1} which is cancelled to create

V_i. For V_i not to be a DMVF, due to Proposition 10.2, we must have another distinct V-path from e_i to v_i in V_{i-1}, $e_i = e_{j_1} \prec v_{j_1} \prec \cdots \prec e_{j_{n'}} \prec v_{i_{n'}} = v_i$. These two non-identical paths form a 1-cycle. Every edge in this cycle except possibly e_i is a matched edge in V_{i-1} and hence participates in a persistence pair by the inductive hypothesis. Then, all edges in the 1-cycle participate in some persistence pair because e_i is also such an edge by assumption. But this is impossible because in any 1-cycle at least one edge has to remain unpaired in persistence. It follows that by cancelling (v_i, e_i), we obtain a DMVF V_i satisfying the inductive hypothesis (i). Also, inductive hypothesis (ii) follows because the new matched pairs in V_i involve edges that were already matched in V_{i-1} and the edge e_i that participates in a persistence pair by assumption. □

The result above holds for vertex–edge pairing in any simplicial complex. Furthermore, using dual graphs, it can be used for edge–triangle pairing in triangulations of 2-manifolds. Given a simplicial 2-complex K whose underlying space is a 2-manifold without boundary, consider the dual graph (1-complex) K^* where each triangle $t \in K$ becomes a vertex $t^* \in K^*$ and two vertices t_1^* and t_2^* are joined with an edge e^* if triangles t_1 and t_2 share an edge e in K.

The following result connects the persistence of a filtration of K and its dual graph K^*.

Proposition 10.4. *Let $\sigma_1, \sigma_2, \ldots, \sigma_n$ be a subsequence of a simplex-wise filtration $\mathcal{F}$ of K consisting of all edges and triangles. An edge–triangle pair (σ_i, σ_j) is a persistence pair for $\mathcal{F}$ if and only if (σ_j^*, σ_i^*) is a persistence pair for the filtration $\sigma_n^*, \sigma_{n-1}^*, \ldots, \sigma_1^*$ of the dual graph K^*.*

PROOF. Recall Proposition 3.8. An edge–triangle persistence pair produced by the filtered boundary matrix D_2 for filtration of K is exactly the same as the triangle–edge persistence pair obtained from the twisted (transposed and reversed) matrix D_2^* by left-to-right column additions. The matrix D_2^* is exactly the filtered boundary matrix of a filtration $\mathcal{F}(K^*)$ of K^* that reverses the subsequence of triangle and edges. Dualizing a triangle t to a vertex t^* and an edge e to an edge e^*, we can view $\mathcal{F}(K^*)$ as a filtration on a 1-complex (graph). Then, applying Theorem 3.6, we get that (t^*, e^*) is indeed a persistence pair for the filtration $\mathcal{F}(K^*)$. □

We can compute a DMVF V^* for K^* by cancelling all persistence pairs as stated in Proposition 10.3. By duality, this also produces a DMVF V for the 2-manifold K. The action of cancelling a vertex–edge pair in K^* can be translated into a cancellation of an edge–triangle pair in K. Combining Propositions 10.3 and 10.4, we obtain the following result.

Theorem 10.5. *Let K be a finite simplicial 2-complex whose underlying space is a 2-manifold without boundary and let $\mathcal{F}$ be a simplex-wise filtration of K (Definition 3.1). Starting from the trivial DMVF where each simplex is critical, one can obtain a DMVF in K by cancelling the vertex–edge and edge–triangle persistence pairs given by $\mathcal{F}$.*

In general, by duality one can apply the above theorem to cancel all persistence pairs between $(d-1)$-simplices and d-simplices in a filtration of a simplicial d-complex where each $(d-1)$-simplex has at most two d-simplices as cofacets. This includes simplicial d-manifolds with boundary. We call a $(d-1)$-simplex a *boundary* if it adjoins exactly one d-simplex. For this extension, one has to introduce a "dummy" vertex in the dual graph that connects to all dual vertices of d-simplices incident to a boundary $(d-1)$-simplex. We leave it as an exercise (Exercise 11).

Unfortunately, it does not extend any further. In particular, the result in Theorem 10.5 does not extend to arbitrary simplicial 2-complexes and hence arbitrary simplicial complexes. The main difficulty arises because such a complex does not admit a dual graph in general. Indeed, there are counterexamples which exhibit that every persistence pair for a filtration of a simplicial 2-complex cannot be cancelled leading to a DMVF. The following dunce hat example exhibits this obstruction.

Dunce Hat
Consider a 2-manifold with boundary which is a cone with apex v and the boundary circle c. Let u be a point on c. Modify the cone by identifying the line segment uv with the circle c. Because of the similarity, the space obtained by this identification is called the dunce hat. Consider a triangulation K of the dunce hat. Notice that the dunce hat and hence $|K|$ is not a 2-manifold. The edges discretizing uv in K have three triangles incident to them. We show that there is no DMVF without any critical edge and triangle for K. The complex K is known to have $\beta_i(K) = 0$ for all $i > 0$ and has two or more triangles adjoining every edge in it. For any filtration of K, there cannot be any edge or triangle that remains unpaired because otherwise that would contradict that $\beta_1(K) = 0$ and $\beta_2(K) = 0$ (Fact 3.9 in Chapter 3). If a DMVF V were possible to be created by cancelling persistence pairs, there would be a finite maximal V-path that cannot be extended any further. Consider such a path π starting at a simplex σ. If σ is a triangle, the edge $\mu^{-1}(\sigma)$ matched with it can be added before it to extend π. If σ is an edge, there is a triangle adjoining σ not in the V-path because at least two triangles adjoin e and the V-path starting at e cannot be cyclic. We can add that triangle to extend π. In both cases, we contradict that π is maximal.

10.2.2 Algorithms

The above results naturally suggest an algorithm for computing a persistence-based DMVF for a simplicial 2-manifold K. We compute the persistence pairs on a chosen filtration $\mathcal{F}$ of K and then cancel them successively as Theorem 10.5 suggests. Both of these tasks can be combined by modifying the well-known Kruskal's algorithm for computing the minimum spanning tree of a graph.

Consider a graph $G = (U, E)$ which can be either the 1-skeleton of a complex K or the dual graph K^* if K is a simplicial 2-manifold. Assume that $u_1, u_2, \ldots, u_k$ and $e_1, e_2, \ldots, e_\ell$ are ordered sequences of vertices and edges in G. For the minimum spanning tree, the sequence of edges are taken in nondecreasing order of their weights. Here we describe the algorithm by assuming any order. Kruskal's algorithm maintains a spanning forest of the vertex set. It brings one edge e at a time in the given order either to join two trees in the current forest or to discover that the edge makes a cycle and hence does not belong to the spanning forest. If the two endpoints of e belong to two different trees in the forest, then it joins those two trees. Otherwise, e connects two vertices in the same tree, creating a cycle. The main computation involves determining if two vertices of an edge belong to the same tree or not. Algorithm 18: PERSDMVF does it by a union–find data structure which maintains the set of vertices of a tree in a single set and two sets are united if an edge joins the two respective trees. This is similar to the FINDSET and UNION operations in the algorithm ZEROPERDG described in Section 3.5.3. All such find and union operations can be done in $O(k + \ell\alpha(\ell))$ time assuming that there are k vertices and ℓ edges in the graph which dominates the overall complexity.

We can incorporate the persistence computation and Morse cancellations simultaneously in the above algorithm with some simple modifications. We process the vertices and edges in their order of the input filtration. Usually, the filtration $\mathcal{F} = \mathcal{F}_f$ is given by a simplex-wise monotone function f as described in Section 3.1.2. We compute the persistence Pers (e) of an edge e as Pers $(e) = |f(e) - f(r)|$ if e pairs with the vertex r and ∞ otherwise.

For a vertex u in the filtration $\mathcal{F}_f$, we do not do anything other than create a new set containing u only. When an edge $e = (u, u')$ comes in, we check if u and u' belong to the same tree by using the union–find data structure. If they do, the edge e is designated as a creator for persistence and as a critical edge in DMVF that is being built on G. Otherwise, we compute Pers (e) after finding the persistence pair for e and at the same time cancel e with its pair in the DMVF as follows. Assume inductively that the current DMVF matches

---

Algorithm 18 PERSDMVF($G, \mathcal{F}_f$)

Input:
A graph G and a filtration $\mathcal{F}_f$ on its n vertices and edges
Output:
A DMVF V and persistence pairs of $\mathcal{F}_f$ which are cancelled for creating V

1: Let $G = (U, E)$ and $\mathcal{F}$ be the input filtration of its n vertices and edges
2: $\mathcal{T} := \{\varnothing\}$; $V := \varnothing \sqcup \varnothing \sqcup \{(U \cup E)\}$; Initialize $\mathcal{U} := U$
3: **for all** $i = 1, \ldots, n$ **do**
4:    **if** $\sigma_i \in \mathcal{F}_f$ is a vertex u **then**
5:       Create a tree T rooted at u; $\mathcal{T} := \mathcal{T} \cup \{T\}$
6:    **else if** $\sigma_i \in F$ is an edge $e = (u, u')$ **then**
7:       **if** $t := \text{FINDSET}(u) = t' := \text{FINDSET}(u')$ **then**
8:          Designate e as creator and critical in V; Pers $(e) := \infty$
9:       **else**
10:          UNION(t,t') updating $\mathcal{U}$
11:          Let T_u and $T_{u'}$ be trees containing u and u'
12:          Find V-paths π_u from u to root r and $\pi_{u'}$ from u' to r' in T_u and $T_{u''}$ respectively
13:          Let r succeed r' in F; Cancel (e, r) considering the V-path π_u and update DMVF V; Pers $(e) := |f(e) - f(r)|$
14:          JOIN($T_u, T_{u'}$) in $\mathcal{T}$
15:       **end if**
16:    **end if**
17: **end for**
18: Output V and persistence pairs with persistence values

---

every vertex other than the roots of the trees to one of its adjacent edges as follows. For a leaf vertex v, consider the path $v = v_1, e_1, \ldots, e_{k-1}, v_k = r$ from v to the root r which consists of matched pairs $(v_1, e_1), \ldots, (v_{k-1}, e_{k-1})$ and the critical vertex r. For the edge $e = (u, u')$, let the roots of the two trees T_u and $T_{u'}$ containing u and u' be r and r', respectively. Assume without loss of generality that r succeeds r' in the input filtration. Then, e pairs with r in persistence because e joins the two components created by r and r' between which r comes later in the filtration. We cancel the persistence pair (r, e) by shifting the matched pairs on the path from u to r as stated in Definition 10.6. We join the two trees T_u and $T_{u'}$ into one tree by calling the routine JOIN. The root of the joined tree becomes r'. Cancelling (r, e) maintains the invariant

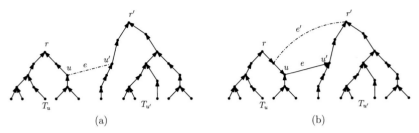

(a)   (b)

Figure 10.3 Illustration for Algorithm 18: PERSDMVF. (a) Destroyer edge $e =$ (u, u') is joining two trees T_u and $T_{u'}$ with roots r and r' respectively. The pair (r, e) is cancelled, reversing the arrows on three edges on the path from r to u'. (b) Edge e' is a creator and does not make any change in the forest.

that every path from the leaf to the root of the new tree remains a V-path. See Figure 10.3 for an illustration.

The costly step in algorithm PERSDMVF is the cancellation step which takes $O(n)$ time and thus incurs a running time $O(n^2)$ in total. However, we observe that all matchings in the final DMVF are made between a node v and the edge e that connects v to its parent, denoted parent(v), in the respective rooted tree and the root remains critical. All nontree edges remain critical. Thus, we can eliminate the cancellation step in PERSDMVF and after computing the final forest we can determine all matched pairs by traversing the trees upward from the leaves to the roots while matching a vertex with the edge visited next in this upward traversal. This matching takes $O(n)$ time. Accounting for the union–find operations, all other steps in PERSDMVF take $O(n\alpha(n))$ time in total. The simplified version Algorithm 19: SIMPLEPERSDMVF incorporates these changes. We have the following result.

Theorem 10.6. *Given a simplicial 1-complex or a simplicial 2-manifold K with n simplices, one can compute*

(a) a DMVF by cancelling all persistence pairs resulting from a given filtration of K in $O(n\alpha(n))$ time;
(b) a DMVF as above when the filtration is induced by a given PL-function on K in $O(n \log n)$ time.

PROOF. We argue for all statements in the theorem when K is a 1-complex. By considering the dual graph K^*, and combining Propositions 10.3 and 10.4, the arguments also hold for K when it is a simplicial 2-manifold. The algorithm SIMPLEPERSDMVF outputs the same as the algorithm PERSDMVF whose correctness follows from Theorem 10.5 because it cancels the persistence pairs exactly as the theorem dictates. The complexity analysis of the

---

Algorithm 19 SIMPLEPERSDMVF($G, \mathcal{F}_f$)

Input:
  A graph G and a filtration $\mathcal{F}_f$ on its n vertices and edges
Output:
  A DMVF V and persistence pairs of $\mathcal{F}_f$ which are cancelled for creating V

1: Let $G = (U, E)$ and $\mathcal{F}_f$ be the input filtration of its n vertices and edges
2: $\mathcal{T} := \{\varnothing\}$; $V := \varnothing$; Initialize $\mathcal{U} := U$
3: **for all** $i = 1, \ldots, n$ **do**
4:    **if** $\sigma_i \in \mathcal{F}_f$ is a vertex u **then**
5:      Create a tree T rooted at u; $\mathcal{T} := \mathcal{T} \cup \{T\}$
6:    **else if** $\sigma_i \in F$ is an edge $e = (u, u')$ **then**
7:      **if** $t := \text{FINDSET}(u) = t' := \text{FINDSET}(u')$ **then**
8:        Designate e as creator and critical in V; Pers $(e) := \infty$
9:      **else**
10:       $\text{UNION}(t, t')$ updating $\mathcal{U}$
11:       Let T_u and $T_{u'}$ be trees containing u and u' with roots r and r'
12:       Let r succeed r' in F; Pers $(e) := |f(e) - f(r)|$
13:       $\text{JOIN}(T_u, T_{u'})$ in $\mathcal{T}$ with edge e
14:     **end if**
15:   **end if**
16: **end for**
17: **for** each tree $T \in \mathcal{T}$ **do**
18:   **for** each node v in T **do**
19:     $e := (v, \text{parent}(v))$, $V := V \sqcup (v, e)$
20:   **end for**
21:   Put the root of T as a critical vertex in V
22: **end for**
23: Output V and persistence pairs with persistence values

---

algorithm SIMPLEPERSDMVF establishes the first statement. For the second statement, given the function values at the vertices of K, we can compute a simplex-wise lower-star filtration (Section 3.5) in $O(n \log n)$ time after sorting these function values. A subsequent application of SIMPLEPERSDMVF on this lower-star filtration provides us the desired DMVF. □

We can modify SIMPLEPERSDMVF slightly to take into account a threshold δ for persistence, that is, we can cancel pairs only with persistence up to δ. To do this, we need a slightly different version of Proposition 10.3.

The cancellation also succeeds if we cancel persistence pairs in the order of their persistence values. The proof of Proposition 10.3 can be adapted for the following proposition.

Proposition 10.7. *Let* (v_1, e_1), (v_2, e_2), ..., (v_n, e_n) *be the sequence of all non-essential persistence pairs of vertices and edges sorted in nondecreasing order of their persistence for a given filtration of* K. *Let* V_0 *be the DMVF in* K *with all simplices being critical. Suppose DMVF* V_{i-1} *can be obtained by cancelling successively* (v_1, e_1), (v_2, e_2), ..., (v_{i-1}, e_{i-1}). *Then,* (v_i, e_i) *can be cancelled in* V_{i-1} *providing a DMVF* V_i *for all* $i \geq 1$.

The modified algorithm proceeds as in SIMPLEPERSDMVF, but we designate those edges *critical* whose persistence is more than δ. Then, before traversing the edges of the trees in the forest $\mathcal{T}$ to output the vertex–edge pairs, we delete all these critical edges from $\mathcal{T}$. This splits the trees in $\mathcal{T}$ and creates more trees. We need to determine the roots of these trees. Observe that, had we done the cancellations as in PERSDMVF, the roots of the trees would have been the vertices that appear the earliest in the filtration among all vertices in the respective trees. So, all trees in $\mathcal{T}$ obtained after deleting all critical edges are rooted at the vertices that appear the earliest in the filtration. Then, the steps 17 to 22 in SIMPLEPERSDMVF compute the vertex–edge matchings into the DMVF from these rooted trees. The new algorithm called PARTIALPERSDMVF modifies step 13 of the algorithm SIMPLEPERSDMVF as:

- **if** Pers $(e) > \delta$ **then** designate e critical in V **endif**; JOIN(T_u,$T_{u'}$)

Also, PARTIALPERSDMVF introduces a step before step 17 of SIMPLEPERSDMVF as:

- Delete all critical edges from $\mathcal{T}$ and create new rooted trees in $\mathcal{T}$ as described

We claim that PARTIALPERSDMVF indeed computes a DMVF guaranteed by Proposition 10.7 (Exercise 9).

Claim 10.1. PARTIALPERSDMVF($\mathcal{F}$,δ) computes a DMVF obtained by cancelling persistence pairs in nondecreasing order of persistence values which do not exceed the input threshold δ.

Let V_δ denote the resulting DMVF after cancelling all *vertex–edge* persistence pairs with persistence at most δ.

Proposition 10.8. *The following statements hold for the output* $\mathcal{T}$ *of the algorithm* PARTIALPERSDMVF *with respect to any* $\delta \geq 0$:

(a) *For each tree* T_i, *its root* r_i *is the only critical simplex in* $V_\delta \cap T_i$. *The collection of these roots corresponds exactly to those vertices whose persistence is bigger than* δ.

(b) *Any edge with* Pers $(e) > \delta$ *remains critical in* V_δ *and cannot be contained in* $\mathcal{T}$.

10.3 Stable and Unstable Manifolds

In Section 1.5.2, we introduced the concept of Morse functions (Definition 1.28). These are smooth functions $f : \mathbb{R}^d \to \mathbb{R}$ satisfying certain conditions. We defined critical points of these functions and analyzed topological structures using the neighborhoods of these critical points. Here, we introduce another well-known structure associated with Morse functions and then draw a parallel between these smooth continuous structures and their discrete counterparts with the discrete Morse functions.

10.3.1 Morse Theory Revisited

For a point $p \in \mathbb{R}^d$, recall that the gradient vector of f at a point p is $\nabla f(p) = [\partial f / \partial x_1 \cdots \partial f / \partial x_d]^{\mathsf{T}}$, which represents the steepest ascending direction of f at p, with its magnitude being the rate of change. An *integral path* of f is a maximal path $\pi : (0, 1) \to \mathbb{R}^d$ where the tangent vector at each point p of this path equals $\nabla f(p)$, which is intuitively a flow path following the steepest ascending direction at any point. Recall that a point $p \in \mathbb{R}^d$ is critical if its gradient vector vanishes, that is, $\nabla f(p) = [0 \cdots 0]^{\mathsf{T}}$. An integral path necessarily "starts" and "ends" at critical points of f; that is, $\lim_{t \to 0} \pi(t) = p$ with $\nabla f(p) = [0 \cdots 0]^{\mathsf{T}}$, and $\lim_{t \to 1} \pi(t) = q$ with $\nabla f(q) = [0 \cdots 0]^{\mathsf{T}}$. See Figure 10.4, where we show the graph of a function $f : \mathbb{R}^2 \to \mathbb{R}$, and there is an integral path from a minimum v to a maximum t_2 and also to a saddle point e_2.

For a critical point p, the union of p and all the points from integral lines flowing into p is referred to as the *stable manifold of p*. Similarly, for a critical point q, the union of q and all the points on integral lines starting from q is called the *unstable manifold of q*. The unstable manifold of a minimum p intuitively corresponds to the basin/valley around p in the terrain of f. The 1-unstable manifold of an index $(d - 1)$-saddle consists of flow paths

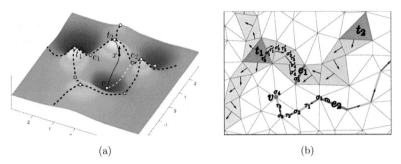

(a)   (b)

Figure 10.4 (Un)stable manifolds for (a) a smooth Morse function and (b) its discrete version (shown partially). (a) Here t_1 and t_2 are maxima (critical triangles in discrete Morse), v is a minimum, and e_1 and e_2 are saddles (critical edges in discrete Morse). The unstable manifold of e_1 flows out of it to t_1 and t_2. On the other hand, its stable manifolds flow out of minima such as v and come to it. These flows work in the opposite direction of "gravity" because if we put a drop of water at x it will flow to v. If we put it on the other side of the mountain ridge, it will flow to another minimum. (b) The flow direction reverses from the smooth case to the discrete case.

connecting this saddle to maxima. These curves intuitively capture "mountain ridges" of the terrain (graph of the function f); see Figure 10.4 for an example. Symmetrically, the stable manifold of a maximum q corresponds to the mountain around q. The 1-stable manifolds consist of a collection of curves connecting minima to 1-saddles, corresponding intuitively to the "valley ridges."

Now, we focus on a graph reconstruction approach using Morse theory. Suppose that a density field $\rho : \Omega \rightarrow \mathbb{R}$ on a domain $\Omega \subseteq \mathbb{R}^d$ is given where ρ concentrates around a hidden geometric graph G embedded in $\mathbb{R}^d$. We want to reconstruct G from ρ. Intuitively, we wish to use the 1-unstable manifolds of saddles (mountain ridges) of the density field ρ to capture the hidden graph.

However, to implement this idea, we will use *discrete Morse theory*, which provides robustness and simplicity due to its combinatorial nature. The cancellations guided by the persistence pairings can help us to remove noise introduced by both discretization and measurement errors. Below, we introduce some concepts necessary for transitioning to the discrete versions of (un)stable manifolds.

10.3.2 (Un)Stable Manifolds in Discrete Morse Vector Fields

The V-paths in a DMVF are analogues to the integral paths in the smooth setting. A *V-path* $\pi : \sigma_0 \prec \sigma_1 \prec \cdots \prec \sigma_{i-1} \prec \sigma_i \prec \sigma_{i+1} \prec \cdots \prec \sigma_k$ is a vertex–edge gradient path if the σ_i alternate between edges and vertices. Similarly, it is an edge–triangle gradient path if they alternate between triangles and edges. Also, we refer to vertex–edge or edge–triangle pairs as gradient vertex–edge and gradient edge–triangle vectors, respectively.

Different from the smooth setting, a maximal V-path may not start or end at critical simplices. However, those that do (i.e., when σ_0 and σ_k are critical simplices) are exactly the critical V-paths. These paths are discrete analogues of maximal integral paths in the smooth setting which "start" and "end" at critical points. One can think of *critical k-simplices* in the discrete Morse setting as *index-k critical points* in the smooth setting as defined in Section 1.5.2. For example, for a function on $\mathbb{R}^2$, critical 0-, 1-, and 2-simplices in the discrete Morse setting correspond to minima, saddles, and maxima in the smooth setting, respectively.

There is one more caveat that one should be aware of. The direction of the integral paths and the V-paths run in the opposite direction by definition: In the smooth setting, function values increase along an integral path, while in the discrete setting, they decrease along a V-path. This means that the stable and unstable manifolds reverse their roles in the two settings; refer to Figure 10.4. For a critical edge e, we define its *stable manifold* to be the union of edge–triangle gradient paths that end at e. Its *unstable manifold* is defined to be the union of vertex–edge gradient paths that begin with e. In the graph reconstruction approach presented below, we use "mountain ridges" for the reconstruction. We have seen that these are 1-unstable manifolds of saddles in the smooth setting and hence correspond to 1-stable manifolds in the discrete gradient fields consisting of triangle–edge paths. Notice that these mountain ridges on a triangulation of a d-manifold correspond to a V-path alternating between d- and $(d-1)$-dimensional simplices. Computationally, however, vertex–edge gradient paths are simpler to handle especially for the Morse cancellations below. Hence in our algorithm below, we negate the density function ρ and consider the function $-\rho$. The algorithm outputs a subset of the 1-*unstable manifolds* that are vertex–edge gradient paths in the discrete setting as the recovered hidden graph.

With the above setup, we have an input function $f \colon V(K) \to \mathbb{R}$ defined at the vertices $V(K)$ of a complex K whose linear extension leads to a PL-function still denoted by $f \colon |K| \to \mathbb{R}$. For computing persistence, we use the lower-star filtration $\mathcal{F}_f$ of f and its simplex-wise version as described in Section 3.1.2.

10.4 Graph Reconstruction

Suppose we have a domain Ω (which will be a cube in $\mathbb{R}^d$) and a density function $\rho \colon \Omega \to \mathbb{R}$ that "concentrates" around a hidden geometric graph $G \subset \Omega$. In the discrete setting, our input will be a triangulation K of Ω and a density

---

Algorithm 20 MORSERECON(K, ρ, δ)

Input:
    A 2-complex K, a vertex function ρ on K, a threshold δ
Output:
    A graph

1: Let $\mathcal{F}$ be a simplex-wise lower-star filtration of K with respect to $f = -\rho$
2: Compute persistence Pers (e) for every edge e for the filtration $\mathcal{F}$
3: Let K^1 be the 1-skeleton of K and $\mathcal{F}^1$ be $\mathcal{F}$ restricted to vertices and edges only
4: Let $\mathcal{T}$ be the forest computed by PARTIALPERSDMVF(K^1,$\mathcal{F}^1$,δ)
5: COLLECTG(K^1,$\mathcal{T}$, Pers ($\cdot$), δ)

---

function given as a PL-function $\rho\colon |K| \to \mathbb{R}$. The algorithm can be easily modified to take a cell complex as input. Our goal is to compute a graph $\hat{G}$ approximating the hidden graph G.

10.4.1 Algorithm

Intuitively, we wish to use "mountain ridges" of the density field to approximate the hidden graph, as Figure 10.6 (later) shows. We compute these ridges as the 1-stable manifolds ("valley ridges") of $f = -\rho$, the negation of the density function. In the discrete setting, these become 1-unstable manifolds consisting of vertex–edge gradient paths in an appropriate DMVF. We compute this DMVF by cancelling vertex–edge persistence pairs whose persistence is at most a threshold δ. The rationale behind this choice is that the small undulations in a 1-unstable manifold caused by noise and discretization need to be ignored by cancellation. The procedure PARTIALPERSDMVF described earlier in Section 10.2.2 achieves this goal. Finally, the union of the 1-unstable manifolds of all remaining high-persistence critical edges is taken as the output graph $\hat{G}$, as outlined in Algorithm 21: COLLECTG. Algorithm 20: MORSERECON presents these steps.

Since we only need 1-unstable manifolds, K is assumed to be a 2-complex. Notice that one only needs to cancel vertex–edge pairs – this is because only vertex–edge gradient vectors contribute to the 1-unstable manifolds, and also new vertex–edge vectors can only be generated while cancelling other vertex–edge pairs.

Let $T_1, T_2, \ldots, T_k$ be the set of trees returned by PARTIALPERSDMVF. The routine COLLECTG outputs the 1-unstable manifold of every edge $e = (u, v)$

Algorithm 21 COLLECTG(K^1,$\mathcal{T}$, Pers $(\cdot)$, δ)

Input:
  A 1-skeleton K^1, a forest $\mathcal{T} \subseteq K^1$, persistence values for edges in K^1, a threshold δ

Output:
  A graph

1: $\hat{G} := \varnothing$
2: **for** every edge $e = (u, v) \in K^1 \setminus \mathcal{T}$ **do**
3:    **if** Pers $(e) > \delta$ **then**
4:       Let $\pi(u)$ and $\pi(v)$ be the two paths from u and v to the roots respectively
5:       Set $\hat{G} := \hat{G} \cup \pi(u) \cup \pi(v) \cup \{e\}$
6:    **end if**
7: **end for**
8: Return $\hat{G}$

---

with Pers $(e) > \delta$, which is simply the union of e and the unique paths from u and v to the root of the tree containing them, respectively.

Notice that we still need to compute the persistence for all edges. If it were only for those edges that pair with vertices, we could have eliminated step 2 in MORSERECON and computed the persistence of these edges in PARTIALPERSDMVF in almost linear time (Theorem 10.6). However, to compute persistence for edges that pair with triangles, we have to use the standard persistence algorithm, whose complexity again depends on the complex K. For example, if K is a simplicial 2-manifold, this can run in $O(n\alpha(n))$ time (Section 3.6 Notes and Exercises); but this time complexity does not hold for general 2-complex K. To take into account this dependence of the time complexity on the type of K, we simply denote the time for computing persistence with Pert(K) in the following theorem.

Theorem 10.9. *The time complexity of the algorithm* MORSERECON *is* $O(\text{Pert}(K))$, *where Pert(K) is the time to compute persistence pairings for K.*

We remark that, for K with n vertices and edges, collecting all 1-unstable manifolds takes $O(n)$ time if one avoids revisiting edges while tracing paths. This $O(n)$ term is subsumed by Pert(K) because there are at least $n/2$ such pairs.

Figure 10.5 Noise model for graph reconstruction.

Consider the DMVF V_δ computed by PARTIALPERSDMVF. Notice that Proposition 10.8(a) implies that for each T_i, any V-path of V_δ starting at a vertex or an edge in T_i terminates at its root r_i. See Figure 10.3 for an example. Hence for any vertex $v \in T_i$, the path $\pi(v)$ computed by COLLECTG is the unique V-path starting at v. This immediately leads to the following result.

Corollary 10.10. *For each critical edge $e = (u, v)$ with* Pers $(e) \geq \delta$, $\pi(u) \cup \pi(v) \cup \{e\}$ *computed by the algorithm* COLLECTG *is the 1-unstable manifold of e in V_δ.*

10.4.2 Noise Model

To establish theoretical guarantees for the graph reconstructed by the algorithm MORSERECON, we assume a noise model for the input. We first describe the noise model in the continuous setting where the domain is the k-dimensional unit cube $\Omega = [0, 1]^k$. We then explain the setup in the discrete setting when the input is a triangulation K of Ω.

Given a connected "true graph" $G \subset \Omega$, consider a ω-neighborhood $G^\omega \subseteq \Omega$, meaning that (i) $G \subseteq G^\omega$, and (ii) for any $x \in G^\omega$, $d(x, G) \leq \omega$ (i.e., G^ω is sandwiched between G and its ω-offset). Given G^ω, we use cl$(\overline{G^\omega}) =$ cl$(\Omega \setminus G^\omega)$ to denote the closure of its complement $\overline{G^\omega} = \Omega \setminus G^\omega$. Figure 10.5 illustrates the noise model in the discrete setting, showing G (red graph) with its ω-neighborhood G^ω (yellow).

Definition 10.7. $((\beta, \nu, \omega)$-approximation) A density function $\rho \colon \Omega \to \mathbb{R}$ is a (β, ν, ω)-*approximation of a connected graph G* if the following conditions hold:

(C-1) There is an ω-neighborhood G^ω of G such that G^ω deformation retract to G.

(C-2) One has $\rho(x) \in [\beta, \beta + \nu]$ for $x \in G^\omega$; and $\rho(x) \in [0, \nu]$ otherwise. Furthermore, $\beta > 2\nu$.

Intuitively, this noise model requires that the density ρ concentrates around the true graph G in the sense that the density is significantly higher inside G^ω than outside; and the density fluctuation inside or outside G^ω is small compared to the density value in G^ω (C-2). Condition (C-1) says that the neighborhood has the same topology as the hidden graph. Such a density field could, for example, be generated as follows: Imagine that there is an ideal density field $f_G : \Omega \to \mathbb{R}$ where $f_G(x) = \beta$ for $x \in G^\omega$ and 0 otherwise. There is a noisy perturbation $g : \Omega \to \mathbb{R}$ whose size is always bounded by $g(x) \in [0, \nu]$ for any $x \in \Omega$. The observed density field $\rho = f_G + g$ is a (β, ν, ω)-approximation of G.

In the discrete setting when we have a triangulation K of Ω, we define an ω-neighborhood G^ω to be a subcomplex of K, that is, $G^\omega \subseteq K$, such that (i) G is contained in the underlying space of G^ω and (ii) for any vertex $v \in V(G^\omega)$, $d(v, G) \leq \omega$. The complex $\mathrm{cl}(\overline{G^\omega}) \subseteq K$ is simply the smallest subcomplex of K that contains all simplices from $K \setminus G^\omega$ (i.e., all simplices **not** in G^ω and their faces). A (β, ν, ω)-approximation of G is extended to this setting by a PL-function $\rho \colon |K| \to \mathbb{R}$ while requiring that the underlying space of G^ω deformation retracts to G as in (C-1), and density conditions in (C-2) are satisfied at vertices of K.

We remark that the noise model is rather limited – in particular, it does not allow significant non-uniform density distribution. However, this is the only case for which theoretical guarantees are known at the moment for a discrete Morse-based reconstruction framework. In practice, the algorithm has often been applied to non-uniform density distributions.

10.4.3 Theoretical Guarantees

In this subsection, we prove results that are applicable to hypercube domains of any dimensions. Recall that V_δ is the discrete gradient field after the cancellation process with threshold δ, where we perform cancellation for *vertex–edge* persistence pairs generated by a simplex-wise filtration induced by the PL-function $f = -\rho$ that negates the density PL-function. At this point, all positive edges, that is, those not paired with vertices, remain critical in V_δ. Some negative edges, that is, those paired with vertices, also remain critical in V_δ – these are exactly the negative edges with persistence bigger than δ. COLLECTG only takes the 1-unstable manifolds of those critical edges (positive or negative) with persistence bigger than δ; so those edges whose persistence is at most δ are ignored.

Input Assumption

Let ρ be an input density field which is a (β, ν, ω)-approximation of a connected graph G, and $\delta \in [\nu, \beta - \nu)$.

Under the above input assumption, let $\hat{G}$ be the output of algorithm MORSERECON(K, ρ, δ). The proof of the following result can be found in [137].

Proposition 10.11. *Under the input assumption, we have the following.*

(a) *There is a single critical vertex left after* MORSERECON *returns, which is in G^ω.*

(b) *Every critical edge considered by* COLLECG *forms a persistence pair with a triangle.*

(c) *Every critical edge considered by* COLLECTG *is in G^ω.*

Theorem 10.12. *Under the input assumption, the output graph satisfies $\hat{G} \subseteq G^\omega$.*

PROOF. Recall that the output graph $\hat{G}$ consists of the union of 1-unstable manifolds of all the edges $e_1^*, \ldots, e_g^*$ with persistence larger than δ — by Proposition 10.11(b) and (c), they are all positive (paired with triangles), and contained inside G^ω. Below we show that other simplices in their 1-unstable manifolds are also contained in G^ω.

Take any $i \in [1, g]$ and consider $e_i^* = (u, v)$. Without loss of generality, consider the critical V-path $\pi : e_i^* \prec (u = u_1) \prec e_1 \prec u_2 \prec \cdots \prec e_s \prec u_{s+1}$. By definition, u_{s+1} is a critical vertex and is necessarily the global minimum v_0 for the density field ρ, which is also contained inside G^ω. We now argue that all simplices in the path π lie inside G^ω. In fact, we argue a stronger statement: First, we say that a gradient vector (v, e) is *crossing* if $v \in G^\omega$ and $e \notin G^\omega$ (i.e., $e \in \mathrm{cl}(\overline{G^\omega})$). Since v is an endpoint of e, this means that the other endpoint of e must lie in $K \setminus G^\omega$. We now prove the following claim, before returning to the proof of the theorem.

Claim 10.2. *During the cancellation with threshold δ in the algorithm* MORSERECON, *no crossing gradient vector is ever produced.*

PROOF. Suppose the claim is not true. Then, let (v, e) be the *first* crossing gradient vector ever produced during the cancellation process. Since we start with a trivial discrete gradient vector field, the creation of (v, e) can only be caused by reversing of some gradient path π' connecting two critical simplices v' and e' while we are performing cancellation for the persistence pair (v', e'). Obviously, Pers $(e') \leq \delta$ because otherwise cancellation would not have been performed. On the other hand, due to our (β, ν, ω) noise model and the choice of δ, it must be that either both $v', e' \in G^\omega$ or both $v', e' \in K \setminus G^\omega$ – as otherwise, the persistence of this pair will be larger than $\beta - \nu > \delta$.

Now consider the V-path π' connecting e' and v' in the current discrete gradient vector field V'. The path π' begins and ends with simplices that are either both in G^ω or both outside G^ω and also it has simplices both inside and outside G^ω. It follows that the path π' contains a gradient vector (v'', e'') going in the opposite direction crossing inside/outside, that is, $v'' \in G^\omega$ and $e'' \notin G^\omega$. In other words, it must contain a crossing gradient vector. This, however, contradicts our assumption that (v, e) is the first crossing gradient vector. Hence, the assumption is wrong and no crossing gradient vector can ever be created. □

As there is no crossing gradient vector during and after cancellation, it follows that π, which is one piece of the 1-unstable manifold of the critical edge e_i^*, has to be contained inside G^ω. The same argument works for the other piece of the 1-unstable manifold of e_i^* which starts from the other endpoint of e_i^*. Since this holds for any $i \in [1, g]$, the theorem follows. □

The previous theorem shows that $\hat{G}$ is geometrically close to G. Next we show that they are also close in topology.

Proposition 10.13. *Under the input assumption, $\hat{G}$ is homotopy equivalent to G.*

PROOF. First we show that $\hat{G}$ is connected. Then, we show that $\hat{G}$ has the same first Betti number as that of G, which implies the claim, as any two connected graphs in $\mathbb{R}^k$ with the same first Betti number are homotopy equivalent. Suppose that $\hat{G}$ has at least two components. These two components should come from two trees in the forest computed by PARTIALPERSDMVF. The roots, say r and r', of these two trees must reside in G^ω due to Claim 10.2 and Proposition 10.11(c). Furthermore, the supporting complex of G^ω is connected because it contains the connected graph G. It follows that there is a path connecting r and r' within G^ω. All vertices and edges in G^ω appear earlier than other vertices and edges in the filtration that PARTIALPERSDMVF works on. These two facts mean that the first edge which connects the two trees rooted at r and r' resides in G^ω. This edge has a persistence less than δ and should be included in the reconstruction by MORSERECON. It follows that COLLECTG returns 1-unstable manifolds of edges ending at a common root of the tree containing both r and r'. In other words, $\hat{G}$ cannot have two components as assumed.

The underlying space of ω-neighborhood G^{ω} of G deformation retracts to G by definition. Observe that, by our noise model, G^{ω} is a sublevel set in the filtration that determines the persistence pairs. This sublevel set, being homotopy equivalent to G, must contain exactly g positive edges where g is the first Betti number of G. Each of these positive edges pairs with a triangle in $\overline{G^{\omega}}$. Therefore, $\mathrm{Pers}\,(e) > \delta$ for each of the g positive edges in G^{ω}. By our earlier results, these are exactly the edges that will be considered by procedure COLLECTG. Our algorithm constructs $\hat{G}$ by adding these g positive edges to the spanning tree each of which adds a new cycle. Thus, $\hat{G}$ has first Betti number g as well, thus proving the proposition.   □

We have already proved that $\hat{G}$ is contained in G^{ω}. This fact along with Proposition 10.13 can be used to argue that any deformation retraction taking (underlying space) G^{ω} to G also takes $\hat{G}$ to a subset $G' \subseteq G$ where G' and G have the same first Betti number. In what follows, we use G^{ω} to denote also its underlying space.

Theorem 10.14. *Let $H : G^{\omega} \times [0, 1] \to G^{\omega}$ be any deformation retraction so that $H(G^{\omega}, 1) = G$. Then, the restriction $H|_{\hat{G}} : \hat{G} \times [0, 1] \to G^{\omega}$ is a homotopy from the embedding $\hat{G}$ to $G' \subseteq G$ where G and G' have the same first Betti number.*

PROOF. The fact that $H|_{\hat{G}}(\cdot, \ell)$ is continuous for any $\ell \in [0, 1]$ is obvious from the continuity of H. The only thing that needs to be shown is that $G' := H|_{\hat{G}}(\hat{G}, 1)$ has the same first Betti number as that of G. We observe that a cycle in $\hat{G}$ created by a positive edge e along with the paths to the root of the spanning tree is also nontrivial in G^{ω} because this is a cycle created by adding the edge e during persistence filtration and the cycle created by the edge e is not destroyed in G^{ω}. Therefore, a cycle basis for $\mathsf{H}_1(\hat{G})$ is also a homology basis for $\mathsf{H}_1(G^{\omega})$. Since the map $H(\cdot, 1) : G^{\omega} \to G$ is a homotopy equivalence, it induces an isomorphism in the respective homology groups; in particular, a basis in $\mathsf{H}_1(G^{\omega})$ is mapped bijectively to a basis in $\mathsf{H}_1(G)$. Therefore, the image $G' = H|_{\hat{G}}(\hat{G}, 1)$ must have a basis of cardinality $g = \beta_1(\hat{G}) = \beta_1(G^{\omega}) = \beta_1(G)$, proving that $\beta_1(G') = \beta_1(G)$.   □

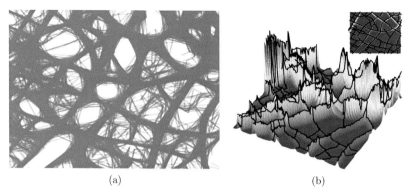

(a)   (b)

Figure 10.6 Road network reconstruction [295]. (a) Input GPS traces. (b) Terrain corresponding to the graph of the density function computed from input GPS traces. Black lines are the output of algorithm MORSERECON, which captures the "mountain ridges" of the terrain, corresponding to the reconstructed road network. The upper right inset is a top view of the terrain.

10.5 Applications

10.5.1 Road Network

Robust and efficient automatic road network reconstruction from Global Positioning System (GPS) traces and satellite images is an important task in Geographic Information System (GIS) data analysis and applications. The Morse-based approach can help in reconstructing the road network in both cases in a conceptually simple and clean manner. The framework provides a meaningful and robust way to remove noise because it is based on the concept of persistent homology. Intuitively, reconstruction of a road network from noisy data is tantamount to reconstructing a graph from a noisy function on a 2D domain. One needs to eliminate noise and at the same time preserve the signal. Persistent homology and discrete Morse theory help address both of these aspects. We can simply use the graph reconstruction algorithm detailed in the previous section for this road network recovery.

GPS Trajectories
Here the input is a set of GPS traces, and the goal is to reconstruct the underlying road network automatically from these traces. The input set of GPS traces can be converted into a density map $\rho : \Omega \to \mathbb{R}$ defined on the planar domain $\Omega = [0, 1] \times [0, 1]$. We then use our graph reconstruction algorithm MORSERECON to recover the "mountain ridges" of the density field; see Figure 10.6.

(a)　　　(b)　　　(c)

Figure 10.7 Road network reconstruction with editing [136]. (a) Red points (minima) are added; red branches are newly reconstructed for the Athens map (black curves are original reconstruction, blue curves are input GPS traces). (b) We also add blue triangles as maxima to capture many missing loops. (c) Top: an example to show that adding extra triangles as maxima will capture more loops Bottom: Berlin with adding both branches and loops.

In Figure 10.7, we show the reconstructed road network after improving the discrete Morse-based output graphs with an *editing strategy* [136]. After the automatic reconstruction, the user can observe the missing branches and can recover them by artificially making a vertex near the tip of each such branch a minimum. This forces a 1-unstable manifold from a saddle edge to each of these minima. Similarly, if a distinct loop in the network is missing, the user can artificially make a triangle in the center of the loop a maximum, which forces the loop to be detected.

Satellite Images

In this case, we combine the Morse-based graph reconstruction with a neural network framework to recover the road network from input satellite images. First, we feed the grayscale values of the input satellite image as a density function to MORSERECON. The output graphs from a set of images are used to train a convolutional neural network (CNN), which outputs an image aiming to capture only the foreground (roads) in the satellite images. After training this CNN, we feed the original satellite images to it to obtain a set of hopefully "cleaner" images. These cleaned images are again fed to MORSERECON to output a graph which can again be used to further train the CNN. Repeated use of this reconstruct-and-train step cleans the noise considerably. In Figure 2(f) in the Prelude, we show an example of the output of this strategy. Notice that this strategy eliminates the need for curating the satellite images manually for creating training samples.

10.5.2 Neuron Network

To understand neuronal circuitry in the brain, a first step is often to reconstruct the neuronal cell morphology and cell connectivity from microscopic neuroanatomical image data. Earlier work often focused on single neuron reconstruction from high-resolution images of a specific region in the brain. With the advancement of imaging techniques, whole brain imaging data are becoming more and more common. Robust and efficient methods that can segment and reconstruct neurons and/or connectivities from such images are highly desirable.

The discrete Morse-based graph reconstruction algorithms have been applied to both fronts. Neuron cells have tree morphology and can commonly be modeled as a rooted tree, where the root of the tree locates in the soma (cell body) of the neuron. In Figure 10.8, we show the reconstructed neuron morphology by applying the discrete Morse algorithm directly to an Olfactory Projection Fibers dataset (specifically, the OP-2 dataset) from the DIADEM challenge [143]. Specifically, the input is an image stack acquired

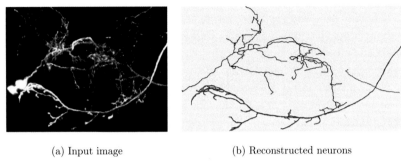

(a) Input image   (b) Reconstructed neurons

Figure 10.8 Discrete Morse-based neuron morphology reconstruction from [294] Images courtesy of S. Wang et al. (2018, figure 13).

by the two-channel confocal microscopy method. In the approach proposed in [294], after some preprocessing, the discrete Morse-based algorithm is applied to the 3D volumetric data to construct a graph skeleton. A tree extraction-based algorithm is then applied to extract a tree structure from the graph output.

The discrete Morse-based graph reconstruction algorithm can also be used in a more sophisticated manner to handle more challenging data. Indeed, a new neural network framework is proposed in [16] to combine the reconstructed Morse graph as a topological prior with a UNet-like [269] neural network architecture for cell process segmentation from various neuroanatomical image data. Intuitively, while UNet has been quite successful in image segmentation, such approaches lack a global view (e.g., connectivity) of the structure behind the segmented signal. Consequently, the output can contain broken pieces for noisy images, and features such as junction nodes in the input signal can be particularly challenging to recover. On the other hand, while the discrete Morse-based graph reconstruction algorithm is particularly effective in capturing global graph structures, it may produce many false positives. The framework proposed in [16], called DM++, uses the output from discrete Morse as a separate channel of input, and co-trains it together with the output of a specific UNet-like architecture called ALBU [61] so as to use these two inputs to complement each other. See Figure 10.9. In particular, the UNet output helps to remove false positives from the discrete Morse output, while the Morse graph output helps to obtain better connectivity.

10.6 Notes and Exercises

Forman [161] developed the discrete analogue of the classical Morse theory in mathematics. This analogy is exemplified by the following fact. Let C_p denote the p-th chain group formed by the p-dimensional critical cells in a discrete

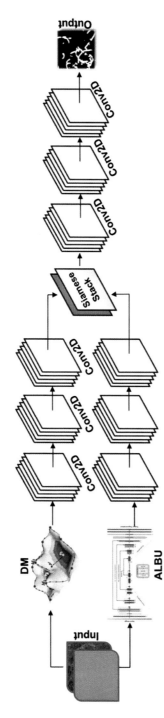

Figure 10.9 The DM++ framework proposed by [16], which combines both the discrete Morse output (DM) with a standard neural network-based output (ALBU) together via a Siamese neural network stack so as to use these two inputs to augment each other and obtain better connected final segmentation. Image courtesy of S. Banerjee et al. (2020, figure 2b).

Morse vector field. It means that C_p is a free abelian group with critical p-cells forming a basis assuming $\mathbb{Z}_2$-additions. For a critical cell c_p, define the boundary operator $\partial_p c_p = \sum_i (m_p \bmod 2) c^i_{p-1}$ where c^i_p is a critical $(p-1)$-cell reachable by m_p number of V-paths from c_p. Extending the boundary operator to the chains we get the boundary homomorphism $\partial_p : C_p \to C_{p-1}$. One can verify that $\partial_{p-1} \circ \partial_p = 0$ (Exercise 12) thus leading to a valid *discrete Morse chain complex*. Naturally, we get a homology group H_p from this construction. It turns out that this homology group is isomorphic to the homology group of the complex on which the DMVF is defined.

Several researchers brought the concept to the area of topological data analysis [22, 214, 224, 236]. King et al. [214] presented an algorithm to produce a discrete Morse function on a complex from a given real-valued function on its vertices. Bauer et al. [22] showed that persistence pairs can be cancelled in order of their persistence values for any simplicial 2-manifolds. They also gave an $O(n \log n)$-time algorithm for cancelling pairs that have persistence below a given threshold. The cancellation algorithm and its analysis in this chapter follow this result though with a slightly different presentation. This cancellation does not generalize to simplicial 2-complexes and beyond as we have illustrated. Mischaikow and Nanda [236] proposed Morse cancellation as a tool to simplify an input complex before computing persistence pairs. The combinatorial view of the vector field given by the discrete Morse theory has recently been extended to dynamical systems; see, for example, [38, 239].

Starting with Lewiner et al. [224], several researchers proposed discrete Morse theory for applications in visualization and image processing. Gyulassy et al. [181], Delgado-Friedrichs et al. [117] and Robins et al. [267] used discrete Morse theory in conjunction with persistence-based cancellations for processing images and analyzing features for, for example porous solids. Sousbie [282] proposed using the theory for detecting filamentary structures in data for cosmic webs. These works proposed using cancellations as long as they are permitted, acknowledging the fact that all cancellations in a 2- or 3-complex may not be possible. Wang et al. [295] proposed to use discrete Morse complexes to compute unstable 1-manifolds as an output for a road network from GPS data. Using unstable 1-manifolds in a discrete Morse complex defined on a triangulation in $\mathbb{R}^2$ to capture the hidden road network was proposed in that paper. Ultimately, this proposed approach was implemented with a simplified algorithm and a proof of guarantee in [137]. The material in Section 10.4 is taken from that paper. The application to road network reconstruction from GPS trajectories and satellite images in Section 10.5 appeared in [136] and [138], respectively. The application to neuron imaging data is taken from [16, 294].

Exercises

1. A Hasse diagram of a simplicial complex K is a directed graph that has a vertex v_σ for every simplex σ in K and a directed edge from v_σ to $v_{\sigma'}$ if and only if σ' is a cofacet of σ. Let M be a matching in K. Modify the Hasse diagram by reversing every edge that is directed from v_σ to $v_{\sigma'}$ and (σ', σ) is in the matching M. Show that M induces a DMVF if and only if the modified Hasse diagram does not have any directed cycle.

2. Let f be a Morse function defined on a simplicial complex K. We say K collapses to K' if there is a simplex σ with a single cofacet σ' and $K' = K \setminus \{\sigma, \sigma'\}$. Let $K_a \subseteq K$ be the subcomplex where $K_a = \{\sigma \mid f(\sigma) \leq a\}$. Show that there is a series of collapses (possibly empty) that brings K_a to K_b for any $b \leq a$ if there is no critical simplex with function value c where $b < c < a$.

3. Call a V-path *extendible* if it can be extended by a simplex at any of the two ends.

   (a) Show an example of a non-extendible V-path that is not critical.

   (b) Show that every non-extendible V-path in a simplicial 2-manifold without boundary must have at least one critical simplex.

4. Show that a discrete Morse function defines a Morse matching.

5. Let K be a simplicial Möbius strip with all its vertices on the boundary. Design a DMVF on K so that there is only one critical edge and only one critical vertex and no critical triangle.

6. Prove that two V-paths that meet must have a common suffix.

7. Show the following:

   (a) The strong Morse inequality implies the weak Morse inequality in Proposition 10.1.

   (b) A matching which is not Morse may not satisfy Morse inequalities as in Proposition 10.1 but always satisfies the equality $c_p - c_{p-1} + \cdots \pm c_0 = \beta_p - \beta_{p-1} + \cdots \pm \beta_0$ for a p-dimensional complex K.

8. Consider a filtration of a simplicial complex K embedded in $\mathbb{R}^3$. We want to create a DMVF where all persistent triangle–tetrahedra pairs with persistence less than a threshold can be cancelled. Show that this is always possible. Write an algorithm to compute the stable manifolds for each of the critical tetrahedra in the resulting DMVF.

9. Prove Claim 10.1.

10. We propose a different version of PARTIALPERSDMVF by changing only step 13 of SIMPLEPERSDMVF as:

    • **if** Pers $(e) > \delta$ **then** designate e critical in V **else** JOIN$(T_u, T_{u'})$ **endif**

Prove that this simple modification produces the same DMVF as the PARTIALPERSDMVF described in the text.

11. Let K be a simplicial d-complex that has every $(d-1)$-simplex incident to at most two d-simplices. Extend Theorem 10.5 to prove that all persistence pairs between $(d-1)$-simplices and d-simplices arising from a filtration of K can be cancelled.

12. Prove $\partial_{p-1} \circ \partial_p = 0$ for the boundary operator defined for chain groups of critical cells as described for the discrete Morse chain complex in the notes above.

11

Multiparameter Persistence and Decomposition

In previous chapters, we have considered filtrations that are parameterized by a single parameter such as $\mathbb{Z}$ or $\mathbb{R}$. Naturally, they give rise to a one-parameter persistence module. In this chapter, we generalize the concept and consider persistence modules that are parameterized by one or more parameters such as $\mathbb{Z}^d$ or $\mathbb{R}^d$. They are called multiparameter persistence modules in general. Multiparameter persistence modules naturally arise from filtrations that are parameterized by multiple values such as the one shown in Figure 11.1 over two parameters.

The classical algorithm of Edelsbrunner et al. [152] presented in Chapter 3 provides a unique decomposition of the one-parameter persistence module over $\mathbb{Z}$ implicitly generated by an input simplicial filtration. Similarly, a multiparameter persistence module M over the grid $\mathbb{Z}^d$ can be implicitly given by an input multiparameter finite simplicial filtration and we look for computing a decomposition (Definition 11.10) $M \cong \bigoplus_i M^i$. The modules M^i are the counterparts of bars in the one-parameter case and are called *indecomposables*. These indecomposables are more complicated and cannot be completely characterized as in the one-parameter case. Nonetheless, for finitely generated persistence modules defined over $\mathbb{Z}^d$, their existence is guaranteed by the Krull–Schmidt theorem [10]. Figure 11.2 illustrates indecomposables of some modules.

An algorithm for decomposing a multiparameter persistence module can be derived from the so-called Meataxe algorithm which applies to much more general modules than we consider in topological data analysis (TDA) at the expense of high computational cost. Sacrificing this generality and still encompassing a large class of modules that appear in TDA, we can design a much more efficient algorithm. Specifically, we present an algorithm that can decompose a finitely presented module with a time complexity that is much better than the Meataxe algorithm, though we lose the generality as the module needs to be *distinctly graded* as explained later.

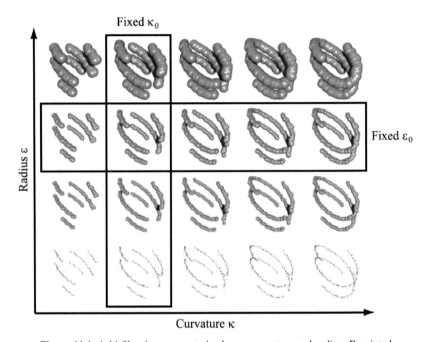

Figure 11.1 A bi-filtration parameterized over curvature and radius. Reprinted by permission from Springer Nature: Springer Nature, *Discrete & Computational Geometry*, "The Theory of Multidimensional Persistence," G. Carlsson et al. [65], © 2009.

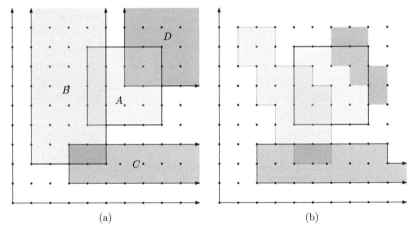

(a)   (b)

Figure 11.2 Decomposition of a finitely generated two-parameter persistence module. (a) Rectangle decomposable module: each indecomposable is supported by either a bounded (A) or unbounded (B and C); rectangle D is a free module. (b) Interval decomposable module: each indecomposable is supported over a 2D interval (defined in the next chapter).

$$\Bbbk \xrightarrow{\binom{1}{0}} \Bbbk^2 \xrightarrow{\binom{1\ 1}{0\ 0}} \Bbbk^2 \xrightarrow{(\,1\ \ 1\,)} \Bbbk \xrightarrow{(1)} \Bbbk$$

$$g_1 \longrightarrow tg_1, g_2 \longrightarrow t^2 g_1, tg_2, g_3 \longrightarrow t^3 g_1, t^2 g_2, tg_3 \longrightarrow t^4 g_1, t^3 g_2, t^2 g_3$$

$$0 = t^2 g_1 + t g_2 \qquad 0 = t^3 g_1 + t^2 g_2 \qquad 0 = t^4 g_1 + t^3 g_2$$
$$0 = t^2 g_2 + t g_3 \qquad 0 = t^3 g_2 + t^2 g_3$$

	(2)	(3)
(0)	1	
(1)	1	1
(2)		1

Figure 11.3 Costly presentation (top) versus graded presentation (bottom, right). The top chain can be summarized by three generators g_1, g_2, and g_3 at grades (0), (1), and (2), respectively, and two relations $0 = t^2 g_1 + t g_2$ and $0 = t_2 g_2 + t g_3$ at grades (2) and (3), respectively (Definition 11.5). The grades of the generators and relations are given by the first times they appear in the chain. Finally, this information can be summarized succinctly by the presentation matrix on the right.

For measuring algorithmic efficiency, it is imperative to specify how the input module is presented. Assuming an index set of size s and vector spaces of dimension $O(m)$, a one-parameter persistence module can be presented by a set of matrices of dimension $O(m) \times O(m)$ each representing a linear map $M_i \to M_{i+1}$ between two consecutive vector spaces M_i and M_{i+1}. This input format is costly as it takes $O(sm^2)$ space ($O(m^2)$-size matrix for each index) and also does not appear to offer any benefit in time complexity for computing the bars. An alternative presentation is obtained by considering the persistence module as a graded module over a polynomial ring $\Bbbk[t]$ and presenting it with the so-called *generators* $\{g_i\}$ of the module and *relations* $\{\sum_i \alpha_i g_i = 0 \mid \alpha_i \in \Bbbk[t]\}$ among them. A presentation matrix encoding the relations in terms of the generators characterizes the module completely. Then, a matrix reduction algorithm akin to the persistence algorithm MATPERSISTENCE from Chapter 3 provides the desired decomposition.[1] Figure 11.3 illustrates the advantage of this presentation over the other costly presentation. In practice, when the one-parameter persistence module is given by an implicit simplicial filtration, one can apply the matrix reduction algorithm directly on a boundary matrix rather than first computing a presentation matrix from it and then decomposing it. If there are $O(n)$ simplices constituting the filtration, the algorithm runs in $O(n^3)$ time with simple matrix reductions and in $O(n^\omega)$ time with more sophisticated matrix multiplication techniques, where $\omega < 2.373$ is the exponent for matrix multiplication.

The Meataxe algorithm for multiparameter persistence modules follows the costly approach analogous to that in the one-parameter case that expects the presentation of each individual linear map explicitly. In particular, it expects the input d-parameter module M over a finite subset of $\mathbb{Z}^d$ to be given as a large matrix in $\mathbf{k}^{D \times D}$ with entries in a fixed field $\mathbf{k} = \mathbb{Z}_q$, where D is the sum

---

[1] The persistence algorithm takes a filtration as input whereas here a module is presented with input matrices.

of the dimensions of the vector spaces over all points in $\mathbb{Z}^d$ supporting M. The time complexity of the Meataxe algorithm is $O(D^6 \log q)$ [196]. In general, D might be quite large. It is not clear what is the most efficient way to transform an input that specifies generators and relations (or a simplicial filtration) to a representation matrix required by the Meataxe algorithm. A naive approach is to consider the minimal subgrid in $\mathbb{Z}^d$ that supports the nontrivial maps. In the worst case, with N being the total number of generators and relations, one has to consider $O\left(\binom{N}{d}\right) = O(N^d)$ grid points in $\mathbb{Z}^d$ each with a vector space of dimension $O(N)$. Therefore, $D = O(N^{d+1})$, giving a worst-case time complexity of $O(N^{6(d+1)} \log q)$. Even allowing approximation, the algorithm runs in $O(N^{3(d+1)} \log q)$ time [197].

In this chapter, we take the alternative approach where the module is treated as a finitely presented graded module over multivariate polynomial ring $R = \mathbf{k}[t_1, \ldots, t_d]$ [107] and presented with a set of generators and relations graded appropriately. Given a presentation matrix encoding relations with generators, our algorithm computes a diagonalization of the matrix giving a presentation of each indecomposable into which the input module decomposes. Compared to the one-parameter case, we have to overcome two main barriers for computing the indecomposables. First, we need to allow row operations along with column operations for reducing the input matrix. In the one-parameter case, row operations become redundant because column operations already produce the bars. Second, unlike in the one-parameter case, we cannot allow all left-to-right column or bottom-to-top row operations for the matrix reduction because the parameter space $\mathbb{Z}^d$, $d > 1$, unlike $\mathbb{Z}$ only induces a partial order on these operations. These two difficulties are overcome by an incremental approach combined with a linearization trick. Given a presentation matrix with a total of $O(N)$ generators and relations that are graded distinctly, the algorithm runs in $O(N^{2\omega+1})$ time. Surprisingly, the complexity does not depend on the parameter d.

Computing the presentation matrix from a multiparameter simplicial filtration is not easy. For d-parameter filtrations with n simplices, a presentation matrix of size $O(n^{d-1}) \times O(n^{d-1})$ can be computed in $O(n^{d+1})$ time by adapting an algorithm of Skryzalin [279] as described presented in [141]. We will not present this construction here. Instead, we focus on two cases: two-parameter persistence modules where the homology groups could be multi-dimensional; and d-parameter persistence modules where the homology group is only zero-dimensional. For these two cases, we can compute the presentation matrices more efficiently. For the two-parameter case, Lesnick and Wright [223] give an efficient $O(n^3)$ algorithm for computing a presentation matrix from an input filtration. In this case, N, the total number of generators and relations, is $O(n)$. For the zeroth homology groups, presentation matrices are given by the boundary matrices straightforwardly as detailed in Section 11.5.2 giving $N = O(n)$.

11.1 Multiparameter Persistence Modules

We define persistence modules in this chapter differently using the definition of graded modules in algebra. Graded module structures provide an appropriate framework for defining the multiparameter persistence, in particular, for the decomposition algorithm that we present. Also, navigating between the simplicial filtration and the module induced by it becomes natural with the graded module structure.

11.1.1 Persistence Modules as Graded Modules

First, we recall the definition of modules from Section 2.4.1. It requires a ring. We consider a module where the ring R is taken as the polynomial ring.

Definition 11.1. (Polynomial ring) Given a variable t and a field $\mathbf{k}$, the set of polynomials given by

$$\mathbf{k}[t] = \{a_0 + a_1 t + a_2 t^2 + \cdots + a_n t^n \mid n \geq 0, a_i \in \mathbf{k}\}$$

forms a ring with the usual polynomial addition and multiplication operations. The definition can be extended to multivariate polynomials:

$$\mathbf{k}[t] = \mathbf{k}[t_1, \ldots, t_k] = \left\{ \sum_{i_1, \ldots, i_k} a_{i_1, \ldots, i_k} t_1^{i_1} \cdots t_j^{i_j} \cdots t_k^{i_k} \mid i_1, \ldots, i_k \geq 0, a_{i_1, \ldots, i_k} \in \mathbf{k} \right\}.$$

We use *polynomial ring* to define multiparameter persistence modules. Specifically, let $R = \mathbf{k}[t_1, \ldots, t_d]$ be the d-variate polynomial ring for some $d \in \mathbb{Z}_+$ with $\mathbf{k}$ being a field. Throughout this chapter, we assume coefficients are in $\mathbf{k}$. Hence homology groups are vector spaces.

Definition 11.2. (Graded module) A $\mathbb{Z}^d$-*graded R-module* (*graded module* in brief) is an R-module M that is a direct sum of $\mathbf{k}$-vector spaces $M_{\mathbf{u}}$ indexed by $\mathbf{u} = (u_1, u_2, \ldots, u_d) \in \mathbb{Z}^d$, that is, $M = \bigoplus_{\mathbf{u}} M_{\mathbf{u}}$, such that the ring action satisfies that for all i, and for all $\mathbf{u} \in \mathbb{Z}^d$, $t_i \cdot M_{\mathbf{u}} \subseteq M_{\mathbf{u}+e_i}$, where $\{e_i\}_{i=1}^d$ is the standard basis in $\mathbb{Z}^d$. The indices $\mathbf{u} \in \mathbb{Z}^d$ are called *grades*.

Another interpretation of graded module is that, for each $\mathbf{u} \in \mathbb{Z}^d$, the action of t_i on $M_{\mathbf{u}}$ determines a linear map $t_i \bullet : M_{\mathbf{u}} \to M_{\mathbf{u}+e_i}$ by $(t_i \bullet)(m) = t_i \cdot$

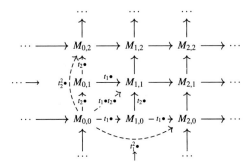

Figure 11.4 A graded two-parameter module. All subdiagrams of maps and compositions of maps are commutative.

m. So, we can also describe a graded module equivalently as a collection of vector spaces $\{M_\mathbf{u}\}_{\mathbf{u} \in \mathbb{Z}^d}$ with a collection of linear maps $\{t_i \bullet : M_\mathbf{u} \to M_{\mathbf{u}+e_i}$, $\forall i, \forall \mathbf{u}\}$, where the commutative property $(t_j \bullet) \circ (t_i \bullet) = (t_i \bullet) \circ (t_j \bullet)$ holds. The commutative diagram in Figure 11.4 shows a graded module for $d = 2$, also called a bigraded module.

Definition 11.3. (Graded module R) There is a special *graded module M* where $M_\mathbf{u}$ is the **k**-vector space generated by $\mathbf{t^u} = t_1^{u_1} t_2^{u_2} \cdots t_d^{u_d}$ and the ring action is given by the ring R. We denote it with R, not to be confused with the ring R which is used to define it.

Before we introduce persistence modules as instances of graded modules, we extend the notion of simplicial filtration to the multiparameter framework.

Definition 11.4. (d-Parameter filtration) A *(d-parameter) simplicial filtration* is a family of simplicial complexes $\{X_\mathbf{u}\}_{\mathbf{u} \in \mathbb{Z}^d}$ such that for each grade $\mathbf{u} \in \mathbb{Z}^d$ and each $i = 1, \ldots, d$, $X_\mathbf{u} \subseteq X_{\mathbf{u}+e_i}$.

Figure 11.5 (top) shows an example of a two-parameter filtration and various graded modules associated with it (bottom). The module resulting with the homology group in Figure 11.5(c) is a persistence module. Also shown are other graded modules of chain groups (Figure 11.5a) and boundary groups (Figure 11.5b).

Definition 11.5. (d-Parameter persistence module) We call a $\mathbb{Z}^d$-graded R-module M a *d-parameter persistence module* when $M_\mathbf{u}$ for each $\mathbf{u} \in \mathbb{Z}^\mathbf{d}$ is a homology group defined over a field and the linear maps corresponding to ring actions among them are induced by inclusions in a d-parameter simplicial filtration. We call M *finitely generated* if there exists a finite set of elements

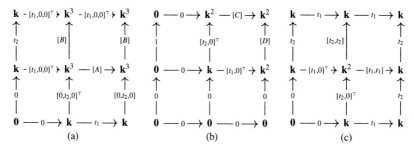

Figure 11.5 (top) An example of a two-parameter simplicial filtration. Each square box indicates what is the current (filtered) simplicial complex at the bottom left grid point of the box. (bottom) We show different modules considering different abelian groups arising out of the complexes with the ring actions on the arrows (see Section 11.5 for details). (a) The module of the zeroth chain groups C_0, $A = \begin{pmatrix} t_1 & 0 & 0 \\ 0 & t_1 & 0 \\ 0 & 0 & t_1 \end{pmatrix}$ and $B = \begin{pmatrix} t_2 & 0 & 0 \\ 0 & t_2 & 0 \\ 0 & 0 & t_2 \end{pmatrix}$. (b) The module of the zeroth boundary groups B_0, $C = \begin{pmatrix} t_1 & 0 \\ 0 & t_1 \end{pmatrix}$ and $D = \begin{pmatrix} t_2 & 0 \\ 0 & t_2 \end{pmatrix}$. (c) The module of the zeroth homology groups H_0; it has one connected component in zeroth homology groups at grades except $(0, 0)$ and $(1, 1)$, and has two connected components at grade $(1, 1)$.

$\{g_1, \ldots, g_n\} \subseteq M$ such that each element $m \in M$ can be written as an R-linear combination of these elements, that is, $m = \sum_{i=1}^{n} \alpha_i g_i$ with $\alpha_i \in R$. We call this set $\{g_i\}$ a *generating set* or *generators* of M. A generating set is called *minimal* if its cardinality is minimal among all generating sets. The R-linear combinations $\sum_{i=1}^{n} \alpha_i g_i$ that are zero are called *relations*. We will see later that a module can be represented by a set of generators and relations.

In this exposition, we assume that all modules are finitely generated. Such modules always admit a minimal generating set. In our example in Figure 11.5, the vertex set $\{v_b, v_r, v_g\}$ is a minimal generating set for the module of zero-dimensional homology groups.

Definition 11.6. (Morphism) A *graded module morphism*, called *morphism* in short, between two graded modules M and N is defined as an R-linear map

$f : M \to N$ preserving grades: $f(M_\mathbf{u}) \subseteq N_\mathbf{u}$, for all $\mathbf{u} \in \mathbb{Z}^d$. Equivalently, it can also be described as a collection of linear maps $\{f_\mathbf{u} : M_\mathbf{u} \to N_\mathbf{u}\}$ which gives the following commutative diagram for each $\mathbf{u}$ and i.

$$
\begin{array}{ccc}
M_\mathbf{u} & \xrightarrow{t_i} & M_{\mathbf{u}+e_i} \\
{\scriptstyle f_\mathbf{u}}\downarrow & & \downarrow{\scriptstyle f_{\mathbf{u}+e_i}} \\
N_\mathbf{u} & \xrightarrow{t_i} & N_{\mathbf{u}+e_i}
\end{array}
$$

Two graded modules M and N are isomorphic if there exist two morphisms $f : M \to N$ and $g : N \to M$ such that $g \circ f$ and $f \circ g$ are identity maps. We denote it as $M \cong N$.

Definition 11.7. (Shifted module) For a graded module M and some $\mathbf{u} \in \mathbb{Z}^d$, define a *shifted graded module* $M_{\to\mathbf{u}}$ by setting $(M_{\to\mathbf{u}})_\mathbf{v} = M_{\mathbf{v}-\mathbf{u}}$ for each $\mathbf{v}$.

Definition 11.8. (Free module) We say a graded module is *free* if it is isomorphic to the direct sum of a collection of R^j, denoted as $\bigoplus_j R^j$, where each $R^j = R_{\to\mathbf{u}_j}$ for some $\mathbf{u}_j \in \mathbb{Z}^d$. Here R is the special graded module in Definition 11.3.

Definition 11.9. (Homogeneous element) We say an element $m \in M$ is *homogeneous* if $m \in M_\mathbf{u}$ for some $\mathbf{u} \in \mathbb{Z}^d$. We denote $\mathrm{gr}(m) = \mathbf{u}$ as the *grade* of such a homogeneous element. To emphasize the grade of a homogeneous element, we also write $m^{\mathrm{gr}(m)} := m$.

A minimal generating set of a free module is called a *basis*. We usually further require that all the elements (generators) in a basis are homogeneous. For a free module $F \cong \bigoplus_j R^j$ such a basis exists. Specifically, $\{e_j : j = 1, 2, \ldots\}$ is a homogeneous basis of F, where e_j indicates the multiplicative identity in R^j. The generating set $\{e_j : j = 1, 2, \ldots\}$ is often referred to as the standard basis of $\bigoplus_j R^j = \langle e_j : j = 1, 2, \ldots \rangle$.

11.2 Presentations of Persistence Modules

Definition 11.10. (Decomposition) For a finitely generated graded module M, we call $M \cong \bigoplus M^i$ a *decomposition* of M for some collection of modules $\{M^i\}$. We say M is *indecomposable* if $M \cong M^1 \oplus M^2 \implies M^1 = \mathbf{0}$ or $M^2 = \mathbf{0}$ where $\mathbf{0}$ denotes a trivial module. By the Krull–Schmidt theorem [10], there exists an essentially unique (up to permutation and isomorphism) decomposition $M \cong \bigoplus M^i$ with every M^i being indecomposable. We call it the *total decomposition* of M.

For example, the free module R in Definition 11.3 is generated by $\langle e_1^{(0,0)}\rangle$ and the free module $R_{\rightarrow(0,1)} \oplus R_{\rightarrow(1,0)}$ is generated by $\langle e_1^{(0,1)}, e_2^{(1,0)}\rangle$. A free module M generated by $\langle e_j^{\mathbf{u}_j} : j = 1, 2, \ldots \rangle$ has a (total) decomposition $M \cong \bigoplus_j R_{\rightarrow \mathbf{u}_j}$.

Definition 11.11. (Isomorphic morphisms) Two morphisms $f : M \rightarrow N$ and $f' : M' \rightarrow N'$ are *isomorphic*, denoted as $f \cong f'$, if there exist isomorphisms $g : M \rightarrow M'$ and $h : N \rightarrow N'$ such that the following diagram commutes.

$$
\begin{array}{ccc}
M & \xrightarrow{\;f\;} & N \\
{\scriptstyle g}\downarrow{\scriptstyle \cong} & & {\scriptstyle h}\downarrow{\scriptstyle \cong} \\
M' & \xrightarrow{\;f'\;} & N'
\end{array}
$$

Essentially, like isomorphic modules, two isomorphic morphisms can be considered the same. For two morphisms $f_1 : M^1 \rightarrow N^1$ and $f_2 : M^2 \rightarrow N^2$, there exists a canonical morphism $g : M^1 \oplus M^2 \rightarrow N^1 \oplus N^2$, $g(m_1, m_2) = (f_1(m_1), f_2(m_2))$, which is essentially uniquely determined by f_1 and f_2 and is denoted as $f_1 \oplus f_2$. A module is trivial if it has only the element 0 at every grade. We denote a trivial morphism by $\mathbb{0} : \mathbf{0} \rightarrow \mathbf{0}$. Analogous to the decomposition of a module, we can also define a decomposition of a morphism.

Definition 11.12. (Morphism decomposition) A morphism f is indecomposable if $f \simeq f_1 \oplus f_2 \implies f_1$ or f_2 is the trivial morphism $\mathbb{0} : \mathbf{0} \rightarrow \mathbf{0}$. We call $f \cong \bigoplus f_i$ a *decomposition* of f. If each f_i is indecomposable, we call it a *total decomposition* of f.

Like decompositions of modules, the total decomposition of a morphism is also essentially unique.

11.2.1 Presentation and Its Decomposition

To study total decompositions of persistence modules that are treated as graded modules, we draw upon the idea of *presentations* of graded modules and build a bridge between the decompositions of persistence modules and the corresponding presentations. The decompositions of presentations can be transformed to a matrix reduction problem with possibly nontrivial constraints which we will introduce in Section 11.3. We first state a result saying that there are one-to-one correspondences between persistence modules, presentations, and presentation matrices. Recall that, by assumption, all modules are finitely generated. A graded module, hence a persistence module, accommodates a description called its *presentation* that aids finding its decomposition. We remind the reader that a sequence of maps is *exact* if the image of one map equals the kernel of the next map.

Definition 11.13. (Presentation) A *presentation* of a graded module H is an *exact* sequence
$$F^1 \xrightarrow{f} F^0 \xrightarrow{g} H \to 0 \quad \text{where } F^1 \text{ and } F^0 \text{ are free.}$$
We call f a presentation map. We say a graded module H is *finitely presented* if there exists a presentation of H with both F^1 and F^0 being finitely generated.

The exactness of the sequence implies that $\operatorname{im} f = \ker g$ and $\operatorname{im} g = H$. The double arrows on the second map in the sequence signifies the surjection of g. It follows that $\operatorname{coker} f \cong H$ and the presentation is determined by the presentation map f.

Remark 11.1. *Presentations of a given graded module are not unique. However, there exists an essentially unique (up to isomorphism) presentation f of a graded module in the sense that any presentation f' of that module can be written as $f' \cong f \oplus f''$ with $\operatorname{coker} f'' = 0$. We call this unique presentation the minimal presentation. See more details of the construction and properties of minimal presentation in [141].*

Definition 11.14. (Presentation matrix) Given a presentation $F^1 \xrightarrow{f} F^0 \to H$, fixed bases of F^1 (relations) and F^0 (generators) provide a matrix form $[f]$ of the presentation map f, which we call a *presentation matrix* of H. It has entries in R. In the special case that H is a free module with F^1 being a zero module, we define the presentation matrix $[f]$ of H to be a *null column matrix* with matrix size $\ell \times 0$ for some $\ell \in \mathbb{N}$.

In Figure 11.6, we illustrate the presentation matrix of the persistence module H_0 consisting of zero-dimensional homology groups induced by the filtration shown in Figure 11.5. We will see later that, in this case, f equals the boundary morphism $\partial_1 : C_1 \to C_0$ whose columns are edges and rows are vertices. For example, the red edge e_r whose grade is $(1, 1)$ has two boundary vertices v_b, the blue vertex with grade $(0, 1)$, and v_r, the red vertex with grade $(1, 0)$. To bring v_b to grade $(1, 1)$, we need to multiply by the polynomial t_1. Similarly, to bring v_r to grade $(1, 1)$, we need to multiply by t_2. The corresponding entries in the column of e_r are t_1 and t_2, respectively, indicated by shaded boxes. Actual matrices are shown later in Example 11.1.

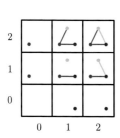
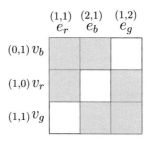

Figure 11.6 The presentation matrix of the module H_0 consisting of zero-dimensional homology groups for the example in Figure 11.5. The boxes in the matrix containing nonzero entries are shaded.

An important property of a graded module H is that a decomposition of its presentation f corresponds to a decomposition of H itself. The decomposition of f can be computed by *diagonalizing* its presentation matrix $[f]$. Informally, a diagonalization of a matrix $\mathbf{A}$ is an equivalent matrix $\mathbf{A}'$ in the following form (see formal Definition 11.15 later):

$$
\mathbf{A}' = \begin{bmatrix} \mathbf{A}_1 & 0 & \cdots & 0 \\ 0 & \mathbf{A}_2 & \cdots & 0 \\ \vdots & \vdots & \ddots & \vdots \\ 0 & 0 & \cdots & \mathbf{A}_k \end{bmatrix}.
$$

All nonzero entries are in the $\mathbf{A}_i$ and we write $\mathbf{A} \cong \bigoplus \mathbf{A}_i$. It is not hard to see that for a map $f \cong \bigoplus f_i$, there is a corresponding diagonalization $[f] \cong \bigoplus [f_i]$. With these definitions, we have the following theorem that motivates the decomposition algorithm (see proof in [141]).

Theorem 11.1. *There are one-to-one correspondences between the following three structures arising from a minimal presentation map $f : F^1 \to F^0$ of a graded module H, and its presentation matrix $[f]$:*

(a) a decomposition of the graded module $H \cong \bigoplus H^i$;
(b) a decomposition of the presentation map $f \cong \bigoplus f_i$;
(c) a diagonalization of the presentation matrix $[f] \cong \bigoplus [f]_i$.

Remark 11.2. *From Theorem 11.1, we can see that there exist an essentially unique total decomposition of a presentation map and an essentially unique total diagonalization of the presentation matrix of H which correspond to an essentially unique total decomposition of H (up to permutation, isomorphism, and trivial summands). In practice, we might be given a presentation which is not necessarily minimal. One way to handle this case is to compute the minimal presentation of the given presentation first. For two-parameter modules, this* *can be done by an algorithm presented in [223]. The other choice is to compute the decomposition of the given presentation directly, which is sufficient to get the decomposition of the module thanks to the following proposition.*

Proposition 11.2. *Let f be any presentation (not necessarily minimal) of a graded module H. The following statements hold:*

(a) for any decomposition of $H \cong \bigoplus H^i$, there exists a decomposition of $f \cong \oplus f_i$ such that $\operatorname{coker} f_i = H^i$, for all i;
(b) the total decomposition of H follows from the total decomposition of f.

Remark 11.3. *By Remark 11.1, any presentation f can be written as $f \cong f^* \oplus f'$ with f^* being the minimal presentation and $\operatorname{coker} f' = \mathbf{0}$. Furthermore, f' can be written as $f' \cong g \oplus h$ where g is an identity map and h is a zero map. The corresponding matrix form is $[f] \cong [f^*] \oplus [g] \oplus [h]$ with $[g]$ being an identity submatrix and $[h]$ being an empty matrix representing a collection of zero columns. Therefore, one can easily read these trivial parts from the result of matrix diagonalization. See the following diagram for an illustration.*

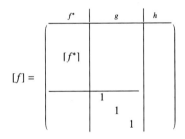

11.3 Presentation Matrix: Diagonalization and Simplification

Our aim is to compute a total diagonalization of a presentation matrix over $\mathbb{Z}_2$. Here we formally define some notation used in the diagonalization. All (graded) modules are assumed to be finitely presented and we take $\mathbf{k} = \mathbb{Z}_2$ for simplicity, though the method can be generalized for any finite field (Exercise 9). We have observed that a total decomposition of a module can be achieved by computing a total decomposition of its presentation f. This in turn requires a total diagonalization of the presentation matrix $[f]$. Here we formally define some notation about the diagonalization.

Given an $\ell \times m$ matrix $\mathbf{A} = [\mathbf{A}_{i,j}]$, with row indices $\mathsf{Row}(\mathbf{A}) = [\ell] := \{1, 2, \ldots, \ell\}$ and column indices $\mathsf{Col}(\mathbf{A}) = [m] := \{1, 2, \ldots, m\}$, we define an *index block* B of $\mathbf{A}$ as a pair $\left[\mathsf{Row}(B), \mathsf{Col}(B)\right]$ with $\mathsf{Row}(B) \subseteq \mathsf{Row}(\mathbf{A})$ and $\mathsf{Col}(B) \subseteq \mathsf{Col}(\mathbf{A})$. We say an index pair (i, j) is in B if $i \in \mathsf{Row}(B)$ and $j \in \mathsf{Col}(B)$, denoted as $(i, j) \in B$. We denote a *block* of $\mathbf{A}$ on B as the matrix restricted to the index block B, that is, $[\mathbf{A}_{i,j}]_{(i,j)\in B}$, denoted as $\mathbf{A}|_B$. We call B the index of the block $\mathbf{A}|_B$. We abuse the notation $\mathsf{Row}(\mathbf{A}|_B) :=$ $\mathsf{Row}(B)$ and $\mathsf{Col}(\mathbf{A}|_B) := \mathsf{Col}(B)$. For example, the ith row $r_i = \mathbf{A}_{i,*} =$ $\mathbf{A}|_{[\{i\},\mathsf{Col}(\mathbf{A})]}$ and the jth column $c_j = \mathbf{A}_{*,j} = \mathbf{A}|_{[\mathsf{Row}(\mathbf{A}),\{j\}]}$ are blocks with indices $\left[\{i\}, \mathsf{Col}(\mathbf{A})\right]$ and $\left[\mathsf{Row}(\mathbf{A}), \{j\}\right]$, respectively. Specifically, $\left[\varnothing, \{j\}\right]$ represents an index block of a single column j and $[\{i\}, \varnothing]$ represents an index block of a single row i. We call $[\varnothing, \varnothing]$ the empty index block.

A matrix can have multiple equivalent forms for the same morphism they represent. We use $\mathbf{A}' \sim \mathbf{A}$ to denote the equivalence of matrices. One fact about equivalent matrices is that they can be obtained from one another by row and column operations introduced later (Chapter 5 in [109]).

Definition 11.15. (Diagonalization) A matrix $\mathbf{A}' \sim \mathbf{A}$ is called a *diagonalization of* $\mathbf{A}$ with a set of non-empty disjoint index blocks $\mathcal{B} = \{B_1, B_2, \ldots, B_k\}$ if the rows and columns of $\mathbf{A}$ are partitioned into these blocks, that is, $\mathsf{Row}(\mathbf{A}) = \coprod_i \mathsf{Row}(B_i)$ and $\mathsf{Col}(\mathbf{A}) = \coprod_i \mathsf{Col}(B_i)$, and all the nonzero entries of $\mathbf{A}'$ have indices in some B_i ($\coprod_i$ denotes disjoint union). We write $\mathbf{A}' = \bigoplus_{B_i \in \mathcal{B}} \mathbf{A}'|_{B_i}$. We say $\mathbf{A}' = \bigoplus_{B_i \in \mathcal{B}} \mathbf{A}'|_{B_i}$ is *total* if no block in this diagonalization can be diagonalized further into smaller non-empty blocks. That means, for each block $\mathbf{A}'|_{B_i}$, there is no nontrivial diagonalization. Specifically, when $\mathbf{A}$ is a null column matrix (the presentation matrix of a free module), we say that $\mathbf{A}$ is itself a total diagonalization with index blocks $\{[\{i\}, \varnothing] \mid i \in \mathsf{Row}(\mathbf{A})\}$.

Note that each non-empty matrix $\mathbf{A}$ has a trivial diagonalization with the set of index blocks being the singleton $\{(\mathsf{Row}(\mathbf{A}), \mathsf{Col}(\mathbf{A}))\}$. Guaranteed by the Krull–Schmidt theorem [10], all total diagonalizations are unique up to permutations of their rows and columns, and equivalent transformation within each block. The total diagonalization of $\mathbf{A}$ is denoted generically as $\mathbf{A}^*$. All total diagonalizations of $\mathbf{A}$ have the same set of index blocks unique up to permutations of rows and columns. See Figure 11.7 for an illustration of a diagonalized matrix.

11.3.1 Simplification

First we want to transform the diagonalization problem to an equivalent problem that involves matrices with a simpler form. The idea is to simplify the presentation matrix to have entries only in $\mathbf{k}$ which is taken as $\mathbb{Z}_2$. There is a

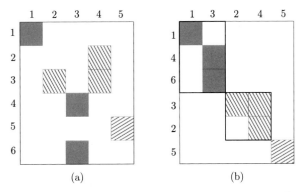

Figure 11.7 (a) A nontrivial diagonalization where the locations of nonzero entries are patterned and the patterns for all such entries in the same block are the same. (b) The same matrix with permutation of columns and rows to bring entries of a block into adjacent locations. The three index blocks are: $((1, 4, 6)(1, 3))$, $((2, 3)(2, 4))$, and $((5)(5))$.

correspondence between diagonalizations of the original presentation matrix and certain constrained diagonalizations of the corresponding transformed matrix.

We first make some observations about the homogeneous property of presentation maps and presentation matrices. Equivalent matrices actually represent isomorphic presentations $f' \cong f$ that admit the commutative diagram

$$\begin{array}{ccc} F^1 & \xrightarrow{\ f\ } & F^0 \\ {\scriptstyle h^1}\downarrow{\scriptstyle\cong} & & {\scriptstyle\cong}\downarrow{\scriptstyle h^0} \\ F^1 & \xrightarrow{\ f'\ } & F^0 \end{array}$$

where h^1 and h^0 are endomorphisms on F^1 and F^0, respectively. The endomorphisms are realized by basis changes between corresponding presentation matrices $[f] \cong [f']$. Since all morphisms between graded modules are required to be homogeneous (preserve grades) by definition, we can use homogeneous bases (all the basis elements chosen are homogeneous elements[2]) for F^0 and F^1 to represent matrices. Let $F^0 = \langle g_1, \ldots, g_\ell \rangle$ and $F^1 = \langle s_1, \ldots, s_m \rangle$ where g_i and s_i are homogeneous elements for every i. With this choice, we can consider only equivalent presentation matrices under homogeneous basis changes. Each entry $[f]_{i,j}$ is also homogeneous. That means $[f]_{i,j} = t_1^{u_1} t_2^{u_2} \cdots t_d^{u_d}$ where $(u_1, u_2, \ldots, u_d) = \mathrm{gr}(s_j) - \mathrm{gr}(g_i)$. Writing

---

[2] Recall that an element $m \in M$ is homogeneous with grade $\mathrm{gr}(m) = \mathbf{u}$ for some $\mathbf{u} \in \mathbb{Z}^d$ if $m \in M_{\mathbf{u}}$.

$\mathbf{u} = (u_1, u_2, \ldots, u_d)$ and $\mathbf{t^u} = t_1^{u_1} t_2^{u_2} \cdots t_d^{u_d}$, we get $[f]_{i,j} = \mathbf{t^u}$ where $\mathbf{u} = \mathrm{gr}(s_j) - \mathrm{gr}(g_i)$ is called the grade of $[f]_{i,j}$. We call such a presentation matrix a *homogeneous presentation matrix*.

For example, given $F^0 = \langle g_1^{(1,1)}, g_2^{(2,2)} \rangle$, the basis change $g_2^{(2,2)} \leftarrow g_2^{(2,2)} + g_1^{(1,1)}$ is not homogeneous since $g_2^{(2,2)} + g_1^{(1,1)}$ is not a homogeneous element. However, $g_2^{(2,2)} \leftarrow g_2^{(2,2)} + \mathbf{t}^{(1,1)} g_1^{(1,1)}$ is a homogeneous change with $\mathrm{gr}(g_2^{(2,2)} + \mathbf{t}^{(1,1)} g_1^{(1,1)}) = \mathrm{gr}(g_2^{(2,2)}) = (2,2)$, which results in a new homogeneous basis, $\{g_1^{(1,1)}, g_2^{(2,2)} + \mathbf{t}^{(1,1)} g_1^{(1,1)}\}$. Homogeneous basis changes always result in homogeneous bases.

Let $[f]$ be a homogeneous presentation matrix of $f \colon F^1 \to F^0$ with bases $F^0 = \langle g_1, \ldots, g_\ell \rangle$ and $F^1 = \langle s_1, \ldots, s_m \rangle$. We extend the notation of grading to every row r_i and every column c_j from the basis elements g_i and s_j they represent, respectively, that is, $\mathrm{gr}(r_i) := \mathrm{gr}(g_i)$ and $\mathrm{gr}(c_j) := \mathrm{gr}(s_j)$. We define a strict partial order $<_{\mathrm{gr}}$ on rows $\{r_i\}$ by asserting $r_i <_{\mathrm{gr}} r_j$ if and only if $\mathrm{gr}(r_i) < \mathrm{gr}(r_j)$. Similarly, we define a strict partial order on columns $\{c_j\}$.

For such a homogeneous presentation matrix $[f]$, we aim to diagonalize it totally by homogeneous change of basis while trying to zero out entries by column and row operations that include additions and scalar multiplication of columns and rows as done in well-known Gaussian elimination. We have the following observations:

1. $\mathrm{gr}([f]_{i,j}) = \mathrm{gr}(c_j) - \mathrm{gr}(r_i)$;
2. a nonzero entry $[f]_{i,j}$ can only be zeroed out by column operations from columns $c_k <_{\mathrm{gr}} c_j$ or by row operations from rows $r_\ell >_{\mathrm{gr}} r_i$.

Observation 2 indicates which subset of column and row operations is sufficient to zero out the entry $[f]_{i,j}$. We restate the diagonalization problem as follows:

Given an $n \times m$ homogeneous presentation matrix $\mathbf{A} = [f]$ consisting of entries in $\mathbf{k}[t_1, \ldots, t_d]$ with grading on rows and columns, find a total diagonalization of $\mathbf{A}$ under the following *admissible* row and column operations:

- multiply a row or column by nonzero $\alpha \in \mathbf{k}$ (for $\mathbf{k} = \mathbb{Z}_2$, we can ignore these operations);
- for two rows r_i and r_j with $j \neq i$ and $r_j <_{\mathrm{gr}} r_i$, set $r_j \leftarrow r_j + \mathbf{t^u} \cdot r_i$ where $\mathbf{u} = \mathrm{gr}(r_i) - \mathrm{gr}(r_j)$;
- for two columns c_i and c_j with $j \neq i$ and $c_i <_{\mathrm{gr}} c_j$, set $c_j \leftarrow c_j + \mathbf{t^v} \cdot c_i$ where $\mathbf{v} = \mathrm{gr}(c_j) - \mathrm{gr}(c_i)$.

The above operations realize all possible homogeneous basis changes. That means that any homogeneous presentation matrix can be realized by a combination of the above operations.

In fact, the values of nonzero entries in the matrix are redundant under the homogeneous property $\mathrm{gr}(\mathbf{A}_{i,j}) = \mathrm{gr}(c_j) - \mathrm{gr}(r_i)$ given by observation 1. So, we can further simplify the matrix by replacing all the nonzero entries with their $\mathbf{k}$-coefficients. For example, we can replace $2 \cdot \mathbf{t}^{\mathbf{u}}$ with 2. What really matters are the partial orders defined by the grading of rows and columns. With our assumption of $\mathbf{k} = \mathbb{Z}_2$, all nonzero entries are replaced with 1. Based on the above observations, we further simplify the diagonalization problem to be the one as follows.

Given a $\mathbf{k}$-valued matrix $\mathbf{A}$ with a partial order on rows and columns, find a total diagonalization $\mathbf{A}^* \sim \mathbf{A}$ with the following *admissible operations*:

- multiply a row or column by nonzero $\alpha \in \mathbf{k}$ (for $\mathbf{k} = \mathbb{Z}_2$, we can ignore these operations);
- adding c_i to c_j only if $j \neq i$ and $\mathrm{gr}(c_i) < \mathrm{gr}(c_j)$; denoted as $c_i \to c_j$;
- adding r_k to r_l only if $l \neq k$ and $\mathrm{gr}(r_\ell) < \mathrm{gr}(r_k)$; denoted as $r_k \to r_l$.

The assumption of $\mathbf{k} = \mathbb{Z}_2$ allows us to ignore the first set of multiplication operations on the binary matrix obtained after transformation. We denote the set of all admissible column and row operations as follows:

$\mathsf{Colop} = \{(i, j) \mid c_i \to c_j \text{ is an admissible column operation}\};$

$\mathsf{Rowop} = \{(k, l) \mid r_k \to r_l \text{ is an admissible row operation}\}.$

Under the assumption that no two columns nor rows have the same grades, Colop and Rowop are closed under transitive relation.

Proposition 11.3. *If* $(i, j), (j, k) \in \mathsf{Colop}$ *(*Rowop*) then* $(i, k) \in \mathsf{Colop}$ *(*Rowop*).*

Given a solution of the diagonalization problem in the simplified form, one can reconstruct a solution of the original problem on the presentation matrix by reversing the above process of simplification. We will illustrate it by running the algorithm on the working example in Figure 11.5 at the end of this section. The matrix reduction we employ for diagonalization may be viewed as a *generalized matrix reduction* because the matrix is reduced under constrained operations Colop and Rowop which might be a nontrivial subset of all basic operations.

Remark 11.4. *There are two extreme but trivial cases: (i) There are no $<_{\mathrm{gr}}$-comparable pairs of rows and columns. In this case, $\mathsf{Colop} = \mathsf{Rowop} = \varnothing$ and the original matrix is a trivial solution. (ii) All pairs of rows and all pairs of columns are $<_{\mathrm{gr}}$-comparable. Or equivalently, both Colop and Rowop are totally ordered. In this case, one can apply the traditional matrix reduction algorithm to reduce the matrix to a diagonal matrix with all nonzero*

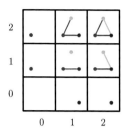
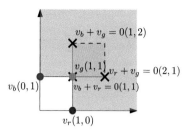

Figure 11.8 The persistence module corresponding to the presentation matrix $[\partial_1]$ shown in Example 11.1. The generators are given by the three vertices with grades $(0, 1), (1, 0)$, and $(1, 1)$ and the relations are given by the edges with grades $(1, 1)$, $(1, 2)$, and $(2, 1)$.

blocks being 1×1 minors. This is also the case for the one-parameter persistence module if one further applies row reduction after column reduction. Note that row reductions are not necessary for reading out persistence information because it essentially does not change the persistence information. However, in multiparameter cases, both column and row reductions are necessary to obtain a diagonalization from which the persistence information can be read. From this viewpoint, the algorithm we present can be thought of as a generalization of the traditional persistence algorithm.

Example 11.1. *Consider our working example in Figure 11.5. One can see later in Section 11.5.2 that the presentation matrix of this example can be chosen to be the same as the matrix of the boundary morphism $\partial_1 : \mathsf{C}_1 \to \mathsf{C}_0$. With fixed bases $\mathsf{C}_0 = \langle v_b^{(0,1)}, v_r^{(1,0)}, v_g^{(1,1)} \rangle$ and $\mathsf{C}_1 = \langle e_r^{(1,1)}, e_b^{(1,2)}, e_g^{(2,1)} \rangle$, this presentation matrix $[\partial_1]$ and the corresponding binary matrix $\mathbf{A}$ can be written as follows (recall that superscripts indicate the grades):*

$$
\begin{array}{cccc}
[\partial_1] & e_r^{(1,1)} & e_b^{(1,2)} & e_g^{(2,1)} \\
v_b^{(0,1)} & \left(\begin{matrix} \mathbf{t}^{(1,0)} & \mathbf{t}^{(1,1)} & 0 \\ v_r^{(1,0)} & \mathbf{t}^{(0,1)} & 0 & \mathbf{t}^{(1,1)} \\ v_g^{(1,1)} & 0 & \mathbf{t}^{(0,1)} & \mathbf{t}^{(1,0)} \end{matrix} \right. &
\end{array}
$$

$$
\begin{array}{cccc}
\mathbf{A} & c_1^{(1,1)} & c_2^{(1,2)} & c_3^{(2,1)} \\
r_1^{(0,1)} & 1 & 1 & 0 \\
r_2^{(1,0)} & 1 & 0 & 1 \\
r_3^{(1,1)} & 0 & 1 & 1
\end{array}.
$$

Four admissible operations are: $r_3 \to r_1, r_3 \to r_2, c_1 \to c_2$, and $c_1 \to c_3$. Figure 11.8 shows the persistence module H_0 whose presentation matrix is $[\partial_1]$.

11.4 Total Diagonalization Algorithm

Assume that no two columns nor rows have the same grades. Without this assumption, the problem of total diagonalization becomes more complicated.

At this point, we do not know how to extend the algorithm to overcome this limitation. However, the algorithm introduced below can still compute a correct diagonalization (not necessarily total) by applying the trick of adding small enough perturbations to tied grades (considering $\mathbb{Z}^d \subseteq \mathbb{R}^d$) to reduce the case to the one satisfying our assumption. Furthermore, this diagonalization in fact coincides with a total diagonalization of some persistence module which is arbitrarily close to the original persistence module under a well-known metric called interleaving distance which we discuss in the next chapter. In practice, the persistence module usually arises from a simplicial filtration as shown in our working example. The assumption of distinct grading of the columns and rows is automatically satisfied if at most one simplex is introduced at each grade in the filtration.

Let $\mathbf{A}$ be the presentation matrix whose total diagonalization we are seeking. We order the rows and columns of the matrix $\mathbf{A}$ according to any order that extends the partial order on the grades to a total order, for example, dictionary

---

Algorithm 22 TOTDIAGONALIZE($\mathbf{A}$)

Input:
   $\mathbf{A}$ = input matrix treated as a global variable whose columns and rows are totally ordered respecting some fixed partial order given by the grading
Output:
   A total diagonalization $\mathbf{A}^*$ with index blocks $\mathcal{B}^*$

1: $\mathcal{B}^{(0)} \leftarrow \{B_i^{(0)} := \big[\{i\}, \varnothing\big] \mid i \in \mathsf{Row}(\mathbf{A})\}$
2: **for** $t \leftarrow 1$ to $m := |\mathsf{Col}(\mathbf{A})|$ **do**
3:    $B_0^{(t)} \leftarrow \big[\varnothing, \{t\}\big]$
4:    **for** each $B_i^{(t-1)} \in \mathcal{B}^{(t-1)}$ **do**
5:       $T := \big[\mathsf{Row}(B_i^{(t-1)}), \mathsf{Col}(\mathbf{A}_{\leq t}) \setminus \mathsf{Col}(B_i^{(t-1)})\big]$
6:       **if** BLOCKREDUCE (T) == false **then**
7:          $B_i^{(t)} \leftarrow B_i^{(t-1)} \oplus B_0^{(t)}$; * update B_i by appending t *\
8:       **else**
9:          $B_i^{(t)} \leftarrow B_i^{(t-1)}$; * B_i remains unchanged *\
10:         * $\mathbf{A}$ and c_t are updated in BLOCKREDUCE when it returns true *\
11:      **end if**
12:   **end for**
13:   $\mathcal{B}^{(t)} \leftarrow \{B_i^{(t)}\}$ with all $B_i^{(t)}$ containing t merged as one block
14: **end for**
15: Return $(\mathbf{A}, \mathcal{B}^{(m)})$

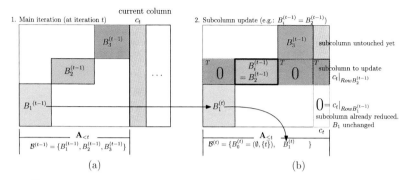

Figure 11.9 (a) Matrix $\mathbf{A}$ at the beginning of iteration t with $\mathbf{A}_{<t}$ being totally diagonalized with three index blocks $\mathcal{B}^{(t-1)} = \{B_1^{(t-1)}, B_2^{(t-1)}, B_3^{(t-1)}\}$. (b) A subcolumn update step: $c_t|_{\mathsf{Row}B_1^{(t-1)}}$ has already been reduced to zero. So, $B_1^{(t)} = B_1^{(t-1)}$ is added into $\mathcal{B}^{(t)}$. White regions including $c_t|_{\mathsf{Row}B_1^{(t-1)}}$ must be preserved afterward. Now for $i = 2$, we attempt to reduce purple subcolumn $c_t|_{\mathsf{Row}B_2^{(t-1)}}$. We extend it to block on $T := \left[\mathsf{Row}(B_2^{(t-1)}), (\mathsf{Col}(A_{\leq t}) \backslash \mathsf{Col}(B_2^{(t-1)}))\right]$ (colored purple) and try to reduce it in BLOCKREDUCE.

order. We fix the indices $\mathsf{Row}(\mathbf{A}) = \{1, 2, \ldots, \ell\}$ and $\mathsf{Col}(\mathbf{A}) = \{1, 2, \ldots, m\}$ according to this order. With this ordering, observe that, for each admissible column operation $c_i \to c_j$, we have $i < j$, and for each admissible row operation $r_l \to r_k$, we have $l > k$.

For any column c_t, let $\mathbf{A}_{\leq t} := \mathbf{A}|_C$ denote the left submatrix on $C = \left[\mathsf{Row}(\mathbf{A}), \{j \in \mathsf{Col}(\mathbf{A}) \mid j \leq t\}\right]$ and $\mathbf{A}_{<t}$ denote its stricter version obtained by excluding column c_t from $\mathbf{A}_{\leq t}$. Our algorithm starts with the finest decomposition that puts every free module given by each generator (rows) into a separate block and then combines them incrementally as we process the relations (columns). The main idea of our algorithm is presented in Algorithm 22: TOTDIAGONALIZE, which runs as follows (see Figure 11.9 for an illustration).

1. **Initialization**: Initialize the collection of index blocks $\mathcal{B}^{(0)} := \{B_i^{(0)} := \left[\{i\}, \varnothing\right] \mid i \in \mathsf{Row}(\mathbf{A})\}$, for the total diagonalization of the null column matrix $\mathbf{A}_{\leq 0}$.

2. **Main iteration**: Process $\mathbf{A}$ from left to right incrementally by introducing a column c_t and considering left submatrices $\mathbf{A}_{\leq t}$ for $t = 1, 2, \ldots, m$. We update and maintain the collection of index blocks $\mathcal{B}^{(t)} \leftarrow \{B_i^{(t)}\}$ for the current submatrix $\mathbf{A}_{\leq t}$ in each iteration by using column and block updates stated below. Here we use upper index $(\cdot)^{(t)}$ to emphasize the iteration t.

(a) **Subcolumn update**: Partition the column c_t into subcolumns,

$$c_t|_{\mathsf{Row}B_i^{(t-1)}} := \mathbf{A}_{[\mathsf{Row}(B_i^{(t-1)}),\,\{t\}]},$$

one for the set of rows $\mathsf{Row}(B_i^{(t-1)})$ for each block from the previous iteration. We process each such subcolumn $c_t|_{\mathsf{Row}B_i^{(t-1)}}$ one by one, checking whether there exists a sequence of admissible operations that are able to reduce the subcolumn to zero while *preserving the prior* as defined below.

Definition 11.16. (Prior) We say a *prior* with respect to a subcolumn $c_t|_{\mathsf{Row}B_i^{(t-1)}}$ is the left submatrix $A_{<t}$ and subcolumns $c_t|_{\mathsf{Row}B_j^{(t-1)}}$ for all $j < i$.

Prior preservation means that the operations together *change neither* $\mathbf{A}_{<t}$ *nor other subcolumns* $c_t|_{\mathsf{Row}B_j^{(t-1)}}$ *for every* $j < i$. If such operations exist, we apply them on the current $\mathbf{A}$ to get an equivalent matrix with the subcolumn $c_t|_{\mathsf{Row}B_i^{(t-1)}}$ being zeroed out and we set $B_i^{(t)} \leftarrow B_i^{(t-1)}$. Otherwise, we leave the matrix $\mathbf{A}$ unchanged and add the column index t to those of $B_i^{(t-1)}$, that is, we set $B_i^{(t)} \leftarrow \left[\mathsf{Row}(B_i^{(t-1)}), \mathsf{Col}(B_i^{(t-1)}) \cup \{t\}\right]$. After processing every subcolumn $c_t|_{\mathsf{Row}B_i^{(t-1)}}$ one by one, all index blocks $B_i^{(t)}$ containing column index t are merged into one single index block. At the end of iteration t, we get an equivalent matrix $\mathbf{A}$ with $\mathbf{A}_{\leq t}$ being totally diagonalized with index blocks $\mathcal{B}^{(t)}$.

(b) **Block reduce**: To update the entries of each subcolumn of c_t described in 2(a), we propose a block reduction Algorithm 24: BLOCKREDUCE to compute the correct entries. Given $T := \left[\mathsf{Row}(B_i^{(t-1)}), (\mathsf{Col}(\mathbf{A}_{\leq t}) \setminus \mathsf{Col}(B_i^{(t-1)}))\right]$, this routine checks whether the block T can be zeroed out by some collection of admissible operations. If so, c_t does not join the block $B_i^{(t)}$ and $\mathbf{A}$ is updated with these operations.

For two index blocks B_1 and B_2, we denote the merging $B_1 \oplus B_2$ of these two index blocks as an index block $\left[\mathsf{Row}(B_1) \cup \mathsf{Row}(B_2), \mathsf{Col}(B_1) \cup \mathsf{Col}(B_2)\right]$. In the following algorithm, we treat the given matrix $\mathbf{A}$ as a global variable which can be visited and modified anywhere by every subroutine called. Consequently, every time we update values on $\mathbf{A}$ by some operations, these operations are applied to the latest $\mathbf{A}$.

The outer loop is the incremental step for main iteration introducing a new column c_t which updates the diagonalization of $\mathbf{A}_{\leq t}$ from the last iteration. The inner loop corresponds to block updates which checks the intersection of the current column and the rows of each previous block one by one.

Remark 11.5. *The algorithm* TOTDIAGONALIZE *does not require the input presentation matrix to be minimal. As indicated in Remark 11.3, the trivial parts result in either identity blocks or single column blocks like* $[\varnothing, \{j\}]$. *Such a single column block corresponds to a zero morphism and is not merged with any other blocks. Therefore,* c_j *is a zero column. For a single row block* $[\{i\}, \varnothing]$ *which is not merged with any other blocks,* r_i *is a zero row vector. It represents a free indecomposable submodule in the total decomposition of the input persistence module.*

We first prove the correctness of TOTDIAGONALIZE assuming that the BLOCKREDUCE routine works as claimed, namely, it checks if a subcolumn of the current column c_t can be zeroed out while preserving the prior, that is, without changing the left submatrix from the previous iteration and also the other subcolumns of c_t that have already been zeroed out.

Proposition 11.4. *At the end of each iteration* t, $\mathbf{A}_{\leq t}$ *is a total diagonalization.*

PROOF. We prove it by induction on t. For the base case $t = 0$, it follows trivially by definition. Now assume $\mathbf{A}^{(t-1)}$ is the matrix we get at the end of iteration $(t - 1)$ with $\mathbf{A}^{(t-1)}_{\leq t-1}$ being totally diagonalized. That means $\mathbf{A}^{(t-1)}_{\leq t-1} = \mathbf{A}^*_{\leq t-1}$ where $\mathbf{A} = \mathbf{A}^{(0)}$ is the original given matrix. For contradiction, assume at the end of iteration t that the matrix we get, $\mathbf{A}^{(t)}$, has left submatrix $\mathbf{A}^{(t)}_{\leq t}$ which is not totally diagonalized. That means some index block $B \in \mathcal{B}^{(t)}$ can be decomposed further. Observe that such B must contain t because all other index blocks (not containing t) in $\mathcal{B}^{(t)}$ are also in $\mathcal{B}^{(t-1)}$ which cannot be decomposed further by our inductive assumption. We denote this index block containing t as B_t. Let $\mathbf{A}'$ be the equivalent matrix of $\mathbf{A}^{(t)}$ such that $\mathbf{A}'_{\leq t}$ is a total diagonalization with index blocks $\mathcal{B}'$. Let F be an equivalent transformation from $\mathbf{A}^{(t)}$ to $\mathbf{A}'$, which decomposes B_t into at least two distinct index blocks of $\mathcal{B}'$, say $B_0, B_1, \ldots$. Only one of them contains t, say B_0. Then B_1 consists of only indices that are from $\mathbf{A}_{\leq t-1}$, which means B_1 equals some index block $B_i \in \mathcal{B}^{(t-1)}$. Therefore, the transformation F gives a sequence of admissible operations which can reduce the subcolumn $c_t|_{\mathsf{Row}(B_i)}$ to zero in $\mathbf{A}^{(t)}$. Starting with this sequence of admissible operations, we construct another sequence of admissible operations which further keeps $\mathbf{A}^{(t)}_{\leq t-1}$ unchanged to reach the contradiction. Note that $\mathbf{A}^{(t)}_{\leq t-1} = \mathbf{A}^{(t-1)}_{\leq t-1}$.

Observe that all index blocks of $\mathcal{B}'$ other than B_0 are also index blocks in $\mathcal{B}^{(t-1)}$, that is, $\mathcal{B}' \setminus \{B_0\} \subseteq \mathcal{B}^{(t-1)}$. For B_0, it can be written as $B_0 = \bigoplus_{B_j \in \mathcal{B}^{(t-1)} \setminus \mathcal{B}'} B_j \oplus [\varnothing, \{t\}]$. Let B_a be the merge of index blocks that are in $\mathbf{A}^{(t-1)}$ and also in $\mathbf{A}'$, and B_b be the merge of the rest of the index blocks of

$\mathbf{A}^{(t-1)}$, that is, $B_a = \bigoplus_{B_j \in \mathcal{B}' \cap \mathcal{B}^{(t-1)}} B_j$ and $B_b = \bigoplus_{B_j \in \mathcal{B}^{(t-1)} \setminus \mathcal{B}'} B_j$. Then B_a and B_b can be viewed as a coarser decomposition on $\mathbf{A}^{(t-1)}_{\leq t-1}$ and also on $\mathbf{A}'_{\leq t-1}$. By taking restrictions, we have $\mathbf{A}'|_{B_a} \sim \mathbf{A}^{(t-1)}|_{B_a}$ with equivalent transformation F_a and $\mathbf{A}'|_{B_b} \sim \mathbf{A}^{(t-1)}|_{B_b}$ with equivalent transformation F_b. Then F_a gives a sequence of admissible operations with indices in B_a, and F_b gives a sequence of admissible operations with indices in B_b. By applying these operations on $\mathbf{A}'$, we can transform $\mathbf{A}'_{\leq t-1}$ to $\mathbf{A}^{(t-1)}_{\leq t-1}$ with subcolumn [Row$(\mathbf{A}) \setminus$ Row(B_0), $\{t\}$] unchanged, which consists of the subcolumns that have already been reduced to zero. Combining all admissible operations from the three transformations F, F_a, and F_b together, we get a sequence of admissible operations that reduce subcolumn [Row(B_i), $\{t\}$] to zero without changing $\mathbf{A}^{(t)}_{\leq t}$ and also those subcolumns which have already been reduced. But then BLOCKREDUCE would have returned "true" signaling that B_i should not be merged with any other block required to form the block B_t, reaching a contradiction.   □

Now we design the BLOCKREDUCE subroutine as required. With the requirement of prior preservation, observe that reducing the subcolumn $c_t|_{\mathsf{Row}B}$ for some $B \in \mathcal{B}^{(t-1)}$ is the same as reducing $T = $ [Row(B), $($Col$(\mathbf{A}_{\leq t}) \setminus$ Col$(B))$] called the *target block* (see Figure 11.9b). The main idea of BLOCKREDUCE is to consider a specific subset of admissible operations called *independent operations*. Within $\mathbf{A}_{\leq t}$, these operations only change entries in T and this change is independent of their order of application. The BLOCKREDUCE subroutine is designed to search for a sequence of admissible operations within this subset and reduce T with it, if it exists. Clearly, the prior is preserved with these operations. The only thing we need to ensure is that searching within the set of independent operations is sufficient. That means, if there exists a sequence of admissible operations that can reduce T to 0 and meanwhile preserves the prior, then we can always find one such sequence with only independent operations. This is what we show next.

Consider the following matrices for each admissible operation. For each admissible column operation $c_i \to c_j$, let

$$\mathbf{Y}^{i,j} := \mathbf{A} \cdot [\delta_{i,j}],$$

where $[\delta_{i,j}]$ is the $m \times m$ square matrix with only one nonzero entry at (i, j). Observe that $\mathbf{A} \cdot [\delta_{i,j}]$ is a matrix with the only nonzero column at j with entries copied from c_i in $\mathbf{A}$. Similarly, for each admissible row operation $r_l \to r_k$, let $[\delta_{k,l}]$ be the $\ell \times \ell$ matrix with only nonzero entry at (k, l), and let

$$\mathbf{X}^{k,l} := [\delta_{k,l}] \cdot \mathbf{A}.$$

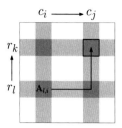

Figure 11.10 We can see that $[\delta_{k,l}]\mathbf{A}[\delta_{i,j}]$ is a matrix with the only nonzero entry at (k, j) being a copy of $\mathbf{A}_{l,i}$.

Application of a column operation $c_i \to c_j$ can be viewed as updating $\mathbf{A}$ to $\mathbf{A}{\cdot}(\mathbf{I} + [\delta_{i,j}]) = \mathbf{A} + \mathbf{Y}^{i,j}$. A similar observation holds for row operations as well. For a target block $T = [\mathsf{Row}(B), \mathsf{Col}(\mathbf{A}_{\leq t})\backslash\mathsf{Col}(B)]$ defined on some $B \in \mathcal{B}^{(t-1)}$, we say an admissible column (row) operation, $c_i \to c_j$ (resp. $r_l \to r_k$) is *independent* on T if $i \notin \mathsf{Col}(T), j \in \mathsf{Col}(T)$ (resp. $l \notin \mathsf{Row}(T), k \in \mathsf{Row}(T)$). Briefly, we just call such operations *independent operations* if T is clear from the context.

We have the following two observations about independent operations that are important. The first one follows from the definition that $T = [\mathsf{Row}(B), \mathsf{Col}(\mathbf{A}_{\leq t}) \backslash \mathsf{Col}(B)]$. The second one is proved below.

Observation 11.1. *Within $\mathbf{A}_{\leq t}$, an independent column or row operation only changes entries on T.*

Observation 11.2. *For any independent column operation $c_i \to c_j$ and row operation $r_l \to r_k$, we have $[\delta_{k,l}]{\cdot}\mathbf{A}{\cdot}[\delta_{i,j}] = 0$. Or, equivalently,*

$$(\mathbf{I} + [\delta_{k,l}]){\cdot}\mathbf{A}{\cdot}(\mathbf{I} + [\delta_{i,j}]) = \mathbf{A} + [\delta_{k,l}]\mathbf{A} + \mathbf{A}[\delta_{i,j}] = \mathbf{A} + \mathbf{X}^{k,l} + \mathbf{Y}^{i,j}. \quad (11.1)$$

PROOF. We have $[\delta_{k,l}]{\cdot}\mathbf{A}{\cdot}[\delta_{i,j}] = \mathbf{A}_{l,i}[\delta_{k,j}]$ (see Figure 11.10 for an illustration). By definitions of independence and T, we have $l \notin \mathsf{Row}(B), i \in \mathsf{Col}(B)$. That means they are the row index and column index from different blocks. Therefore, $\mathbf{A}_{l,i} = 0$. □

The following proposition reveals why we are after the independent operations.

Proposition 11.5. *The target block $\mathbf{A}|_T$ can be reduced to 0 while preserving the prior if and only if $\mathbf{A}|_T$ can be written as a linear combination of independent operations. That is,*

Algorithm 23 COLREDUCE(S, *c*)

Input:
    Source matrix **S** and target column *c* to reduce
Output:
    Reduced column *c* with **S**

1: **S′** ← [**S**|*c*]
2: Call MATPERSISTENCE(**S′**);
3: **return** *c* along with indices of columns in **S** used for reduction of *c*

$$\mathbf{A}|_T = \sum_{\substack{l\notin\mathrm{Row}(T)\\k\in\mathrm{Row}(T)}} \alpha_{k,l}\mathbf{X}^{k,l}|_T + \sum_{\substack{i\notin\mathrm{Col}(T)\\j\in\mathrm{Col}(T)}} \beta_{i,j}\mathbf{Y}^{i,j}|_T,$$

where the $\alpha_{k,l}$ *and* $\beta_{i,j}$ *are coefficients in* $\mathbf{k} = \mathbb{Z}_2$.

The full proof can be seen in [141]. Here, we give some intuitive explanation. Reducing the target block $\mathbf{A}|_T$ to 0 is equivalent to finding matrices **P** and **Q** encoding sequences of admissible row operations and admissible column operations, respectively, so that $\mathbf{PAQ}|_T = 0$.

For the "if" direction, we can build $\mathbf{P} = \mathbf{I} + \sum \alpha_{k,l}[\delta_{k,l}]$ and $\mathbf{Q} = \mathbf{I} + \sum \beta_{i,j}[\delta_{i,j}]$ with binary coefficients $\alpha_{k,l}$ and $\beta_{i,j}$ given in Eq. (11.5). Then using Observations 11.1 and 11.2, one can show **PAQ** indeed reduces $\mathbf{A}|_T$ to 0 with the prior being preserved. This provides the proof for the "if" direction.

For the "only if" direction, as long as we show that the existence of a transformation reducing $\mathbf{A}|_T$ to 0 implies the existence of a transformation reducing $\mathbf{A}|_T$ to 0 by independent operations, we are done. This is formally proved in [141].

We can view $\mathbf{A}|_T$, $\mathbf{Y}^{i,j}|_T$, and $\mathbf{X}^{k,l}|_T$ as binary vectors in the same $|T|$-dimensional space. Proposition 11.5 tells us that it is sufficient to check if $\mathbf{A}|_T$ can be a linear combination of the vectors corresponding to a set of independent operations. So, we first *linearize* each of the matrices $\mathbf{Y}^{i,j}|_T$, $\mathbf{X}^{k,l}|_T$, and $\mathbf{A}|_T$ to a column vector as described later (see Figure 11.11). Then, we check if $\mathbf{A}|_T$ is in the span of the $\mathbf{Y}^{i,j}|_T$ and $\mathbf{X}^{k,l}|_T$. This is done by collecting all vectors $\mathbf{X}^{i,j}|_T$ and $\mathbf{Y}^{k,l}|_T$ into a matrix *S* called the *source matrix* (Figure 11.11c) and then reducing the vector $c := \mathbf{A}|_T$ with *S* by some standard matrix reduction algorithm with left-to-right column additions, which we have seen before in Section 3.3.1 for computing persistence. This routine is presented in Algorithm 23: COLREDUCE (**S**, *c*) which reduces the column *c* with respect to the

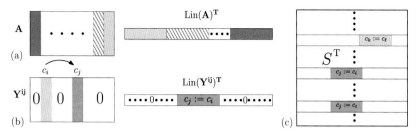

Figure 11.11 (a) Matrix $\mathbf{A}$ is linearized to the vector $\mathsf{Lin}(\mathbf{A})$ in the middle. (b) The column operation $c_i \to c_j$ is captured by $\mathbf{Y}^{ij}$ whose linearization is illustrated in the middle. (c) Source matrix S combining all operations (row operations not shown). In the figure, $(\cdot)^{\mathsf{T}}$ denotes transposed matrices.

input matrix $\mathbf{S}$ by reducing the matrix $[\mathbf{S}|c]$ altogether by MATPERSISTENCE in Section 3.3.1.

If $c = \mathbf{A}|_T$ can be reduced to 0, we apply the corresponding independent operations to update $\mathbf{A}$. Observe that all column operations used in reducing $\mathbf{A}|_T$ together only change the subcolumn $c_t|_{\mathsf{Row}(B)}$ while row operations may change $\mathbf{A}$ to the right of the column t. We say this procedure *reduces c with* $\mathbf{S}$.

Fact 11.1. *There exists a set of column operations adding a column only to its right such that the matrix* $[\mathbf{S}|c]$ *is reduced to* $[\mathbf{S'}|0]$ *if and only if* COLREDUCE($\mathbf{S}$, c) *returns a zero vector.*

Now we describe the linearization used in in Algorithm 24: BLOCKRE-DUCE. We fix a linear order $\leq_{\mathsf{Lin}}$ on the set of matrix indices, $\mathsf{Row}(\mathbf{A}) \times \mathsf{Col}(\mathbf{A})$, as follows: $(i, j) \leq_{\mathsf{Lin}} (i', j')$ if $j > j'$ or $j = j', i < i'$. Explicitly, we linearly order the indices as

$$((1, m), (2, m), \ldots, (\ell, n), (1, m - 1), (2, m - 1), \ldots).$$

For any index block B, let $\mathsf{Lin}(\mathbf{A}|_B)$ be the vector of dimension $|\mathsf{Col}(B)| \cdot |\mathsf{Row}(B)|$ obtained by linearizing $\mathbf{A}|_B$ to a vector in the above linear order on the indices.

Proposition 11.6. *The target block on T can be reduced to zero in $\mathbf{A}$ while preserving the prior if and only if* BLOCKREDUCE(T) *returns true.*

Time Complexity
First we analyze the time complexity of TOTDIAGONALIZE assuming that the input matrix has size $\ell \times m$. Clearly, $\max\{\ell, m\} = O(N)$ where N is the total number of generators and relations. For each of $O(N)$ columns, we attempt to zero out every subcolumn with row indices coinciding with each block B of

---

Algorithm 24 BLOCKREDUCE(T)

Input:
Index of target block T to be reduced; Given matrix $\mathbf{A}$ is assumed to be a global variable
Output:
A boolean to indicate whether $\mathbf{A}|_T$ can be reduced and reduced block $\mathbf{A}|_T$ if possible

1: Compute $c := \mathsf{Lin}(\mathbf{A}|_T)$ and initialize empty matrix $\mathbf{S}$
2: **for** each admissible column operation $c_i \to c_j$ with $i \notin \mathsf{Col}(T), j \in \mathsf{Col}(T)$ **do**
3:    Compute $\mathbf{Y}^{i,j}|_T := (\mathbf{A}\cdot[\delta_{i,j}])|_T$ and $y^{i,j} = \mathsf{Lin}(\mathbf{Y}^{i,j}|_T)$; update $\mathbf{S} \leftarrow [\mathbf{S}|y^{i,j}]$
4: **end for**
5: **for** each admissible row operation $r_l \to r_k$ with $l \notin \mathsf{Row}(T), k \in \mathsf{Row}(T)$ **do**
6:    Compute $\mathbf{X}^{k,l}|_T := ([\delta_{k,l}]\cdot\mathbf{A})|_T$ and $x^{k,l} := \mathsf{Lin}(\mathbf{X}^{k,l}|_T)$; update $\mathbf{S} \leftarrow [\mathbf{S}|x^{k,l}]$
7: **end for**
8: COLREDUCE $(\mathbf{S}, c)$ returns indices of the $y^{i,j}$ and $x^{k,l}$ used to reduce c (if possible)
9: For every returned index of $y^{i,j}$ or $x^{k,l}$ apply $c_i \to c_j$ or $r_l \to r_k$ to transform $\mathbf{A}$
10: Return $\mathbf{A}|_T == 0$

---

the previously determined $O(N)$ blocks. Let B have N_B rows. Then, the block T has N_B rows and $O(N)$ columns.

To zero out a subcolumn, we create a source matrix out of T which has size $O(NN_B) \times O(N^2)$ because each of $O\left(\binom{N}{2}\right)$ possible operations is converted to a column of size $O(NN_B)$ in the source matrix. The source matrix $\mathbf{S}$ with the target vector c can be reduced with an efficient algorithm [57, 200] in $O(a + N^2(NN_B)^{\omega-1})$ time, where a is the total number of nonzero elements in $[\mathbf{S}|c]$ and $\omega \in [2, 2.373)$ is the exponent for matrix multiplication. We have $a = O(NN_B \cdot N^2) = O(N^3 N_B)$. Therefore, for each block B we spend $O(N^3 N_B + N^2(NN_B)^{\omega-1})$ time. Then, observing that $\sum_{B \in \mathcal{B}} N_B = O(N)$, for each column we spend a total time of

$$\sum_{B \in \mathcal{B}} O(N^3 N_B + N^2(NN_B)^{\omega-1}) = O\left(N^4 + N^{\omega+1}\sum_{B \in \mathcal{B}} N_B^{\omega-1}\right)$$
$$= O(N^4 + N^{2\omega}) = O(N^{2\omega}).$$

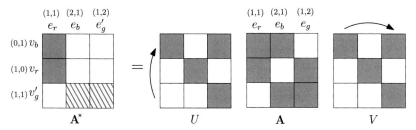

Figure 11.12 Diagonalizing the binary matrix given in Example 11.1. It can be viewed as multiplying the original matrix $\mathbf{A}$ with a left matrix U that represents the row operation and a right matrix V that represents the column operations.

Therefore, accounting for all of the $O(N)$ columns, the total time for decomposition takes $O(N^{2\omega+1})$ time.

We finish this analysis by commenting that one can build the presentation matrix from a given simplicial filtration consisting of n simplices leading to the following cases: (i) For zeroth homology, the boundary matrix ∂_1 can be taken as the presentation matrix, giving $N = O(n)$ and a total time complexity of $O(n^{2\omega+1})$. (ii) For two-parameter case, $N = O(n)$ and presentations can be computed in $O(n^3)$ time, giving a total time complexity of $O(n^{2\omega+1})$. (iii) For the d-parameter case, $N = O(n^{d-1})$ and a presentation matrix can be computed in $O(n^{d+1})$ time, giving a total time complexity of $O(n^{(2\omega+1)(d-1)})$. We discuss the details in Section 11.5.

11.4.1 Running TOTDIAGONALIZE on the Working Example in Figure 11.5

Example 11.2. *Consider the binary matrix after simplification as illustrated in Example 11.1.*

$$
\begin{array}{c}
\mathbf{A} \\
r_1^{(0,1)} \\
r_2^{(1,0)} \\
r_3^{(1,1)}
\end{array}
\begin{array}{ccc}
c_1^{(1,1)} & c_2^{(1,2)} & c_3^{(2,1)} \\
\left(\begin{array}{ccc}
1 & 1 & 0 \\
1 & 0 & 1 \\
0 & 1 & 1
\end{array}\right)
\end{array}
$$

This has four admissible operations: $r_3 \to r_1, r_3 \to r_2, c_1 \to c_2$, and $c_1 \to c_3$. Diagonalizing $\mathbf{A}$ reduces to finding appropriate row and column operations, as Figure 11.12 illustrates.

Before the first iteration, $\mathcal{B}$ is initialized to be $\mathcal{B} = \{B_1 = (\{1\}, \varnothing), B_2 = (\{2\}, \varnothing), B_3 = (\{3\}, \varnothing)\}$. In the first iteration when $t = 1$, we have block $B_0 = (\varnothing, \{1\})$ for column c_1. For $B_1 = (\{1\}, \varnothing)$, the target block we hope to zero out is $T = (\{1\}, \{1\})$. So we call BLOCKREDUCE(T) to check if $\mathbf{A}|_T$

can be zeroed out and update the entries on T according to the results of BLOCKREDUCE(T). *There is only one admissible operation from outside of T into it, namely, $r_3 \to r_1$. The target vector $c = \mathsf{Lin}(A|_T)$ and the source matrix* $S = \{\mathsf{Lin}(([\delta_{1,3}]A)|_T)\}$ *are*

$$
\begin{array}{c c c}
S & \mathsf{Lin}(([\delta_{1,3}]A)|_T) & c=\mathsf{Lin}(A|_T) \\
\begin{bmatrix} & 0 & \end{bmatrix} & & 1 \end{array}.
$$

The result of COLREDUCE(S, c) *stays the same as its input. That means we cannot reduce c at all. Therefore,* BLOCKREDUCE(T, t) *returns* FALSE *and nothing is updated in the original matrix.*

It is not surprising that the matrix remains the same because the only admissible operation that can affect T does not change any entries in T at all. So there is nothing one can do to reduce it, which results in merging $B_1 \oplus B_0 = (\{1\}, \{1\})$. Similarly, for B_2 with $T = (\{2\}, \{1\})$, the only admissible operation $r_3 \to r_2$ does not change anything in T. Therefore, the matrix does not change and B_2 is merged with $B_1 \oplus B_0$, which results in the block $(\{1, 2\}, \{1\})$. For B_3 with $T = (\{3\}, \{1\})$, there is no admissible operation. So the matrix does not change. But $A|_T = A|_{(\{3\},\{1\})} = 0$. That means BLOCKRE-DUCE *returns* TRUE. *Therefore, we do not merge B_3. In summary, B_0, B_1, and B_2 are merged to be one block $(\{1, 2\}, \{1\})$ in the first iteration. So after the first iteration, there are two index blocks in $\mathcal{B}^{(1)}$: $(\{1, 2\}, \{1\})$ and $(\{3\}, \varnothing)$.*

In the second iteration $t = 2$, we process the second column c_2. Now $B_1 = (\{1, 2\}, \{1\})$, $B_2 = (\{3\}, \varnothing)$, and $B_0 = (\varnothing, \{2\})$. For the block $B_1 = (\{1, 2\}, \{1\})$, the target block we hope to zero out is $T = (\{1, 2\}, \{2\})$. There are three admissible operations from outside of T into T, $r_3 \to r_1, r_3 \to r_2$, and $c_1 \to c_2$. BLOCKREDUCE(T) *constructs the target vector $c = \mathsf{Lin}(A|_T)$ and the source matrix $S = \{\mathsf{Lin}(([\delta_{1,3}]A)|_T), \mathsf{Lin}(([\delta_{2,3}]A)|_T), \mathsf{Lin}((A[\delta_{1,2}])|_T)\}$ illustrated as follows:*

$$
\begin{array}{c c c c c}
S & \mathsf{Lin}(([\delta_{1,3}]A)|_T) & \mathsf{Lin}([(\delta_{2,3}]A)|_T) & \mathsf{Lin}((A[\delta_{1,2}])|_T) & c=\mathsf{Lin}(A|_T) \\
\begin{bmatrix} 1 & 0 & 1 \\ 0 & 1 & 1 \end{bmatrix} & & & & \begin{matrix} 1 \\ 0 \end{matrix} \end{array}.
$$

The result of COLREDUCE(S, c) *is*

$$
\begin{array}{cc}
S & c \\
\begin{bmatrix} 1 & 0 & 0 \\ 0 & 1 & 0 \end{bmatrix} & \begin{matrix} 0 \\ 0 \end{matrix} \end{array}.
$$

So the BLOCKREDUCE *updates* $\mathbf{A}|_T$ *to get the updated matrix*

$$
\begin{array}{c}
\mathbf{A}' \\
r_1^{(0,1)} + r_3^{(1,1)} \\
r_2^{(1,0)} \\
r_3^{(1,1)}
\end{array}
\begin{array}{ccc}
c_1^{(1,1)} & c_2^{(1,2)} & c_3^{(2,1)}
\end{array}
\left(
\begin{array}{ccc}
1 & 0 & 1 \\
1 & 0 & 1 \\
0 & 1 & 1
\end{array}
\right)
$$

and returns TRUE *since* $\mathbf{A}'|_T == 0$. *Therefore, we do not merge* B_1. *We continue to check for the block* $B_2 = (\{3\}, \varnothing)$ *and* $T = (\{3\}, \{1, 2\})$, *whether* $\mathbf{A}'|_T$ *can be reduced to zero. There is no admissible operation for this block at all. Therefore, the matrix stays the same and* BLOCKREDUCE *returns* FALSE. *We merge* $B_2 \oplus B_0 = (\{3\}, \{2\})$.

Continuing the process for the last column c_3 *in the third iteration* $t = 3$, *we see that* $B_1 = (\{1, 2\}, \{1\})$, $B_2 = (\{3\}, \{2\})$, *and* $B_0 = (\varnothing, \{3\})$. *For the block* $B_1 = (\{1, 2\}, \{1\})$, *the target block we hope to zero out is* $T = (\{1, 2\}, \{2, 3\})$. *There are four admissible operations from outside of* T *into* T: $r_3 \rightarrow r_1$, $r_3 \rightarrow r_2$, $c_1 \rightarrow c_2$, *and* $c_1 \rightarrow c_3$. BLOCKREDUCE(T) *constructs the target vector* $c = \mathsf{Lin}(\mathbf{A}|_T)$ *and the source matrix* $\mathbf{S} = \{\mathsf{Lin}(([\delta_{1,3}]\mathbf{A})|_T), \mathsf{Lin}(([\delta_{2,3}]\mathbf{A})|_T), \mathsf{Lin}((\mathbf{A}[\delta_{1,2}])|_T)\}, \mathsf{Lin}((\mathbf{A}[\delta_{1,3}])|_T)\}$ *illustrated as follows:*

$\mathbf{S}$	$\mathsf{Lin}(([\delta_{1,3}]\mathbf{A})\|_T)$	$\mathsf{Lin}([(\delta_{2,3}]\mathbf{A})\|_T)$	$\mathsf{Lin}((\mathbf{A}[\delta_{1,2}])\|_T)$	$\mathsf{Lin}((\mathbf{A}[\delta_{1,3}])\|_T)$	$c=\mathsf{Lin}(\mathbf{A}\|_T)$
	1	0	0	1	1
	0	1	0	1	1
	1	0	1	0	0
	0	1	1	0	0

The result of COLREDUCE$(\mathbf{S}, c)$ *is*

$$
\begin{array}{cc}
\mathbf{S} & c
\end{array}
\left[
\begin{array}{cccc|c}
1 & 0 & 1 & 0 & 0 \\
0 & 1 & 1 & 0 & 0 \\
1 & 0 & 0 & 0 & 0 \\
0 & 1 & 0 & 0 & 0
\end{array}
\right]
$$

So the BLOCKREDUCE *updates* $\mathbf{A}|_T$ *to get the following updated matrix:*

$$
\begin{array}{c}
\mathbf{A}' \\
r_1^{(0,1)} \\
r_2^{(1,0)} + r_3^{(1,1)} \\
r_3^{(1,1)}
\end{array}
\begin{array}{ccc}
c_1^{(1,1)} & c_2^{(1,2)} + c_1^{(1,1)} & c_3^{(2,1)}
\end{array}
\left(
\begin{array}{ccc}
1 & 0 & 0 \\
1 & 0 & 0 \\
0 & 1 & 1
\end{array}
\right)
$$

and returns TRUE *since* $\mathbf{A}'|_T == 0$. *Therefore, we do not merge* B_1 *with any other block. We continue to check for the block* $B_2 = (\{3\}, \{2\})$ *and* $T = (\{3\}, \{1, 3\})$ *whether* $\mathbf{A}'|_T$ *can be reduced to zero. There is no admissible operation for this block at all. Therefore, the matrix stays the same and* BLOCKREDUCE *returns* FALSE. *We merge* $B_2 \oplus B_0 = (\{3\}, \{2, 3\})$.

Finally, the algorithm returns the matrix $\mathbf{A}'$ *shown above as the final result. It is the correct total diagonalization with two index blocks in* $\mathcal{B}^{\mathbf{A}^*}$: $B_1 = (\{1, 2\}, \{1\})$ *and* $B_2 = (\{3\}, \{2, 3\})$. *An examination of* COLREDUCE$(\mathbf{S}, c)$ *in all three iterations over columns reveals that the entire matrix* $\mathbf{A}$ *is updated by operations* $r_3 \to r_2$ *and* $c_1 \to c_2$. *We can further transform it back to the original form of the presentation matrix* $[\partial_1]$. *Observe that a row addition* $r_i \leftarrow r_i + r_j$ *reverts to a basis change in the opposite direction:*

$$
[\partial_1] \quad
\begin{array}{c}
 \\
v_b^{(0,1)} \\
v_r^{(1,0)} \\
v_g^{(1,1)}
\end{array}
\begin{array}{ccc}
e_r^{(1,1)} & e_b^{(1,2)} & e_g^{(2,1)} \\
\left(\mathbf{t}^{(1,0)} \right. & \mathbf{t}^{(1,1)} & 0 \\
\mathbf{t}^{(0,1)} & 0 & \mathbf{t}^{(1,1)} \\
0 & \mathbf{t}^{(0,1)} & \left. \mathbf{t}^{(1,0)} \right)
\end{array}
$$

$$
\Longrightarrow
$$

$$
[\partial_1]^* \quad
\begin{array}{c}
 \\
v_b^{(0,1)} \\
v_r^{(1,0)} \\
v_g^{(1,1)} + \mathbf{t}^{(0,1)} v_r^{(1,0)}
\end{array}
\begin{array}{ccc}
e_r^{(1,1)} & e_b^{(1,2)} + \mathbf{t}^{(0,1)} e_r^{(1,1)} & e_g^{(2,1)} \\
\left(\mathbf{t}^{(1,0)} \right. & 0 & 0 \\
\mathbf{t}^{(0,1)} & 0 & 0 \\
0 & \mathbf{t}^{(0,1)} & \left. \mathbf{t}^{(1,0)} \right)
\end{array}
$$

11.5 Computing Presentations

Now that we know how to decompose a presentation by diagonalizing its matrix form, we describe how to construct and compute these matrices in this section. For a persistence module H_p with p-th homology groups, we consider a presentation $\mathsf{C}_{p+1} \to \mathsf{Z}_p \twoheadrightarrow \mathsf{H}_p \to 0$ where C_{p+1} is a graded module of $(p + 1)$-chains and Z_p is a graded module of p-cycles which we describe now. Recall that a (d-parameter) *simplicial filtration* is a family of simplicial complexes $\{X_{\mathbf{u}}\}_{\mathbf{u} \in \mathbb{Z}^d}$ such that for each grade $\mathbf{u} \in \mathbb{Z}^d$ and each $i = 1, \ldots, d$, $X_{\mathbf{u}} \subseteq X_{\mathbf{u}+e_i}$.

11.5.1 Graded Chain, Cycle, and Boundary Modules

We obtain a simplicial chain complex $(\mathsf{C}.(X_{\mathbf{u}}), \partial.)$ for each $X_{\mathbf{u}}$ in the given simplicial filtration. For each comparable pair in the grading $\mathbf{u} \leq \mathbf{v} \in \mathbb{Z}^d$, a family of inclusion maps $\mathsf{C}.(X_{\mathbf{u}}) \hookrightarrow \mathsf{C}.(X_{\mathbf{v}})$ is induced by the canonical inclusion $X_{\mathbf{u}} \hookrightarrow X_{\mathbf{v}}$ giving rise to the following diagram.

$$
\begin{array}{ccccccccc}
\mathsf{C}.(X_{\mathbf{u}}): & \cdots \xrightarrow{\partial_{p+2}} & \mathsf{C}_{p+1}(X_{\mathbf{u}}) & \xrightarrow{\partial_{p+1}} & \mathsf{C}_p(X_{\mathbf{u}}) & \xrightarrow{\partial_p} & \mathsf{C}_{p-1}(X_{\mathbf{u}}) & \xrightarrow{\partial_{p-1}} & \cdots \\
& & \big\downarrow & & \big\downarrow & & \big\downarrow & & \\
\mathsf{C}.(X_{\mathbf{v}}): & \cdots \xrightarrow{\partial_{p+2}} & \mathsf{C}_{p+1}(X_{\mathbf{v}}) & \xrightarrow{\partial_{p+1}} & \mathsf{C}_p(X_{\mathbf{v}}) & \xrightarrow{\partial_p} & \mathsf{C}_{p-1}(X_{\mathbf{v}}) & \xrightarrow{\partial_{p-1}} & \cdots
\end{array}
$$

For each chain complex $\mathsf{C}.(X_{\mathbf{u}})$, we have the cycle spaces $\mathsf{Z}_p(X_{\mathbf{u}})$ and boundary spaces $\mathsf{B}_p(X_{\mathbf{u}})$ as kernels and images of boundary maps ∂_p, respectively, and the homology group $\mathsf{H}_p(X_{\mathbf{u}}) = \mathsf{Z}_p(X_{\mathbf{u}})/\mathsf{B}_p(X_{\mathbf{u}})$ as the cokernel of the inclusion maps $\mathsf{B}_p(X_{\mathbf{u}}) \hookrightarrow \mathsf{Z}_p(X_{\mathbf{u}})$. In line with category theory we use the notation im, ker, and coker for indicating both the modules of kernel, image, and cokernel and the corresponding morphisms uniquely determined by their constructions.[3] We obtain the following commutative diagram.

$$
\begin{array}{ccc}
\mathsf{B}_p(X_{\mathbf{u}}) & \longrightarrow \mathsf{Z}_p(X_{\mathbf{u}}) \; -\text{coker} \twoheadrightarrow & \mathsf{H}_p(X_{\mathbf{u}}) \\[4pt]
{\scriptstyle \mathrm{im}\,\partial_{p+1}} \nearrow & \quad \curvearrowleft {\scriptstyle \ker \partial_p} & \searrow \\[4pt]
\cdots \mathsf{C}_{p+1}(X_{\mathbf{u}}) & \xrightarrow{\quad\quad\quad \partial_{p+1} \quad\quad\quad} & \mathsf{C}_p(X_{\mathbf{u}}) \; \cdots
\end{array}
$$

In the language of graded modules, for each p, the family of vector spaces and linear maps (inclusions) $(\{\mathsf{C}_p(X_{\mathbf{u}})\}_{\mathbf{u} \in \mathbb{Z}^d}, \{\mathsf{C}_p(X_{\mathbf{u}}) \hookrightarrow \mathsf{C}_p(X_{\mathbf{v}})\}_{\mathbf{u} \le \mathbf{v}})$ can be summarized as a $\mathbb{Z}^d$-graded R-module:

$$
\mathsf{C}_p(X) := \bigoplus_{\mathbf{u} \in \mathbb{Z}^d} \mathsf{C}_p(X_{\mathbf{u}}), \text{ with the ring action } t_i \cdot \mathsf{C}_p(X_{\mathbf{u}}) : \mathsf{C}_p(X_{\mathbf{u}})
$$
$$
\hookrightarrow \mathsf{C}_p(X_{\mathbf{u}+e_i}) \; \forall\, i, \; \forall\, \mathbf{u}.
$$

That is, the ring R acts as the linear map (inclusion) between pairs of vector spaces in $\mathsf{C}_p(X.)$ with comparable grades. It is not too hard to check that this $\mathsf{C}_p(X.)$ is indeed a graded module. Each p-chain in a chain space $\mathsf{C}_p(X_{\mathbf{u}})$ is a homogeneous element with grade $\mathbf{u}$.

Then we have a chain complex of graded modules $(\mathsf{C}_*(X), \partial_*)$ where $\partial_* : \mathsf{C}_*(X) \to \mathsf{C}_{*-1}(X)$ is the boundary morphism given by $\partial_* \triangleq \bigoplus_{\mathbf{u} \in \mathbb{Z}^d} \partial_{*,\mathbf{u}}$ with $\partial_{*,\mathbf{u}} : \mathsf{C}_*(X_{\mathbf{u}}) \to \mathsf{C}_{*-1}(X_{\mathbf{u}})$ being the boundary map on $\mathsf{C}_*(X_{\mathbf{u}})$.

The kernel and image of a graded module morphism are also graded modules as submodules of domain and codomain, respectively whereas the

---

[3] For example, $\ker \partial_p$ denotes the inclusion of Z_p into C_p.

cokernel is a quotient module of the codomain. They can also be defined grade-wise in the expected way:

for $f \colon M \to N$, $(\ker f)_{\mathbf{u}} = \ker f_{\mathbf{u}}$, $(\operatorname{im} f)_{\mathbf{u}} = \operatorname{im} f_{\mathbf{u}}$, $(\operatorname{coker} f)_{\mathbf{u}} = \operatorname{coker} f_{\mathbf{u}}$.

All the linear maps are naturally induced from the original linear maps in M and N. In our chain complex cases, the kernel and image of the boundary morphism $\partial_p \colon \mathsf{C}_p(X) \to \mathsf{C}_{p-1}(X)$ are the family of cycle spaces $\mathsf{Z}_p(X)$ and the family of boundary spaces $\mathsf{B}_{p-1}(X)$, respectively, with linear maps induced by inclusions. Also, from the inclusion induced morphism $\mathsf{B}_p(X) \hookrightarrow \mathsf{Z}_p(X)$, we have the cokernel module $\mathsf{H}_p(X)$, consisting of homology groups $\bigoplus_{\mathbf{u} \in \mathbb{Z}^d} \mathsf{H}_p(X_{\mathbf{u}})$ and linear maps induced from inclusion maps $X_{\mathbf{u}} \hookrightarrow X_{\mathbf{v}}$ for each comparable pair $\mathbf{u} \le \mathbf{v}$. This $\mathsf{H}_p(X)$ is the *persistence module* M which we decompose. Classical persistence modules arising from a filtration of a simplicial complex over $\mathbb{Z}$ is an example of a one-parameter persistence module where the action $t_1 \cdot M_{\mathbf{u}} \subseteq M_{\mathbf{u}+e_1}$ signifies the linear map $M_{\mathbf{u}} \to M_{\mathbf{v}}$ between homology groups induced by the inclusion of the complex at $\mathbf{u}$ into the complex at $\mathbf{v} = \mathbf{u} + e_1$.

In our case, we have the chain complex of graded modules and induced homology groups which can be succinctly described by the following diagram.

$$
\begin{array}{ccccccccc}
& \mathsf{B}_p(X) \hookrightarrow \mathsf{Z}_p(X) \twoheadrightarrow \mathsf{H}_p(X) & & \mathsf{B}_{p-1}(X) \hookrightarrow \mathsf{Z}_{p-1}(X) \twoheadrightarrow \mathsf{H}_{p-1}(X) \\
\operatorname{im}\partial_{p+1} & & \ker(\partial_p) & \operatorname{im}\partial_p & & \ker\partial_{p-1} \\
\cdots \mathsf{C}_{p+1}(X) \xrightarrow{\quad \partial_{p+1} \quad} \mathsf{C}_p(X) & & \xrightarrow{\quad \partial_p \quad} \mathsf{C}_{p-1}(X) \cdots
\end{array}
$$

An Assumption

We always assume that the simplicial filtration is 1-*critical*, which means that each simplex has a unique earliest birth time. For the case which is not 1-critical, called multi-critical, one may utilize the *mapping telescope*, a standard algebraic construction [186], which transforms a multi-critical filtration to a 1-critical one. However, notice that this transformation increases the input size depending on the multiplicity of the incomparable birth times of the simplices. For 1-critical filtrations, each module C_p is free. With a fixed basis for each free module C_p, a concrete matrix $[\partial_p]$ for each boundary morphism ∂_p based on the chosen bases can be constructed.

With this input, we discuss our strategies for different cases that depend on two parameters: d, the number of parameters of filtration function, and p, the dimension of the homology groups in the persistence modules.

Note that a presentation gives an exact sequence $F^1 \to F^0 \twoheadrightarrow H \to \mathbf{0}$. To reveal further details of a presentation of H, we recognize that it respects the commutative diagram

$$
\begin{array}{ccc}
 & Y^1 & \\
 & \nearrow^{\text{im} f^1} \quad \searrow^{\text{ker} f^0} & \\
F^1 \xrightarrow{\quad f^1 \quad} & F^0 & \xrightarrow{\ f^0 = \text{coker} f^1\ }\hspace{-4pt}\twoheadrightarrow H
\end{array}
$$

where $Y^1 \hookrightarrow F^0$ is the kernel of f^0. With this diagram being commutative, all maps in this diagram are essentially determined by the presentation map f^1. We call the surjective map $f^0 \colon F^0 \to H$ the *generating map*, and $Y^1 = \ker f^0$ the *first syzygy module* of H.

11.5.2 Multiparameter Filtration, Zero-Dimensional Homology

In this case, $p = 0$ and $d > 0$. We obtain a presentation matrix straightforwardly with the observation that the module Z_0 of cycle spaces coincides with the module C_0 of chain spaces.

1. Presentation $\mathsf{C}_1 \xrightarrow{\ \partial_1\ } \mathsf{C}_0 \xrightarrow{\ \text{coker}\partial_1\ }\hspace{-4pt}\twoheadrightarrow \mathsf{H}_0$.
2. Presentation matrix $= [\partial_1]$ is given as part of the input.

Justification
For $p = 0$, the cycle module $\mathsf{Z}_0 = \mathsf{C}_0$ is a free module. So we have the presentation of H_0 as claimed. It is easy to check that $\partial_1 \colon \mathsf{C}_1 \to \mathsf{C}_0$ is a presentation of H_0 since both C_1 and C_0 are free modules. With the standard basis of chain modules C_p, we have a presentation matrix $[\partial_1]$ as the valid input to our decomposition algorithm.

The zeroth homology in our working example (Figure 11.5) corresponds to this case. The presentation matrix is the same as the matrix of boundary morphism ∂_1.

11.5.3 Two-Parameter Filtration, Multi-Dimensional Homology

In this case, $d = 2$ and $p \geq 0$. Lesnick and Wright [223] present an algorithm to compute a presentation, in fact a minimal presentation, for this case. When $d = 2$, by Hilbert's syzygy theorem [191], the kernel of a morphism between two free graded modules is always free. This implies that the canonical surjective map $\mathsf{Z}_p \twoheadrightarrow \mathsf{H}_p$ from free module Z_p can be naturally chosen as a generating map in the presentation of H_p. In this case we have the following.

1. Presentation $\mathsf{C}_{p+1} \xrightarrow{\ \bar{\partial}_{p+1}\ } \mathsf{Z}_p \xrightarrow{\ \text{coker}\bar{\partial}_{p+1}\ }\hspace{-4pt}\twoheadrightarrow \mathsf{H}_p$ where $\bar{\partial}_{p+1}$ is the induced map from the diagram:
2. Presentation matrix $= [\bar{\partial}_{p+1}]$ is constructed as follows:

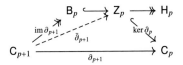

- Compute a basis $G(\mathsf{Z}_p)$ for the free module Z_p where $G(\mathsf{Z}_p)$ is presented as a set of generators in the basis of C_p. This can be done by an algorithm in [223]. Take $G(\mathsf{Z}_p)$ as the row basis of the presentation matrix $[\bar{\partial}_{p+1}]$.
- Present $\mathrm{im}\,\partial_{p+1}$ in the basis of $G(\mathsf{Z}_p)$ to get the presentation matrix $[\bar{\partial}_{p+1}]$ of the induced map as follows. Originally, $\mathrm{im}\,\partial_{p+1}$ is presented in the basis of C_p through the given matrix $[\partial_{p+1}]$. One needs to rewrite each column of $[\partial_{p+1}]$ in the basis $G(\mathsf{Z}_p)$ computed in the previous step. This can be done as follows. Let $[G(\mathsf{Z}_p)]$ denote the matrix presenting basis elements in $G(\mathsf{Z}_p)$ in the basis of C_p. Let c be any column vector in $[\partial_{p+1}]$. We reduce c to a zero vector by the matrix $[G(\mathsf{Z}_p)]$ and note the columns that are added to c. These columns provide the necessary presentation of c in the basis $G(\mathsf{Z}_p)$. This reduction can be done by the persistent algorithm described in Chapter 3.

Justification

Unlike the $p = 0$ case, for $p > 0$, we just know Z_p is a (proper) submodule of C_p, which means that Z_p is not necessarily equal to the free module C_p. However, fortunately for $d = 2$, the module Z_p is free, and we have an efficient algorithm to compute a basis of Z_p as the kernel of the boundary map $\partial_p : \mathsf{C}_p \to \mathsf{C}_{p-1}$. Then, we can construct the following presentation of H_p.

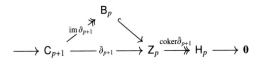

Here the $\bar{\partial}_{p+1}$ is an induced map from ∂_{p+1}. With a fixed basis on Z_p and standard basis of C_{p+1}, we rewrite the presentation matrix $[\partial_{p+1}]$ to get $[\bar{\partial}_{p+1}]$, which constitutes a valid input to our decomposition algorithm.

11.5.4 d-Parameter $(d > 2)$ Filtration, Multi-Dimensional Homology

The above construction of presentation matrix cannot be extended straightforwardly to d-parameter $(d > 2)$ persistence modules. Unlike the case in $d \le 2$, the cycle module Z is not necessarily free when $d > 2$. The issue caused by nonfree Z is that, if we use the same presentation matrix as we did

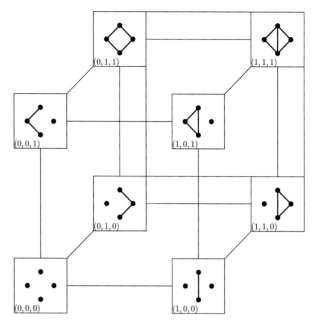

Figure 11.13 An example of a filtration of a simplicial complex for $d = 3$ with nonfree $\mathbf{Z}$ when $p = 1$. The three cycles at gradings $(0, 1, 1)$, $(1, 0, 1)$, and $(1, 1, 0)$ are three generators in $\mathbf{Z}_1$. However, at grading $(1, 1, 1)$, the earliest time these three cycles exist simultaneously, there is a relation among these three generators.

in the previous case with free $\mathbf{Z}$, we may lose some relations coming from the inner relations of a generating set of $\mathbf{Z}$. One can fix this problem by adding these inner relations into the presentation matrix as detailed in [141]. It is more complicated and we skip it here.

Figure 11.13 shows a simple example of a filtration of a simplicial complex whose persistence module H_p for $p = 1$ is a quotient module of non-free module $\mathbf{Z}$. The module H_1 is generated by three 1-cycles presented as $g_1^{(0,1,1)}, g_2^{(1,0,1)}$, and $g_3^{(1,1,0)}$. But when they appear together in $(1, 1, 1)$, there is a relation between these three: $\mathbf{t}^{(1,0,0)}g_1^{(0,1,1)} + \mathbf{t}^{(0,1,0)}g_2^{(1,0,1)} + \mathbf{t}^{(0,0,1)}g_3^{(1,1,0)} = 0$. Although im $\partial_1 = 0$, we still have a nontrivial relation from $\mathbf{Z}$. So, we have $\mathsf{H}_1 = \langle g_1^{(0,1,1)}, g_2^{(1,0,1)}, g_3^{(1,1,0)} : s^{(1,1,1)} = \mathbf{t}^{(1,0,0)}g_1^{(0,1,1)} + \mathbf{t}^{(0,1,0)}g_2^{(1,0,1)} + \mathbf{t}^{(0,0,1)}g_3^{(1,1,0)}\rangle$. The presentation matrix turns out to be the following:

$$
\begin{array}{c}
 \\
g_1^{(0,1,1)} \\
g_2^{(1,0,1)} \\
g_3^{(1,1,0)}
\end{array}
\begin{array}{c}
s^{(1,1,1)} \\
\left(
\begin{array}{c}
\mathbf{t}^{(1,0,0)} \\
\mathbf{t}^{(0,1,0)} \\
\mathbf{t}^{(0,0,1)}
\end{array}
\right).
\end{array}
$$

11.5.5 Time Complexity

Now we consider the time complexity for computing presentation and decomposition together. Let n be the size of the input filtration, that is, the total number of simplices obtained by counting at most one new simplex at a grid point of $\mathbb{Z}^d$. We consider three different cases as before.

Multiparameter, Zeroth Homology
In this case, the presentation matrix $[\partial_1]$ where $\partial_1 : \mathsf{C}_1 \to \mathsf{C}_0$ has size $O(n) \times O(n)$, that is, $N = O(n)$. Therefore, the total time complexity for this case is $O(n^{2\omega+1})$.

Two-Parameter, Multi-Dimensional Homology
In this case, as described in Section 11.5.3, first we compute a basis $G(\mathsf{Z}_p)$ that is presented in the basis of C_p. This is done by the algorithm of Lesnick and Wright [223] which runs in $O(n^3)$ time. Using $[G(\mathsf{Z}_p)]$, we compute the presentation matrix $[\bar{\partial}_{p+1}]$ as described in Section 11.5.3. This can be done in $O(n^3)$ time assuming that $G(\mathsf{Z}_p)$ has at most $O(n)$ elements. The presentation matrix is decomposed with TOTDIAGONALIZE as in the previous case. However, to claim that it runs in $O(n^{2\omega+1})$ time, one needs to ensure that the basis $G(\mathsf{Z}_p)$ has $O(n)$ elements. This follows from the fact that Z_p, being a free submodule of C_p, cannot have a rank larger than that of C_p.

In summary, the total time complexity in this case becomes $O(n^3) + O(n^{2\omega+1}) = O(n^{2\omega+1})$.

d-Parameter ($d > 2$), Multi-Dimensional Homology
For d-parameter persistence modules where $d \geq 2$ (subsumes the previous case), an algorithm running in time $O(n^{d+1})$ that produces a presentation matrix of dimensions $O(n^{d-1}) \times O(n^{d-1})$ can be designed using a result of Skryzalin [279] as described in [141]. Plugging $N = O(n^{d-1})$ and taking the computation of presentation matrix into consideration, we get a time complexity bound of $O(n^{d+1}) + O(n^{(2\omega+1)(d-1)}) = O(n^{(2\omega+1)(d-1)})$.

11.6 Invariants

For a given persistence module, it is useful to compute invariants that in some sense summarize the information contained in them. Ideally, these invariants should characterize the input module completely, meaning that the two invariants should be equal *if and only if* the modules are isomorphic. Persistence diagrams for one-parameter tame persistence modules are such invariants. For multiparameter persistence modules, no such complete invariants exist that are finite and hence computable. However, we can still aim for invariants that are computable and characterize the modules in some limited sense, meaning that these invariants remain equal for isomorphic modules though they may not differentiate non-isomorphic modules. Of course, their effectiveness in practice is determined by their discriminative power. We present two such invariants below: the first one, *rank invariant* was suggested in [65]; whereas the second one, *graded Betti number* was brought to TDA by [215] and studied further in [222].

11.6.1 Rank Invariants

Assume that the input graded module M is finitely generated as before and additionally finitely supported. For this we need to define the support of M.

Definition 11.17. (Support) Let M be a $\mathbb{Z}^d$-graded module. Its *support* is defined as the graph $\mathrm{supp}(M) = (V, E \subseteq V \times V)$ where a node $\mathbf{v} \in V$ if and only if $M_{\mathbf{v}} \neq \mathbf{0}$ and an edge $(\mathbf{u}, \mathbf{v}) \in E$ if and only if (i) $\mathbf{u} < \mathbf{v}$ and there is no $\mathbf{s} \in \mathbb{Z}^d$ satisfying $\mathbf{u} < \mathbf{s} < \mathbf{v}$, and (ii) $\mathrm{rank}(M_{\mathbf{u}} \to M_{\mathbf{v}}) \neq 0$. We say M is *finitely supported* if $\mathrm{supp}(M)$ is finite.

Fact 11.2. *The support* $\mathrm{supp}(M)$ *is disconnected if there exist two grades* $\mathbf{u} < \mathbf{v}$ *in* $\mathrm{supp}(M)$ *so that* $\mathrm{rank}(M_{\mathbf{u}} \to M_{\mathbf{v}}) = 0$.

For a finitely generated and finitely supported module M, we can compute a finite number of ranks of linear maps which collectively form the *rank invariant* of M. For two grades $\mathbf{u} \not\leq \mathbf{v}$, the linear maps between $M_{\mathbf{u}}$ and $M_{\mathbf{v}}$ are not defined. In the following definitions, we take them as zero maps.

Definition 11.18. (Rank invariant) Let $r_{\mathbf{uv}}(M) = \mathrm{rank}(M_{\mathbf{u}} \to M_{\mathbf{v}})$ for any pair $\mathbf{u}, \mathbf{v} \in \mathrm{supp}(M)$. The collection $\{r_{\mathbf{uv}}(M)\}_{\mathbf{u}, \mathbf{v} \in \mathrm{supp}(M)}$ is called the *rank invariant* of M.

Fact 11.3. *The rank invariant of a one-parameter module is a complete invariant. For a one-parameter persistence module* H_p, *it is given by persistent Betti numbers* $\beta_p^{i,j}$ *as defined in Definition 3.4.*

Although in the one-parameter case, the rank invariant provides complete information about the module, it does not do so for multiparameter persistence modules. For example, it cannot provide information about the "birth" and "death" of generators. This information can be deduced from a wider collection of rank invariant data called *multirank invariant* where we compute ranks of the linear maps between vector spaces at multiple grades. Multirank invariant is still not a complete invariant.

Definition 11.19. (Multirank invariant) The collection $\{r_{\mathbf{UV}}(M)\}$ for every pair $\mathbf{U} \subseteq \operatorname{supp}(M)$ and $\mathbf{V} \subseteq \operatorname{supp}(M)$ where $r_{\mathbf{UV}}(M) = \operatorname{rank}\left(\bigoplus_{\mathbf{u} \in \mathbf{U}} M_{\mathbf{u}} \to \bigoplus_{\mathbf{v} \in \mathbf{V}} M_{\mathbf{v}} \right)$ is called the *multirank invariant* of M.

We can retrieve information about the birth and death of generators from the multirank. For a grade $\mathbf{u}$, define its immediate predecessors $P_{\mathbf{u}}$ and immediate successors $S_{\mathbf{u}}$ as

$$P_{\mathbf{u}} = \{\mathbf{u}' \in \operatorname{supp}(M) \mid \mathbf{u}' < \mathbf{u} \text{ and } \not\exists\, \mathbf{u}'' \text{ with } \mathbf{u}' < \mathbf{u}'' < \mathbf{u}\},$$
$$S_{\mathbf{u}} = \{\mathbf{u}' \in \operatorname{supp}(M) \mid \mathbf{u}' > \mathbf{u} \text{ and } \not\exists\, \mathbf{u}'' \text{ with } \mathbf{u} < \mathbf{u}'' < \mathbf{u}'\}.$$

Fact 11.4. *We know the following.*

(a) We have that m generators get born at grade $\mathbf{u}$ if and only if $\operatorname{coker}\left(\bigoplus_{\mathbf{u}' \in P_{\mathbf{u}}} M_{\mathbf{u}'} \to M_{\mathbf{u}} \right)$ has dimension m.
(b) We have that m generators die leaving grade $\mathbf{u}$ if and only if $\ker\left(M_{\mathbf{u}} \to \bigoplus_{\mathbf{u}' \in S_{\mathbf{u}}} M_{\mathbf{u}'} \right)$ has dimension m.

Although multirank invariants cannot characterize multiparameter persistence modules completely in general, they do so for the special case of interval decomposable modules. We will describe these modules in detail in the next chapter. Here we introduce them briefly.

We call $I \subseteq \operatorname{supp}(M)$ an *interval* if I is connected and for every $\mathbf{u}, \mathbf{v} \in I$, if $\mathbf{u} < \mathbf{w} < \mathbf{v}$, then $\mathbf{w} \in I$. We call a persistence module with support on an interval an *interval module* if $M_{\mathbf{u}}$ is one-dimensional for each vertex $\mathbf{u} \in \operatorname{supp}(M)$. A persistence module M is called *interval decomposable* if there is a decomposition $M = \bigoplus M^i$ where each M^i is an interval module.

Fact 11.5. *Two interval decomposable modules are isomorphic if and only if they have the same multirank invariants.*

11.6.2 Graded Betti Numbers and Blockcodes

For one-parameter persistence modules, the barcodes provide a complete invariant. For multiparameter persistence, we first introduce an invariant called *graded Betti number*, which we refine further to define *persistent graded Betti numbers* as a generalization of persistence diagrams. The decomposition of a module also allows us to define *blockcodes* as a generalization of barcodes. Both of them depend on the ideas of free resolution and graded Betti numbers which are well studied in commutative algebra and were first introduced in topological data analysis by Knudson [215].

Table 11.1 *All the nonzero graded Betti numbers* $\beta_{i,\mathbf{u}}$ *are listed. Empty entries are all zeros.*

β^M	(1,0)	(0,1)	(1,1)	(2,1)	(1,2)	(2,2)	$\cdots$
β_0	1	1	1				
β_1			1	1	1		
β_2						1	
$\beta_{\geq 3}$							

Definition 11.20. (Free resolution) For a graded module M, a *free resolution* $\mathcal{F} \to M$ is an exact sequence

$$\cdots \longrightarrow F^2 \xrightarrow{f^2} F^1 \xrightarrow{f^1} F^0 \xrightarrow{f^0} M \longrightarrow 0$$

where each F^i is a free graded R-module.

Now we observe that a free resolution can be obtained as an extension of a free presentation. Consider a free presentation of M as depicted below.

If the presentation map f^1 has nontrivial kernel, we can find a nontrivial map $f^2 \colon F^2 \to F^1$ with $\operatorname{im} f^2 = \ker f^1$, which implies $\operatorname{coker} f^2 \cong \operatorname{im} f^1 = \ker f^0 = Y^1$. Therefore, f^2 is in fact a presentation map of the module Y^1 which is the so-called first syzygy module of M (named after Hilbert's famous syzygy theorem [191]). We can keep doing this to get $f^3, f^4, \ldots$ by constructing presentation maps on higher-order syzygy modules $Y^2, Y^3, \ldots$ of M, which results in a diagram depicted below, which gives a free resolution of M.

Free resolution is not unique. However, there exists an essentially unique minimal free resolution in the sense that any free resolution can be obtained by summing the minimal free resolution with a free resolution of a trivial module. Below we give a construction to build a minimal free resolution from a minimal free presentation. The proof that it indeed creates a minimal free resolution can be found in [50, 268].

Construction of Minimal Free Resolution

Choose a minimal set of homogeneous generators $g_1, \ldots, g_n$ of M. Let $F^0 = \bigoplus_{i=1}^n R_{\to \text{gr}(g_i)}$ with standard basis $e_1^{\text{gr}(g_1)}, \ldots, e_n^{\text{gr}(g_n)}$ of F^0. The homogeneous R-map $f^0 \colon F^0 \to M$ is determined by $f^0(e_i) = g_i$. Now the first syzygy module of M, $Y^1 \xhookrightarrow{\ker f^0} F^0$, is again a finitely generated graded R-module. We choose a minimal set of homogeneous generators $y_1, \ldots, y_m$ of Y_1 and let $F^1 = \bigoplus_{j=1}^m R_{\to \text{gr}(y_j)}$ with standard basis $e_1^{/\text{gr}(y_1)}, \ldots, e_m^{/\text{gr}(y_m)}$ of F^1. The homogeneous R-map $f^1 \colon F^1 \to F^0$ is determined by $f^1(e_j') = y_j$. By repeating this procedure for $Y_2 = \ker f^1$ and moving backward further, one gets a graded free resolution of M.

Definition 11.21. (Graded Betti numbers) Let F^j be a free module in the minimal free resolution of a graded module M. Let $\beta_{j,\mathbf{u}}^M$ be the multiplicity of each grade $\mathbf{u} \in \mathbb{Z}^d$ in the multiset consisting of the grades of homogeneous basis elements for F^j. Then, the mapping $\beta_{(-,-)}^M \colon \mathbb{Z}_{\geq 0} \times \mathbb{Z}^d \to \mathbb{Z}_{\geq 0}$ is an invariant called the *graded Betti numbers* of M.

For example, the graded Betti numbers of the persistence module for our working example in Figure 11.5 are listed in Table 11.1.

Definition 11.22. (Persistent graded Betti numbers) Let $M \cong \bigoplus M^i$ be a total decomposition of a graded module M. We have for each indecomposable M^i, the refined graded Betti numbers $\beta^{M^i} - \{\beta_{j,\mathbf{u}}^{M^i} \mid j \in \mathbb{N}, \mathbf{u} \in \mathbb{Z}^d\}$. We call the set $\mathcal{PB}(M) := \{\beta^{M^i}\}$ the *persistent graded Betti numbers* of M.

For the working example in Figure 11.5, the persistent graded Betti numbers are given in Table 11.2.

One way to summarize the information of graded Betti numbers is to use the Hilbert function, which is also called the dimension function [140], defined as

$$\text{dm}M : \mathbb{Z}^d \to \mathbb{Z}_{\geq 0} \quad \text{dm}M(\mathbf{u}) = \dim(M_{\mathbf{u}}).$$

Fact 11.6. *There is a relation between the graded Betti numbers and dimension function of a persistence module as follows:*

$$\text{for all } \mathbf{u} \in \mathbb{Z}^d, \ \text{dm}M(\mathbf{u}) = \sum_{\mathbf{v} \leq \mathbf{u}} \sum_j (-1)^j \beta_{j,\mathbf{v}}.$$

Then for each indecomposable M^i, we have the dimension function $\text{dm}M^i$ related to persistent graded Betti numbers restricted to M^i.

Definition 11.23. (Blockcode) The set of dimension functions $\mathcal{B}_{\text{dm}}(M) := \{\text{dm}M^i\}$ is called the *blockcode* of M.

Table 11.2 *Persistence grades* $\mathcal{PB}(M) = \{\beta^{M^1}, \beta^{M^2}\}$. *Only nonzero entries are listed. Empty entries are all zeros.*

β^{M^1}	(1,0)	(0,1)	(1,1)	(2,1)	(1,2)	(2,2)	$\cdots$
β_0	1	1					
β_1			1				
$\beta_{\geq 2}$							

β^{M^2}	(1,0)	(0,1)	(1,1)	(2,1)	(1,2)	(2,2)	$\cdots$
β_0			1				
β_1				1	1		
β_2						1	
$\beta_{\geq 3}$							

For our working example, the dimension functions of indecomposable summands M^1 and M^2 are (see Figure 11.14 for the visualization):

$$\mathrm{dm}M^1(\mathbf{u}) = \begin{cases} 1 & \text{if } \mathbf{u} \geq (1,0) \text{ or } \mathbf{u} \geq (0,1), \\ 0 & \text{otherwise,} \end{cases} \qquad \mathrm{dm}M^2(\mathbf{u}) = \begin{cases} 1 & \text{if } \mathbf{u} = (1,1), \\ 0 & \text{otherwise.} \end{cases} \quad (11.2)$$

We can read out some useful information from dimension functions on each indecomposable. We take the dimension functions of our working example as an example. For $\mathrm{dm}M^1$, two connected components are born at the two left-bottom corners of the purple region. They are merged together immediately when they meet at grade $(1, 1)$. After that, they persist forever as one connected component. For $\mathrm{dm}M^2$, one connected component is born at the left-bottom corner of the square green region. Later at the grades of left-top corner and right-bottom corner of the green region, it is merged with some other connected component with smaller grades of birth. Therefore, it only persists within this green region.

In general, both persistent graded Betti numbers and blockcodes are not sufficient to classify multiparameter persistence modules, which means they are not complete invariants. As indicated in [64], there is no complete discrete invariant for multiparameter persistence modules. However, interestingly, these two invariants are indeed complete invariants for interval decomposable modules like this example, which we will study in the next chapter.

11.7 Notes and Exercises

In one of the first extensions of the persistence algorithm for the one-parameter case, the authors in [9] presented a matrix reduction-based algorithm which applies to a very special case of commutative ladder C_n for $n \leq 4$ defined on a subgrid of $\mathbb{Z}^2$. This matrix construction and the algorithm are very different

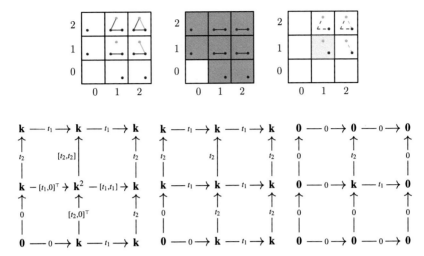

Figure 11.14 (top) Two-parameter simplicial filtration for our working example in Figure 11.5. $\mathrm{dm}M^1$ and $\mathrm{dm}M^2$: each colored square represents a one-dimensional vector space $\mathbf{k}$ and each white square represents a zero-dimensional vector space. In the middle picture M^1 is generated by $v_b^{0,1}$ and $v_r^{1,0}$ which are drawn as a blue dot, and a red dot, respectively. They are merged at $(1, 1)$ by the red edge e_r. In the right picture, M^2 is generated by $v_g^{(1,1)} + \mathbf{t}^{(0,1)}v_r^{1,0}$ which is represented by the combination of the green circle and the red circle together at $(1, 1)$. After this point $(1, 1)$, the generator is reduced to zero by relation of e_g starting at $(2, 1)$, represented by the green dashed line segment, and by relation of $e_b + \mathbf{t}^{(0,1)}e_r$ starting at $(1, 2)$, represented by the blue dashed line segment connected with the red dashed line segment.

from the one presented here. This algorithm may not terminate if the input does not satisfy the stated assumption.

We have already mentioned that the Meataxe algorithm [252] known in the computational algebra community can be used for more general modules and hence for persistence modules. The main advantage of this algorithm is that it applies to general persistence modules, but a major disadvantage is that it runs very slowly. Even allowing approximation, the algorithm [197] runs in $O(N^{3(d+1)} \log q)$ time (or $O(N^{4(d+1)} \log q)$ as conjectured in [196] because of some special cases mentioned in [197]), where N is the number of generators and relations in the input module that is defined with polynomial ring $\mathbb{Z}_q[t_1 t_2 \ldots t_d]$.

Under a suitable finiteness condition, the fact that persistence modules are indeed finitely presented graded modules over multivariate polynomials was first recognized by Carlsson et al. [64, 65] and Knudson [215] and further studied by Lesnick and Wright [221, 223]. The graded module structure studied in algebraic geometry and commutative algebra [155, 232] encodes a lot of information and thus can be leveraged for designing efficient algorithms. Lesnick and Wright [223] leveraged this fact to design an efficient algorithm for computing minimal presentations for two-parameter persistence modules from an input two-parameter simplicial filtration. Recognizing the power of expressing graded modules in terms of presentations, Dey and Xin [141] proposed the decomposition algorithm using matrix equivalents of presentations and their direct sums. The materials in this chapter are mostly taken from that paper. This decomposition algorithm can be viewed as a generalization of the classical persistence algorithm for the one-parameter case, though the matrix reduction technique is more involved because it has to accommodate constraints on grades. The algorithm in [141] handled these constraints using the technique of matrix linearization as described in Section 11.4.

As a generalization of the one-parameter persistence algorithm, it is expected that the algorithm in [141] can be interpreted as computing invariants such as persistence diagrams or barcodes. A roadblock to this goal is that d-parameter persistence modules do not have complete discrete invariants for $d \geq 2$ [65, 221]. Consequently, one needs to invent other invariants suitable for multiparameter persistence modules. The rank invariants and multirank invariants described in Section 11.6.1 serve this purpose. There is a related notion of generalized persistence diagram introduced by Patel [255] and further studied in [213].

One natural approach taking advantage of a decomposition algorithm would be to consider the decomposition and take the discrete invariants in each indecomposable component. This gives invariants which may not be complete but still contain rich information. We mentioned two interpretations of the output of the algorithm presented in this chapter as two different invariants: *persistent graded Betti numbers* as a generalization of persistence diagrams and *blockcodes* as a generalization of barcodes. The persistent graded Betti numbers are linked to the graded Betti numbers studied in commutative algebra brought to TDA by [215]. The bigraded Betti numbers are further studied in [223]. By constructing the free resolution of a persistence module, we can compute its graded Betti numbers and then decompose them according to each indecomposable module, which results in the presistent graded Betti numbers. For each indecomposable, we apply the dimension function [140], which is also known as the Hilbert function in commutative algebra, to summarize the graded Betti numbers for each indecomposable module. This constitutes a blockcode for the indecomposable module of the persistence module. The blockcode is a good vehicle for visualizing lower-dimensional persistence modules such as two- or three-parameter persistence modules. For details on these invariants, see [141].

Exercises

1. Using the matrix diagonalization algorithm as described in this chapter, devise an algorithm to compute a minimal presentation of a two-parameter persistence module given by a simplicial filtration over $\mathbb{Z}^2$.

2. Give an example of a two-parameter simplicial filtration over $\mathbb{Z}^2$ at least one of whose decomposables is not free.

3. Give an example of a two-parameter simplicial filtration over $\mathbb{Z}^2$ at least one of whose decomposables does not have all of its nontrivial vector spaces over the grades being isomorphic.

4. Give an example of a two-parameter persistence module M with three generators and relations that have the following properties:
   (a) M is indecomposable,
   (b) M has two indecomposables, and
   (c) M has three indecomposables.

5. Prove that the cycle module Z_p arising from a two-parameter simplicial filtration is always free.

6. Design a polynomial-time algorithm for computing the decomposition of the persistence module induced by a given simplicial filtration over $\mathbb{Z}^2$ when a simplex can be a generator at different grades.

7. Let $\mathbf{A}$ be a presentation matrix with n generators and relations whose grades are distinct and totally ordered. Design an $O(n^3)$-time algorithm to decompose $\mathbf{A}$. Interpret the types of each indecomposable in such a case.

8. The algorithm TOTDIAGONALIZE has been written assuming that the field of the polynomial ring is $\mathbb{Z}_2$. Write it for a general finite field.

9. Give an example of two non-isomorphic two-parameter persistence modules which have the same rank invariant.

10. Design an efficient algorithm to compute the rank invariant of a module from the simplicial filtration inducing it.

11. Prove that a two-parameter persistence module M is an interval (see Section 11.6.1) if and only if supp(M) is connected and each $M_{\mathbf{u}}$ for $\mathbf{u} \in$ supp(M) has dimension one.

12. Suppose that a two-parameter persistence module is given by a presentation matrix. Design an algorithm to determine if M is an interval or not without decomposing the input matrix. (Hint: Consider computing graded Betti numbers from the grades of the rows and columns of the matrix.)

13. Show that for a finitely presented (finite number of generators and relations) graded module M, there exist two interval decomposable graded modules M^1 and M^2 so that the rank invariants (Definition 11.18) satisfy $r_{\mathbf{uv}}(M) = r_{\mathbf{uv}}(M^1) - r_{\mathbf{uv}}(M^2)$ for every $\mathbf{u}, \mathbf{v} \in$ supp(M). Given a presentation matrix for M, compute such M^1 and M^2 efficiently.

14. Write a pseudocode for the construction of a minimal free resolution given in Section 11.6.2. Analyze its complexity.

12
Multiparameter Persistence and Distances

We have seen that persistence modules are important objects of study in topological data analysis in that they serve as an intermediate between the raw input data and the output summarization with persistence diagrams. For the one-parameter case, the distances between modules can be computed from the bottleneck distances between the corresponding persistence diagrams. For multiparameter persistence modules, we already saw in Chapter 11 that the indecomposables which are analogues to bars in the one-parameter case are more complicated. So, defining distances between persistence modules in terms of indecomposables also becomes more complicated. However, we need a distance or distance-like notion between persistence modules to compare the input data inducing them. Figure 12.1 shows an output of RIVET software [222] that implemented the so-called matching distance between two-parameter persistence modules. In this chapter, we describe some of these distances proposed in the literature and algorithms for computing them efficiently (polynomial time).

The interleaving distance d_I between one-parameter persistence modules as defined in Chapter 3 provides a useful means to compare them. Fortunately, for one-parameter persistence modules, they can be computed exactly by computing the bottleneck distance d_b between their persistence diagrams thanks to the isometry theorem [221] (see also [23, 80]). Chapter 3 gives a polynomial-time algorithm $O(n^{1.5} \log n)$ for computing bottleneck distance. The status, however, is not so well settled for multiparameter persistence modules.

One of the difficulties facing the definition and computation of distances among multiparameter persistence modules is the fact that their indecomposables do not have a finite characterization as indicated in Chapter 11. Even for finitely generated modules, this is true, though a unique decomposition is guaranteed by the Krull–Schmidt theorem [10]. Despite this difficulty, one can define an interleaving distance d_I for multiparameter persistence modules

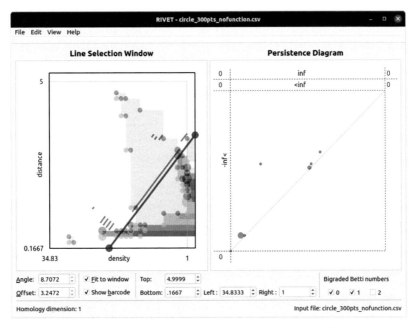

Figure 12.1 A two-parameter module is sliced by lines that provide matching distance between two modules as we explain in Section 12.3. This figure is an output of RIVET software due to [222]. Courtesy of M. Lesnick and M. Wright (2015, figure 3).

which can be viewed as an extension of the interleaving distance defined for one-parameter persistence modules. Shown by Lesnick [221], this distance is the most fundamental one because it is the most discriminative distance among persistence modules that is also stable with respect to functions or simplicial filtrations that give rise to the modules. Unfortunately, it turns out that computing d_I for n-parameter persistence modules and even approximating it within a factor less than 3 is NP-hard for $n \geq 2$. For a special case of modules called *interval modules*, d_I can be computed in polynomial time. In Section 12.2, we introduce the interleaving distance for multiparameter persistence modules. We follow it with a polynomial-time algorithm [140] in Section 12.4.3 which computes d_I for two-parameter interval modules.

To circumvent the problem of computing interleaving distances, several other distances have been proposed in the literature that are computable in polynomial time and bound the interleaving distance from either above or below, but not both in the general case. Given the NP-hardness of approximating interleaving distance, there cannot exist any polynomial-time computable distance that bounds d_I from both above and below within a constant factor of 3 unless

P = NP. The *matching distance* d_m as defined in Section 12.3 bounds d_I from below, that is, $d_m \leq d_I$, and it can be computed in polynomial time.

Finally, in Section 12.4, we extend the definition of the bottleneck distance to multiparameter persistence modules. Extending the concept from the one-parameter case, one can define d_b as the supremum of the pairwise interleaving distances between indecomposables under an optimal matching. Then, straightforwardly, $d_I \leq d_b$, but the converse is not necessarily true. It is known that no lower bound for d_I in terms of d_b may exist even for a special class of two-parameter persistence modules called interval decomposable modules [47]. However, d_b can be useful as a reasonable upper bound to d_I. Unfortunately, a polynomial-time algorithm for computing d_b is not known for general persistence modules. For some persistence modules whose indecomposables have constant description, such as block decomposable modules, one can compute d_b in polynomial time simply because the interleaving distance between any two modules with constant description cannot take more than $O(1)$ time.

In Section 12.4, we consider a special class of persistence modules whose indecomposables are intervals and present a polynomial-time algorithm for computing d_b for them. These are modules whose indecomposables are supported by "staircase" polyhedra. Our algorithm assumes that all indecomposables are given and computes d_b exactly for two-parameter interval decomposable modules. Although the algorithm can be extended to persistence modules with larger number of parameters, we choose to present it only for the two-parameter case for simplicity while not losing the essential ingredients for the general case. The indecomposables required as input can be computed by the decomposition algorithm presented in the previous chapter (Chapter 11).

12.1 Persistence Modules from Categorical Viewpoint

In this chapter we define the persistence modules as categorical structures which are different from the graded structures used in the previous chapter. Other than introducing a different viewpoint of persistence modules, we do so because this definition becomes more amenable to defining distances. Thanks to representation theory [65, 107, 215], these two notions coincide when the modules are finitely generated in the graded module definition (Definition 11.5) and are of finite type (Definition 12.5) in the categorical definition. Let us recall the definition in the one-parameter case. A persistence module M parameterized over $A = \mathbb{Z}$ or $\mathbb{R}$ is defined by a sequence of vector spaces M_x, $x \in A$, with linear maps $\rho_{x,y} : M_x \to M_y$ so that $\rho_{x,x}$ is the identity for every

$x \in A$, and for all $x, y, z \in A$ with $x \leq y \leq z$, one has $\rho_{x,z} = \rho_{y,z} \circ \rho_{x,y}$. These conditions can be formulated using category theory.

Definition 12.1. (Category) A *category* $\mathcal{C}$ is a set of objects $\mathrm{Obj}\,\mathcal{C}$ with a set of morphisms $\hom(x, y)$ for every pair of elements $x, y \in \mathrm{Obj}\,\mathcal{C}$ where:

- for every $x \in \mathrm{Obj}\,\mathcal{C}$, there is a special identity morphism $\mathbb{1}_x \in \hom(x, x)$;
- if $f \in \hom(x, y)$ and $g \in \hom(y, z)$, then $g \circ f \in \hom(x, z)$;
- for homomorphisms f, g, and h, the compositions wherever defined are associative, that is, $(f \circ g) \circ h = f \circ (g \circ h)$; and
- $\mathbb{1}_x \circ f_{x,y} = f_{x,y}$ and $f_{x,y} \circ \mathbb{1}_y = f_{x,y}$ for every pair $x, y \in \mathrm{Obj}\,\mathcal{C}$.

All sets form a category **Set** with functions between them playing the role of morphisms. Topological spaces form a category **Top** with continuous maps between them being the morphisms. Vector spaces form the category **Vec** with linear maps between them being the morphisms. A poset $\mathbb{P}$ forms a category with every pair $x, y \in \mathbb{P}$ admitting at most one morphism; $\hom(x, y)$ has one element if $x \leq y$ and is empty otherwise. Such a category is called a *thin* category in the literature, for which composition rules take trivial form.

Definition 12.2. (Functor) A *functor* between two categories $\mathcal{C}$ and $\mathcal{D}$ is an assignment $F : \mathcal{C} \to \mathcal{D}$ satisfying the following conditions:

- for every $x \in \mathrm{Obj}\,\mathcal{C}$, $F(x) \in \mathrm{Obj}\,\mathcal{D}$;
- for every morphism $f \in \hom(x, y)$, $F(f) \in \hom(F(x), F(y))$;
- F respects composition, that is, $F(f \circ g) = F(f) \circ F(g)$; and
- F preserves identity morphisms, that is, $F(\mathbb{1}_x) = \mathbb{1}_{F(x)}$ for every $x \in \mathrm{Obj}\,\mathcal{C}$.

One can observe that the one-parameter persistence module is a functor from the category of totally ordered set of $\mathbb{Z}$ (or $\mathbb{R}$) to the category of **Vec**. Homology groups with a field coefficient for topological spaces provide a functor from category **Top** to the category of vectors spaces **Vec**. We can define maps between functors themselves.

Definition 12.3. (Natural transformation) Given two functors $F, G \colon \mathcal{C} \to \mathcal{D}$, a *natural transformation* η from F to G, denoted as $\eta \colon F \Longrightarrow G$, is a family of morphisms $\{\eta_x \colon F(x) \to G(x)\}$ for every $x \in \mathrm{Obj}\,\mathcal{C}$ so that the following diagram commutes.

$$F(x) \xrightarrow{F(\rho)} F(y)$$

$$\downarrow \eta_x \qquad \qquad \downarrow \eta_y$$

$$G(x) \xrightarrow{G(\rho)} G(y)$$

Let **k** be a field, **Vec** be the category of vector spaces over **k**, and **vec** be the subcategory of finite-dimensional vector spaces. As usual, for simplicity, we assume $\mathbf{k} = \mathbb{Z}_2$.

Definition 12.4. (Persistence module) Let $\mathbb{P}$ be a poset category. A $\mathbb{P}$-*indexed persistence module* is a functor $M : \mathbb{P} \to \mathbf{Vec}$. If M takes values in **vec**, we say M is *pointwise finite-dimensional* (p.f.d.). The $\mathbb{P}$-indexed persistence modules themselves form another category where the natural transformations between functors constitute the morphisms.

Definition 12.5. (Finite type) A $\mathbb{P}$-indexed persistence module M is said to have *finite type* if M is p.f.d. and all morphisms $M(x \leq y)$ are isomorphisms outside a finite subset of $\mathbb{P}$.

Here we consider the poset category to be $\mathbb{R}^d$ with the standard partial order and all modules to be of finite type. We call $\mathbb{R}^d$-indexed persistence modules as d-parameter modules in short. The reader can recognize that this is a shift from our assumption in the last chapter where we considered $\mathbb{Z}^d$-indexed modules. The category of d-parameter modules in this chapter is denoted as $\mathbb{R}^d$-**mod**. For a d-parameter module $M \in \mathbb{R}^d$-**mod**, we use notation $M_x := M(x)$ and $\rho^M_{x \to y} := M(x \leq y)$.

Definition 12.6. (Shift) For any $\delta \in \mathbb{R}$, we denote $\vec{\delta} = (\delta, \ldots, \delta) = \delta \cdot \vec{e}$, where $\vec{e} = (e_1, e_2, \ldots, e_d)$ with $\{e_i\}_{i=1}^d$ being the standard basis of $\mathbb{R}^d$. We define a shift functor $(\cdot)_{\to \delta} : \mathbb{R}^d$-**mod** $\to \mathbb{R}^d$-**mod** where $M_{\to \delta} := (\cdot)_{\to \delta}(M)$ is given by $M_{\to \delta}(x) = M(x + \vec{\delta})$ and $M_{\to \delta}(x \leq y) = M(x + \vec{\delta} \leq y + \vec{\delta})$. In other words, $M_{\to \delta}$ is the module M shifted diagonally by $\vec{\delta}$.

12.2 Interleaving Distance

The following definition of interleaving adapts the original definition designed for one-parameter modules in [77, 80] to d-parameter modules.

Definition 12.7. (Interleaving) For two d-parameter persistence modules M and N, and $\delta \geq 0$, a δ-*interleaving* between M and N are two families of

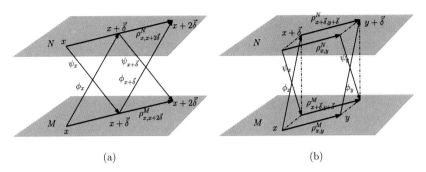

Figure 12.2 (a) Triangular commutativity and (b) rectangular commutativity.

linear maps $\{\phi_x : M_x \rightarrow N_{x+\vec{\delta}}\}_{x \in \mathbb{R}^d}$ and $\{\psi_x : N_x \rightarrow M_{x+\vec{\delta}}\}_{x \in \mathbb{R}^d}$ satisfying the following two conditions (see Figure 12.2):

- for all $x \in \mathbb{R}^n$, $\rho^M_{x \rightarrow x+2\vec{\delta}} = \psi_{x+\vec{\delta}} \circ \phi_x$ and $\rho^N_{x \rightarrow x+2\vec{\delta}} = \phi_{x+\vec{\delta}} \circ \psi_x$;
- for all $x \leq y \in \mathbb{R}^n$, $\phi_y \circ \rho^M_{x \rightarrow y} = \rho^N_{x+\vec{\delta} \rightarrow y+\vec{\delta}} \circ \phi_x$ and $\psi_y \circ \rho^N_{x \rightarrow y} = \rho^M_{x+\vec{\delta} \rightarrow y+\vec{\delta}} \circ \psi_x$.

If such a δ-interleaving exists, we say M and N are δ-*interleaved*. We call the first condition *triangular commutativity* and the second condition *rectangular commutativity*.

Definition 12.8. (Interleaving distance) The *interleaving distance* between modules M and N is defined as $\mathsf{d}_I(M, N) = \inf_\delta \{M$ and N are δ-interleaved$\}$. We say M and N are ∞-interleaved if they are not δ-interleaved for any $\delta \in \mathbb{R}^+$, and assign $\mathsf{d}_I(M, N) = \infty$.

The following computational hardness result from [33] is stated assuming that the input modules are represented with the graded matrices as in Chapter 11. As we mentioned before, these modules coincide with the category of modules of finite type.

Theorem 12.1. *Given two modules M and N given by graded matrix representations, the problem of computing a real r so that $\mathsf{d}_I(M, N) \leq r < 3\mathsf{d}_I(M, N)$ is NP-hard.*

12.3 Matching Distance

The matching distance between two persistence modules M and N draws upon the idea of taking the restrictions of M and N over lines with positive slopes and then determining the supremum of weighted interleaving distances on these restrictions. It can be defined for d-parameter modules. We are going to describe a polynomial-time algorithm for computing it for two-parameter modules, so for simplicity we define the matching distance for two-parameter modules. Let $\ell\colon sx + t$ denote any line in $\mathbb{R}^2$ with $s > 0$ and let Λ denote the space of all such lines. Define a parameterization $\lambda\colon \mathbb{R} \to \ell$ of ℓ by taking $\lambda(x) = (1/(1 + s^2))(x, sx + t)$. For a line $\ell \in \Lambda$, let $M|_\ell$ denote the restriction of M on ℓ where $M|_\ell(x) = M(\lambda(x))$ with linear maps induced from M. This is a one-parameter persistence module. We define a weight $w(\ell)$ that accounts for projections on one of the two axes depending on the slope:

$$
w(\ell) = \begin{cases} \dfrac{1}{\sqrt{1 + s^2}} & \text{for } s \geq 1, \\[2ex] \dfrac{1}{\sqrt{1 + 1/s^2}} & \text{for } 0 < s < 1. \end{cases}
$$

Definition 12.9. The matching distance $\mathsf{d}_m(M, N)$ between two persistence modules is defined as

$$
\mathsf{d}_m(M, N) = \sup_{\ell \in \Lambda}\{w(\ell) \cdot \mathsf{d}_I(M|_\ell, N|_\ell)\}.
$$

The weight $w(\ell)$ is introduced to make the matching distance stable with respect to the interleaving distance.

12.3.1 Computing Matching Distance

We define a point–line duality in $\mathbb{R}^2$: a line $\ell \subset \mathbb{R}^2$ is dual to a point $\ell^* = (s, t)$ where $\ell\colon y = sx - t$, and a point $p = (s, t)$ is dual to a line $p^*\colon y = sx - t$. The following facts can be deduced from the definition easily (Exercise 4).

Fact 12.1. *We know the following.*

(a) For a point p and a line ℓ, one has $(p^*)^* = p$ and $(\ell^*)^* = \ell$.
(b) If a point p is in a line ℓ, then point ℓ^* is in line p^*.
(c) If a point p is above (below) a line ℓ, then point ℓ^* is above (below) line p^*.

Consider the open half-plane Ω of $\mathbb{R}^2$ where $\Omega = \{x, y \mid x > 0\}$. Let α denote the bijective map between Ω and the space Λ of lines with positive slopes where $\alpha(p) = p^*$.

The representation theory [65, 107, 215] tells us that finitely generated graded modules as defined in Chapter 11 are essentially equivalent to persistence modules as defined in this chapter as long as they are of finite type (Definition 12.5). Then, if a persistence module M is a functor on the poset $\mathbb{P} = \mathbb{R}^2$ or $\mathbb{Z}^2$, we can talk about the grades (elements of $\mathbb{P}$) of a *generating set* of M and the relations which are combinations of generators that become zero. A mindful reader can recognize that these are exactly the grades of the rows and columns of the presentation matrix for M (Definition 11.14).

Given two two-parameter persistence modules M and N, let $\mathrm{gr}(M)$ and $\mathrm{gr}(N)$ denote the grades of all generators and relations in M and N, respectively. Consider the set of lines L dual to the points in $\mathrm{gr}(M) \cup \mathrm{gr}(N)$. These lines together create a line arrangement in Ω which is a partition of Ω into vertices, edges, and faces. The vertices are points where two lines meet, the edges are maximal connected subsets of the lines excluding the vertices, and the faces are maximal connected subsets of Ω excluding the vertices and edges. Let A_0 denote this initial arrangement. We refine this arrangement further later. First, we observe an invariant property of the arrangement for which we need the following definition.

Definition 12.10. (Point pair type) Given two points p and q and a line ℓ, we say (p, q) has the following types with respect to ℓ: (i) *Type-1* if both p and q lie above ℓ; (ii) *Type-2* if both p and q lie below ℓ; (iii) *Type-3* if p lies above and q lies below ℓ; and (iv) *Type-4* if p lies below and q lies above ℓ.

The following proposition follows from Fact 12.1.

Proposition 12.2. *For two points $p, q \in \mathrm{gr}(M) \cup \mathrm{gr}(N)$ and a face $\tau \in A_0$, the type of (p, q) with respect to the line z^* is the same for all $z \in \tau$.*

Our goal is to refine A_0 further to another arrangement A so that for every face $\tau \in A$ the grade points p and q that realize $\mathsf{d}_I(M|_\ell, N|_\ell)$ for every $\ell = z^*$ remain the same for all $z \in \tau$. Toward that goal, we define the push of a grade point.

Definition 12.11. (Push) For a point $p = (p_x, p_y)$ and a line $\ell : y = sx - t$, the *push* $\mathrm{push}(p, \ell)$ is defined as

$$\mathrm{push}(p, \ell) = \begin{cases} (p_x, sp_x - t) & \text{for } p \text{ below } \ell; \\ ((p_y + t)/s, p_y) & \text{for } p \text{ above } \ell. \end{cases}$$

Geometrically, $\mathrm{push}(p, \ell)$ is the intersection of ℓ with the upward ray originating from p in the first case, and with the horizontal ray originating from p in the second case. Figure 12.3 illustrates the two cases.

For $p, q \in \mathbb{R}^2$, let

$$\delta_{p,q}(\ell) = \|\mathrm{push}(p, \ell) - \mathrm{push}(q, \ell)\|_2.$$

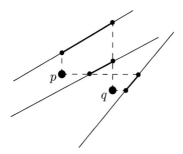

Figure 12.3 Pushes of two points to three lines. Thick segments indicate $\delta_{p,q}$ for the corresponding lines.

Consider the equations

$$\begin{aligned}
\delta_{p,q}(\ell) &= 0 & \text{for } p, q \in \mathrm{gr}(M) \text{ or } p, q \in \mathrm{gr}(N), \\
c_{p,q}\delta_{p,q}(\ell) &= c_{p',q'}\delta_{p',q'}(\ell) & \text{for } p, q, p', q' \in \mathrm{gr}(M) \sqcup \mathrm{gr}(N),
\end{aligned}$$

where

$$c_{p,q} = \begin{cases} \frac{1}{2} & \text{if } p, q \in \mathrm{gr}(M) \text{ or } p, q \in \mathrm{gr}(N), \\ 1 & \text{otherwise.} \end{cases}$$

The following proposition is proved in [208].

Proposition 12.3. *The solution set $z \in \tau$ for a face $\tau \in A_0$ so that $\delta_{p,q}(z^*)$ satisfies the above equations is either empty, the entire face τ, the intersection of a line with τ, or the intersection of two lines with τ.*

Let A be the arrangement of Ω with the lines used to form A_0, the lines stated in the above proposition, and the vertical line $x = 1$.

Proposition 12.4. *Arrangement A is formed with $O(n^4)$ lines where $n = |\mathrm{gr}(M) + \mathrm{gr}(N)|$.*

The next theorem states the main property of A which allows us to consider only finitely many (polynomially bounded) lines ℓ for computing the supremum of $\{\mathrm{d}_I(M|_\ell, N|_\ell)\}$.

Theorem 12.5. *For any face $\tau \in A$, there exists a pair $p, q \in \mathrm{gr}(M) \cup \mathrm{gr}(N)$ so that $c_{p,q}\delta_{p,q}(z^*) = \mathrm{d}_I(M|_{z^*}, N|_{z^*})$ for every $z \in \tau$.*

The above theorem implies that after determining the pair (p, q) for the face $\tau \in A$, we need to compute the $\sup_{z \in \tau} F(z)$ where $F(z) = \mathrm{d}_I(M|_{z^*}, N|_{z^*})$

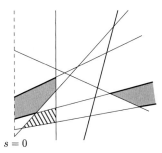

$s = 0$

Figure 12.4 Outer regions are shaded gray whose outer segments are drawn with thickened segments; the hatched region is inner.

because then considering all F over all faces in A gives the global supremum. So, now we focus on how to compute the supremum of F on a face τ.

A *region* is the closure of a face $\tau \in A$ in Ω. A region R is called *inner* if it is bounded and its closure in $\mathbb{R}^2$ does not meet the vertical line $s = 0$. See Figure 12.4. All other regions are called *outer*. An outer region has exactly two edges that are either unbounded or reach the vertical line $s = 0$ in the limit. They are called *outer edges*. It turns out that $\sup F(z)$ is achieved either at a vertex or at the limit point of the outer edges that can be computed easily.

Theorem 12.6. *The supremum* $\sup_{z \in R} F(z)$ *for a region R is realized either at a boundary vertex of R or at the limit point of an outer edge. In the latter case, let p, q be the pair given by Theorem 12.5 for $\tau \subseteq R$. If e is an outer edge and p lies above z^* for any (and all by Proposition 12.2) $z \in \tau$, then* $\sup F$ *restricted to e is given by*

$$\sup F|_e = \begin{cases} |p_x - t| & \text{if line of } e \text{ intersects line } x = 0 \text{ at } t, \\ |q_x + r| & \text{if line of } e \text{ is infinite and has slope } r. \end{cases}$$

The roles of p and q reverse if p lies below z^* for any $z \in \tau$.

We present the entire algorithm in Algorithm 25: MATCHDIST. It is known that this algorithm runs in $O(n^{11})$ time where n is the total number of generators and relations for the two input modules. A more efficient algorithm approximating the matching distance is also known [210].

12.4 Bottleneck Distance

Definition 12.12. (Matching) A *matching* $\mu \colon A \nrightarrow B$ between two multisets A and B is a partial bijection, that is, $\mu \colon A' \to B'$ for some $A' \subseteq A$ and $B' \subseteq B$. We say $\operatorname{im} \mu = B'$ and $\operatorname{coim} \mu = A'$.

Algorithm 25 MATCHDIST(M, N)

Input:
    Two modules M and N with grades of their generators and relations
Output:
    Matching distance between M and N

1: Compute arrangement A as described from $\mathrm{gr}(M) \cup \mathrm{gr}(N)$;
2: Let V be the vertex set of A;
3: Compute maximum $m = \max_{z \in V} F(z^*)$ over all vertices $z \in V$;
4: **for** every outer region R **do**
5:     Pick a point $z \in R$;
6:     Compute the pair $p, q \in \mathrm{gr}(M) \cup \mathrm{gr}(N)$ that realizes $\mathsf{d}_I(M|_{z^*}, N|_{z^*})$;
7:     **if** p is above z^* **then**
8:         **if** e as defined in Theorem 12.6 is infinite **then**
9:             $m := \max(m, q_x + r)$ where r is the slope of e
10:         **else**
11:             $m := \max(m, p_x - t)$ where e meets line $x = 0$ at t
12:         **end if**
13:     **else**
14:         Reverse roles of p and q
15:     **end if**
16: **end for**
17: Return m

For the next definition, we call a d-parameter module M δ-*trivial* if $\rho^M_{x \to x + \vec{\delta}} = 0$ for all $x \in \mathbb{R}^d$.

Definition 12.13. (Bottleneck distance) Let $M \cong \bigoplus_{i=1}^m M_i$ and $N \cong \bigoplus_{j=1}^n N_j$ be two persistence modules, where M_i and N_j are indecomposable submodules of M and N, respectively. Let $I = \{1, \ldots, m\}$ and $J = \{1, \ldots, n\}$. We say M and N are δ-matched for $\delta \geq 0$ if there exists a matching $\mu : I \nrightarrow J$ so that: (i) $i \in I \setminus \mathrm{coim}\,\mu \implies M_i$ is 2δ-trivial, (ii) $j \in J \setminus \mathrm{im}\,\mu \implies N_j$ is 2δ-trivial, and (iii) $i \in \mathrm{coim}\,\mu \implies M_i$ and $N_{\mu(i)}$ are δ-interleaved.

The *bottleneck distance* is defined as

$$\mathsf{d}_b(M, N) = \inf\{\delta \mid M \text{ and } N \text{ are } \delta\text{-matched}\}.$$

The following fact observed in [47] is straightforward from the definition.

Fact 12.2. *One has* $\mathsf{d}_I \leq \mathsf{d}_b$.

12.4.1 Interval Decomposable Modules

We present a polynomial-time algorithm for computing the bottleneck distances for a class of persistence modules called interval decomposable modules which we have seen in the previous chapter (Section 11.6.1). For ease of description, we will describe the algorithm for the two-parameter case, though an extension to the multiparameter case exists.

Persistence modules whose indecomposables are interval modules (Definition 12.15) are called *interval decomposable modules*. To account for the boundaries of free modules, we enrich the poset $\mathbb{R}^d$ by adding points at $\pm\infty$ and consider the poset $\bar{\mathbb{R}}^n = \bar{\mathbb{R}} \times \cdots \times \bar{\mathbb{R}}$ where $\bar{\mathbb{R}} = \mathbb{R} \cup \{\pm\infty\}$ with the usual additional rule $a \pm \infty = \pm\infty$.

Definition 12.14. (Interval) An *interval* is a subset $\varnothing \neq I \subset \bar{\mathbb{R}}^d$ that satisfies the following:

• If $p, q \in I$ and $p \leq r \leq q$, then $r \in I$ (convexity condition).
• If $p, q \in I$, then there exists a sequence $(p = p_0, \ldots, p_m = q) \in I$ for some $m \in \mathbb{N}$ so that for every $i \in [0, k-1]$ either $p_i \leq p_{i+1}$ or $p_i \geq p_{i+1}$ (connectivity condition). We call the sequence $(p = p_0, \ldots, p_m = q)$ a path from p to q (in I).

Let $\bar{I}$ denote the closure of an interval I in the standard topology of $\bar{\mathbb{R}}^d$. The lower and upper boundaries of I are defined as

$$L(I) = \{x = (x_1, \ldots, x_d) \in \bar{I} \mid \forall\ y = (y_1, \ldots, y_d) \text{ with } y_i < x_i\ \forall\ i \implies y \notin I\},$$
$$U(I) = \{x = (x_1, \ldots, x_d) \in \bar{I} \mid \forall\ y = (y_1, \ldots, y_d) \text{ with } y_i > x_i\ \forall\ i \implies y \notin I\}.$$

Let $B(I) = L(I) \cup U(I)$. According to this definition, $\bar{\mathbb{R}}^d$ is an interval with boundary $B(\bar{\mathbb{R}}^d)$ that consists of all the points with at least one coordinate ∞. The vertex set $V(\bar{\mathbb{R}}^d)$ consists of 2^d corner points with coordinates $(\pm\infty, \ldots, \pm\infty)$.

Definition 12.15. (*d*-Parameter interval module) A *d*-parameter *interval persistence module*, or *interval module* in short, is a persistence module M that satisfies the following condition: for an interval $I_M \subseteq \bar{\mathbb{R}}^d$, called the interval of M,

$$M_x = \begin{cases} \mathbb{k} & \text{if } x \in I_M, \\ 0 & \text{otherwise,} \end{cases} \qquad \rho_{x \to y}^M = \begin{cases} \mathbb{1} & \text{if } x, y \in I_M, \\ \mathbb{0} & \text{otherwise,} \end{cases}$$

where $\mathbb{1}$ and $\mathbb{0}$ denote the identity and zero maps, respectively.

It is known that an interval module is indecomposable [47].

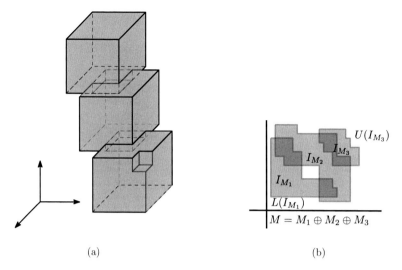

(a)   (b)

Figure 12.5 (a) Interval in $\mathbb{R}^3$ and (b) intervals in $\mathbb{R}^2$.

Definition 12.16. (Interval decomposable module) A d-parameter *interval decomposable module* is a persistence module that can be decomposed into interval modules.

Definition 12.17. (Rectangle) A k-dimensional *rectangle*, $0 \le k \le d$, or k-*rectangle*, in $\mathbb{R}^d$ is a set $I = [a_1, b_1] \times \cdots \times [a_d, b_d], a_i, b_i \in \bar{\mathbb{R}}$, such that there exists a size-k index set $\Lambda \subseteq [d]$ where for all $i \in \Lambda, a_i \ne b_i$, and for all $j \in [d] \setminus \Lambda, a_j = b_j$.

A 0-rectangle is a vertex. A 1-rectangle is an edge. Note that a rectangle is an example of an interval.

We say an interval $I \subseteq \bar{\mathbb{R}}^d$ is *discretely presented* if it is a finite union of d-rectangles. We also require that the boundary of the interval is a $(d-1)$-manifold. A facet of I is a $(d-1)$-dimensional subset $f = \hat{f} \cap L \subseteq \bar{\mathbb{R}}^d$ where $\hat{f} = \{x_i = c\}$ is a hyperplane at some standard direction $\vec{e}_i$ in $\mathbb{R}^d$ and L is either $L(I)$ or $U(I)$. We denote the facet set as $F(I)$ and the union of all of their vertices as $V(I)$. So the boundary of I is the union of facets. And the vertices of each facet form a subset of $V(I)$. Figure 12.5(a) and (b) show intervals in $\mathbb{R}^3$ and $\mathbb{R}^2$, respectively.

For the two-parameter case, a discretely presented interval $I \subseteq \bar{\mathbb{R}}^2$ has boundary consisting of a finite set of horizontal and vertical line segments called edges, with endpoints called vertices, which satisfy the following conditions: (i) every vertex is incident to either a single horizontal edge or a vertical edge, and (ii) no vertex appears in the interior of an edge. We denote the set of edges and vertices with $E(I)$ and $V(I)$, respectively.

We say a d-parameter interval decomposable module is finitely presented if it can be decomposed into finitely many interval modules whose intervals are discretely presented (see Figure 12.5[b]). They belong to the finitely presented persistence modules as defined in Chapter 11. In the following, we focus on finitely presented interval decomposable modules.

For an interval module M, let $\overline{M}$ be the interval module defined on the closure $\overline{I_M}$. To avoid complication in this exposition, we assume that every interval module has closed intervals, which is justified by the following proposition (Exercise 8).

Proposition 12.7. *One has* $\mathsf{d}_I(M, N) = \mathsf{d}_I(\overline{M}, \overline{N})$.

12.4.2 Bottleneck Distance for Two-Parameter Interval Decomposable Modules

We present an algorithm for two-parameter interval decomposable persistence modules, though most of our definitions and claims in this section apply to general d-parameter persistence modules. They are stated and proved in the general setting wherever applicable.

Given the intervals of the indecomposables (interval modules) as input, an approach based on bipartite graph matching is presented in Section 3.2.1 for computing the bottleneck distance $\mathsf{d}_b(M, N)$ between two one-parameter persistence modules M and N. This approach constructs a bipartite graph G out of the intervals of M and N and their pairwise interleaving distances, including the distances to zero modules. If these distance computations take $O(C)$ time in total, then the algorithm for computing d_b takes time $O(m^{5/2} \log m + C)$, where M and N together have m indecomposables altogether. Observe that the term $m^{5/2}$ in the complexity comes from the bipartite matching. Although this could be avoided in the one-parameter case taking advantage of the two-dimensional geometry of the persistence diagrams, we cannot do this here for determining matching among indecomposables according to Definition 12.13. Given indecomposables (say, computed by the algorithm in Chapter 11 or Meataxe [252]), this approach is readily extensible to d-parameter modules if one can compute the interleaving distance between any pair of indecomposables, including the zero modules. To this end, we present an algorithm to compute the interleaving distance between two two-parameter interval modules M_i and N_j with t_i and t_j vertices, respectively, on their intervals in $O((t_i + t_j) \log(t_i + t_j))$ time. This gives a total time of $O(m^{5/2} \log m + \sum_{i,j}(t_i + t_j)$

$\log(t_i + t_j)) = O(m^{5/2} \log m + t^2 \log t)$, where t is the total number of vertices over all input intervals.

Now we focus on computing the interleaving distance between two given intervals. Given intervals I_M and I_N with t vertices, the algorithm searches for a value δ so that there exist two families of linear maps from M to $N_{\to \delta}$ and from N to $M_{\to \delta}$, respectively, which satisfy both triangular and square commutativity. The search is done with a binary probing: For a chosen δ from a candidate set of $O(t)$ values, the algorithm determines the direction of the search by checking two conditions called *trivializability* and *validity* on the intersections of modules M and N.

Definition 12.18. (Intersection module) For two interval modules M and N with intervals I_M and I_N, respectively, let $I_Q = I_M \cap I_N$, which is a disjoint union of intervals, $\coprod I_{Q_i}$. The *intersection module* Q of M and N is $Q = \bigoplus Q_i$, where Q_i is the interval module with interval I_{Q_i}. That is,

$$Q_x = \begin{cases} \mathbf{k} & \text{if } x \in I_M \cap I_N, \\ \mathbf{0} & \text{otherwise,} \end{cases} \quad \text{and for } x \leq y, \quad \rho^Q_{x \to y} = \begin{cases} \mathbb{1} & \text{if } x, y \in I_M \cap I_N, \\ \mathbb{0} & \text{otherwise.} \end{cases}$$

From the definition we can see that the support of Q, $\mathrm{supp}(Q)$, is $I_M \cap I_N$. We call each Q_i an intersection component of M and N. Write $I := I_{Q_i}$ and consider $\phi \colon M \to N$ to be any morphism. The following proposition says that ϕ is constant on I.

Proposition 12.8. *One has $\phi|_I \equiv a \cdot \mathbb{1}$ for some $a \in \mathbf{k}$.*

PROOF. For any $x, y \in I$, consider a path $(x = p_0, p_1, p_2, \ldots, p_{2m}, p_{2m+1} = y)$ in I from x *to* y and the commutative diagrams below for $p_i \leq p_{i+1}$ (left) and $p_i \geq p_{i+1}$(right), respectively.

$$\begin{array}{ccc} M_{p_i} & \xrightarrow{\mathbb{1}} & M_{p_{i+1}} \\ \phi_{p_i} \downarrow & & \downarrow \phi_{p_{i+1}} \\ N_{p_i} & \xrightarrow{\mathbb{1}} & N_{p_{i+1}} \end{array} \qquad \begin{array}{ccc} M_{p_i} & \xleftarrow{\mathbb{1}} & M_{p_{i+1}} \\ \phi_{p_i} \downarrow & & \downarrow \phi_{p_{i+1}} \\ N_{p_i} & \xleftarrow{\mathbb{1}} & N_{p_{i+1}} \end{array}$$

Observe that $\phi_{p_i} = \phi_{p_{i+1}}$ in both cases due to the commutativity. Inducting on i, we get that $\phi(x) = \phi(y)$. □

Definition 12.19. (Valid intersection) An intersection component Q_i is (M, N)-*valid* if for each $x \in I_{Q_i}$ the following two conditions hold (see Figure 12.6):

- $y \leq x$ and $y \in I_M \implies y \in I_N$, and
- $z \geq x$ and $z \in I_N \implies z \in I_M$.

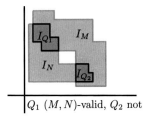

Q_1 (M, N)-valid, Q_2 not

Figure 12.6 Examples of a valid intersection and an invalid intersection.

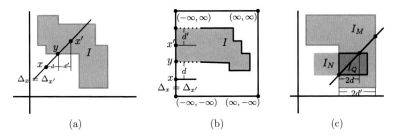

(a)   (b)   (c)

Figure 12.7 Here, (a) $d = \mathrm{dl}(x, I)$, $y = \pi_I(x)$, $d' = \mathrm{dl}(x', L(I))$; (b) $d = \mathrm{dl}(x, I)$ and $d' = \mathrm{dl}(x', U(I))$ are defined on the left edge of $B(\bar{\mathbb{R}}^2)$; and (c) Q is $d'_{(M,N)}$- and $d_{(N,M)}$-trivializable.

Proposition 12.9. *Let $\{Q_i\}$ be a set of intersection components of M and N with intervals $\{I_{Q_i}\}$. Let $\{\phi_x\}\colon M \to N$ be the family of linear maps defined as $\phi_x = \mathbb{1}$ for all $x \in I_{Q_i}$ and $\phi_x = \mathbb{0}$ otherwise. Then ϕ is a morphism if and only if every Q_i is (M, N)-valid.*

Definition 12.20. (Diagonal projection and distance) Let I be an interval and $x \in \bar{\mathbb{R}}^n$. Let $\Delta_x = \{x + \bar{\alpha} \mid \alpha \in \mathbb{R}\}$ denote the line called *diagonal* with slope 1 that passes through x. We define the *distance* (see Figure 12.7)

$$\mathrm{dl}(x, I) = \begin{cases} \min_{y \in \Delta_x \cap I}\{\mathsf{d}_\infty(x, y) := |x - y|_\infty\} & \text{if } \Delta_x \cap I \neq \varnothing, \\ +\infty & \text{otherwise.} \end{cases}$$

In case $\Delta_x \cap I \neq \varnothing$, define $\pi_I(x)$, called the *projection point* of x on I, to be the point $y \in \Delta_x \cap I$ where $\mathrm{dl}(x, I) = \mathsf{d}_\infty(x, y)$.

Note that for all $\alpha \in \mathbb{R}$, we have $\pm\infty + \alpha = \pm\infty$. Therefore, for $x \in V(\bar{\mathbb{R}}^n)$, the line collapses to a single point. In that case, $\mathrm{dl}(x, I) \neq +\infty$ if and only if $x \in I$, which means $\pi_I(x) = x$.

Notice that the upper and lower boundaries of an interval are also intervals by definition. With this understanding, the following properties of dl are obvious from the above definition.

Fact 12.3. *One has the following.*

(a) For any $x \in I_M$,

$$\text{dl}(x, U(I_M)) = \sup_{\delta \in \bar{\mathbb{R}}} \{x + \vec{\delta} \in I_M\} \text{ and } \text{dl}(x, L(I_M)) = \sup_{\delta \in \bar{\mathbb{R}}} \{x - \vec{\delta} \in I_M\}.$$

(b) Let $L = L(I_M)$ or $U(I_M)$ and let x and x' be two points such that $\pi_L(x)$ and $\pi_L(x')$ both exist. If x and x' are on the same facet or the same diagonal line, then $|\text{dl}(x, L) - \text{dl}(x', L)| \le d_\infty(x, x')$.

We now set $VL(I) := V(I) \cap L(I)$, $EL(I) := E(I) \cap L(I)$, $VU(I) := V(I) \cap U(I)$, and $EU(I) := E(I) \cap U(I)$.

Proposition 12.10. *For an intersection component Q of M and N with interval I, the following conditions are equivalent:*

(a) Q is (M, N)-valid;
(b) $L(I) \subseteq L(I_M)$ and $U(I) \subseteq U(I_N)$;
(c) $VL(I) \subseteq L(I_M)$ and $VU(I) \subseteq U(I_N)$.

Definition 12.21. (Trivializable intersection) Let Q be a connected component of the intersection of two modules M and N. For each point $x \in I_Q$, define

$$\mathsf{d}_{triv}^{(M,N)}(x) = \max\{\text{dl}(x, U(I_M))/2, \text{dl}(x, L(I_N))/2)\}.$$

For $\delta \ge 0$, we say a point x is $\delta_{(M,N)}$-*trivializable* if $\mathsf{d}_{triv}^{(M,N)}(x) < \delta$. We say an intersection component Q is $\delta_{(M,N)}$-*trivializable* if each point in I_Q is $\delta_{(M,N)}$-*trivializable* (Figure 12.7). We also denote $\mathsf{d}_{triv}^{(M,N)}(I_Q) := \sup_{x \in I_Q} \{\mathsf{d}_{triv}^{(M,N)}(x)\}$.

The following proposition discretizes the search for trivializability.

Proposition 12.11. *An intersection component Q is $\delta_{(M,N)}$-trivializable if and only if every vertex of Q is $\delta_{(M,N)}$-trivializable.*

Recall that for two modules to be δ-interleaved, we need two families of linear maps satisfying both triangular commutativity and square commutativity. For a given δ, Theorem 12.14 below provides criteria which ensure that such linear maps exist. In the algorithm, we then will make sure that these criteria are verified.

Given an interval module M and the diagonal line Δ_x for any $x \in \bar{\mathbb{R}}^d$, there is a one-parameter persistence module $M|_{\Delta_x}$ which is the functor restricted on the poset Δ_x as a subcategory of $\bar{\mathbb{R}}^d$. We call it a *one-dimensional slice* of M along Δ_x. Define

$$\delta^* = \inf_{\delta \in \bar{\mathbb{R}}} \{\delta : \forall x \in \bar{\mathbb{R}}^d, M|_{\Delta_x} \text{ and } N|_{\Delta_x} \text{ are } \delta\text{-interleaved}\}.$$

Equivalently we have $\delta^* = \sup_{x \in \bar{\mathbb{R}}^n} \{ \mathsf{d}_I(M|_{\Delta_x}, N|_{\Delta_x}) \}$. We have the following proposition and corollary from the equivalent definitions of δ^*.

Proposition 12.12. *For two interval modules M and N and $\delta > \delta^* \in \mathbb{R}^+$, there exist two families of linear maps $\phi = \{\phi_x : M_x \to N_{(x+\delta)}\}$ and $\psi = \{\psi_x : N_x \to M_{(x+\delta)}\}$ such that for each $x \in \bar{\mathbb{R}}^d$, the one-dimensional slices $M|_{\Delta_x}$ and $N|_{\Delta_x}$ are δ-interleaved by the linear maps $\phi|_{\Delta_x}$ and $\psi|_{\Delta_x}$.*

Corollary 12.13. *One has that $\mathsf{d}_I(M, N) \geq \delta^*$.*

Theorem 12.14. *For two interval modules M and N, $\mathsf{d}_I(M, N) \leq \delta$ if and only if the following two conditions are satisfied:*

(a) $\delta \geq \delta^*$;
(b) for all $\delta' > \delta$, each intersection component of M and $N_{\to \delta'}$ is either $(M, N_{\to \delta'})$-valid or $\delta_{(M, N_{\to \delta'})}$-trivializable, and each intersection component of $M_{\to \delta'}$ and N is either $(N, M_{\to \delta'})$-valid or $\delta_{(N, M_{\to \delta'})}$-trivializable.

PROOF. Note that $\mathsf{d}_I(M, N) \leq \delta$ if and only if for all $\delta' > \delta$, M and N is δ'-interleaved.

The "only if" direction: Given M and N are δ-interleaved. Part (a) follows from Corollary 12.13 directly. For part (b), by definition of interleaving, for all $\delta' > \delta$, we have two families of linear maps $\{\phi_x\}$ and $\{\psi_x\}$ which satisfy both triangular and square commutativities. Let the morphisms between the two persistence modules constituted by these two families of linear maps be $\phi = \{\phi_x\}$ and $\psi = \{\psi_x\}$, respectively. For each intersection component Q of M and $N_{\to \delta'}$ with interval $I := I_Q$, consider the restriction $\phi|_I$. By Proposition 12.8, $\phi|_I$ is constant, that is, $\phi|_I \equiv \mathbb{0}$ or $\mathbb{1}$. If $\phi|_I \equiv \mathbb{1}$, by Proposition 12.9, Q is $(M, N_{\to \delta'})$-valid. If $\phi|_I \equiv \mathbb{0}$, by the triangular commutativity of ϕ, we have that $\rho^M_{x \to x + 2\vec{\delta}'} = \psi_{x + \vec{\delta}'} \circ \phi_x = \mathbb{0}$ for each point $x \in I$. That means $x + 2\vec{\delta}' \notin I_M$. By Fact 12.3(a), $\mathrm{dl}(x, U(I_M))/2 < \delta'$. Similarly, $\rho^N_{x - \vec{\delta}' \to x + \vec{\delta}'} = \phi_x \circ \psi_{x - \vec{\delta}'} = 0 \implies x - \vec{\delta}' \notin I_N$, which is the same as saying that $x - 2\vec{\delta}' \notin I_{N_{\to \delta'}}$. By Fact 12.3(a), $\mathrm{dl}(x, L(I_{N_{\to \delta'}}))/2 < \delta'$. So for all $x \in I$, we have $\mathsf{d}^{(M, N_{\to \delta'})}_{triv}(x) < \delta'$. This means Q is $\delta'_{(M, N_{\to \delta'})}$-trivializable. A similar statement holds for intersection components of $M_{\to \delta'}$ and N.

The "if" direction: We construct two families of linear maps $\{\phi_x\}$ and $\{\psi_x\}$ as follows. On the interval $I := I_{Q_i}$ of each intersection component Q_i of M and $N_{\to \delta'}$, set $\phi|_I \equiv \mathbb{1}$ if Q_i is $(M, N_{\to \delta'})$-valid and $\phi|_I \equiv \mathbb{0}$ otherwise.

Set $\phi_x \equiv \mathbb{0}$ for all x not in the interval of any intersection component. Similarly, construct $\{\psi_x\}$. Note that, by Proposition 12.9, $\phi := \{\phi_x\}$ is a morphism between M and $N_{\to\delta'}$, and $\psi := \{\psi_x\}$ is a morphism between N and $M_{\to\delta'}$. Hence, they satisfy the square commutativity. We show that they also satisfy the triangular commutativity.

We claim that for all $x \in I_M$, $\rho^M_{x\to x+2\vec{\delta'}} = \mathbb{1} \implies x + \vec{\delta'} \in I_N$ and a similar statement holds for I_N. From the condition that $\delta' > \delta \geq \delta^*$ and by Proposition 12.12, we know that there exist two families of linear maps satisfying triangular commutativity everywhere, especially on the pair of one-parameter persistence modules $M|_{\Delta_x}$ and $N|_{\Delta_x}$. From triangular commutativity, we know that for all $x \in I_M$ with $\rho^M_{x\to x+2\vec{\delta'}} = \mathbb{1}$, $x + \vec{\delta'} \in I_N$, since otherwise one cannot construct a δ-interleaving between $M|_{\Delta_x}$ and $N|_{\Delta_x}$. So we get our claim.

Now for each $x \in I_M$ with $\rho^M_{x\to x+2\vec{\delta'}} = \mathbb{1}$, we have $\mathrm{dl}(x, U(I_M))/2 \geq \delta'$ by Fact 12.3, and $x + \vec{\delta'} \in I_N$ by our claim. This implies that $x \in I_M \cap I_{N\to\delta'}$ is a point in an interval of an intersection component Q_x of M and $N_{\to\delta'}$ which is not $\delta'_{(M,N_{\to\delta'})}$-*trivializable*. Hence, it is $(M, N_{\to\delta'})$-*valid* by the assumption. So, by our construction of ϕ on valid intersection components, $\phi_x = \mathbb{1}$. Symmetrically, we have that $x + \vec{\delta'} \in I_N \cap I_{M\to\delta'}$ is a point in an interval of an intersection component of N and $M_{\to\delta'}$ which is not $\delta'_{(N,M_{\to\delta'})}$-*trivializable* since $\mathrm{dl}(x + \vec{\delta'}, L(I_M))/2 \geq \delta'$. So by our construction of ψ on valid intersection components, $\psi_{x+\vec{\delta'}} = \mathbb{1}$. Then, we have $\rho^M_{x\to x+2\vec{\delta'}} = \psi_{x+\vec{\delta'}} \circ \phi_x$ for every nonzero linear map $\rho^M_{x\to x+2\vec{\delta'}}$. The statement also holds for any nonzero linear map $\rho^N_{x\to x+2\vec{\delta'}}$. Therefore, the triangular commutativity holds. $\qquad\square$

Note that the above proof provides a construction of the interleaving maps for any specific δ' if it exists. Furthermore, the interleaving distance $\mathsf{d}_I(M, N)$ is the infimum of all δ' satisfying the two conditions in the theorem, which means $\mathsf{d}_I(M, N)$ is the infimum of all $\delta' \geq \delta^*$ satisfying Theorem 12.14(b).

12.4.3 Algorithm to Compute d_I for Intervals

In practice, we cannot verify all those infinitely many values $\delta' > \delta^*$. But we propose a finite candidate set of potentially possible interleaving distance values and prove later that our final target, the interleaving distance, is always contained in this finite set. Surprisingly, the size of the candidate set is only $O(n)$ with respect to the number of vertices for two-parameter interval modules.

Based on our results, we propose a search algorithm for computing the interleaving distance $d_I(M, N)$ for interval modules M and N.

Definition 12.22. (Candidate set) For two interval modules M and N, and for each point x in $I_M \cup I_N$, let

$$D(x) = \{\text{dl}(x, L(I_M)), \text{dl}(x, L(I_N)), \text{dl}(x, U(I_M)), \text{dl}(x, U(I_N))\}.$$

Define the *candidate set*

$$S = \{d \mid d \in D(x) \text{ or } 2d \in D(x) \text{ for some vertex } x \in V(I_M) \cup V(I_N)\}.$$

Also, let

$$S_{\geq \delta} := \{d \mid d \geq \delta, d \in S\}.$$

Algorithm 26 INTERLEAVING(I_M, I_N)

Input:
  I_M and I_N with t vertices in total
Output:
  $d_I(M, N)$

1: Compute the candidate set S and let ϵ be half of the smallest difference between any two numbers in S * $O(l)$ time *\
2: Compute δ^*; Let $\delta = \delta^*$ * $O(t)$ time *\
3: Let $\delta^* = \delta_0, \delta_1, \ldots, \delta_k$ be numbers in $S_{\geq \delta^*}$ in nondecreasing order * $O(t \log t)$ time *\
4: $\ell := 0; u = k;$
5: **while** $\ell < u$ * $O(\log t)$ probes *\ **do**
6:   $i := \lfloor (u + \ell)/2 \rfloor; \delta := \delta_i; \delta' := \delta + \varepsilon;$
7:   Compute intersections $\mathcal{Q} := \{I_M \cap I_{N \to \delta'}\} \cup \{I_N \cap I_{M \to \delta'}\}$ * $O(t)$ time *\
8:   **if** every $Q \in \mathcal{Q}$ is valid or trivializable according to Theorem 12.14 * $O(t)$ time *\ **then**
9:     $u := i$
10:   **else**
11:     $\ell := i$
12:   **end if**
13: **end while**
14: Output δ

In Algorithm 26: INTERLEAVING, the following generic task of computing the *diagonal span* is performed for several steps. Let L and U be any
two chains of vertical and horizontal edges that are both x- and y-monotone.
Assume that L and U have at most t vertices. Then, for a set X of $O(t)$ points
in L, one can compute the intersection of Δ_x with U for every $x \in X$ in $O(t)$
total time. The idea is to first compute by a binary search a point x in X so that
Δ_x intersects U if at all. Then, for other points in X, traverse from x in both
directions while searching for the intersections of the diagonal line with U in
lock steps.

Now we analyze the complexity of the algorithm INTERLEAVING. The candidate set, by definition, has $O(t)$ values which can be computed in $O(t)$
time by the diagonal span procedure. By Proposition 12.15, δ^* is in S and
can be determined by computing the interleaving distances $\mathsf{d}_I(M|_{\Delta_x}, N|_{\Delta_x})$
for modules indexed by diagonal lines passing through $O(t)$ vertices of I_M
and I_N. This can be done in $O(t)$ time by the diagonal span procedure. Once
we determine δ^*, we perform a binary search (`while` loop) with $O(\log t)$
probes for $\delta = \mathsf{d}_I(M, N)$ in the truncated set $S_{\delta \geq \delta^*}$ to satisfy the first condition of Theorem 12.14. Intersections between two polygons I_M and I_N
bounded by x- and y-monotone chains can be computed in $O(t)$ time by a
simple traversal of the boundaries. The validity and trivializability of each
intersection component can be determined in time linear in the number of its
vertices due to Proposition 12.10 and Proposition 12.11, respectively. Since
the total number of intersection points is $O(t)$, the validity check takes $O(t)$
time in total. The check for trivializabilty also takes $O(t)$ time if one uses
the diagonal span procedure. So the total time complexity of the algorithm is
$O(t \log t)$.

Proposition 12.15 below says that δ^* is determined by a vertex in I_M or I_N
and $\delta^* \in S$.

Proposition 12.15. *The following hold:*

(a) $\delta^* = \max_{x \in V(I_M) \cup V(I_N)} \{\mathsf{d}_I(M|_{\Delta_x}, N|_{\Delta_x})\}$,
(b) $\delta^* \in S$.

The correctness of the algorithm INTERLEAVING already follows from Theorem 12.14 as long as the candidate set contains the distance $\mathsf{d}_I(M, N)$. This
is indeed true as shown in [140].

Theorem 12.16. *One has* $\mathsf{d}_I(M, N) \in S$.

Remark 12.1. *Our main theorem and algorithm consider the persistence
modules defined on* $\mathbb{R}^2$. *For a persistence module defined on a finite or discrete*

poset like $\mathbb{Z}^2$, one can extend it to a persistence module M on $\mathbb{R}^2$ to apply our theorem and algorithm. This extension is achieved by assuming that all morphisms outside the given persistence module are isomorphisms and $M_{x \to -\infty} = 0$ if it is not given otherwise. The reader can draw the analogy between this extension and the one we had for one-parameter persistence modules (Remark 3.3).

12.5 Notes and Exercises

We already mentioned in Chapter 3 that for one-parameter persistence modules, Chazal et al. [77] showed that the bottleneck distance is bounded from above by the interleaving distance d_I; see also [46, 53, 115] for further generalizations. Lesnick [221] established the isometry theorem which showed that indeed $d_I = d_b$. Consequently, d_I for one-parameter persistence modules can be computed exactly by efficient algorithms known for computing d_b. In Section 3.2.1, we present an algorithm for computing d_b from two given persistence diagrams.

Lesnick defined the interleaving distance for multiparameter persistence modules, and proved its stability and universality [221]. Specifically, he established that the interleaving distance between persistence modules is the best discriminating distance between modules having the property of stability. It is straightforward to observe that $d_I \leq d_b$. For some special cases, results in the reverse direction exist. Botnan and Lesnick [47] proved that, for the special class of two-parameter persistence modules, called block decomposable modules, $d_b \leq \frac{5}{2} d_I$. The support of each indecomposable in such modules consists of the intersection of a bounded or unbounded axis-parallel rectangle with the upper half-plane supported by the diagonal line $x_1 = x_2$. Bjerkevik [32] improved this result to $d_b \leq d_I$, thereby extending the isometry theorem $d_I = d_b$ to two-parameter block decomposable persistence modules.

Interestingly, a zigzag persistence module (Chapter 4) can be mapped to a block decomposable module [47]. Therefore, one can define an interleaving and a bottleneck distance between two zigzag persistence modules by the same distances on their respective block decomposable modules. Suppose that M_1 and M_2 denote the block decomposable modules corresponding to two zigzag filtrations $\mathcal{F}_1$ and $\mathcal{F}_2$, respectively. Bjerkevik's result implies that $d_b(\mathrm{Dgm}_p(\mathcal{F}_1), \mathrm{Dgm}_p(\mathcal{F}_2)) \leq 2d_b(M_1, M_2) = 2d_I(M_1, M_2)$. The factor of 2 comes due to the difference between how distances to a null module are computed in one-parameter and two-parameter cases. It is important to note that the bottleneck distance d_b for persistence diagrams here takes into account the types of the bars as described in Section 4.3. This means, while matching the bars for computing this distance, only bars of similar types are matched.

A similar conclusion can also be derived for the bottleneck distance between the levelset persistence diagrams of Reeb graphs. Mapping the zeroth levelset zigzag modules $\mathcal{Z}_f$ and $\mathcal{Z}_g$ of two Reeb graphs $(\mathbb{F}, f)$ and $(\mathbb{G}, g)$ to block decomposable modules M_f and M_g, respectively, one gets that $d_b(\mathrm{Dgm}_0(\mathcal{Z}_f), \mathrm{Dgm}_0(\mathcal{Z}_g)) \leq 2d_b(M_f, M_g) = 2d_I(M_f, M_g)$. The interleaving distance $d_I(M_f, M_g)$ between block decomposable modules is bounded from above by (not necessarily equal to) the interleaving distance between Reeb graphs given by Definition 7.6, that is, $d_I(M_f, M_g) \leq d_I(\mathbb{F}, \mathbb{G})$.

Bjerkevik also extended his result to rectangle decomposable d-parameter modules (indecomposables are supported on bounded or unbounded rectangles). Specifically, he showed that $d_b \leq (2d-1)d_I$ for rectangle decomposable d-parameter modules and $d_b \leq (d-1)d_I$ for free d-parameter modules. He gave an example for exactness of this bound when $d = 2$.

Multiparameter matching distance d_m introduced in [71] provides a lower bound to interleaving distance [217]. This matching distance can be approximated within any error threshold by algorithms proposed in [29, 72]. But it cannot provide an upper bound like d_b. The algorithm for computing d_m exactly as presented in Section 12.3 is taken from [208]. The complexity of this algorithm is rather high. To address this issue, an approximation algorithm with better time complexity has been proposed in [210] which builds on the result in [29].

For free, block, rectangle, and triangular decomposable modules, one can compute d_b by computing pairwise interleaving distances between indecomposables in constant time because they have a description of constant complexity. Due to the results mentioned earlier, d_I can be estimated within constant or dimension-dependent factors by computing d_b for these modules. On the other hand, Botnan and Lesnick [47] observed that even for interval decomposable modules, d_b cannot approximate d_I by any constant factor approximation.

Bjerkevik et al. [33] showed that computing interleaving distance for two-parameter interval decomposable persistence modules as considered in this chapter is NP-hard. Worse, it cannot be approximated within a factor of 3 in polynomial time. In this context, the fact that d_b does not approximate d_I within any factor for two-parameter interval decomposable modules [47] turns out to be a boon in disguise because otherwise a polynomial-time algorithm for computing it by the algorithm as presented in Section 12.4 would not have existed. This algorithm is taken from [140] whose extension to multiparameter persistence modules is available on arXiv.

Exercises

1. Show that d_I and d_b are pseudometrics on the space of finitely generated multiparameter persistence modules. Show that if the grades of generators and relations of the modules do not coincide, both become metrics.

2. Give an example of two persistence modules M and N for which $\mathsf{d}_m(M, N) = 0$ but $\mathsf{d}_I(M, N) \neq 0$.

3. Prove that $\mathsf{d}_I \leq \mathsf{d}_b$ and $\mathsf{d}_m \leq \mathsf{d}_I$.

4. Prove Fact 12.1 for point–line duality.

5. The algorithm MATCHDIST computes d_m in $O(n^{11})$ time where n is the total number of generators and relations with which the input modules are described. Design an algorithm for computing d_m that runs in $O(n^{11})$ time.

6. Consider the matching distance d_m between two interval modules. Compute d_m in this case in $O(n^4)$ time.

7. Given an interval decomposable persistence module $M \in \mathbb{R}^d\text{-}\mathbf{mod}$ and the subcategory $\mathbb{B} \subseteq \mathbb{R}^d\text{-}\mathbf{mod}$ of rectangle decomposable modules, let M^* denote an optimal approximation of M with a module in $\mathbb{B}$ with respect to the bottleneck distance d_b, that is, $M^* = \operatorname{argmin}_{M' \in \mathbb{B}} \mathsf{d}_b(M, M')$. Show that if $M = \bigoplus M^i$, then $M^* = \bigoplus M^{i*}$.

8. Prove Proposition 12.7.

9. For two points $x, y \in \mathbb{R}^2$, the ℓ_∞ distance between x and y is given by $\ell_\infty(x, y) = \max\{x_1 - y_1, x_2 - y_2\}$. Given a nonnegative real $\delta \geq 0$, we can define an ℓ_∞ δ-ball centered at a point $x \in \mathbb{R}^2$ as $\sqcup_\delta(x) = \{x' \in \mathbb{R}^2 : \ell_\infty(x, x') \leq \delta\}$. We can further extend this idea to a set $I \in \mathbb{R}^2$ as $I^{+\delta} = \bigcup_{x \in I} \square_\delta(x)$, which is the union of all ℓ_∞ δ-balls centered at all points in I. For two intervals $I, J \subset \mathbb{R}^2$, the ℓ_∞ Hausdorff distance is defined as $\mathsf{d}_H(I, J) = \inf_\delta\{I \subseteq J^{+\delta}, J \subseteq I^{+\delta}\}$. Show that:

(a) for two interval modules M and N, we have $\mathsf{d}_I(M, N) \leq \mathsf{d}_H(I_M, I_N)$;

(b) $\mathsf{d}_I(M, N) \lneq \mathsf{d}_H(I_M, I_N)$ strictly.

(Hint: Show that $\mathsf{d}_H(I_M, I_N) \geq \delta^*$ and for all $\delta \leq \mathsf{d}_H(I_M, I_N)$ each intersection component between M and $N_{\to \delta}$, and between N and $M_{\to \delta}$ is valid.)

13
Topological Persistence and Machine Learning

Machine learning (ML) has been a prevailing technique for data analysis. Naturally, researchers in the past few years have explored ways to combine machine learning techniques with topological data analysis (TDA) techniques. In previous chapters we have introduced various topological structures and algorithms for computing them. In this chapter, we give two examples of combining topological ideas with machine learning approaches. Note that this chapter is not intended to be a survey of such TDA + ML approaches, given that this is a very active and rapidly evolving field.

We have seen that persistent homology, in some sense, encodes the "shape" of data. Thus, it is natural to use persistent homology to map potentially complex input data (e.g., a point set or a graph) to a feature representation (persistence diagram). In particular, a simple persistence-based feature vectorization and data analysis framework can be as follows: Given a collection C of objects (e.g., a set of images, a collection of graphs, etc.), apply the persistent homology to map each object to a persistence diagram representation. Thus, objects in the input collection are now mapped to a set of points in the space of persistence diagrams. Different types of input data can all be now mapped to a common feature space: the space of persistence diagrams. Equipping this space with appropriate metric structures, one can then carry out downstream data analysis tasks on C in the space of persistence diagrams. In Section 13.1, we further elaborate on this framework, by describing several methods to assign a nice metric or kernel on the space of persistence diagrams.

One way to further incorporate topological information into the machine learning framework is by using a "topological loss function." In particular, as topology provides a language to describe global properties of a space, it can help a machine learning task at hand by allowing one to inject topological constraints or priors. This usually leads to optimizing a "topological function" over certain persistence diagrams. An example is given in Figure 13.1, taken from

389

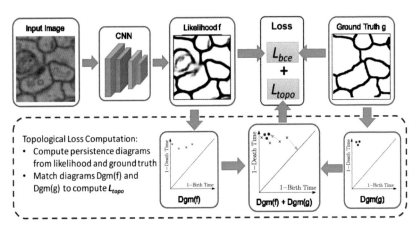

Figure 13.1 The high-level neural network framework, where topological information of the segmented image (captured via persistent homology) is used to help
train the neural network for better segmentation. Reprinted by permission from X.
Hu et al. (2019, figure 2) [199].

[199], where a term representing the topological quality of the output segmented images is added as part of the loss function to help improve topology of
the segmented foregrounds. In Section 13.2, we give another example of how
to use a "topological function" and describe how to address the key challenge
of differentiating such a topological loss function when it involves persistent
homology-based information.

In this book, we have focused mainly on the mathematical structures and
algorithmic/computational aspects involving TDA. However, we note that
there have been important developments in the statistical treatment of topological summaries, which are crucial for quantification of uncertainty, noise,
and convergence of topological summaries computed from sampled data. In
concluding this book, we provide a very brief description of some such developments in Section 13.3 at the end of this chapter. Interested readers can follow
the references given within this section for further details.

13.1 Feature Vectorization of Persistence Diagrams

The space of persistence diagrams equipped with the bottleneck distance (or
the p-th Wasserstein distance) introduced in previous chapters lacks (e.g., inner
product) structure, which can pose challenges when used within a machine
learning framework. To address this issue, in the past few years, starting
with the persistence landscapes [51], a series of methods have been developed to map persistence diagrams to a (finite- or infinite-dimensional) vector
space or a Hilbert space. (A Hilbert space is a vector space equipped with an

inner product, and it is a complete metric space with respect to the distance induced by the inner product.) This can be done explicitly, or by defining a positive (semi)definite kernel for persistence diagrams. Below we briefly introduce some of them. In what follows, for simplicity, let $\mathbb{D}$ denote the space of *bounded and finite* persistence diagrams. Some of the results require only finite persistence diagrams, where the total number of points other than the points from the diagonal in a diagram (including multiplicity) is finite. However, for simplicity of presentation, we assume that diagrams are also bounded within a finite box.

13.1.1 Persistence Landscape

Persistence landscape was introduced in [51], aiming to make persistence-based summaries easier for statistical analysis via mapping persistence diagrams to a function space.

Definition 13.1. (Persistence landscape) Given a finite persistence diagram $D = \{(b_i, d_i)\}_{i \in [1,n]}$ from $\mathbb{D}$, the *persistence landscape with respect to D is a* function $\lambda_D \colon \mathbb{N} \times \mathbb{R} \to \mathbb{R}$ where

$$\lambda_D(k, t) := k\text{-th largest value of } [\min\{t - b_i, d_i - t\}]_+ \text{ for } i \in [1, n].$$

Here, $[c]_+ = max(c, 0)$.

For a fixed k, $\lambda_D(k, \cdot) : \mathbb{R} \to \mathbb{R}$ is a function on $\mathbb{R}$. In particular, one can think of each persistent point (b_i, d_i) giving rise to a triangle whose upper boundary is traced out by points

$$\big\{(t, [\min\{t - b_i, d_i - t\}]_+) \mid t \in \mathbb{R}\big\};$$

see Figure 13.2. There are n such triangles, and the function $\lambda_D(k, \cdot)$ is the *k-th upper envelope* in the arrangement formed by the union of these triangles, which intuitively are points on the boundary of the k-th layer of these triangles.

The persistence landscape maps the persistence diagrams to a linear function space. The p-norm on persistence landscapes is defined as

$$\|\lambda_D\|_p^p = \sum_{k=1}^{\infty} \|\lambda_D(k, \cdot)\|_p^p.$$

Given two persistence diagrams D_1 and D_2, their *p-landscape distance* is defined by

$$\Lambda_p(D_1, D_2) = \|\lambda_{D_1} - \lambda_{D_2}\|_p. \tag{13.1}$$

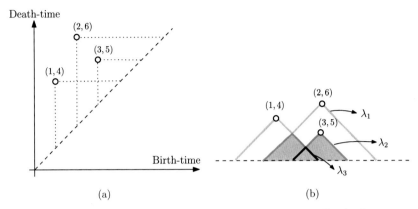

Figure 13.2 (a) A persistence diagram D and (b) its corresponding landscape functions, where $\lambda_k := \lambda_D(k, \cdot)$ for $k = 1, 2$, and 3.

Note that for any $k > n$, $\lambda_D(k, \cdot) \equiv 0$. One can recognize that persistence landscapes for finite persistence diagrams lie in the so-called L^p-space $\mathcal{L}^p(\mathbb{N} \times \mathbb{R})$.[1] If $p = 2$, then this is a Hilbert space. Given a set of persistence diagrams, one can compute their mean or carry out other statistical analysis in $\mathcal{L}^p(\mathbb{N} \times \mathbb{R})$. For example, given a set of ℓ finite diagrams $D_1, \ldots, D_\ell \in \mathbb{D}$, one can define the *mean landscape* $\overline{\lambda}$ of their corresponding landscapes $\lambda_{D_1}, \ldots, \lambda_{D_\ell}$ to be

$$\overline{\lambda}(k, t) = \frac{1}{\ell} \sum_{i=1}^{\ell} \lambda_{D_i}(k, t).$$

The following claim states that the map from the space of finite persistence diagrams $\mathbb{D}$ to the space of persistence landscapes is injective, and this map is lossless in terms of the information encoded in the persistence diagram.

Claim 13.1. *Given a persistence diagram D, let λ_D be its persistence landscape. Then from λ_D one can uniquely recover the persistence diagram D.*

However, a function $\lambda\colon \mathbb{N} \times \mathbb{R} \to \mathbb{R}$ may not be the image of any valid persistence diagram. For example, the mean landscape introduced above may not be the image of any persistence diagram.

Finally, in addition to being injective, under appropriate norms, the map from persistence diagram to persistence landscape is also stable (1-Lipschitz with respect to the bottleneck distance between persistence diagrams):

---

[1] For $1 \leq p < \infty$, $L^p(X)$ is defined as $L^p(X) := \{f : X \to \mathbb{R} \mid \|f\|_p \leq +\infty\}$. For example, $L^2(\mathbb{R}^d)$ is the space of standard square-integrable functions on $\mathbb{R}^d$. Then $\mathcal{L}^p(X)$ is defined as $\mathcal{L}^p(X) = L^p(X)/\sim$ where $f \sim g$ if $\|f - g\|_p = 0$.

Theorem 13.1. *For persistence diagrams* D *and* D', $\Lambda_\infty(D, D') \leq d_B(D, D')$.

Additional stability results for Λ_p are given in [51], relating it to the p-th Wasserstein distance for persistence diagrams, or for the case where the persistence diagrams are induced by tame Lipschitz functions.

13.1.2 Persistence Scale Space Kernel (PSSK)

In the previous subsection, we introduced a way to map persistence diagrams into a function space $\mathcal{L}^p(\mathbb{N}, \mathbb{R})$ (which is a Hilbert space when $p = 2$). One can also map persistence diagrams to a so-called reproducing kernel Hilbert space via the use of kernels. The work of [263] is the first of a line of work defining a (positive semidefinite) kernel on persistence diagrams.

Definition 13.2. (Positive, negative semidefinite kernel) Given a topological space X, a function $\mathsf{k} : X \times X \to \mathbb{R}$ is a *positive semidefinite kernel* if it is symmetric and for any integer $n > 0$, any $x_1, \ldots, x_n \in X$, and any $a_1, \ldots, a_n \in \mathbb{R}$ with $\sum_i a_i = 0$, it holds that $\sum_{i,j} a_i a_j \mathsf{k}(x_i, x_j) \geq 0$. Analogously, k is a *negative semidefinite kernel* if it is symmetric and for any integer $n > 0$, any $x_1, \ldots, x_n \in X$, and any $a_1, \ldots, a_n \in \mathbb{R}$ with $\sum_i a_i = 0$, it holds that $\sum_{i,j} a_i a_j \mathsf{k}(x_i, x_j) \leq 0$.

Now, given a set X and a Hilbert space $\mathcal{H}$ of real-valued functions on X, the *evaluation functional* over $\mathcal{H}$ is a linear functional that evaluates each function in f at a point x: that is, given x, $L_x : \mathcal{H} \to \mathbb{R}$ is defined as $L_x(f) = f(x)$ for any $f \in \mathcal{H}$. The Hilbert space $\mathcal{H}$ is called a *reproducing kernel Hilbert space (RKHS)* if L_x is continuous for all $x \in X$. It is known that given a positive semidefinite kernel k, there is a unique RKHS $\mathcal{H}_\mathsf{k}$ such that $\mathsf{k}(x, y) = \langle \mathsf{k}(\cdot, x), \mathsf{k}(\cdot, y) \rangle_{\mathcal{H}_\mathsf{k}}$. We call k the *reproducing kernel* for $\mathcal{H}_\mathsf{k}$. From now on, we simply use "kernel" to refer to a positive semidefinite kernel. See [271] for more detailed discussions of kernels, RKHS, and related concepts.

Equivalently, a kernel can be thought of as the inner product $k(x, y) = \langle \Phi(x), \Phi(y) \rangle_{\mathcal{H}}$ after mapping X to some Hilbert space $\mathcal{H}$ via a *feature map* $\Phi : X \to \mathcal{H}$. With this inner product, one can also further induce a pseudometric[2] by

$$\mathsf{d}_\mathsf{k}^2(x, y) := \mathsf{k}(x, x) + \mathsf{k}(y, y) - 2\mathsf{k}(x, y), \text{ or equivalently, } \mathsf{d}_\mathsf{k}(x, y) = \|x - y\|_{\mathcal{H}}.$$

Many machine learning pipelines directly use kernels and its associate inner product structure.

---

[2] Recall that different from a metric, for a pseudometric, $\mathsf{d}(x, y) = 0$ may not imply that $x = y$. All other conditions for a metric hold for a pseudometric.

The work of [263] constructs the following persistence scale space kernel (PSSK) by defining an explicit feature map. Let $\Omega = \{x = (x_1, x_2) \in \mathbb{R}^2 \mid x_2 \geq x_1\}$ denote the subspace of $\mathbb{R}^2$ on or above the diagonal.[3] Recall the L^2-space $\mathcal{L}^2(\Omega)$, which is a Hilbert space.

Definition 13.3. (Persistence scale space kernel (PSSK)) Define the feature map $\Phi_\sigma : \mathbb{D} \to \mathcal{L}^2(\Omega)$ at scale $\sigma > 0$ as follows. For a persistence diagram $D \in \mathbb{D}$ and $x \in \Omega$, set

$$\Phi_\sigma(D)(x) = \frac{1}{4\pi\sigma} \sum_{y \in D} [e^{-\|x-y\|^2/4\sigma} - e^{-\|x-\bar{y}\|^2/4\sigma}],$$

where $\bar{y} = (y_2, y_1)$ if $y = (y_1, y_2)$ (i.e., $\bar{y}$ is the reflection of y across the diagonal). This feature map induces the following *persistence scale space kernel (PSSK)* $k_\sigma : \mathbb{D} \times \mathbb{D} \to \mathbb{R}$ using the inner product structure on $\mathcal{L}^2(\Omega)$: given two diagrams $D, E \in \mathbb{D}$, we have

$$k_\sigma(D, E) = \langle \Phi_\sigma(D), \Phi_\sigma(E) \rangle_{\mathcal{L}^2(\Omega)}$$
$$= \frac{1}{8\pi\sigma} \sum_{y \in D; \, z \in E} [e^{-\|y-z\|^2/8\sigma} - e^{-\|y-\bar{z}\|^2/8\sigma}]. \quad (13.2)$$

In other words, a persistence diagram is now mapped to a function $\Phi_\sigma(D) : \Omega \to \mathbb{R}$ under the feature map Φ_σ. By construction, the PSSK is positive definite. Now consider the distance induced by the PSSK:

$$\|\Phi_\sigma(D) - \Phi_\sigma(E)\|_{\mathcal{L}^2(\Omega)} = \sqrt{k_\sigma(D, D) + k_\sigma(E, E) - 2k_\sigma(D, E)}$$

This distance is stable in the sense that the feature map Φ_σ is Lipschitz with respect to the 1-Wasserstein distance:

Theorem 13.2. *Given two persistence diagrams $D, E \in \mathbb{D}$, we have*

$$\|\Phi_\sigma(D) - \Phi_\sigma(E)\|_{\mathcal{L}^2(\Omega)} \leq \frac{1}{2\pi\sigma} \mathsf{d}_{W,1}(D, E).$$

13.1.3 Persistence Images

Let $D \in \mathbb{D}$ be a finite persistence diagram. We set $T : \mathbb{R}^2 \to \mathbb{R}^2$ to be the linear transformation where for each $(x, y) \in \mathbb{R}^2$, $T(x, y) = (x, y - x)$. Let $T(D)$ denote the transformed diagram of D. Let $\phi_u : \mathbb{R}^2 \to \mathbb{R}$ be a differentiable probability distribution with mean $u \in \mathbb{R}^2$ (e.g., the normalized Gaussian where for any $z \in \mathbb{R}^2$, $\phi_u(z) = (1/(2\pi\tau^2))e^{-\|z-u\|^2/2\tau^2}$). We now define the persistence images introduced in [4].

---

[3] Often in the literature, one assumes that the standard persistent homology is considered where the birth time is smaller than or equal to the death time in the filtration. Several of the kernels introduced here, including PSSK and PWGK, assume that persistence diagrams lie in Ω.

Definition 13.4. (Persistence image) Let $\omega \colon \mathbb{R}^2 \to \mathbb{R}$ be a nonnegative weight function for $\mathbb{R}^2$. Given a persistence diagram D, its *persistence surface* $\mu_D \colon \mathbb{R}^2 \to \mathbb{R}$ (w.r.t. ω) is defined as

$$\mu_D(z) := \sum_{u \in T(D)} \omega(u)\phi_u(z), \quad \text{for any } z \in \mathbb{R}^2. \tag{13.3}$$

The persistence image is a discretization of the persistence surface. Specifically, fix a grid on a rectangular region in the plane with a collection P of N rectangles (pixels). The *persistence image* of a persistence diagram D is $I_D = \{\, I[\mathsf{p}] \,\}_{\mathsf{p} \in \mathsf{P}}$ which consists of N numbers (i.e., a vector in $\mathbb{R}^N$), one for each pixel p in the grid P with $I_D[\mathsf{p}] := \int_{\mathsf{p}} \rho_D \, dx \, dy$.

We remark that the weight function ω in constructing the persistence surface allows points in the persistence diagrams to have different contributions in the final representation. A natural choice of $\omega(u)$ could be the persistence $|b - d|$ of point $u = (b, d)$.

The persistence image can be viewed as a vector in $\mathbb{R}^N$. One could then compute the distance between two persistence diagrams D and E by the L_2-distance $\|I_D - I_E\|_2$ between their persistence images (vectors) I_D and I_E. Other L_p-norms can also be used.

Persistence images are shown to be stable with respect to the 1-Wasserstein distance between persistence diagrams [4]. As an example, below we state the stability result for the special case where the persistence surfaces are generated using the normalized Gaussian distribution $\phi_u \colon \mathbb{R}^2 \to \mathbb{R}$ defined via $\phi_u(z) = (1/(2\pi\sigma^2))e^{-\|z-u\|_2^2/2\sigma^2}$ for any $z \in \mathbb{R}^2$. See [4] for stability results for the general cases.

Theorem 13.3. *Suppose persistence images are computed with the normalized Gaussian distribution with variance σ^2 and weight function $\omega : \mathbb{R}^2 \to \mathbb{R}$. Then the persistence images are stable with respect to the 1-Wasserstein distance between persistence diagrams. More precisely, given two finite and bounded persistence diagrams D and E, we have*

$$\|I_D - I_E\|_1 \leq \left(\sqrt{5}\,|\nabla\omega| + \sqrt{\frac{10}{\pi}}\,\frac{\|\omega\|_\infty}{\sigma} \right) \cdot \mathsf{d}_{W,1}(D, E).$$

Here, $\nabla\omega$ stands for the gradient of ω, and $|\nabla\omega| = \sup_{z \in \mathbb{R}^2} \|\nabla\omega\|_2$ is the maximum norm of the gradient vector of ω at any point in $\mathbb{R}^2$. The same upper bound holds for $\|I_D - I_E\|_2$ and $\|I_D - I_E\|_\infty$ as well.

13.1.4 Persistence Weighted Gaussian Kernel (PWGK)

The work of [216] proposes to first embed each persistence diagram to an RKHS. Using the representation of persistence diagrams in this RKHS, one can further put another kernel on top of them to obtain a final kernel for persistence diagrams.

In particular, the first step of embedding persistence diagrams to an RKHS is achieved by kernel embedding for (signed) measures. Recall that given a kernel k, there is a unique RKHS $\mathcal{H}_k$ associated to k where k is its reproducing kernel. Now given a locally compact Hausdorff space X, let $C_0(X)$ denote the space of continuous functions vanishing at infinity. A kernel k on X is called a C_0-kernel if $k(\cdot, x)$ is in $C_0(X)$ for any $x \in \Omega$. It turns out that if k is a C_0-kernel, then its associated RKHS $\mathcal{H}_k$ is a subspace of $C_0(X)$; we further call k C_0-universal if it is a C_0-kernel and $\mathcal{H}_k$ is dense in $C_0(X)$. For example, the d-dimensional Gaussian kernel $k_G(x, y) = (1/(2\pi \tau^2))e^{-\|x-y\|^2/2\tau^2}$ is C_0-universal on $\mathbb{R}^d$ [283]. Recall that $\Omega = \{x = (x_1, x_2) \in \mathbb{R}^2 \mid x_2 \geq x_1\}$ denotes the subspace of $\mathbb{R}^2$ on or above the diagonal.

Definition 13.5. (Persistence weighted kernel) Let $k \colon \Omega \to \mathbb{R}$ be a C_0-universal kernel on Ω (e.g., a Gaussian), and $\omega \colon \Omega \to \mathbb{R}^+$ a strictly positive (weight) function on Ω. The following feature map $\Psi_{k,\omega} \colon \mathbb{D} \to \mathcal{H}_k$ maps each persistence diagram $D \in \mathbb{D}$ to the RKHS $\mathcal{H}_k$ associated to k:

$$\Psi_{k,\omega}(D) = \sum_{x \in D} \omega(x)k(\cdot, x).$$

This feature map induces the following *persistence weighted kernel (PWK)* $K_{k,\omega} \colon \mathbb{D} \times \mathbb{D} \to \mathbb{R}$:

$$K_{k,\omega}(D, E) = \langle \Psi_{k,\omega}(D), \Psi_{k,\omega}(E) \rangle_{\mathcal{H}_k} = \sum_{x \in D;\ y \in E} \omega(x)\omega(y)k(x, y).$$

$$(13.4)$$

The intuition of the above feature map is as follows. Given a persistence diagram D, it can be viewed as a discrete measure $\mu_D^\omega := \sum_{x \in D} \omega(x)\delta_x$, where $\omega \colon \mathbb{R}^2 \to \mathbb{R}$ is a weight function, and δ_x is the Dirac measure at x. (Similar to persistence images, the use of the weight function ω allows different points in the birth–death plane to have different influence.) The map $\Psi_{k,w}(D)$ is essentially the kernel mean embedding of distributions (with persistence diagrams viewed as discrete measures) to the RKHS. It is known that if the kernel k is C_0-universal, then this embedding is in fact injective [283], and hence the resulting induced distance $\| \Psi_{k,w}(D) - \Psi_{k,w}(E) \|_{\mathcal{H}_k}$ is a proper metric (instead of a pseudometric).

An Alternative Construction

An equivalent construction for Eq. (13.4) is as follows. Treat a persistence diagram D as an unweighted discrete measure $\mu_D = \sum_{x \in D} \delta_x$. Given a kernel k, consider an ω-weighted version of it:

$$k^\omega(x, y) := \omega(x) \cdot \omega(y) \cdot k(x, y).$$

This weighted kernel k^ω is still positive semidefinite for strictly positive weight function $\omega : \Omega \to \mathbb{R}^+$. Let $\mathcal{H}_{k^\omega}$ denote its associated RKHS. Then the map

$$\Psi_{k^\omega}(D) := \sum_{x \in D} \omega(x)\omega(\cdot)k(\cdot, x)$$

defines a valid feature map $\Psi_{k^\omega} : \mathbb{D} \to \mathcal{H}_{k^\omega}$ to the RKHS $\mathcal{H}_{k^\omega}$. It is shown in [216] that the induced inner product $K_{k^\omega} : \mathbb{D} \times \mathbb{D} \to \mathbb{R}$ by this feature map equals the inner product in Eq. (13.4):

$$K_{k^\omega}(D, E) = \langle \Psi_{k^\omega}(D), \Psi_{k^\omega}(E) \rangle_{\mathcal{H}_{k^\omega}} = \sum_{x \in D; y \in E} \omega(x)\omega(y)k(x, y) = K_{k,w}(D, E). \quad (13.5)$$

Persistence Weighted Gaussian Kernel (PWGK)

There are different choices for the weight function ω and the kernel k. For example, given a persistence point $x = (b, d)$, let $pers(x) = |d - b|$. Then we can set the weight function to be

$$\omega_{arc}(x) = \arctan(C \cdot pers(x)^p), \quad \text{where } C \text{ is a constant and } p \in \mathbb{Z}_{>0}.$$

We can also choose the kernel k to be the 2D (unnormalized) Gaussian kernel $k_G(x, y) = e^{-\|x-y\|^2/2\tau^2}$. Then the weighted kernel $k_G^{\omega_{arc}}(x, y) = \omega_{arc}(x)\omega_{arc}(y)k_G(x, y)$ is referred to as *the persistence weighted Gaussian kernel (PWGK)*. Stability results of the PWGK-induced distance with respect to the bottleneck distance d_B and the 1-Wasserstein distance $\mathsf{d}_{W,1}$ on persistence diagrams are shown in [216], with bounds depending on the weight function ω and the kernel k_G. The precise statements are somewhat involved, so we omit the details here. We remark that the stability with respect to the bottleneck distance is provided, which is usually harder to obtain than for the Wasserstein distance for such vectorizations of persistence diagrams.

Finally, now that the persistence diagrams are embedded in an RKHS, one can directly use the associated inner product and kernel for machine learning pipelines. One can also further put another kernel based on the RKHS representation of persistence diagrams. Indeed, the persistence weighted kernel in Eq. (13.4) is equivalent to putting a linear kernel on the RKHS $\mathcal{H}_k$. We can also consider using a nonlinear kernel, say the Gaussian kernel, on the

RKHS $\mathcal{H}_k$, and obtain yet another kernel on persistence diagrams, called the (k, ω)-Gaussian kernel:[4]

$$K_{k,w}^G(D, E) = \exp\left(-\frac{1}{2\tau^2}\|\Psi_{k,w}(D) - \Psi_{k,w}(E)\|_{\mathcal{H}_k}^2\right).$$

13.1.5 Sliced Wasserstein Kernel

Instead of using feature maps, one can also construct a kernel for persistence diagrams directly, and we now describe such an approach taken in [68]. This requires a positive semidefinite kernel (recall Definition 13.2). One way to construct a positive definite kernel is to exponentiate a negative (semi)definite kernel: see the following result [25].[5]

Theorem 13.4. *Given X and $\phi\colon X \times X \to \mathbb{R}$, the kernel ϕ is negative semidefinite if and only if $e^{-t\phi}$ is positive semidefinite for all $t > 0$.*

In what follows, we construct the so-called *sliced Wasserstein distance* d_{SW} for persistence diagrams, which is shown to be negative definite. We then use it to construct the sliced Wasserstein kernel following the above theorem.

Specifically, let μ and ν be two (unnormalized) nonnegative measures on the real line, such that the total mass $\mu(\mathbb{R})$ equals $\nu(\mathbb{R})$, and they are bounded. Let $\Pi(\mu, \nu)$ denote the set of measures on $\mathbb{R}^2$ with marginals μ and ν. Consider

$$\mathcal{W}(\mu, \nu) = \inf_{P \in \Pi(\mu,\nu)} \iint_{\mathbb{R}\times\mathbb{R}} |x - y| \cdot dP(x, y), \tag{13.6}$$

which is simply the 1-Wasserstein distance between measures μ and ν. In the following definition, $\mathbb{S}^1$ denotes the unit circle in the plane.

Definition 13.6. (Sliced Wasserstein distance) Given a unit vector $\theta \in \mathbb{S}^1 \subseteq \mathbb{R}^2$, let $L(\theta)$ denote the line $\{\lambda\theta \mid \lambda \in \mathbb{R}\}$. Let $\pi_\theta\colon \mathbb{R}^2 \to L(\theta)$ be the orthogonal projection of the plane onto $L(\theta)$. Given two persistence diagrams D and E, set $\mu_D^\theta := \sum_{p \in D} \delta_{\pi_\theta(p)}$ and $\overline{\mu}_D^\theta := \sum_{p \in D} \delta_{\pi_\theta \circ \pi_\Delta(p)}$, where $\pi_\Delta\colon \mathbb{R}^2 \to \Delta$ is the orthogonal projection onto the diagonal $\Delta = \{(x, x) \mid x \in \mathbb{R}\}$. Set μ_E^θ and $\overline{\mu}_E^\theta$ in a symmetric manner. Then the *sliced Wasserstein distance* between D and E is defined as

$$\mathsf{d}_{SW}(D, E) := \frac{1}{2\pi} \int_{\mathbb{S}^1} \mathcal{W}(\mu_D^\theta + \overline{\mu}_E^\theta, \ \mu_E^\theta + \overline{\mu}_D^\theta)\, d\theta.$$

---

[4] In the work of [216], it also sometimes refers to $\Psi_{k_G,w_{arc}}^G$ as the *persistence weighted Gaussian kernel*.

[5] In [25], the use of positive (negative) definite kernel is the same as our positive (negative) semidefinite kernel.

In the above definition, the sums $\mu_D^\theta + \overline{\mu}_E^\theta$ and $\mu_E^\theta + \overline{\mu}_D^\theta$ ensure the resulting two measures have the same total mass.

Proposition 13.5. *One has that* d_{SW} *is negative semidefinite on* $\mathbb{D}$ *where* $\mathbb{D}$ *is the space of bounded and finite persistence diagrams.*

Combining the above proposition with Theorem 13.4, we can now define the positive semidefinite *sliced Wasserstein kernel* k_{SW} on $\mathbb{D}$ as

$$k_{SW}(D, E) := e^{-d_{SW}(D,E)/2\sigma^2}, \quad \text{for } \sigma > 0. \qquad (13.7)$$

The sliced Wasserstein distance is not only stable, but also strongly equivalent to the 1-Wasserstein distance $d_{W,1}$ on bounded persistence diagrams in the following sense.

Theorem 13.6. *Let* $\mathbb{D}_N$ *be the set of bounded persistence diagrams with cardinalities at most N. For any* $D, E \in \mathbb{D}_N$, *one has*

$$\frac{d_{W,1}(D, E)}{4N(4N-1)+2} \le d_{SW}(D, E) \le 2\sqrt{2} \cdot d_{W,1}(D, E).$$

13.1.6 Persistence Fisher Kernel

The construction of the persistence Fisher (PF) kernel, proposed by [219], uses a similar idea as the sliced Wasserstein (SW) distance in the sense that it will also leverage Theorem 13.4 to construct a positive definite kernel on persistence diagrams. However, it uses the Fisher information metric from information geometry (usually used for probability measures) to derive the kernel. First, given a persistence diagram D, we map it to a function $\mu_D : \mathbb{R}^2 \to \mathbb{R}^+ \cup \{0\}$ as follows:

$$\mu_D(x) := \frac{1}{Z} \sum_{z \in D} \phi_{G,\sigma}(x, u),$$

where $\phi_{G,\sigma}(x, u) = e^{-\|x-u\|^2/2\sigma}$ and $Z = \int_{\mathbb{R}^2} \sum_{u \in D} \phi_{G,\sigma}(x, u)\,dx.$

This function is similar to the *persistence surface* used in [4]. Recall that Δ denotes the diagonal in the plane. Given a diagram D, let $D_\Delta := \{\pi_\Delta(u) \mid u \in D\}$ where π_Δ denotes the orthogonal projection onto the diagonal Δ.

Definition 13.7. (Persistence Fisher (PF) kernel) Given two persistence diagrams D and E, the *Fisher information metric* between their corresponding persistence surfaces μ_D and μ_E is defined as

$$\mathsf{d}_{FIM}(D, E) := \mathsf{d}_{FIM}(\mu_{D \cup E_\Delta}, \mu_{E \cup D_\Delta})$$

$$= \arccos\left(\int_{\mathbb{R}^2} \sqrt{\mu_{D \cup E_\Delta}(x)\mu_{E \cup D_\Delta}(x)}\, dx\right).$$

The *persistence Fisher (PF) kernel* for persistence diagrams is then defined as

$$k_{PF}(D, E) := e^{-t \cdot \mathsf{d}_{FIM}(D, E)}, \quad \text{for some } t > 0.$$

Note that similar to the sliced Wasserstein distance, the use of $D \cup E_\Delta$ (resp. $E \cup D_\Delta$) is to address the issue that D and E may have different cardinality.

Proposition 13.7. *The function* $(\mathsf{d}_{FIM} - \tau)$ *is negative definite on the set of bounded and finite persistence diagrams* $\mathbb{D}$ *for any* $\tau \geq \pi/2$.

By the above result and Theorem 13.4, we have that $e^{-t(\mathsf{d}_{FIM} - \tau)}$ is positive definite for $t > 0$ and $\tau \geq \pi/2$. Furthermore, by definition, we can rewrite the persistence Fisher kernel as

$$k_{PF}(D, E) = e^{-t \cdot \mathsf{d}_{FIM}(D, E)} = \alpha \cdot e^{-t \cdot (\mathsf{d}_{FIM}(D, E) - \tau)},$$

$$\text{where } \tau \geq \frac{\pi}{2} \text{ and } \alpha = e^{-t\tau} > 0.$$

As $\alpha > 0$ is a fixed constant, it then follows that:

Corollary 13.8. *The persistence Fisher kernel* k_{PF} *is positive definite on* $\mathbb{D}$.

The work of [219] provides interesting analysis of the eigensystem of the integral operator induced by k_{PF}. Furthermore, both persistence Fisher kernel and sliced Wasserstein kernel are infinitely divisible. This could bring computational advantages when using them in kernel machines. The PSSK and PWGK do not have this property.

We remark that there are other vectorization approaches of persistence diagrams developed. Very recently, there have also been several pieces of work on learning the representation of persistence diagrams in *an end-to-end manner* using labeled data. We will mention some of these works in the bibliographical notes later (Section 13.4).

13.2 Optimizing Topological Loss Functions

Topology provides a language to describe the global properties of a space. One could envision adding topological constraints or priors for a machine learning task at hand. This usually leads to optimizing a "topological function" over certain persistence diagrams. To motivate this, in Section 13.2.1 we give an example where one aims to regularize the topological complexity of a classifier that leads to a topological loss function. We then describe how a resulting topological function can be optimized in Section 13.2.2. We briefly discuss some other recent work on injecting topological constraints/loss in machine learning pipelines at the end of this section.

13.2.1 Topological Regularizer

We describe the work of [95], which uses persistent homology to regularize classifiers as an example to illustrate the occurrence of topological functions. For simplicity, we consider the binary classification problem, and assume that the domain X (where the input data is sampled from) is a d-dimensional hypercube. A classifier function is a smooth scalar function $f : X \to \mathbb{R}$, which provides the prediction for a training/testing data point $x \in X$ by evaluating $\mathrm{sign}(f(x))$. In other words, the *classification boundary* (separating the positive and negative classification regions), denoted by S_f, is simply the 0-level set (i.e., the levelset at function value 0):

$$S_f = f^{-1}(0) = \{x \in X \mid f(x) = 0\}.$$

See Figure 13.3(a) for an example, where the classification boundary S_f consists of the U-shaped curve and two closed loops.

The classifier may have unnecessary details that overfit the input data, and one way to address this is via regularizing (constraining) the properties of f (e.g., requiring that it is smooth). The work of [95] proposed to regularize the "topological simplicity" of a classifier. In the example of Figure 13.3(a), there are three components (zeroth homological features) in S_f. To develop a notion of "topological complexity" of the classification boundary, it is desirable to quantify the "robustness" of these topological features. To do so, we will need to use the information of the entire classifier function f beyond just the 0-level set; see Figure 13.3(b). Notice that, while the two small components in S_f are of similar size in S_f, intuitively, it takes less perturbation of the classifier function f to remove the left component. In particular, one could push down the saddle point q_1 so that this component is merged with the large component in the levelset S_f and thus reduces the zeroth Betti number of S_f. See Figure 13.3(c) and (d). The perturbation required to do so in terms of the maximum changes in the function values is less than what is required for pushing q_2 or p_2 to remove the right component.

Hence the "robustness" of features within the levelset S_f depends on information of f beyond just S_f. To this end, one can do the following. Let $\mathrm{Dgm} f$ be the levelset zigzag persistence diagram of f. Set

$$\Pi_{S_f} := \{(b, d) \in \mathrm{Dgm} f \mid b \leq 0; d \geq 0\}.$$

See Figure 13.4 for an illustration. Intuitively, points in Π_{S_f} are those persistent features whose lifetime passes through the 0-level set S_f. There is a

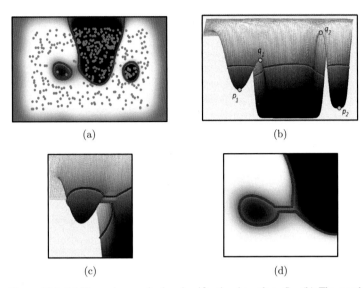

(a)   (b)

(c)   (d)

Figure 13.3 (a) The red curve is the classification boundary S_f. (b) The graph of the classifier function f, with S_f (the levelset at value 0) marked in red. (c) Pushing the saddle q_1 down to remove this left component in S_f as shown in (d). Image taken from [95].

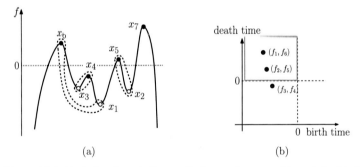

(a)   (b)

Figure 13.4 (a) A function $f : \mathbb{R} \rightarrow \mathbb{R}$. Its persistence pairings (of critical points) are marked by the dotted curves: $\{(x_1, x_6), (x_2, x_5), (x_3, x_4), \ldots\}$. The corresponding persistence diagram is shown in (b). The set Π_{S_f} consists of all points within the solid rectangle; that is, $\Pi_{S_f} = \{(f_1, f_6), (f_2, f_5), \ldots\}$, where $f_i = f(x_i)$ for $i \in [1, 6]$. Note that (f_3, f_4) is not in Π_{S_f} as the interval $[f_3, f_4]$ does not contain 0. Image taken from [95].

one-to-one correspondence between the "topological features" in S_f and the points in Π_{S_f} (this can be made more precise via the persistent cycles concept introduced in Definition 5.7 of Chapter 5), and one can view a point $(b, d) \in \Pi_{S_f}$ as the lifetime of its corresponding feature in the 0-level set S_f. The *robustness* of the feature corresponding to point $c = (b, d)$ is then defined as $\rho(c) = \min\{|b|, |d|\}$. Intuitively, this is the least amount of function perturbation in terms of the L_∞-norm needed to remove this feature from S_f (i.e., to push persistent point c out of the set Π_{S_f}). One can then define the topological complexity (topological penalty) for the classifier f as

$$L_{topo}(f) := \sum_{c \in \Pi_{S_f}} \rho^2(c).$$

In practice, suppose, for example, that we have the supervised setting where we are given a set of points $X_n = \{x_1, \ldots, x_n\}$ with class labels $\{y_1, \ldots, y_n\}$. Assume the classifier f_ω is parameterized by $\omega \in \mathbb{R}^m$. We can combine the topological penalty L_{topo} with any standard loss function to define a final loss function, for example,

$$L(f_\omega, X_n) = \sum_{x_i \in X_n} \ell(f_\omega(x_i), y_i) + \lambda L_{topo}(f_\omega), \qquad (13.8)$$

where the first term represents standard loss and $\ell(\cdot, \cdot)$ could be cross-entropy loss, hinge loss and so on.

Finally, to optimize $L(f_\omega, X_n)$ with respect to ω (so as to learn the best classifier f_ω), we can do (stochastic) gradient descent, and thus need to compute the gradient for $L_{topo}(f_\omega)$. To this end, we approximate the domain $\mathbb{X}$ by taking a certain simplicial complex K spanned by samples X_n. In [95], only the zeroth topological information of the classification boundary S_f is used. Hence one only needs the 1-skeleton of K. In the implementation of [95], that is then simply taken as the k-nearest-neighbor graph spanned by input samples X_n. One then uses the approach to be described shortly in Section 13.2.2 to compute gradients of this loss function, which is a persistence-based topological function.

13.2.2 Gradients of a Persistence-Based Topological Function

For simplicity, we describe the setting where we are given an input *topological function* which incorporates the persistence induced by a PL-function on a simplicial complex.

Specifically, given a simplicial complex domain K and a PL-function $f \colon |K| \to \mathbb{R}$, let $\mathrm{Dgm}\, f$ denote the (sublevel set, superlevel set, union of them, or levelset zigzag) persistence diagram induced by f. Now suppose that the function f is parameterized by some m-dimensional parameter $\omega \in \mathbb{R}^m$, and denote the corresponding function by $f_\omega \colon |K| \to \mathbb{R}$. Its resulting persistence diagram is denoted by $\mathrm{Dgm}\, f_\omega$. In the exposition below, we sometimes omit the subscript ω from f_ω for simplicity.

Recall that $\mathrm{Dgm}\,f$ consists of a multiset set of points $\{(b_i, d_i)\}_{i \in I_f}$, where I_f is an index set. Suppose we have a *persistence-based topological function*,

$$\mathcal{T}(\omega) := \mathcal{T}(\mathrm{Dgm}\,f_\omega) = \mathcal{T}\big(\{(b_i, d_i)\}_{i \in I_{f_\omega}}\big);$$

for example, $\mathcal{T}(\omega)$ could be $L_{topo}(f_\omega)$ introduced in the previous section.

To optimize the topological function $\mathcal{T}(\omega)$, one may need to compute gradient of $\mathcal{T}$ with respect to the parameter ω. Applying the chain rule, this means that one needs to be able to compute $\partial b_i/\partial \omega$ and $\partial d_i/\partial \omega$ for certain points (b_i, d_i) in the persistence diagram. (Terms such as $\partial \mathcal{T}/\partial b_i$ can be computed easily if the analytic form of $\mathcal{T}$ with respect to the b_i and d_i are given; again, consider $L_{topo}(f)$ from the previous section as an example.) Intuitively, this requires the "inverse" of the map which maps f_ω to its persistence diagram $\mathrm{Dgm}\,f_\omega$. This inverse in general does not exist. However, assuming that f_ω is a PL-function defined on K, then it turns out that one can map the b_i and d_i back to vertices of K, and this map is locally constant if all vertices of K have distinct function values.

More specifically, suppose $\mathrm{Dgm}\,f$ is generated by the persistent homology of the sublevel set filtration induced by f. Recall as described in Section 3.5.2, from the algorithmic point of view, that the sublevel set filtration is simulated by the so-called lower-star filtration of K. Using the notation from Section 3.5.2, let V_i be the first i vertices of V, sorted in nondecreasing order of their function values, and $K_i = \bigcup_{j \leq i} \mathrm{Lst}(v_i)$ the set of all simplices spanned by vertices in V_i (i.e., by vertices whose function value is at most $f(v_i)$). The sublevel set filtration $\mathcal{F}$ is constructed by adding v_i and all simplices in its lower star in increasing order of i; recall Eq. (3.10). Furthermore, recall that (Theorem 3.16) each persistent point in the diagram $\mathrm{Dgm}\,f$ is in fact of the form $(b_i, d_i) = (f(v_{\ell_i}), f(v_{r_i}))$ such that the pairing function $\mu_f^{\ell_i, r_i} > 0$, and vertices v_{ℓ_i} and v_{r_i} are both homological critical points for the PL-function f. We use the map $\rho: \mathrm{Dgm}\,f \to V \times V$ to denote this correspondence[6] with $\rho(b_i, d_i) = (v_{\ell_i}, v_{r_i})$. We will also abuse notation slightly and write $\rho(b_i) = v_{\ell_i}$ and $\rho(d_i) = v_{r_i}$. In other words, birth and death points in the persistence diagram $\mathrm{Dgm}\,f$ can be mapped back to unique vertices in the vertex set of K.

This gives us a map $\xi: \mathbb{R}^m \to 2^{V \times V}$ that maps any parameter $\omega \in \mathbb{R}^m$ to a collection of pairs $\xi(\omega) := \rho(\mathrm{Dgm}\,f_\omega) \subseteq V \times V$. Assume that as the parameter $\omega \in \mathbb{R}^m$ changes, the function f_ω changes continuously (w.r.t. the L_∞-norm on the function space). It then follows that its persistence diagram $\mathrm{Dgm}\,f_\omega$ also changes continuously due to the stability result of persistence [101]. The image of $\mathrm{Dgm}\,f_\omega$ under ρ_ω also changes, although not necessarily

---

[6] Note that while formulated differently, this map is the same as the one used in [258].

continuously. Nevertheless, for a PL-function f_ω, this image set stays *fixed* (constant) within a small neighborhood of ω if f_ω is "nice." More specifically, we have the following.

Proposition 13.9. *Suppose* $f_\omega \colon |K| \to \mathbb{R}$ *is a PL-function with distinct values on all vertices V of K, and K is a finite simplicial complex. Then there exists a neighborhood of ω in the parameter space such that ξ remains constant within this neighborhood; that is, the image set $\xi(\omega) = \rho_\omega(\mathrm{Dgm}\, f_\omega)$ remains the same for all parameters within this neighborhood.*

Recall that $b_i = f_\omega(\rho_\omega(b_i)) = f_\omega(v_{\ell_i})$. It follows that, if the conditions on f_ω in Proposition 13.9 hold, then within a sufficiently small neighborhood of ω, even though b_i moves continuously, the identity of v_{ℓ_i} remains the same, and $b_i = f_\omega(v_{\ell_i})$ as ω varies within this neighborhood. Hence we have

$$\frac{\partial b_i}{\partial \omega} = \frac{\partial f_\omega(\rho_\omega(b_i))}{\partial \omega} = \frac{\partial f_\omega(v_{\ell_i})}{\partial \omega} = \partial f_\omega(v_{\ell_i})/\partial \omega.$$

The derivative $\partial d_i/\partial \omega$ can be computed in an analogous manner by $\partial f_\omega(v_{r_i})/\partial \omega$. This in turn leads to the computation of the derivative $\partial \mathcal{T}/\partial \omega$ for the persistence-based topological function $\mathcal{T}(\omega)$.

13.3 Statistical Treatment of Topological Summaries

This book has been focused on the mathematical structures and computational aspects of various topological objects useful for topological data analysis. Topological methods help us to map an input object to its topological summaries, and thus it is natural to compute statistics or perform statistical analysis of a collection of objects over their topological summaries. In this penultimate section of the book, we briefly mention some developments regarding stochastic and statistical aspects of topological summaries. We note that while the main content of this book does not focus on them, these are important topics for the development of topological data analysis, for example, leading to more rigorous quantification of uncertainty, noise, consistency, and so on.

Performing Statistics on Space of Persistence Diagrams
One key objective in data analysis is to model and quantify variations in data, such as computing the mean or variance of a collection of data. Given the power of persistent homology in mapping an input complex object to its persistence diagram summary, it is natural to ask whether we can compute the mean or variance in the space of persistence diagrams. This question was first studied in [231], and to answer it, one needs to study the properties of the space of persistence diagrams equipped with certain metrics. To state the results, we

first need to refine the definition of Wasserstein distance of persistence diagrams (Definition 3.10) to allow different norms for measuring the distance between two points in the persistence diagram. The definition below assumes that we take the general view where a persistence diagram includes infinitely many copies of the diagonal.

Definition 13.8. $((p, q)$-Wasserstein distance) Let P and Q be two persistence diagrams. The *(p, q)-Wasserstein distance* between these two diagrams is

$$
\mathsf{d}^p_{W,q}(\mathsf{P}, \mathsf{Q}) := \inf_{\Pi:\mathsf{P}\to\mathsf{Q}} \left[\sum_{x\in\mathsf{P}} \|x - \Pi(x)\|^q_p \right]^{1/q}, \qquad (13.9)
$$

where Π ranges over all bijections from P to Q.

Note that our q-Wasserstein distance introduced in Definition 3.10 is simply $\mathsf{d}^\infty_{W,q}$ under this definition.

Now, let $\mathsf{D}_\varnothing$ denote the *trivial persistence diagram* which contains only infinite copies of the diagonal.

Definition 13.9. (Space of persistence diagrams) Given p and q, the *space of persistence diagrams* $\mathbb{D}^p_q$ consists of all persistence diagrams within a finite distance to the trivial persistence diagram $D_\varnothing$; that is,

$$
\mathbb{D}^p_q := \left\{ \mathsf{P} \mid \mathsf{d}^p_{W,q}(\mathsf{P}, D_\varnothing) \prec \infty \right\}.
$$

In what follows, for simplicity, we abuse the notation slightly and let $\mathbb{D}^p_q$ denote the metric space $(\mathbb{D}^p_q, \mathsf{d}^p_{W,q})$ equipped with $\mathsf{d}^p_{W,q}$. It is shown in [231] that $\mathbb{D}^\infty_q$ is a so-called Polish (i.e., complete and separable) space, and probability measures can be defined. It is later shown that more can be said about the space $\mathbb{D}^2_2$ (Theorem 2.5 in [291]), which is a nonnegatively curved Alexandrov space (i.e., a geodesic space with curvature bounded from below by zero).

Furthermore, in both cases, the concepts of "mean" and "variance" can be introduced using the notion of the Fréchet function. Specifically, in what follows, we use $\mathcal{D}$ to denote either $\mathbb{D}^\infty_q$ and $\mathbb{D}^2_2$ with metric $\mathsf{d}_\mathcal{D}$ being the corresponding metric $\mathsf{d}^\infty_{W,q}$ or $\mathsf{d}^2_{W,2}$. We will consider probability measures defined on $(\mathcal{D}, \mathcal{B}(\mathcal{D}))$, where $\mathcal{B}(\mathcal{D})$ is the Borel σ-algebra on $\mathcal{D}$.

Definition 13.10. (Fréchet function, variance, and expectation) Given a probability distribution ρ on $(\mathcal{D}, \mathcal{B}(\mathcal{D}))$, its *Fréchet function* $\mathcal{F}_\rho : \mathcal{D} \to \mathbb{R}$ is defined as, for any $X \in \mathcal{D}$,

$$
\mathcal{F}_\rho(X) := \int_\mathcal{D} \mathsf{d}^2_\mathcal{D}(X, Y) \, d\rho(Y). \qquad (13.10)
$$

The *Fréchet variance of* ρ is defined as the quantity

$$Var_\rho := \inf_{X \in \mathcal{D}} \mathcal{F}_\rho(X),$$

while the set at which this variance is obtained, that is,

$$\mathbb{E}(\rho) = \{X \mid \mathcal{F}_\rho(X) = Var_\rho\},$$

is called the *Fréchet expectation*, or alternatively, the *Fréchet mean set* of ρ.

Often in the literature, one uses the Fréchet mean to refer to an element in the Fréchet mean set. Intuitively, the Fréchet function is a generalization of the arithmetic mean in the sense that it minimizes the sum of the square distances to all points in the distribution. If the input is a collection of persistence diagrams $\Omega = \{D_1, D_2, \ldots, D_m\}$, then we can talk about the mean of this collection as the mean of the discrete measure $\rho_\Omega = (1/m) \sum_{i=1}^{m} \delta_{D_i}$ induced by them, where δ_X is the Dirac measure centered at $X \in \mathcal{D}$.

In general, it is not clear whether the Fréchet mean even exists. However, for the space $\mathcal{D}$ as defined above, it is shown [231, 291] that the Fréchet mean set is not empty under mild conditions on the distribution.

Theorem 13.10. *Let* ρ *be a probability measure on* $(\mathcal{D}, \mathcal{B}(\mathcal{D}))$ *with a finite second moment, that is,* $\mathcal{F}_\rho(X) < \infty$ *for any* $X \in \mathcal{D}$. *If* ρ *has compact support, then* $\mathbb{E}(\rho) \neq \varnothing$.

In the case when $\mathcal{D} = \mathbb{D}_2^2$, leveraging the property of $\mathbb{D}_2^2$, Turner et al. developed an iterative algorithm to compute a local minimum of the Fréchet functional [291]. The computational question for Fréchet mean, however, remains open. We also note that, in general, the Fréchet mean is not unique. This becomes undesirable for example when tracking the mean of a set of varying persistence diagrams. To address this issue, a modified concept of *probabilistic Fréchet mean* is proposed in [240], which intuitively is a probabilistic measure on $\mathbb{D}_2^2$, and the authors show how to use this to build useful statistics on the vineyard (the time-varying persistence diagrams).

Other Statistical Analyses of Topological Summaries

Another line of statistical treatment of topological objects concerns relating the estimation of them when input data is assumed to be sampled from a target space/distribution. For example, a common setting is that suppose we observe an independent and identically distributed (i.i.d.) sample $x_1, \ldots, x_n \sim \mathcal{P}$ drawn from a distribution $\mathcal{P}$ supported on a compact set $X \subset \mathbb{R}^d$. One can then ask questions of how to relate the topological summaries estimated from these samples to that of X (when appropriate), whether such estimates converge, or how to compute the confidence interval (set), and so on. We will not review the results here, as that would require careful description of the models used. We refer the reader to the nice survey by Wasserman in [297], which discussed the statistical estimation for various topological objects, including (hierarchical) clustering (related to merge trees), persistence diagrams, and ridge estimation. We will just mention that in the context of persistence diagram and its variants (e.g., persistence landscapes), there has been work to analyze their concentration and convergence behavior as the number of samples n tends to infinity for different settings [84, 85], or to obtain the confidence set for them via bootstrapping or subsampling [34, 83, 160].

The inference and estimation of topological information has been discussed earlier in Chapter 6; however, we have assumed that the samples are deterministic there. Also note that as the distribution $\mathcal{P}$ (where input points are sampled from) deviates further from the true distribution we are interested in, the standard construction based on the Rips or Čech complexes to approximate the sublevel sets of the distance field (recall Definition 6.7 in Section 6.3.1) is no longer appropriate. Instead, one now needs to use more robust notions of "distance field." To this end, an elegant concept called *distance to measures* (DTM) has been proposed [79], which has many nice properties and can lead to more robust topology inferences; see, for example, [82]. An alternative is to use kernel distance as proposed in [257].

Finally, we note that there has also been a line of work to study topological properties (e.g., Betti numbers, or the largest persistence in the persistence diagram) of random simplicial complexes [35–37, 202–204]. We will not describe this interesting line of work in this book.

13.4 Bibliographical Notes

The persistence landscape proposed by Bubenik [51] is perhaps the first to map persistence diagrams into a function space, which may often be taken to be a Banach space or a Hilbert space, so as to facilitate statistical analysis (e.g., computing the mean) of persistence-based summaries. The persistence scale space kernel (PSSK), the persistence images, persistence weighted Gaussian kernel, sliced Wasserstein kernel, and persistence Fisher kernel described in this chapter are introduced in [263], [4], [216], [68], and [219], respectively. Note that this chapter does not aim to provide a complete survey on vectorization frameworks for persistence diagrams; and in addition to those presented in Section 13.1, there are other similar approaches such as [1, 142, 205, 251]. As mentioned in Section 13.3, there has also been work exploring how to perform statistical analysis in the space of persistence diagrams equipped with standard bottleneck distance or Wasserstein distance; see, for example, [34, 52, 85, 160, 231, 240, 291], and more details in Section 13.3.

Recently, there has been a line of work to learn the representation of persistence diagrams in an end-to-end manner using labeled data. The work by Hofer et al. [194] was one of the first along this direction, and the authors developed a neural network layer for this purpose (and the work is later refined in [195]). Later, an alternative layer (based on the Deep Sets architecture [303]) was proposed and developed in [67], which provides a rather general and versatile framework to learn the vector representation for persistence diagrams. For example, the representation learning can be based on several existing vectorizations or kernelizations of persistence diagrams, including the persistence landscapes, persistence surfaces, and sliced Wasserstein. In a related work [305], learning the best representation based on persistence images is formulated as an optimization problem and solved directly via (stochastic) gradient descents. The resulting learned vector representations can be combined simply with SVM (support vector machine) for different tasks such as graph classification.

Differentiating a function involving persistence has been independently proposed and studied in several works from different communities, first in [165] for continuation of point clouds, then in [258] for continuous shape matching, and in [95] for topological regularization of classifiers. The gradients computation of a persistence-based topological function presented in Section 13.2.2 follows mostly from the discussion in [258]. The general topological optimization framework is rather general and powerful. Several recent works apply such ideas to different stages of machine learning applications. For example, [100, 199] used topological loss terms to help enforce a topological prior on an *individual input object* for deep-learning-based image segmentation. The work of [100] assumed certain prior knowledge of the topology of the segmented images. Instead of assuming this prior knowledge, [199] proposed to learn to segment with correct topology by using a topological loss function to help ensure that the topology of segmented images is the same as the ground truth for labeled images. The potential applications of these ideas has been further broadened in [49], where the authors introduced and developed a *topological layer* for function-induced persistence and for distance-based filtration-induced persistence. Such a persistence layer idea is further developed in [212] using the persistence landscape representation of general filtrations. Instead of having a topological constraint on an individual input data point, one can also consider using it for the *latent space* behind data. For example, [193, 237] applied such ideas with auto-encoders.

There has also been some recent work on using (persistent) homology to help characterize the complexity of a neural network (or its training process). For example, [264] proposed the so-called *neural persistence*, to characterize the structural complexity of neural networks. Others [28, 180] proposed to measure the capacity of an architecture by the topological complexity of the classifiers it can produce. The authors of [168] proposed to study the topology of the activation networks (neural networks with node activation for specific inputs), and used such patterns to help understand adversarial examples. Finally, [244] studied the change of the topology of the transformed data space across different layers of a deep neural network. While, overall, exploration in this direction is still in the initial stage, these are exciting ideas and there is much potential in using topological tools to understand neural networks.

References

[1] Aaron Aadcock, Erik Carlsson, and Gunnar Carlsson. The ring of algebraic functions on persistence barcodes. *Homology, Homotopy Appl.*, 18:381–402, 2016.

[2] Michal Adamaszek and Henry Adams. The Vietoris–Rips complexes of a circle. *Pacific J. Math.*, 290:1–40, 2017.

[3] Michal Adamaszek, Henry Adams, Ellen Gasparovic, et al. On homotopy types of Vietoris–Rips complexes of metric gluings. *J. Appl. Comput. Topol.*, 4: 425–454, 2020.

[4] Henry Adams, Tegan Emerson, Michael Kirby, et al. Persistence images: a stable vector representation of persistent homology. *J. Mach. Learn. Res.*, 18:218–252, 2017.

[5] Pankaj K. Agarwal, Herbert Edelsbrunner, John Harer, and Yusu Wang. Extreme elevation on a 2-manifold. *Discrete Comput. Geom.*, 36(4):553–572, 2006.

[6] Pankaj K. Agarwal, Kyle Fox, Abhinandan Nath, Anastasios Sidiropoulos, and Yusu Wang. Computing the Gromov–Hausdorff distance for metric trees. *ACM Trans. Algorithms*, 14(2):24:1–24:20, 2018.

[7] Paul Aleksandroff. Über den allgemeinen Dimensionsbegriff und seine Beziehungen zur elementaren geometrischen Anschauung. *Math. Ann.*, 98: 617–635, 1928.

[8] Nina Amenta, Marshall W. Bern, and David Eppstein. The crust and the beta-skeleton: combinatorial curve reconstruction. *Graph. Models Image Process.*, 60(2):125–135, 1998.

[9] Hideto Asashiba, Emerson G. Escolar, Yasuaki Hiraoka, and Hiroshi Takeuchi. Matrix method for persistence modules on commutative ladders of finite type. *J. Ind. Appl. Math.*, 36(1):97–130, 2019.

[10] Michael Atiyah. On the Krull–Schmidt theorem with application to sheaves. *Bull. Soc. Math. France*, 84:307–317, 1956.

[11] Dominique Attali, Herbert Edelsbrunner, and Yuriy Mileyko. Weak witnesses for Delaunay triangulations of submanifolds. In *Proc. ACM Sympos. on Solid and Physical Modeling*, pages 143–150, 2007.

[12] Dominique Attali, André Lieutier, and David Salinas. Efficient data structure for representing and simplifying simplicial complexes in high dimensions. In *Proc. 27th Annu. Sympos. on Computational Geometry*, pages 501–509, 2011.

[13] Maurice Auslander. Representation theory of Artin algebras II. *Commun. Alg.*, 1(4):269–310, 1974.

[14] Maurice Auslander and David Buchsbaum. *Groups, Rings, Modules*. Dover Publications, 2014.

[15] Aravindakshan Babu. Zigzag coarsenings, mapper stability and gene-network analyses. PhD thesis, Stanford University, 2013.

[16] Samik Banerjee, Lucas Magee, Dingkang Wang, et al. Semantic segmentation of microscopic neuroanatomical data by combining topological priors with encoder–decoder deep networks. *Nature Mach. Intell.*, 2:585–594, 2020.

[17] Jonathan Ariel Barmak and Elias Gabriel Minian. Strong homotopy types, nerves and collapses. *Discrete Comput. Geom.*, 47(2):301–328, 2012.

[18] Saugata Basu and Negin Karisani. Efficient simplicial replacement of semialgebraic sets and applications. *CoRR*, arXiv:2009.13365, 2020.

[19] Ulrich Bauer. Ripser: efficient computation of Vietoris–Rips persistence barcodes. *CoRR*, arXiv:1908.02518, 2019.

[20] Ulrich Bauer, Xiaoyin Ge, and Yusu Wang. Measuring distance between Reeb graphs. In *Proc. 30th Annu. Sympos. on Computational Geometry*, pages 464–473, 2014.

[21] Ulrich Bauer, Claudia Landi, and Facundo Mémoli. The Reeb graph edit distance is universal. In *Proc. 36th Internat. Sympos. on Computational Geometry*, pages 15:1–15:16, 2020.

[22] Ulrich Bauer, Carsten Lange, and Max Wardetzky. Optimal topological simplification of discrete functions on surfaces. *Discrete Comput. Geom.*, 47(2): 347–377, 2012.

[23] Ulrich Bauer and Michael Lesnick. Induced matchings of barcodes and the algebraic stability of persistence. In *Proc. 13th Annu. Sympos. on Computational Geometry*, pages 355–364, 2014.

[24] Ulrich Bauer, Elizabeth Munch, and Yusu Wang. Strong equivalence of the interleaving and functional distortion metrics for Reeb graphs. In *Proc. 31st Annu. Sympos. on Computational Geometry*, pages 461–475, 2015.

[25] Christian Berg, Jens P. R. Christensen, and Paul Ressel. *Harmonic Analysis on Semigroups: Theory of Positive Definite and Related Functions*. Springer, 1984.

[26] Marshall W. Bern, David Eppstein, Pankaj K. Agarwal, et al. Emerging challenges in computational topology. *CoRR*, arXiv:cs/9909001, 1999.

[27] Dimitris Bertsimas and John N. Tsitsiklis. *Introduction to Linear Optimization*. Athena Scientific, 1997.

[28] Monica Bianchini and Franco Scarselli. On the complexity of neural network classifiers: a comparison between shallow and deep architectures. *IEEE Trans. Neural Networks Learn. Syst.*, 25(8):1553–1565, 2014.

[29] Silvia Biasotti, Andrea Cerri, Patrizio Frosini, and Daniela Giorgi. A new algorithm for computing the 2-dimensional matching distance between size functions. *Pattern Recogn. Lett.*, 32(14):1735–1746, 2011.

[30] Silvia Biasotti, Bianca Falcidieno, and Michela Spagnuolo. Extended Reeb graphs for surface understanding and description. In *Proc. 9th Internat. Conf. on Discrete Geometry for Computer Imagery*, pages 185–197, 2000.

[31] Silvia Biasotti, Daniela Giorgi, Michela Spagnuolo, and Bianca Falcidieno. Reeb graphs for shape analysis and applications. *Theor. Comput. Sci.*, 392 (1–3):5–22, 2008.

[32] Håvard Bjerkevik. Stability of higher-dimensional interval decomposable persistence modules. *CoRR*, arXiv:1609.02086, 2020.

[33] Håvard Bjerkevik, Magnus Botnan, and Michael Kerber. Computing the interleaving distance is NP-hard. *Found. Comput. Math.*, 20:1237–1271, 2019.

[34] Ander J. Blumberg, Itamar Gal, Michael A. Mandell, and Matthew Pancia. Robust statistics, hypothesis testing, and confidence intervals for persistent homology on metric measure spaces. *Found. Comput. Math.*, 14:745–789, 2014.

[35] Omer Bobrowski and Matthew Kahle. Topology of random geometric complexes: a survey. *J. Appl. Comput. Topol.*, 1(3):331–364, 2018.

[36] Omer Bobrowski, Matthew Kahle, and Primoz Skraba. Maximally persistent cycles in random geometric complexes. *Ann. Appl. Probab.*, 27(4):2032–2060, 2017.

[37] Omer Bobrowski and Primoz Skraba. Homological percolation and the Euler characteristic. *Phys. Rev. E*, 101:032304, 2020.

[38] Erik Boczko, William D. Kalies, and Konstantin Mischaikow. Polygonal approximation of flows. *Topol. Appl.*, 154:2501–2520, 2007.

[39] Jean-Daniel Boissonnat, Frédéric Chazal, and Mariette Yvinec. *Geometric and Topological Inference*. Cambridge Texts in Applied Mathematics. Cambridge University Press, 2018.

[40] Jean-Daniel Boissonnat, Leonidas J. Guibas, and Steve Y. Oudot. Manifold reconstruction in arbitrary dimensions using witness complexes. In *Proc. 23rd Annu. Sympos. on Computational Geometry*, pages 194–203, 2007.

[41] Jean-Daniel Boissonnat and Siddharth Pritam. Edge collapse and persistence of flag complexes. In *Proc. 36th Internat. Sympos. on Computational Geometry*, volume 164 of *LIPIcs*, pages 19:1–19:15, 2020.

[42] Jean-Daniel Boissonnat, Siddharth Pritam, and Divyansh Pareek. Strong collapse for persistence. In *Proc. 26th Annu. European Sympos. on Algorithms*, volume 112 of *LIPIcs*, pages 67:1–67:13, 2018.

[43] Glencora Borradaile, Erin Wolf Chambers, Kyle Fox, and Amir Nayyeri. Minimum cycle and homology bases of surface-embedded graphs. *J. Comput. Geom.*, 8(2):58–79, 2017.

[44] Glencora Borradaile, William Maxwell, and Amir Nayyeri. Minimum bounded chains and minimum homologous chains in embedded simplicial complexes. In *Proc. 36th Internat. Sympos. on Computational Geometry*, volume 164 of *LIPIcs*, pages 21:1–21:15, 2020.

[45] Karol Borsuk. On the imbedding of systems of compacta in simplicial complexes. *Fundam. Math.*, 35:217–234, 1948.

[46] Magnus Botnan, Justin Curry, and Elizabeth Munch. A relative theory of interleavings. *CoRR*, arXiv:2004.14286, 2020.

[47] Magnus Botnan and Michael Lesnick. Algebraic stability of zigzag persistence modules. *Alg. Geom. Topol.*, 18:3133–3204, 2018.

[48] Stephane Bressan, Jingyan Li, Shiquan Ren, and Jie Wu. The embedded homology of hypergraphs and applications. *Asian J. Math.*, 23(3):479–500, 2019.

[49] Rickard Brüel-Gabrielsson, Bradley J. Nelson, Anjan Dwaraknath, Primoz Skraba, Leonidas J. Guibas, and Gunnar Carlsson. A topology layer for machine learning. *CoRR*, arXiv:1905.12200, 2019. Code available at https://github.com/bruel-gabrielsson/TopologyLayer.

[50] Winfried Bruns and H Jürgen Herzog. *Cohen–Macaulay Rings*. Cambridge University Press, 1998.

[51] Peter Bubenik. Statistical topological data analysis using persistence landscapes. *J. Mach. Learn. Res.*, 16(1):77–102, 2015.

[52] Peter Bubenik and Peter T. Kim. A statistical approach to persistent homology. *Homology, Homotopy Appl.*, 9(2):337–362, 2007.

[53] Peter Bubenik and Jonathan Scott. Categorification of persistent homology. *Discrete Comput. Geom.*, 51(3):600–627, 2014.

[54] Peter Bubenik, Jonathan A. Scott, and Donald Stanley. Wasserstein distance for generalized persistence modules and abelian categories. arXiv:1809.09654, 2018.

[55] Michaël Buchet, Frédéric Chazal, Tamal K. Dey, Fengtao Fan, Steve Y. Oudot, and Yusu Wang. Topological analysis of scalar fields with outliers. In *Proc. 31st Annu. Sympos. on Computational Geometry*, pages 827–841, 2015.

[56] Mickaël Buchet, Frédéric Chazal, Steve Y. Oudot, and Donald Sheehy. Efficient and robust persistent homology for measures. In *Proc. 26th Annu. ACM–SIAM Sympos. on Discrete Algorithms*, pages 168–180, 2015.

[57] James R. Bunch and John E. Hopcroft. Triangular factorization and inversion by fast matrix multiplication. *Math. Comput.*, 28(125):231–236, 1974.

[58] Dmitri Burago, Yuri Burago, and Sergei Ivanov. *A Course in Metric Geometry*. AMS Graduate Studies in Mathematics, vol. 33. American Mathematical Society, 2001.

[59] Dan Burghelea and Tamal K. Dey. Topological persistence for circle-valued maps. *Discrete Comput. Geom.*, 50(1):69–98, 2013.

[60] Oleksiy Busaryev, Sergio Cabello, Chao Chen, Tamal K. Dey, and Yusu Wang. Annotating simplices with a homology basis and its applications. In *Algorithm Theory – 13th Scandinavian Sympos. Workshops*, pages 189–200, 2012.

[61] Alexander Buslaev, Selim S. Seferbekov, Vladimir Iglovikov, and Alexey Shvets. Fully convolutional network for automatic road extraction from satellite imagery. In *Computer Vision and Pattern Recognition Conf. Workshops*, pages 207–210, 2018.

[62] Gunnar Carlsson and Vin de Silva. Zigzag persistence. *Found. Comput. Math.*, 10(4):367–405, 2010.

[63] Gunnar Carlsson, Vin de Silva, and Dmitriy Morozov. Zigzag persistent homology and real-valued functions. In *Proc. 26th Annu. Sympos. on Computational Geometry*, pages 247–256, 2009.

[64] Gunnar Carlsson, Gurjeet Singh, and Afra Zomorodian. Computing multidimensional persistence. In *Proc. Internat. Sympos. on Algorithms and Computation*, pages 730–739. Springer, 2009.

[65] Gunnar Carlsson and Afra Zomorodian. The theory of multidimensional persistence. *Discrete Comput. Geom.*, 42(1):71–93, 2009.

[66] Hamish Carr, Jack Snoeyink, and Ulrike Axen. Computing contour trees in all dimensions. *Comput. Geom.: Theory Appl.*, 24(2):75–94, 2003.

[67] Mathieu Carrière, Frédéric Chazal, Yuichi Ike, Théo Lacombe, Martin Royer, and Yuhei Umeda. Perslay: a neural network layer for persistence diagrams and new graph topological signatures. In *Proc. 23rd Internat. Conf. on Artificial Intelligence and Statistics*, volume 108, pages 2786–2796, 2020.

[68] Mathieu Carrière, Marco Cuturi, and Steve Y. Oudot. Sliced Wasserstein kernel for persistence diagrams. In *Proc. Internat. Conf. on Machine Learning*, pages 664–673, 2017.

[69] Mathieu Carrière and Steve Oudot. Structure and stability of the one-dimensional mapper. *Found. Comput. Math.*, 18(6):1333–1396, 2018.

[70] Nicholas J. Cavanna, Mahmoodreza Jahanseir, and Donald R. Sheehy. A geometric perspective on sparse filtrations. In *Proc. Canadian Conf. on Computational Geometry*, 2015.

[71] Andrea Cerri, Barbara Di Fabio, Massimo Ferri, Patrizio Frosini, and Claudia Landi. Betti numbers in multidimensional persistent homology are stable functions. *Math. Methods Appl. Sci.*, 36(12):1543–1557, 2013.

[72] Andrea Cerri and Patrizio Frosini. A new approximation algorithm for the matching distance in multidimensional persistence. *J. Comput. Math.*, 38: 291–309, 2020.

[73] Erin W. Chambers, Jeff Erickson, and Amir Nayyeri. Minimum cuts and shortest homologous cycles. In *Proc. 25th Annu. Sympos. on Computational Geometry*, pages 377–385, 2009.

[74] Erin W. Chambers, Jeff Erickson, and Amir Nayyeri. Homology flows, cohomology cuts. *SIAM J. Comput.*, 41(6):1605–1634, 2012.

[75] Manoj K. Chari. On discrete Morse functions and combinatorial decompositions. *Discrete Math.*, 217(1–3):101–113, 2000.

[76] Isaac Chavel. *Riemannian Geometry: A Modern Introduction*, 2nd edn. Cambridge University Press, 2006.

[77] Frédéric Chazal, David Cohen-Steiner, Marc Glisse, Leonidas J. Guibas, and Steve Oudot. Proximity of persistence modules and their diagrams. In *Proc. 25th Annu. Sympos. on Computational Geometry*, pages 237–246, 2009.

[78] Frédéric Chazal, David Cohen-Steiner, Leonidas J. Guibas, Facundo Mémoli, and Steve Y. Oudot. Gromov–Hausdorff stable signatures for shapes using persistence. *Comput. Graphics Forum*, 28(5):1393–1403, 2009.

[79] Frédéric Chazal, David Cohen-Steiner, and Quentin Mérigot. Geometric inference for probability distributions. *Found. Comput. Math.*, 11(6):733–751, 2011.

[80] Frédéric Chazal, Vin de Silva, Marc Glisse, and Steve Oudot. The structure and stability of persistence modules. *CoRR*, arXiv:1207.3674, 2012.

[81] Frédéric Chazal, Vin de Silva, and Steve Oudot. Persistence stability for geometric complexes. *Geom. Dedicata*, 173(1):193–214, 2014.

[82] Frédéric Chazal, Brittany Fasy, Fabrizio Lecci, Bertrand Michel, Alessandro Rinaldo, and Larry Wasserman. Robust topological inference: distance to a measure and kernel distance. *J. Mach. Learn. Res.*, 18(159):1–40, 2018.

[83] Frédéric Chazal, Brittany Fasy, Fabrizio Lecci, Alessandro Rinaldo, Aarti Singh, and Larry Wasserman. On the bootstrap for persistence diagrams and landscapes. *Model. Anal. Info. Syst.*, 20(6):96–105, 2013.

[84] Frédéric Chazal, Brittany Terese Fasy, Fabrizio Lecci, Alessandro Rinaldo, and Larry A. Wasserman. Stochastic convergence of persistence landscapes and silhouettes. *J. Comput. Geom.*, 6(2):140–161, 2015.

[85] Frédéric Chazal, Marc Glisse, Catherine Labruère, and Bertrand Michel. Convergence rates for persistence diagram estimation in topological data analysis. *J. Mach. Learn. Res.*, 16(110):3603–3635, 2015.

[86] Frédéric Chazal, Leonidas J. Guibas, Steve Oudot, and Primoz Skraba. Analysis of scalar fields over point cloud data. *Discrete Comput. Geom.*, 46(4):743–775, 2011.

[87] Frédéric Chazal, Ruqi Huang, and Jian Sun. Gromov–Hausdorff approximation of filamentary structures using Reeb-type graphs. *Discrete Comput. Geom.*, 53:621–649, 2015.

[88] Frédéric Chazal and André Lieutier. Weak feature size and persistent homology: computing homology of solids in $\mathbb{R}^n$ from noisy data samples. In *Proc. 21st Annu. Sympos. on Computational Geometry*, pages 255–262, 2005.

[89] Frédéric Chazal and André Lieutier. Stability and computation of topological invariants of solids in $\mathbb{R}^n$. *Discrete Comput. Geom.*, 37(4):601–617, 2007.

[90] Frédéric Chazal and Steve Y. Oudot. Towards persistence-based reconstruction in Euclidean spaces. In *Proc. 24th Annu. Sympos. on Computational Geometry*, pages 232–241, 2008.

[91] Bernard Chazelle. An optimal convex hull algorithm in any fixed dimension. *Discrete Comput. Geom.*, 10:377–409, 1993.

[92] Chao Chen and Daniel Freedman. Measuring and computing natural generators for homology groups. *Comput. Geom.: Theory Appl.*, 43(2):169–181, 2010.

[93] Chao Chen and Daniel Freedman. Hardness results for homology localization. *Discrete Comput. Geom.*, 45(3):425–448, 2011.

[94] Chao Chen and Michael Kerber. An output-sensitive algorithm for persistent homology. *Comput. Geom.: Theory Appl.*, 46(4):435–447, 2013.

[95] Chao Chen, Xiuyan Ni, Qinxun Bai, and Yusu Wang. A topological regularizer for classifiers via persistent homology. In *Proc. 22nd Internat. Conf. on Artificial Intelligence and Statistics*, pages 2573–2582, 2019.

[96] Siu-Wing Cheng, Tamal K. Dey, and Edgar A. Ramos. Manifold reconstruction from point samples. In *Proc. 16th Annu. ACM–SIAM Sympos. on Discrete Algorithms*, pages 1018–1027, 2005.

[97] Siu-Wing Cheng, Tamal K. Dey, and Jonathan R. Shewchuk. *Delaunay Mesh Generation*. CRC Press, 2012.

[98] Samir Chowdhury and Facundo Mémoli. Persistent homology of asymmetric networks: an approach based on Dowker filtrations. *CoRR*, arXiv:1608.05432, 2018.

[99] Samir Chowdhury and Facundo Mémoli. Persistent path homology of directed networks. In *Proc. 29th Annu. ACM–SIAM Sympos. on Discrete Algorithms*, pages 1152–1169. SIAM, 2018.

[100] James R. Clough, Ilkay Oksuz, Nicholas Byrne, Veronika A. Zimmer, Julia A. Schnabel, and Andrew P. King. A topological loss function for deep-learning based image segmentation using persistent homology. *CoRR*, arXiv:1910.01877, 2019.

[101] David Cohen-Steiner, Herbert Edelsbrunner, and John Harer. Stability of persistence diagrams. *Discrete Comput. Geom.*, 37(1):103–120, 2007.

[102] David Cohen-Steiner, Herbert Edelsbrunner, and John Harer. Extending persistence using Poincaré and Lefschetz duality. *Found. Comput. Math.*, 9(1):79–103, 2009.

[103] David Cohen-Steiner, Herbert Edelsbrunner, John Harer, and Yuriy Mileyko. Lipschitz functions have L_p-stable persistence. *Found. Comput. Math.*, 10(2):127–139, 2010.

[104] David Cohen-Steiner, Herbert Edelsbrunner, John Harer, and Dmitriy Morozov. Persistent homology for kernels, images, and cokernels. In *Proc. 20th Annu. ACM–SIAM Sympos. on Discrete Algorithms*, pages 1011–1020, 2009.

[105] David Cohen-Steiner, Herbert Edelsbrunner, and Dmitriy Morozov. Vines and vineyards by updating persistence in linear time. In *Proc. 22nd Annu. Sympos. on Computational Geometry*, pages 119–126, 2006.

[106] Kree Cole-McLaughlin, Herbert Edelsbrunner, John Harer, Vijay Natarajan, and Valerio Pascucci. Loops in Reeb graphs of 2-manifolds. *Discrete Comput. Geom.*, 32(2):231–244, 2004.

[107] René Corbet and Michael Kerber. The representation theorem of persistence revisited and generalized. *J. Appl. Comput. Topol.*, 2(1):1–31, 2018.

[108] Thomas H. Cormen, Charles E. Leiserson, Ronald L. Rivest, and Clifford Stein. *Introduction to Algorithms*, 3rd edn. MIT Press, 2009.

[109] David A. Cox, John Little, and Donal O'Shea. *Using Algebraic Geometry*. Graduate Texts in Mathematics, vol. 185. Springer, 2006.

[110] William Crawley-Boevey. Decomposition of pointwise finite-dimensional persistence modules. *J. Alg. Appl.*, 14(05):1550066, 2015.

[111] Justin Curry. Sheaves, cosheaves and applications. *CoRR*, arXiv:1303.3255, 2013.

[112] Vin de Silva. A weak definition of Delaunay triangulation. *CoRR*, arXiv:cs/0310031, 2003.

[113] Vin de Silva and Gunnar Carlsson. Topological estimation using witness complexes. In *Proc. Sympos. Ion Point-Based Graphics*, 2004.

[114] Vin de Silva, Dmitriy Morozov, and Mikael Vejdemo-Johansson. Dualities in persistent (co)homology. *Inverse Problems*, 27:124003, 2011.

[115] Vin de Silva, Elizabeth Munch, and Amit Patel. Categorified Reeb graphs. *Discrete Comput. Geom.*, 55(4):854–906, 2016.

[116] Cecil Jose A. Delfinado and Herbert Edelsbrunner. An incremental algorithm for Betti numbers of simplicial complexes on the 3-sphere. *Comput. Aided Geom. Design*, 12(7):771–784, 1995.

[117] Olaf Delgado-Friedrichs, Vanessa Robins, and Adrian P. Sheppard. Skeletonization and partitioning of digital images using discrete Morse theory. *IEEE Trans. Pattern Anal. Mach. Intell.*, 37(3):654–666, 2015.

[118] Tamal K. Dey. *Curve and Surface Reconstruction: Algorithms with Mathematical Analysis*. Cambridge Monographs on Applied and Computational Mathematics. Cambridge University Press, 2006.

[119] Tamal K. Dey, Herbert Edelsbrunner, and Sumanta Guha. Computational topology. In *Advances in Discrete and Computational Geometry*. Contemporary Mathematics 223, pages 109–144. American Mathematical Society, 1999.

[120] Tamal K. Dey, Herbert Edelsbrunner, Sumanta Guha, and Dmitry V. Nekhayev. Topology preserving edge contraction. *Publ. Inst. Math. (Beograd)*, 60:23–45, 1999.

[121] Tamal K. Dey, Fengtao Fan, and Yusu Wang. Computing topological persistence for simplicial maps. *CoRR*, arXiv:1208.5018, 2012.

[122] Tamal K. Dey, Fengtao Fan, and Yusu Wang. An efficient computation of handle and tunnel loops via Reeb graphs. *ACM Trans. Graphics*, 32(4):32, 2013.

[123] Tamal K. Dey, Fengtao Fan, and Yusu Wang. Graph induced complex for point data. In *Proc. 29th. Annu. Sympos. on Computational Geometry*, pages 107–116, 2013.

[124] Tamal K. Dey, Fengtao Fan, and Yusu Wang. Computing topological persistence for simplicial maps. In *Proc. 13th Annu. Sympos. on Computational Geometry*, pages 345:345–345:354, 2014.

[125] Tamal K. Dey, Anil N. Hirani, and Bala Krishnamoorthy. Optimal homologous cycles, total unimodularity, and linear programming. *SIAM J. Comput.*, 40(4):1026–1044, 2011.

[126] Tamal K. Dey and Tao Hou. Computing zigzag persistence on graphs in near-linear time. In *Proc. 37th Internat. Sympos. on Computational Geometry*, 2021.

[127] Tamal K. Dey, Tao Hou, and Sayan Mandal. Persistent 1-cycles: definition, computation, and its application. In *Computational Topology in Image Context – 7th Internat. Workshop*, pages 123–136, 2019.

[128] Tamal K. Dey, Tao Hou, and Sayan Mandal. Computing minimal persistent cycles: polynomial and hard cases. In *Proc. ACM–SIAM Sympos. on Discrete Algorithms*, pages 2587–2606, 2020.

[129] Tamal K. Dey, Tianqi Li, and Yusu Wang. Efficient algorithms for computing a minimal homology basis. In *LATIN 2018: Theoretical Informatics – 13th Latin American Sympos.*, pages 376–398, 2018.

[130] Tamal K. Dey, Tianqi Li, and Yusu Wang. An efficient algorithm for 1-dimensional (persistent) path homology. In *Proc. 36th. Internat. Sympos. on Computational Geometry*, pages 36:1–36:15, 2020.

[131] Tamal K. Dey, Facundo Mémoli, and Yusu Wang. Multiscale mapper: topological summarization via codomain covers. In *Proc. 27th Annu. ACM–SIAM Sympos. on Discrete Algorithms*, pages 997–1013, 2016.

[132] Tamal K. Dey, Facundo Mémoli, and Yusu Wang. Topological analysis of nerves, Reeb spaces, mappers, and multiscale mappers. In *Proc. 33rd Internat. Sympos. on Computational Geometry*, pages 36:1–36:16, 2017.

[133] Tamal K. Dey, Dayu Shi, and Yusu Wang. Comparing graphs via persistence distortion. In *Proc. 31st Annu. Sympos. on Computational Geometry*, pages 491–506, 2015.

[134] Tamal K. Dey, Dayu Shi, and Yusu Wang. SimBa: an efficient tool for approximating Rips-filtration persistence via simplicial batch-collapse. In *Proc. 24th Annu. European Sympos. on Algorithms*, volume 57 of *LIPIcs*, pages 35:1–35:16, 2016.

[135] Tamal K. Dey, Jian Sun, and Yusu Wang. Approximating loops in a shortest homology basis from point data. In *Proc. 26th Annu. Sympos. on Computational Geometry*, pages 166–175, 2010.

[136] Tamal K. Dey, Jiayuan Wang, and Yusu Wang. Improved road network reconstruction using discrete Morse theory. In *Proc. 25th ACM SIGSPATIAL Internat. Conf. on Advances in GIS*, pages 58:1–58:4, 2017.

[137] Tamal K. Dey, Jiayuan Wang, and Yusu Wang. Graph reconstruction by discrete Morse theory. In *Proc. 34th Internat. Sympos. on Computational Geometry*, pages 31:1–31:15, 2018.

[138] Tamal K. Dey, Jiayuan Wang, and Yusu Wang. Road network reconstruction from satellite images with machine learning supported by topological methods. In *Proc. 27th ACM SIGSPATIAL Internat. Conf. on Advances in GIS*, pages 520–523, 2019.

[139] Tamal K. Dey and Yusu Wang. Reeb graphs: approximation and persistence. *Discrete Comput. Geom.*, 49(1):46–73, 2013.

[140] Tamal K. Dey and Cheng Xin. Computing bottleneck distance for 2-D interval decomposable modules. In *Proc. 34th Internat. Sympos. on Computational Geometry*, pages 32:1–32:15, 2018.

[141] Tamal K. Dey and Cheng Xin. Generalized persistence algorithm for decomposing multi-parameter persistence modules. *CoRR*, arXiv:1904.03766, 2019.

[142] Barbara Di Fabio and Massimo Ferri. Comparing persistence diagrams through complex vectors. In *Image Analysis and Processing – ICIAP 2015*, pages 294–305, 2015.

[143] DIADEM (Digital Reconstruction of Axonal and Dendritic Morphology) Challenge. http://diademchallenge.org.

[144] Pawel Dlotko, Kathryn Hess, Ran Levi, et al. Topological analysis of the connectome of digital reconstructions of neural microcircuits. *CoRR*, arXiv:1601.01580, 2016.

[145] Harish Doraiswamy and Vijay Natarajan. Efficient output-sensitive construction of Reeb graphs. In *Proc. 19th Internat. Sympos. Algorithms and Computation*, pages 556–567, 2008.

[146] Harish Doraiswamy and Vijay Natarajan. Efficient algorithms for computing Reeb graphs. *Comput. Geom.: Theory Appl.*, 42:606–616, 2009.

[147] Clifford H. Dowker. Homology groups of relations. *Annals of Math.*, 56:84–95, 1952.

[148] Herbert Edelsbrunner. *Geometry and Topology for Mesh Generation*. Cambridge Monographs on Applied and Computational Mathematics, vol. 7. Cambridge University Press, 2001.

[149] Herbert Edelsbrunner and John Harer. *Computational Topology: An Introduction*. American Mathematical Society, 2010.

[150] Herbert Edelsbrunner, John Harer, and Amit K. Patel. Reeb spaces of piecewise linear mappings. In *Proc. 24th Annu. Sympos. on Computational Geometry*, pages 242–250, 2008.

[151] Herbert Edelsbrunner, David G. Kirkpatrick, and Raimund Seidel. On the shape of a set of points in the plane. *IEEE Trans. Info. Theory*, 29(4):551–558, 1983.

[152] Herbert Edelsbrunner, David Letscher, and Afra Zomorodian. Topological persistence and simplification. *Discrete Comput. Geom.*, 28:511–533, 2002.

[153] Herbert Edelsbrunner and Ernst P. Mücke. Three-dimensional alpha shapes. *ACM Trans. Graphics*, 13(1):43–72, 1994.

[154] Alon Efrat, Alon Itai, and Matthew J. Katz. Geometry helps in bottleneck matching and related problems. *Algorithmica*, 31(1):1–28, 2001.

[155] David Eisenbud. *The Geometry of Syzygies: A Second Course in Algebraic Geometry and Commutative Algebra*. Graduate Texts in Mathematics, vol. 229. Springer, 2005.

[156] Jeff Erickson and Kim Whittlesey. Greedy optimal homotopy and homology generators. In *Proc. 16th Annu. ACM–SIAM Sympos. on Discrete Algorithms*, pages 1038–1046, 2005.

[157] Emerson G. Escolar and Yasuaki Hiraoka. Optimal cycles for persistent homology via linear programming. *Optim. Real World*, 13:79–96, 2016.

[158] Barbara Di Fabio and Claudia Landi. Reeb graphs of curves are stable under function perturbations. *Math. Methods Appl. Sci.*, 35:1456–1471, 2012.

[159] Barbara Di Fabio and Claudia Landi. The edit distance for Reeb graphs of surfaces. *Discrete Comput. Geom.*, 55:423–461, 2016.

[160] Brittany Terese Fasy, Fabrizio Lecci, Alessandro Rinaldo, Larry Wasserman, Sivaraman Balakrishnan, and Aarti Singh. Confidence sets for persistence diagrams. *Ann. Statist*, 42(6):2301–2339, 2014.

[161] Robin Forman. Morse theory for cell complexes. *Adv. Math.*, 134:90–145, 1998.

[162] Patrizio Frosini. A distance for similarity classes of submanifolds of a Euclidean space. *Bull. Aust. Math. Soc.*, 42(3):407–415, 1990.

[163] Peter Gabriel. Unzerlegbare Darstellungen I. *Manuscr. Math.*, 6(1):71–103, 1972.

[164] Sylvestre Gallot, Dominique Hulin, and Jacques Lafontaine. *Riemannian Geometry*, 2nd edn. Springer, 1993.

[165] Marcio Gameiro, Yasuaki Hiraoka, and Ippei Obayashi. Continuation of point clouds via persistence diagrams. *Physica D· Nonlin. Phenom.*, 334:118–132, 2016.

[166] Ellen Gasparovic, Maria Gommel, Emilie Purvine, et al. The relationship between the intrinsic Čech and persistence distortion distances for metric graphs. *J. Comput. Geom.*, 10(1):477–499, 2019.

[167] Xiaoyin Ge, Issam Safa, Mikhail Belkin, and Yusu Wang. Data skeletonization via Reeb graphs. In *Proc. 25th Annu. Conf. on Neural Information Processing Systems*, pages 837–845, 2011.

[168] Thomas Gebhart, Paul Schrater, and Alan Hylton. Characterizing the shape of activation space in deep neural networks. *CoRR*, arXiv:1901.09496, 2019.

[169] Loukas Georgiadis, Robert Endre Tarjan, and Renato Fonseca Werneck. Design of data structures for mergeable trees. In *Proc. 17th Annu. ACM–SIAM Sympos. on Discrete Algorithms*, pages 394–403, 2006.

[170] Robert Ghrist. *Elementary Applied Topology*. CreateSpace Independent Publishing Platform, 2014.

[171] Alexander Grigor'yan, Yong Lin, Yuri Muranov, and Shing-Tung Yau. Homologies of path complexes and digraphs. *CoRR*, arXiv:1207.2834, 2012.

[172] Alexander Grigor'yan, Yong Lin, Yuri Muranov, and Shing-Tung Yau. Homotopy theory for digraphs. *CoRR*, arXiv:1407.0234, 2014.

[173] Alexander Grigor'yan, Yong Lin, Yuri Muranov, and Shing-Tung Yau. Cohomology of digraphs and (undirected) graphs. *Asian J. Math.*, 19(5):887–931, 2015.

[174] Alexander Grigor'yan, Yuri Muranov, and Shing-Tung Yau. Homologies of digraphs and Künneth formulas. *Commun. Anal. Geom.*, 25(5):969–1018, 2017.

[175] Mikhail Gromov. Groups of polynomial growth and expanding maps (with an appendix by Jacques Tits). *Publ. Math. Inst. Hautes Études Sci.*, 53(1):53–78, 1981.

[176] Mikhail Gromov. Hyperbolic groups. In *Essays in Group Theory*. Mathematical Sciences Research Institute Publications, vol. 8, pages 75–263. Springer, 1987.

[177] Karsten Grove. Critical point theory for distance functions. *Proc. Sympos. Pure Math.*, 54(3):357–385, 1993.

[178] Leonidas J. Guibas and Steve Y. Oudot. Reconstructing using witness complexes. *Discrete Comput. Geom.*, 30:325–356, 2008.

[179] Victor Guillemin and Alan Pollack. *Differential Topology*. Prentice Hall, 1974.

[180] William H. Guss and Ruslan Salakhutdinov. On characterizing the capacity of neural networks using algebraic topology. *CoRR*, arXiv:1802.04443, 2018.

[181] Attila Gyulassy, Natallia Kotava, Mark Kim, Charles Hansen, Hans Hagen, and Valerio Pascucci. Direct feature visualization using Morse–Smale complexes. *IEEE Trans. Visualiz. Comput. Graphics*, 18(9):1549–1562, 2012.

[182] Sariel Har-Peled and Manor Mendel. Fast construction of nets in low-dimensional metrics and their applications. *SIAM J. Comput.*, 35(5):1148–1184, 2006.

[183] Frank Harary. *Graph Theory*. Addison Wesley Series in Mathematics. Addison-Wesley, 1971.

[184] William Harvey, In-Hee Park, Oliver Rübel, et al. A collaborative visual analytics suite for protein folding research. *J. Mol. Graph. Model.*, 53:59–71, 2014.

[185] William Harvey, Raphael Wenger, and Yusu Wang. A randomized $O(m \log m)$ time algorithm for computing Reeb graph of arbitrary simplicial complexes. In *Proc. 25th Annu. ACM Sympos. on Computational Geometry*, pages 267–276, 2010.

[186] Allen Hatcher. *Algebraic Topology*. Cambridge University Press, 2002.

[187] Jean-Claude Hausmann. On the Vietoris–Rips complexes and a cohomology theory for metric spaces. *Ann. Math. Stud.*, 138:175–188, 1995.

[188] John Hershberger and Jack Snoeyink. Computing minimum length paths of a given homotopy class. *Comput. Geom.: Theory Appl.*, 4:63–97, 1994.

[189] Franck Hétroy and Dominique Attali. Topological quadrangulations of closed triangulated surfaces using the Reeb graph. *Graph. Models*, 65(1–3):131–148, 2003.

[190] Masaki Hilaga, Yoshihisa Shinagawa, Taku Kohmura, and Tosiyasu L. Kunii. Topology matching for fully automatic similarity estimation of 3D shapes. In *Proc. 28th Annu. Conf. on Computer Graphics and Interactive Techniques*, pages 203–212, 2001.

[191] David Hilbert. Über die Theorie der algebraischen Formen. *Math. Ann.*, 36: 473–530, 1890.

[192] Yasuaki Hiraoka, Takenobu Nakamura, Akihiko Hirata, Emerson G. Escolar, Kaname Matsue, and Yasumasa Nishiura. Hierarchical structures of amorphous solids characterized by persistent homology. *Proc. Natl. Acad. Sci.*, 113(26):7035–7040, 2016.

[193] Christoph Hofer, Roland Kwitt, Marc Niethammer, and Mandar Dixit. Connectivity-optimized representation learning via persistent homology. In *Proc. 36th Internat. Conf. on Machine Learning Proceedings of Machine Learning Research*, vol. 97, pages 2751–2760, 2019.

[194] Christoph Hofer, Roland Kwitt, Marc Niethammer, and Andreas Uhl. Deep learning with topological signatures. In *Proc. 31st Internat. Conf. on Neural Information Processing Systems*, pages 1634–1644, 2017.

[195] Christoph D. Hofer, Roland Kwitt, and Marc Niethammer. Learning representations of persistence barcodes. *J. Mach. Learn. Res.*, 20(126):1–45, 2019.

[196] Derek F. Holt. The Meataxe as a tool in computational group theory. *The Atlas of Finite Groups – Ten Years On*. London Mathematical Society Lecture Note Series, no. 249, pages 74–81. Cambridge University Press, 1998.

[197] Derek F. Holt and Sarah Rees. Testing modules for irreducibility. *J. Aust. Math. Soc.*, 57(1):1–16, 1994.

[198] John E. Hopcroft and Richard M. Karp. An $n^{5/2}$ algorithm for maximum matchings in bipartite graphs. *SIAM J. Comput.*, 2(4):225–231, 1973.

[199] Xiaoling Hu, Fuxin Li, Dimitris Samaras, and Chao Chen. Topology-preserving deep image segmentation. In *Proc. 33rd Annu. Conf. on Neural Information Processing Systems*, pages 5658–5669, 2019.

[200] Oscar H. Ibarra, Shlomo Moran, and Roger Hui. A generalization of the fast LUP matrix decomposition algorithm and applications. *J. Algorithms*, 3(1): 45 – 56, 1982.

[201] Arthur F. Veinott Jr. and George B. Dantzig. Integral extreme points. *SIAM Rev.*, 10(3):371–372, 1968.

[202] Matthew Kahle. Topology of random clique complexes. *Discrete Math.*, 309(6):1658–1671, 2009.

[203] Matthew Kahle. Sharp vanishing thresholds for cohomology of random flag complexes. *Annals of Math.*, 179(3):1085–1107, 2014.

[204] Matthew Kahle and Elizabeth Meckes. Limit theorems for Betti numbers of random simplicial complexes. *Homology, Homotopy Appl.*, 15(1):343–374, 2013.

[205] Sara Kališnik. Tropical coordinates on the space of persistence barcodes. *Found. Comput. Math.*, 19:101–129, 2019.

[206] Lida Kanari, Paweł Dłotko, Martina Scolamiero, et al. A topological representation of branching neuronal morphologies. *Neuroinformatics*, 16(1):3–13, 2018.

[207] Pizzanu Kanongchaiyos and Yoshihisa Shinagawa. Articulated Reeb graphs for interactive skeleton animation. In *Multimedia Modeling: Modeling Multimedia Information and Systems*, pages 451–467. World Scientific, 2000.

[208] Michael Kerber, Michael Lesnick, and Steve Oudot. Exact computation of the matching distance on 2-parameter persistence modules. In *Proc. 35th Internat. Sympos. on Computational Geometry*, volume 129 of *LIPIcs*, pages 46:1–46:15, 2019.

[209] Michael Kerber, Dmitriy Morozov, and Arnur Nigmetov. Geometry helps to compare persistence diagrams. *J. Exp. Algorithmics*, 22(1):1–4, 2017.

[210] Michael Kerber and Arnur Nigmetov. Efficient approximation of the matching distance for 2-parameter persistence. *CoRR*, arXiv:1912.05826, 2019.

[211] Michael Kerber and Hannah Schreiber. Barcodes of towers and a streaming algorithm for persistent homology. *Discrete Comput. Geom.*, 61(4):852–879, 2019.

[212] Kwangho Kim, Jisu Kim, Manzil Zaheer, Joon Kim, and Frédéric Chazal, Larry Wasserman. PLLay: efficient topological layer based on persistent landscapes. In *Proc. 33rd Annu. Conf. on Neural Information Processing Systems*, pages 15965–15977, 2020.

[213] Woojin Kim and Facundo Mémoli. Generalized persistence diagrams for persistence modules over posets. *CoRR*, arXiv:1810.11517, 2018.

[214] Henry King, Kevin P. Knudson, and Neza Mramor. Generating discrete Morse functions from point data. *Exp. Math.*, 14(4):435–444, 2005.

[215] Kevin P. Knudson. A refinement of multi-dimensional persistence. *CoRR*, arXiv:0706.2608, 2007.

[216] Genki Kusano, Kenji Fukumizu, and Yasuaki Hiraoka. Kernel method for persistence diagrams via kernel embedding and weight factor. *J. Mach. Learn. Res.*, 18(189):1–41, 2018.

[217] Claudia Landi. The rank invariant stability via interleavings. *CoRR*, arXiv:1412.3374, 2014.

[218] Janko Latschev. Vietoris–Rips complexes of metric spaces near a closed Riemannian manifold. *Arch. Math.*, 77(6):522–528, 2001.

[219] Tam Le and Makoto Yamada. Persistence Fisher kernel: a Riemannian manifold kernel for persistence diagrams. In *Proc. Conf. on Neural Information Processing Systems*, pages 10028–10039, 2018.

[220] Jean Leray. Sur la forme des espaces topologiques et sur les points fixes des représentations. *J. Math. Pure Appl.*, 24:95–167, 1945.

[221] Michael Lesnick. The theory of the interleaving distance on multidimensional persistence modules. *Found. Comput. Math.*, 15(3):613–650, 2015.

[222] Michael Lesnick and Matthew Wright. Interactive visualization of 2-d persistence modules. *CoRR*, arXiv:1512.00180, 2015.

[223] Michael Lesnick and Matthew Wright. Computing minimal presentations and Betti numbers of 2-parameter persistent homology. *CoRR*, arXiv:1902.05708, 2019.

[224] Thomas Lewiner, Hélio Lopes, and Geovan Tavares. Applications of Forman's discrete Morse theory to topology visualization and mesh compression. *IEEE Trans. Visualiz. Comput. Graphics*, 10(5):499–508, 2004.

[225] Li Li, Wei-Yi Cheng, Benjamin S. Glicksberg, et al. Identification of type 2 diabetes subgroups through topological analysis of patient similarity. *Science: Transl. Med.*, 7(311):311ra174, 2015.

[226] André Lieutier. Any open bounded subset of R^n has the same homotopy type as its medial axis. *Comput.-Aided Design*, 36(11):1029–1046, 2004.

[227] Yong Lin, Linyuan Lu, and Shing-Tung Yau. Ricci curvature of graphs. *Tohoku Math. J.*, 63(4):605–627, 2011.

[228] Pek Y. Lum, Gurjeet Singh, Alan Lehman, et al. Extracting insights from the shape of complex data using topology. *Sci. Rep.*, 3:1236, 2013.

[229] Clément Maria and Steve Y. Oudot. Zigzag persistence via reflections and transpositions. In *Proc. 26th Annu. ACM–SIAM Sympos. on Discrete Algorithms*, pages 181–199, 2015.

[230] Paolo Masulli and Alessandro E. P. Villa. The topology of the directed clique complex as a network invariant. *SpringerPlus*, 5(1):388, 2016.

[231] Yuriy Mileyko, Sayan Mukherjee, and John Harer. Probability measures on the space of persistence diagrams. *Inverse Problems*, 27(12):124007, 2011.

[232] Ezra Miller and Bernd Sturmfels. *Combinatorial Commutative Algebra*. Springer, 2004.

[233] John W. Milnor. *Topology from a Differentiable Viewpoint*. Virginia University Press, 1965.

[234] John W. Milnor. *Morse Theory*, 5th edn. Annals of Mathematics Studies. Princeton University Press, 1973.

[235] Nikola Milosavljević, Dmitriy Morozov, and Primoz Skraba. Zigzag persistent homology in matrix multiplication time. In *Proc. 27th Annu. Sympos. on Computational Geometry*, pages 216–225, 2011.

[236] Konstantin Mischaikow and Vidit Nanda. Morse theory for filtrations and efficient computation of persistent homology. *Discrete Comput. Geom.*, 50(2): 330–353, 2013.

[237] Michael Moor, Max Horn, Bastian Rieck, and Karsten Borgwardt. Topological autoencoders. *CoRR*, arXiv:1906.00722, 2019.

[238] Dmitriy Morozov, Kenes Beketayev, and Gunther H. Weber. Interleaving distance between merge trees. In *Workshop on Topological Methods in Data Analysis and Visualization: Theory, Algorithms and Applications*, 2013.

[239] Marian Mrozek. Conley–Morse–Forman theory for combinatorial multivector fields on Lefschetz complexes. *Found. Comput. Math.*, 17(6):1585–1633, 2017.

[240] Elizabeth Munch, Katharine Turner, Paul Bendich, Sayan Mukherjee, Jonathan Mattingly, and John Harer. Probabilistic Fréchet means for time varying persistence diagrams. *Electron. J. Statist.*, 9(1):1173–1204, 2015.

[241] Elizabeth Munch and Bei Wang. Convergence between categorical representations of Reeb space and mapper. In *Proc. 32nd Internat. Sympos. on Computational Geometry*, volume 51 of *LIPIcs*, pages 53:1–53:16, 2016.

[242] James R. Munkres. *Elements of Algebraic Topology*. Addison-Wesley, 1984.

[243] James R. Munkres. *Topology*, 2nd edn. Prentice Hall, 2000.

[244] Gregory Naitzat, Andrey Zhitnikov, and Lek-Heng Lim. Topology of deep neural networks. *J. Mach. Learn. Res.*, 21:184:1–184:40, 2020.

[245] Monica Nicolau, Arnold J. Levine, and Gunnar Carlsson. Topology based data analysis identifies a subgroup of breast cancers with a unique mutational profile and excellent survival. *Proc. Natl. Acad. Sci.*, 108(17):7265–7270, 2011.

[246] Partha Niyogi, Stephen Smale, and Shmuel Weinberger. Finding the homology of submanifolds with high confidence from random samples. *Discrete Comput. Geom.*, 39(1–3):419–441, 2008.

[247] Partha Niyogi, Stephen Smale, and Shmuel Weinberger. A topological view of unsupervised learning from noisy data. *SIAM J. Comput.*, 40(3):646–663, 2011.

[248] Ippei Obayashi. Volume-optimal cycle: tightest representative cycle of a generator in persistent homology. *SIAM J. Appl. Alg. Geom.*, 2(4):508–534, 2018.

[249] James B. Orlin. Max flows in $O(nm)$ time, or better. In *Proc. 45th Annu. ACM Sympos. on Theory of Computing*, pages 765–774, 2013.

[250] Steve Oudot. *Persistence Theory: From Quiver Representations to Data Analysis*. AMS Mathematical Surveys and Monographs, vol. 209. American Mathematical Society, 2015.

[251] Deepti Pachauri, Chris Hinrichs, Moo K. Chung, Sterling C. Johnson, and Vikas Singh. Topology-based kernels with application to inference problems in Alzheimer's disease. *IEEE Trans. Med. Imaging*, 30(10):1760–1770, 2011.

[252] Richard A. Parker. The computer calculation of modular characters (the Meataxe). In *Computational Group Theory: Proceedings of the London Mathematical Society Symposium on Computational Group Theory*, pages 267–274. Academic Press, 1984.

[253] Salman Parsa. A deterministic $O(m \log m)$ time algorithm for the Reeb graph. *Discrete Comput. Geom.*, 49(4):864–878, 2013.

[254] Valerio Pascucci, Giorgio Scorzelli, Peer-Timo Bremer, and Ajith Mascarenhas. Robust on-line computation of Reeb graphs: simplicity and speed. *ACM Trans. Graphics*, 26(3):58, 2007.

[255] Amit Patel. Generalized persistence diagrams. *J. Appl. Comput. Topol.*, 1: 397–419, 2018.

[256] Giovanni Petri, Martina Scolamiero, Irene Donato, and Francesco Vaccarino. Topological strata of weighted complex networks. *PLoS One*, 8:1–8, 06 2013.

[257] Jeff M. Phillips, Bei Wang, and Yan Zheng. Geometric inference on kernel density estimates. In *Proc. 31st Internat. Sympos. on Computational Geometry*, volume 34 of *LIPIcs*, pages 857–871, 2015.

[258] Adrien Poulenard, Primoz Skraba, and Maks Ovsjanikov. Topological function optimization for continuous shape matching. *Comput. Graphics Forum*, 37(5):13–25, 2018.

[259] Victor V. Prasolov. *Elements of Combinatorial and Differential Topology*. Graduate Studies in Mathematics, vol. 74. American Mathematical Society, 2006.

[260] Raúl Rabadán and Andrew J. Blumberg. *Topological Data Analysis for Genomics and Evolution: Topology in Biology*. Cambridge University Press, 2019.

[261] Geoge Reeb. Sur les points singuliers d'une forme de Pfaff complètement intégrable ou d'une fonction numérique. *C. R. Hebdom. Séances Acad. Sci.*, 222:847–849, 1946.

[262] Michael W. Reimann, Max Nolte, Martina Scolamiero, et al. Cliques of neurons bound into cavities provide a missing link between structure and function. *Front. Comput. Neurosci.*, 11:48, 2017.

[263] Jan Reininghaus, Stefan Huber, Ulrich Bauer, and Roland Kwitt. A stable multi-scale kernel for topological machine learning. In *Proc. IEEE Conf. on Computer Vision and Pattern Recognition*, pages 4741–4748, 2015.

[264] Bastian Rieck, Matteo Togninalli, Christian Bock, et al. Neural persistence: a complexity measure for deep neural networks using algebraic topology. In *Proc. Internat. Conf. on Learning Representations*, 2019.

[265] Claus M. Ringel and Hiroyuki Tachikawa. Q-F3 rings. *J. Reine Angew. Math.*, 272:49–72, 1975.

[266] Vanessa Robins. Towards computing homology from finite approximations. *Topology Proc.*, 24(1):503–532, 1999.

[267] Vanessa Robins, Peter J. Wood, and Adrian P. Sheppard. Theory and algorithms for constructing discrete Morse complexes from grayscale digital images. *IEEE Trans. Pattern Anal. Mach. Intell.*, 33(8):1646–1658, 2011.

[268] Tim Römer. On minimal graded free resolutions. *Illinois J. Math.*, 45(2): 1361–1376, 2001.

[269] Olaf Ronneberger, Philipp Fischer, and Thomas Brox. U-net: convolutional networks for biomedical image segmentation. In *Proc. Internat. Conf. on Medical Image Computing and Computer-Assisted Intervention*, pages 234–241. Springer, 2015.

[270] Jim Ruppert. A Delaunay refinement algorithm for quality 2-dimensional mesh generation. *J. Algorithms*, 18:548–585, 1995.

[271] Bernhard Schölkopf and Alexander J. Smola. *Learning with Kernels: Support Vector Machines, Regularization, Optimization, and Beyond*. MIT Press, 1998.

[272] Alexander Schrijver. *Theory of Linear and Integer Programming*. John Wiley & Sons, 1986.

[273] Paul D. Seymour. Decomposition of regular matroids. *J. Combin. Theory B*, 28(3):305–359, 1980.

[274] Donald R. Sheehy. Linear-size approximations to the Vietoris–Rips filtration. In *Proc. 28th. Annu. Sympos. on Computational Geometry*, pages 239–248, 2012.

[275] Donald R. Sheehy. Linear-size approximations to the Vietoris–Rips filtration. *Discrete Comput. Geom.*, 49:778–796, 2013.

[276] Yoshihisa Shinagawa, Tosiyasu L. Kunii, and Yannick L. Kergosien. Surface coding based on Morse theory. *IEEE Comput. Graphics Appl.*, 11(5):66–78, 1991.

[277] Gurjeet Singh, Facundo Mémoli, and Gunnar Carlsson. Topological methods for the analysis of high dimensional data sets and 3D object recognition. In *Proc. Eurographics Sympos. on Point-Based Graphics*, pages 91–100, 2007.

[278] Primoz Skraba and Katharine Turner. Wasserstein stability for persistence diagrams. *CoRR*, arXiv:2006.16824, 2021.

[279] Jacek Skryzalin. Numeric invariants from multidimensional persistence. PhD thesis, Stanford University, 2016.

[280] Daniel D. Sleator and Robert Endre Tarjan. A data structure for dynamic trees. *J. Comput. Syst. Sci.*, 26(3):362–391, 1983.

[281] Henry J. S. Smith. On systems of linear indeterminate equations and congruences. *Philos. Trans. R. Soc. Lond.*, 151:293–326, 1861.

[282] Thierry Sousbie. The persistent cosmic web and its filamentary structure – I. Theory and implementation. *Mon. Not. R. Astron. Soc.*, 414(1):350–383, 2011.

[283] Bharath K. Sriperumbudur, Kenji Fukumizu, and Gert R. G. Lanckriet. Universality, characteristic kernels and RKHS embedding of measures. *J. Mach. Learn. Res.*, 12(70):2389–2410, 2011.

[284] Jian Sun, Maks Ovsjanikov, and Leonidas Guibas. A concise and provably informative multi-scale signature based on heat diffusion. In *Proc. Sympos. on Geometry Processing*, pages 1383–1392, 2009.

[285] Julien Tierny. Reeb graph based 3D shape modeling and applications. PhD thesis, Université des Sciences et Technologies de Lille, 2008.

[286] Julien Tierny. *Topological Data Analysis for Scientific Visualization*. Springer, 2017.

[287] Julien Tierny, Attila Gyulassy, Eddie Simon, and Valerio Pascucci. Loop surgery for volumetric meshes: Reeb graphs reduced to contour trees. *IEEE Trans. Visualiz. Comput. Graphics*, 15(6):1177–1184, 2009.

[288] Brenda Y. Torres, Jose H. M. Oliveira, Ann Thomas Tate, Poonam Rath, Katherine Cumnock, and David S. Schneider. Tracking resilience to infections by mapping disease space. *PLoS Biol.*, 14(4):1–19, 2016.

[289] Elena Farahbakhsh Touli and Yusu Wang. FPT-algorithms for computing Gromov–Hausdorff and interleaving distances between trees. *CoRR*, arXiv:1811.02425, 2018.

[290] Tony Tung and Francis Schmitt. The augmented multiresolution Reeb graph approach for content-based retrieval of 3d shapes. *Internat. J. Shape Model.*, 11(1):91–120, 2005.

[291] Katharine Turner, Yuriy Mileyko, Sayan Mukherjee, and John Harer. Fréchet means for distributions of persistence diagrams. *Discrete Comput. Geom.*, 52(1):44–70, 2014.

[292] Gert Vegter and Chee K. Yap. Computational complexity of combinatorial surfaces. In *Proc. 6th Annu. Sympos. on Computational Geometry*, pages 102–111, 1990.

[293] Leopold Vietoris. Über den höheren Zusammenhang Kompakter Räume und eine Klasse von zusammenhangstreuen Abbildungen. *Math. Ann.*, 97:454–472, 1927.

[294] Suyi Wang, Xu Li, Partha Mitra, and Yusu Wang. Topological skeletonization and tree-summarization of neurons using discrete Morse theory. *CoRR*, arXiv:1805.04997, 2018.

[295] Suyi Wang, Yusu Wang, and Yanjie Li. Efficient map reconstruction and augmentation via topological methods. In *Proc. 23rd SIGSPATIAL Internat. Conf. on Advances in GIS*, pages 25:1–25:10, 2015.

[296] Suyi Wang, Yusu Wang, and Rephael Wenger. The JS-graph of join and split trees. In *Proc. 30th Annu. Sympos. on Computational Geometry*, pages 539–548, 2014.

[297] Larry Wasserman. Topological data analysis. *Annu. Rev. Statist. Appl.*, 5(1): 501–532, 2018.

[298] Carry Webb. Decomposition of graded modules. *Proc. Am. Math. Soc.*, 94(4):565–571, 1985.

[299] Gunther Weber, Peer-Timo Bremer, and Valerio Pascucci. Topological landscapes: a terrain metaphor for scientific data. *IEEE Trans. Visualiz. Comput. Graphics*, 13(6):1416–1423, 2007.

[300] André Weil. Sur les théoréms de de Rham. *Comment. Math. Helv.*, 26:119–145, 1952.

[301] Zoë Wood, Hugues Hoppe, Mathieu Desbrun, and Peter Schröder. Removing excess topology from isosurfaces. *ACM Trans. Graphics*, 23(2):190–208, 2004.

[302] Pengxiang Wu, Chao Chen, Yusu Wang, et al. Optimal topological cycles and their application in cardiac trabeculae restoration. In *Information Processing in Medical Imaging – 25th Internat. Conf.*, IPMI, pages 80–92, 2017.

[303] Manzil Zaheer, Satwik Kottur, Siamak Ravanbakhsh, Barnabas Poczos, Ruslan Salakhutdinov, and Alexander Smola. Deep sets. In *Proc. Advances in Neural Information Processing Systems*, pages 3391–3401, 2017.

[304] Simon Zhang, Mengbai Xiao, and Hao Wang. GPU-accelerated computation of Vietoris–Rips persistence barcodes. In *36th Internat. Sympos. on Computational Geometry*, volume 164 *of LIPIcs*, pages 70:1–70:17, 2020.

[305] Qi Zhao and Yusu Wang. Learning metrics for persistence-based summaries and applications for graph classification. In *Proc. 33rd Annu. Conf. on Neural Information Processing Systems*, pages 9855–9866, 2019.

Index

Printed in the United States
by Baker & Taylor Publisher Services